ARTIFICIAL REEFS IN FISHERIES MANAGEMENT

CRC

MARINE BIOLOGY

SERIES

The late Peter L. Lutz, Founding Editor
David H. Evans, Series Editor

Published Titles

Biology of Marine Birds
E. A. Schreiber and Joanna Burger

Biology of the Spotted Seatrout
Stephen A. Bortone

The Biology of Sea Turtles, Volume II
Peter L. Lutz, John A. Musick, and Jeanette Wyneken

Biology of Sharks and Their Relatives
Jeffrey C. Carrier, John A. Musick, and Michael R. Heithaus

Early Stages of Atlantic Fishes: An Identification Guide for the Western Central North Atlantic
William J. Richards

The Physiology of Fishes, Third Edition
David H. Evans and James B. Claiborne

Biology of the Southern Ocean, Second Edition
George A. Knox

Biology of the Three-Spined Stickleback
Sara Östlund-Nilsson, Ian Mayer, and Felicity Anne Huntingford

Biology and Management of the World Tarpon and Bonefish Fisheries
Jerald S. Ault

Methods in Reproductive Aquaculture: Marine and Freshwater Species
Elsa Cabrita, Vanesa Robles, and Paz Herráez

Sharks and Their Relatives II: Biodiversity, Adaptive Physiology, and Conservation
Jeffrety C. Carrier, John A. Musick, and Michael R. Heithaus

ARTIFICIAL REEFS IN FISHERIES MANAGEMENT

EDITED BY

STEPHEN A. BORTONE

FREDERICO PEREIRA BRANDINI

GIANNA FABI

SHINYA OTAKE

CRC Press
Taylor & Francis Group
Boca Raton London New York

CRC Press is an imprint of the
Taylor & Francis Group, an **informa** business

CRC Press
Taylor & Francis Group
6000 Broken Sound Parkway NW, Suite 300
Boca Raton, FL 33487-2742

Printed in the United States of America on acid-free paper
10 9 8 7 6 5 4 3 2 1

International Standard Book Number: 978-1-4398-2007-0 (Hardback)

Library of Congress Cataloging-in-Publication Data

Artificial reefs in fisheries management / editors, Stephen A. Bortone ... [et al.].
p. cm. -- (Marine biology)
Includes bibliographical references and index.
ISBN 978-1-4398-2007-0 (alk. paper)
1. Artificial reefs. I. Bortone, Stephen A.

SH157.85.A7A774 2011
639.3--dc22 2010048795

Visit the Taylor & Francis Web site at
http://www.taylorandfrancis.com

and the CRC Press Web site at
http://www.crcpress.com

We dedicate this volume to our recently deceased friend and colleague, Dr. David Whitmarsh, Department of Economics, University of Portsmouth, Portsmouth, United Kingdom. David gave us an innovative perspective on artificial reefs through the eyes of an economist.

Contents

Preface

Fishery management has served as a focal point and rationale for much of the research regarding artificial reefs since they were seriously considered scientifically in the mid 1970s. The problem is that there has not been an overwhelming embrace of artificial reefs as an option by fisheries managers in the management of fisheries. While artificial reefs may have much to offer, they have remained of lesser importance in the greater scheme of fisheries management. The Ninth CARAH (Conference of Artificial Reefs and Related Aquatic Habitats) held November 8–13, 2009 in Curitiba, Brazil, provided a venue to bring to the forefront the most recent information regarding artificial reefs. Because of the interest among a number of researchers, a special session entitled "Role of Artificial Reefs in Fisheries Management" was graciously included as part of the conference to more fully explore the use of the application of artificial reefs in fisheries management and issues to be considered when applying artificial reefs to fishery management problems. This book is largely a compilation of the expanded and written versions of many of the contributions to that special session and other sessions under ancillary headings at the Ninth CARAH. Additionally, other individuals who did not participate in the conference, but with interests and expertise aligned with exploring the relationship between artificial reefs and fishery management, also contributed chapters to this book. Chapters are purposefully included from authors from a diversity of geographical areas. This gives the reader the broadest of perspectives and also indicates regional interests and experience with artificial reefs in different parts of the world. No single work on this theme can be considered definitive; nevertheless, this book represents an effort to bring to the forefront the current state of knowledge, on a worldwide basis, regarding artificial reefs and their pragmatic application to furthering fisheries sustainability.

Acknowledgments

The editors thank organizers of the Ninth CARAH for providing a productive forum that was conducive to fully exploring the various perspectives regarding artificial reef applications toward advancing fishery management. Those responsible for the Ninth CARAH (chiefly, Frederico Brandini–Conference Chair, Ilana Rosental Zalmon, Marcelo Vianna, Ricardo Coutinho, Roberto A. Bernades, and Andre Berberi) are commended for convening world experts interested in sharing their research experiences with artificial reefs. In this regard we also thank officers of the host institution at the Federal University of Paraná in Curitiba for making available their excellent conference facilities. We thank the editorial staff at CRC Press (a division of the Taylor & Francis Publishing Company) for their help in making this idea a reality. Specifically at CRC Press, we thank John Sulzycki, Gail Renard, and Kathryn Younce. Last, we are grateful for the special perspective that a long list of fellow artificial reef research colleagues has provided us over the past several decades so that our collective experiences were able to come together in this volume.

Contributors

C. Aliaume
Université Montpellier II
Ecole Polytechnique Universitaire
Montpellier, France

Pablo Arechavala-Lopez
Department of Marine Sciences and Applied Biology
University of Alicante
Alicante, Spain

Just T. Bayle-Sempere
Department of Marine Sciences and Applied Biology
University of Alicante
Alicante, Spain

Pål-Arne Bjørn
NOFIMA
Tromsø, Norway

Stephen A. Bortone
Gulf of Mexico Fishery Management Council
Tampa, Florida, USA

Daniel Shimada Brotto
Department of Health Sciences
Veiga de Almeida University
Rio de Janeiro, Brazil

John F. Caddy
Latina, Italy

Erick R. Baqueiro Cárdenas
Instituto de Ciencia y Tecnología
Universidad Arturo Prat
Puerto Montt, Chile

Hyung Kee Cha
National Fisheries Research and Development Institute
Busan, South Korea

Burak Ali Çiçek
Underwater Research and Imaging Center
Eastern Mediterranean University
Famagusta, North Cyprus, Turkey

Tim Dempster
SINTEF Fisheries and Aquaculture
Trondheim, Norway
and
Department of Zoology
University of Melbourne
Victoria, Australia

F. Ozan Düzbastılar
Faculty of Fisheries
Ege University
Izmir, Turkey

Dor Edelist
Leon Recanati Institute for Maritime Studies and Department of Maritime Civilizations
Leon H. Charney School for Marine Sciences
University of Haifa
Haifa, Israel

Gianna Fabi
National Council of Researchers (CNIR)
Institute of Marine Sciences (ISMAR)
Ancona, Italy

Damian Fernandez-Jover
Department of Marine Sciences and Applied Biology
University of Alicante
Alicante, Spain

Robert Grove
Southern California Edison
Rosemead, California, USA

Benal Gül
Faculty of Fisheries
Ege University
Izmir, Turkey

O. Hashimoto
Fisheries Agency of Japan
Tokyo, Japan

Kenji Hirata
Fisheries Infrastructure Development Center
Tokyo, Japan

Junichi Ito
Fisheries Department
Nagasaki Prefecture Government
Nagasaki, Japan

Yasushi Ito
Japanese Institute of Technology on Fishing Ports, Grounds, and Communities
Tokyo, Japan

David Izquierdo
Department of Marine Sciences and Applied Biology
University of Alicante
Alicante, Spain

Suetoshi Kawaguchi
Fisheries Department
Nagasaki Prefecture Government
Nagasaki, Japan

Chang Gil Kim
West Sea Fisheries Research Institute
National Fisheries Research and Development Institute
Incheon, Korea

G. Lasserre
Institut Halieutique et des Sciences Marines (IH-SM) de l'Université Tuléar
Toliara, Madagascar

Kari L. Lavalli
Division of Natural Sciences and Mathematics
College of General Studies
Boston University
Boston, Massachusetts, USA

Moon Ock Lee
Department of Naval Architecture & Ocean Engineering
Chonnam National University
Yeosu, South Korea

Sung Il Lee
East Sea Fisheries Research Institute
National Fisheries Research and Development Institute
Gangneung, South Korea

Francisco Leitão
National Fisheries and Marine Science Institute
Olhão, Portugal

Altan Lök
Faculty of Fisheries
Ege University
Izmir, Turkey

Koji Mitsuhashi
Fisheries Infrastructure Development Center
Tokyo, Japan

Carlos Costa Monteiro
National Fisheries and Marine Science Institute (INRB, I.P./L-IPIMAR)
Olhão, Portugal

Yoshifumi Nakagawa
Fisheries Infrastructure Development Center
Tokyo, Japan

Miguel Neves Santos
National Fisheries and Marine Science Institute
Olhão, Portugal

Rune Nilsen
NOFIMA
Tromsø, Norway

Tae Gun Oh
National Fisheries Research & Development
South Sea Fisheries Research Institute
Yeosu, South Korea

Takahiro Okano
Fisheries Infrastructure Development Center
Tokyo, Japan

Shinya Otake
Laboratory of Marine Ecosystem Engineering
Fukui Prefectural University
Fukui, Japan

Aytaç Özgül
Faculty of Fisheries
Ege University
Izmir, Turkey

Carlos Valle Pérez
Department of Marine Sciences and Applied Biology
University of Alicante
Alicante, Spain

Sylvain Pioch
Montpellier/Société Egis Eau
Université Montpellier III
Montpellier, France

Tony Pitcher
Fisheries Centre
University of British Columbia
Vancouver, British Columbia, Canada

J.-C. Raynal
Ecole des Hautes Etudes de Sciences Sociales
Paris, France

Giulio Relini
University of Genoa
Genoa, Italy

Pablo Sanchez-Jerez
Department of Marine Sciences and Applied Biology
University of Alicante
Alicante, Spain

Luciano Neves dos Santos
Department of Ecology and Marine Resources
Federal University of Rio de Janeiro State
Rio de Janeiro, Brazil

William Seaman, Jr.
Fisheries and Aquatic Sciences
University of Florida
Gainesville, Florida, USA

Yong Soo Son
East Sea Fisheries Research Institute
National Fisheries Research and Development Institute
Gangneung, South Korea

Alessandra Spagnolo
National Council of Researches (CNR)
Institute of Marine Sciences (ISMAR)
Ancona, Italy

Ehud Spanier
Leon Recanati Institute for Maritime Studies and Department of Maritime Civilizations
Leon H. Charney School for Marine Sciences
University of Haifa
Haifa, Israel

T. Suzuki
Artificial Sea-Mount Institute
Tokyo, Japan

Stephen T. Szedlmayer
Department of Fisheries and Allied Aquacultures
Auburn University
Fairhope, Alabama, USA

Masanori Takeda
Fisheries Infrastructure Development Center
Tokyo, Japan

Ingebrigt Uglem
Norwegian Institute for Nature Research (NINA)
Trondheim, Norway

Ali Ulaş
Faculty of Fisheries
Ege University
İzmir, Turkey

David Whitmarsh
Department of Economics
University of Portsmouth
Portsmouth, United Kingdom

Jae Hyeong Yang
East Sea Fisheries Research Institute
National Fisheries Research and Development Institute
Gangneung, South Korea

Ilana Rosental Zalmon
Centre of Biosciences and Biotechnology
North Rio de Janeiro State University
Campos dos Goytacazes, Brazil

CHAPTER 1

Introduction to the Role Artificial Reefs Play in Fisheries Management

Stephen A. Bortone

CONTENTS

HOW DID WE GET HERE?

In 1974, the first international artificial reef conference was held in Houston, Texas. One of the chief reasons for the gathering was that artificial reefs hold promise in fishery management. In recent years reef building has become popular and scientists from many countries are studying the use of artificial reefs in fishery management (Clark et al., 1974). Including that initial conference, there have been a total of nine international conferences on artificial reefs (CARAH—Conference on Artificial Reefs and Related Artificial Habitats; Bortone, 2006), with the most recent held in Curitiba, Brazil, in 2009. Additionally, there have been a host of other conferences concerned with artificial reefs on a broad variety of fisheries (e.g., Nakamura et al., 1991; Jensen, 1997; Jensen et al., 2000; Secretaria de Pesca, 1992). At each of these conferences (and numerous other conferences, workshops, symposia, and meetings on the artificial reefs; see Bortone, 2006 for a partial review), the topic of the role of artificial reefs in fishery management has been one of the central themes.

Bohnsack and Sutherland (1985) reviewed the state of artificial reef research up until 1984, and Bortone (2006) added to that review by way of a general summary of research trends and status of artificial reef research. It is not without consequence that both reviews indicated a lack of application of artificial reefs in fishery management. This obvious lack of artificial reef utilization in management has occurred in spite of the plethora of research papers presented at each CARAH. Although the total number of oral presentations on artificial reefs offered during these international congregations has declined in recent years (Figure 1.1), any reader would get the impression that there has been a dominance of fishery management as a theme of the world research effort. Indicative among those past research efforts are projects conducted by Miguel Neves dos Santos and his colleagues in the eastern Atlantic Ocean on the Algarve region off southern Portugal (see Chapter 14 for a partial review of these activities). Additionally, there have been numerous artificial reef deployments in

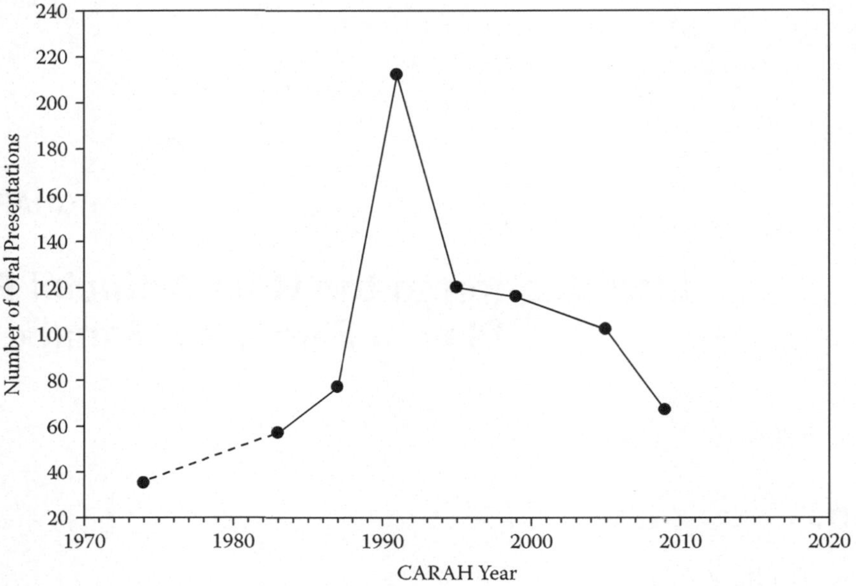

Figure 1.1 Scatter-line plot of the number of oral presentations offered at the CARAH conferences on artificial reefs by year. (*Note*: The number of presentations for 1977 is not available.)

the Mediterranean Sea, for an indirect effect on fisheries through the deployment of artificial reefs to dissuade trawling in certain areas (Relini and Moretti, 1986). In Asia, specifically off Japan's coasts, there have been substantial artificial reef deployments. These were chiefly engineering experiments to demonstrate their efficacy to enhance fisheries as opposed to deployments to assist in a specific fishery management plan (Grove et al., 1991). In the United States, most artificial reef programs are organized at the state, county, and local levels, and these are largely directed toward the recreational fishing sector to improve fishing opportunities but not to specifically help manage fishery resources. An exception to this general statement could be the large permit area off the coast of Alabama (USA) in the northern Gulf of Mexico. Shipp and Bortone (2009) and Gallaway et al. (2009) argued that the increase in the red snapper fishery in the northern Gulf of Mexico may be linked to artificial reef deployments. However, these deployments, though significant in spatial area, were not part of a fishery management plan and the program lacks specific landings and effort objectives. Perhaps one of the best past examples of the use of artificial reefs being objectively used in fishery management has been conducted off Hong Kong by Tony Pitcher and his colleagues (Pitcher et al., 2002). This artificial reef project was specific in its cost–benefit analysis on several levels, and can serve as an exception to the statement that "artificial reefs play no role in fisheries management."

Clearly, there has been activity involving artificial reefs in fishery management but, for the most part, this has been tangential or incidental to the purposeful deployment of artificial reefs with a clearly stated management objective.

WHAT IS THE CURRENT SITUATION?

The application of *casitas* or *pesqueros* (see Chapter 6, this volume) in the spiny lobster fishery off Cuba (and perhaps throughout the Caribbean) is an excellent example of the role that artificial reefs can play in fishery management—even if uncoordinated and without prescribed objectives, in a purely artisanal fishery. Casitas are deployed by fishermen to serve several purposes in the fishery.

Ostensibly, they serve as congregating places for adult spiny lobsters. This facilitates lobster capture by the spiny lobster fishery but may inadvertently increase fishing mortality while leading to higher catches (Polovina, 1991). More interestingly, and perhaps unintentionally, the deployment of casitas has afforded increased survival of juvenile and early settlement stages of spiny lobsters. The casita creates a refuge from predation and thus relieves an early life history bottleneck instituted by predation (e.g., Eggleston et al., 1992). It is important to realize, however, that casita deployment is not institutionalized by any agency charged with fishery management as a formal part of a management strategy to either aid lobster fishermen in the procurement of adults, or as an attempt to decrease natural mortality among juveniles.

Of note, the only significant, purposeful artificial reef deployment that specifically incorporated artificial reefs into a directed and government sanctioned fishery management plan is offered by the study presented in this volume by Kim and his colleagues (Chapter 7) off the coast of South Korea. Briefly, the South Korean government directed the construction of artificial reefs off Korea to increase the stocks of sandfish (*Arctoscopus japonicus*) as a response to indications of declines in that fishery. Reefs were deployed to provide spawning substratum and orientation space for adults. Early indications are that this project is successful in helping to increase sandfish stocks.

The preceding description of the current situation regarding the use of artificial reefs in fishery management is admittedly brief and is not intended to be comprehension. Accordingly, the chapters presented here offer a much more inclusive and encompassing description of the situation. In deference to the opinions of some of my colleagues, I offer the following statement that serves as a hypothesis for this book: Currently, artificial reefs play almost no role in the management of fisheries—anywhere in the world!

WHAT IS FISHERY MANAGEMENT?

One could take exception to the statement, "artificial reefs play almost no role" in the management of fisheries, depending upon the perspective one has of fishery management. Granted, broad considerations of fishery management could inevitably lead one to assume that any deployment (planned or unplanned) of an artificial reef could be construed as affecting fishery management. However, this overly broad usage is without utility in any progression toward actively including artificial reefs as a cognitive option of a fishery management plan. Following are some borrowed definitions and uses of fishery management that emphasize my point in practically excluding artificial reefs from specific involvement in fishery management.

Neilson (1993) considered fishery management as "The manipulation of human interactions with living aquatic resources in a manner that allows humans to gain some sustainable benefit from these resources." Similarly, Ross (1997) offered that fishery management included manipulating human behavior (controlling harvest with regulations), controlling aquatic habitats (pollution), and controlling resources themselves (introduction). Rothschild and Beamish (2009) noted that it in its simplest form fishery management should determine optimum yield and estimate fishing mortality. Moreover, making changes in a fishery requires managing ecosystems, managing habitat, ending overfishing, using a precautionary approach, and rebuilding stocks. In brief, it involves active manipulation by managers. As part of the fishery management process, most fishery management decisions involve quantitative choices: how many, what size, how large an area, how many fishers allowed, how much fishing effort, how much harvest, etc. (Walters and Martell, 2004). Ultimately, fishery management is concerned with applying controls on the current fishery so that the future fishery will be better (Gulland, 1983).

Clearly, there are a host of definitions and perspectives on fishery management but, to synthesize here, the most useful feature of fishery management is that it is based on science. Inherently, scientists seek to explain (i.e., predict with useful certainty) circumstances and conditions under the assumption that there are factors or features that determine (i.e., cause) a result. Without such predictive ability, management becomes a chaotic venture. With it, fisheries cannot only be understood but manipulated to a directed and, hopefully, sustainable condition. To be of use to fishery managers, artificial reefs should be designed, regulated, and prescribed in a way that allows an assessment of their effectiveness as a management option for altering some aspect of the fishery. Moreover, once some aspect of the fishery is altered, the outcome should be predictable and repeatable. It is under these assumptions that I proceed.

WHY SO FEW ARTIFICIAL REEFS IN FISHERY MANAGEMENT?

In one sense, accounting for the lack of artificial reef application in management necessitates circular reasoning (i.e., there are no artificial reefs in fishery management; therefore, there are few examples for managers to follow. Because there are few examples to follow, there are few artificial reefs used in management). While this scenario indicates the most probable reason, there are others. They include a range of attitudes among some fishery scientists and managers toward artificial reefs, from a simple lack of appreciation for the utility of artificial reefs to a strong antireef bias among others. Artificial reefs (or artificial habitats) are generally only given passing attention in fishery biology text books. Moreover, unless the professor offering a course in fishery management has an interest in artificial reefs, they are hardly mentioned in fishery-related course offerings.

There are many other reasons that artificial reefs are not used in fishery management (i.e., skepticism, unfamiliarity, an untested management option, etc.), but there is a single feature that is undoubtedly the most significant reason for their lack of application—the lack of data specific to validating their use. By way of example, Steimle and Meier (1997) surveyed artificial reef managers to determine their needs, and these largely revolved around getting fishery science information that would be directly useful to the application of artificial reefs. While this is similar to the initial, circular reasoning offered earlier, there is more to it; hence, what is behind this apparent shortage of information will be the subject of the remaining portion of this chapter. Generally, scientists do not conduct—and funding agencies do not promote—the research necessary to provide the answers to questions that would help fishery managers justify the deployment of artificial reefs as a management option.

OPPORTUNITY KNOCKS!

The 9th CARAH (Conference on Artificial Reefs and Related Artificial Habitats) held in November 2009 in Curitiba, Brazil, was the platform that served as an impetus for this volume. We will present a compilation of the most recent and up-to-date findings of researchers with interest in artificial reefs. This book also represents a directed attempt to honestly portray and depict the best and most current applications of artificial reefs in fisheries science. The initial chapters (Seaman et al., Szedlmayer, Caddy, and Pioch et al.) give an in-depth summary of the field, as well as introduction on how the features of artificial reefs may be used to glean better life history information on the fishes associated with them. The next set of chapters (Spanier et al., Kim et al., Santos et al., Fabi and Spagnolo, Sanchez et al.) represent a traditional approach to applying artificial reefs to fisheries management. In total, they represent a world survey of artificial reef applications under a variety of circumstances. The last two chapters of this group (Baqueiro and Santos et al.) show the practical applications of fisheries management in chiefly artisanal fisheries. A series of chapters from authors based in Japan and Korea (Ito, Okano et al., Oh et al., and Suzuki and Hashimoto) examine the physical aspects of artificial reef design and how their configuration and placement can facilitate and enhance the application of artificial reefs in fisheries. Lastly, the chapter by Bortone indicates a potential new pathway for research to follow that may help avoid some pitfalls previous researchers have encountered when addressing the issue of artificial reefs in fisheries management.

This volume comes at an opportune time in the history of studies on artificial reefs. Hopefully, it will represent a turning point in the attitude among researchers and natural resource managers to more carefully consider the special features of artificial reefs in resolving fisheries management problems. Let us also hope this book represents merely the first of many steps toward improving the prescribed utilization of artificial reefs as a viable option in many of the world's fisheries in a quest to make more of the world's fisheries sustainable.

REFERENCES

Bohnsack, J.A. and D.L. Sutherland. 1985. Artificial reef research: A review with recommendations for future priorities. *Bull. Mar. Sci.* 37:11–39.

Bortone, S.A. 2006. A perspective of artificial reef research: the past, present, and future. *Bull. Mar. Sci.* 78(1):1–8.

Clark, W., J. Mosely, and R. Stone. 1974. Foreword [sic]. Page 3 (abst.) in L. Colunga and R. Stone (Eds.). *Proc. Int. Conf. Artificial Reefs*, Houston. TAMU-SG-74-103. Texas A & M Sea Grant College, College Station.

Eggleston, D.B., R.N. Lipcius, and D.L. Miller. 1992. Artificial shelters and survival of juvenile Caribbean spiny lobster, *Panulirus argus*—Spatial, habitat, and lobster size effects. *Fish. Bull. U.S.* 90:691–702.

Gallaway, B.J., S.T. Szedlmayer, and W.J. Gazey. 2009. A life history review for red snapper in the Gulf of Mexico with an evaluation of the importance of offshore petroleum platforms and other artificial reefs. *Rev. Fisheries Sci.* 17(1):48–67.

Grove, R.S., C.J. Sonu, and M. Nakamura. 1991. Design and engineering of manufactured habitats for fisheries enhancement. Pages 109–152 in W. Seaman, Jr. and L.M. Sprague. *Artificial Habitats for Marine and Freshwater Fisheries*. Academic Press, San Diego, California.

Gulland, J.A. 1983. *Fish Stock Assessment*. John Wiley & Sons, Chichester, United Kingdom.

Jensen, A.C. (Ed.). 1997. European artificial reef research. *Proc. 1st EARRN Conf.*, Ancona. Pub. Southampton Oceanography Centre, Southampton. 449 pp.

Jensen, A.C., K.J. Collins, and A.P.M. Lockwood (Eds.). 2000. *Artificial Reefs in European Seas*. Kluwer Academic Publishers, Dordrecht. 508 pp.

Nakamura, M., R.S. Gove, and C.J. Soule (Eds.). 1991. Recent advances in aquatic habitat technology. *Japan-U.S. Symp. Artificial Habitats for Fisheries Proceedings*. Tokyo. 345 pp.

Neilson, L.A. 1993. History of inland fisheries management in North America. Pages 3–32 in C.C. Kohler and W.A. Hubert (Eds.). *Inland Fisheries Management in North America*. American Fisheries Society, Bethesda. MD.

Pitcher, T.J., E. Buchary, and T. Hutton. 2002. Forecasting the benefits of no-take human made reefs using spatial ecosystem modeling. Pages S17–S26 in A.C. Jensen (Ed.). *Proc. 7th Int. Conf. Artificial Reefs and Related Aquatic Habitats*, San Remo. *ICES J. Mar. Sci.* 59: (suppl.).

Polovina, J.J. 1991. Fisheries applications and biological impacts of artificial habitats. Pages 154–176 in W. Seaman, Jr. and L.M. Sprague (Eds.). *Artificial Habitats for Marine and Freshwater Fisheries*. Academic Press, San Diego, California.

Relini, G. and S. Moretti. 1986. Artificial reef and *Posidonia* bed protection off Loano (Western Ligurian Riviera). FAO. *Fisheries Rep.* 357:104–109.

Ross, M.R. 1997. *Fisheries Conservation and Management*. Prentice Hall, Upper Saddle River, New Jersey. 374 pages.

Rothschild, B.J. and R.J. Beamish. 2009. On the future of fisheries science. *Fish and Fisheries series*. 31:1–11. Springer.

Secretaria de Pesca (Ed.). 1992. I. Reunion International Sobre Mejoramiento de Habitats Acuaticos para Pesquerieas (Arrecifes Artificiales). 29–30 Octubre de 1992. Manzanillo, Colima, Mexico. Instituto Nacional de la Pesca, Centro Regional de Investigacion Pesquerain Manzanillo, Colima.

Shipp, R.L. and S.A. Bortone. 2009. A perspective of the importance of artificial habitat on the management of red snapper in the Gulf of Mexico. *Rev. Fisheries Sci.* 17(1): 41–47.

Steimle, F.W. and M.H. Meier. 1997. What information do artificial reef managers really want from Fishery Science? *Fisheries* 22(4):6–8.

Walters, C.J. and S.J.D. Martell. 2004. *Fisheries Ecology and Management*. Princeton University Press. Princeton, NJ. 399 pages.

CHAPTER 2

Artificial Reefs as Unifying and Energizing Factors in Future Research and Management of Fisheries and Ecosystems

William Seaman, Jr., Robert Grove, David Whitmarsh, Miguel Neves Santos, Gianna Fabi, Chang Gil Kim, Giulio Relini, and Tony Pitcher

CONTENTS

Artificial reefs have received varied acceptance in over 50 nations for practical purposes such as ecosystem conservation and fishery production. Scientifically, they are used increasingly as platforms for rigorous ecological experimentation. Historically, a case has been made that at least some artificial reef technologies have not been grounded upon sound principles, and that their utilization can harm fish populations and the environment. Can artificial reef deployments, and their associated body of technical knowledge, be considered as established practices and components in fishery science? What is the appropriate role and scope for them in marine ecosystem management? While these questions are debated, advances in the science and applications of this technology nevertheless continue. This chapter concludes that this relatively young field does "add value" to environmental science and management.

Our understanding of artificial reefs needs to be informed by theory and empirical evidence, and this paper adopts both approaches. Bioeconomics provides a frame of reference for analyzing the circumstances in which reef deployment will generate sustainable economic benefits, and gives a theoretical context for the examples that follow. Case studies illustrate a sustained community-based fishery in Portugal, attainment of resource protection and allocation objectives in Italy, success in designing a reef to create habitat for kelp in the United States, and incorporation of reefs into marine ranching in Korea, all at relatively large scales. Together, these findings illustrate shared attributes and trends concerning study and deployment of artificial reefs, which may unify, guide, and foster future investigations and management in various branches of aquatic sciences. For example:

Sustainability of resource utilization is a standard concern that sets reefs in a broader social and environmental context.

Maintaining or enhancing biodiversity is a common goal that extends reef objectives beyond exploitation.

Engaging a broad variety of community stakeholders enhances reef development and assessment efforts.

Using artificial reefs as sites for manipulative experiments affords study designs that could not be conducted at natural habitats.

Promoting multidisciplinary studies offers a powerful approach for addressing complex environmental issues.

Adaptation of fundamental knowledge of species life history requirements to the design of physical structure of reefs combines basic research with the interests of society for applications.

INTRODUCTION

Can artificial reefs and their associated body of scientific knowledge be considered as established practices and components in fishery science? Further, and more generally, what is their appropriate role and scope in marine ecosystem management? While these questions are debated, advances in the science and applications of this technology have continued in at least 50 nations. This progress in turn gives rise to another question, namely, Does this field contribute or "add value" to environmental science and management in general? The aim of this chapter is to stimulate the thoughts, perceptions, and even lines of inquiry of persons and organizations concerned with artificial reefs, regarding their potential limits and how they may serve to unify and energize future research that facilitates management in coastal environments.

Benthic features of the world's coastal oceans artificial reefs are open systems, both ecologically and with respect to human patterns of use and management of natural resources. Thus, the ecological structure and function of these habitats is connected in various ways and levels to adjoining natural ecosystems, such as in providing physical sites for settlement of larvae recruiting from off-reef locations and for foraging around the reef. Artificial reefs usually also are open systems through their use in aquatic resource conservation, exploitation, restoration, and research. This chapter considers the placement and connection of artificial reef science and technology in mainstream fishery science, both by considering some theoretical and bioeconomic aspects and also by focusing on selected, representative situations with sufficient documentation to evaluate their role.

CONTEXT AND APPROACH

We identify five broad sectors that have an interest in the validity of artificial reefs in fishery and ecosystem management:

1. *Science*, in terms of needing to know the rigor and reliability of findings from reef research and be aware of syntheses of ecological, economic, and engineering knowledge
2. *Investors*, both public organizations concerned with expenditure of tax funds for reef projects and businesses considering opportunities for service and profit
3. *Managers of natural resources*, needing to apply best practices for sustainable habitat enhancement or restoration and to balance stakeholder and broader ecological landscape concerns
4. *Reef planners and builders*, concerned with ecology of reefs, impacts of structure, and performance of design in the ocean
5. *Users/consumers*, whose needs concerning marine environments must be reflected in realistic and measurable objectives and success criteria for artificial reefs that lead to cost-effective results.

In fact, if some or all of these sectors can be satisfied through application of reef technologies, would that be evidence for mainstream status for artificial reefs?

Part of the impetus for this chapter came from two projects that are models of long-term study, and which found ecological functional equivalency between artificial and natural reef systems. In the Ligurian Sea of Italy, structures have been placed intentionally since 1970, with over 100 research publications produced (Relini et al., 2007). In contrast, a so-called secondary reef at King Harbor, Redondo Beach, California, was created over 50 years ago by a rock breakwater constructed only for harbor protection, which nonetheless was monitored for fishes beginning in 1974 (Pondella et al., 2002).

In the Ligurian Sea, reefs achieved coastal management aims including protection of habitat from illegal trawling, supply of new microhabitats, and biological production (Relini et al., 2007), while also sustaining fish abundance and diversity and fishing yields comparable to control, natural rocky reefs in the Mediterranean Sea. Meanwhile, at King Harbor, a 25-year database for two well-studied species of Embiotocidae revealed that mean densities on the breakwater exceeded those of natural reefs, whereby for *Embiotoca jacksoni*, the black surfperch, juveniles were at 1.3 individuals per 100 m^2 of breakwater habitat and 0.1 at the control, and adults were at 5.6 (breakwater) versus 3.1 (control). *Rhacochilus vacca*, the pile surfperch, followed a similar pattern, with breakwater densities for juveniles (0.39) and adults (2.4) exceeding values at the control (0.08 and 0.13, respectively) (Pondella et al., 2002). Annual biomass estimates followed a similar pattern for these two species, as did the abundance of the predator *Paralabrax clathratus* (kelp bass). The authors conclude that ecologically a higher carrying capacity exists at the breakwater secondary reef, due to higher physical relief and more three-dimensional space, and that a well-designed artificial reef can enhance fish production without negatively influencing natural reefs.

Table 2.1 Criteria for Evaluating Contribution of Artificial Reef Case Studies to Fishery and Ecosystem Enhancement

Scientific Aspects	Applications Aspects
Project design based on valid scientific principles, including species life histories	Linkage of reefs to larger plan for natural resources
Rigor of research methods including field data acquisition and analysis	Scope and scale of reef deployment at "real-world" level
Pilot studies, as appropriate	Partnerships and leveraging of resources for reef deployment and utilization
Advancement of scholarship through publication	Public awareness of reefs
Training students	Net benefit to stakeholders

To our knowledge, no systematic or comprehensive assessment and synthesis of the role of artificial reefs in fisheries (and increasingly, ecosystem) management has been attempted. This is not surprising. Published articles usually are very subject and geographic specific. Only now are a small number of extensive datasets becoming available, which afford a basis for evaluation of longer-term reef performance, both ecologically and in management. Pondella et al. (2002) discuss the difficulties of determining production rates of fishes on reefs. This chapter does not attempt an exhaustive analysis of reefs in fishery and ecosystem settings. Instead, we turn to situations characterized by extensive science-based planning and documentation.

The discussion below is structured as follows: First, we start with a theoretical section that outlines the bioeconomic implications of artificial reef deployment, highlighting in particular the need for regulatory control of reef-based fishery resources in order to prevent overexploitation. Second, four situations were selected for detailed analysis, namely, the Algarve region, Portugal, possibly the largest artificial reef in Europe; the Marche region, Italy, with one of the world's longest datasets; San Onofre in California, the largest kelp restoration reef in America; and Tongyong, Korea, where a large-scale marine ranching effort includes new benthic structures. They were evaluated for both scientific and applications aspects, identified in Table 2.1. In this manner, the concerns of the five sets of stakeholders identified earlier (i.e., science, investors, managers, planners, users) were addressed.

BIOECONOMIC CONSEQUENCES OF ESTABLISHING ARTIFICIAL HABITATS

Modifying marine habitats through artificial reefs may have widespread social and economic impacts, and the success (or otherwise) of reef deployment is likely to be judged on the benefits to individuals and stakeholder groups. While there is no shortage of scientific and anecdotal evidence that catch rates at artificial reef sites may be higher than elsewhere, and that fishermen's livelihoods may at least temporarily improve by focusing their activities adjacent to such sites, these alone are insufficient to prove that the overall net benefits are positive or indeed sustainable. In fact, since artificial reefs may result in intensified harvesting pressure, there is every reason to suppose that artificial reefs may fail to generate any long-term benefits if fishing effort is not controlled (Garcia, 1990; Willmann, 1990; Milon, 1989, 1991; Ungson et al., 1995; Morton, 1996; Santos et al., 1997; Grossman et al., 1997; Pitcher et al., 2002; Sayer and Wilding, 2002; Watanuki and Gonzales, 2006; Whitmarsh et al., 2008). Bioeconomics provides a useful theoretical framework for exploring this problem and helps unify our understanding of artificial habitats within the wider subject area of environmental management. In the analysis that follows, our aim is to identify the circumstances in which reef deployment will generate sustainable economic benefits, and to demonstrate in principle what needs to be done to ensure that the benefits are captured.

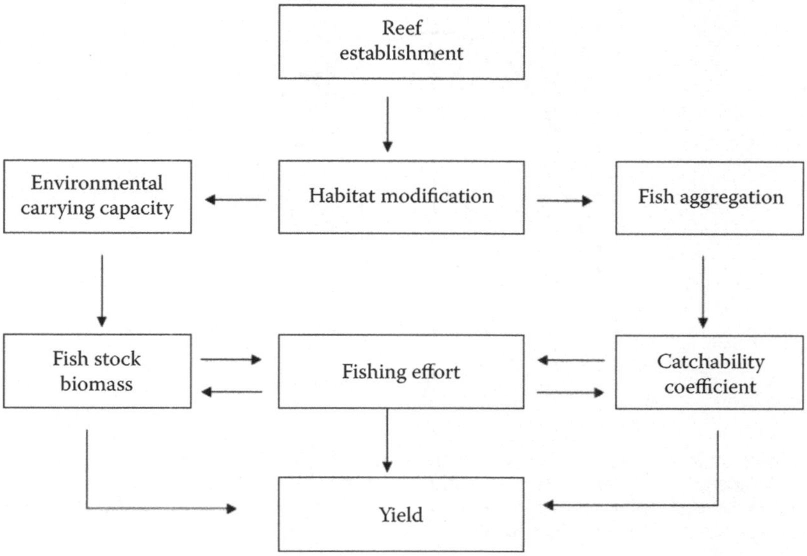

Figure 2.1 Bioeconomic implications of artificial reef deployment. (From Whitmarsh, D. et al. 2008. *Ocean Coast Manage*. 51: 463–468. With permission from Elsevier.)

We may start by considering two scenarios, illustrated schematically in Figure 2.1. In the first, the effect of an artificial reef is purely that of an aggregating device, which by making fish more vulnerable to harvesting results in an increase in the capture efficiency. In the second, the effect of the artificial reef is one in which there is genuine enhancement arising from the creation of additional habitat, resulting in an increase in environmental carrying capacity and thus biomass of the stock. The bioeconomic consequences are shown in Figures 2.2 and 2.3, where in each case the shift in the yield function has an equivalent and corresponding effect on the revenue curves. Under scenario 1 (aggregation), the yield–effort curve for the fishery is shifted horizontally to the left, while under

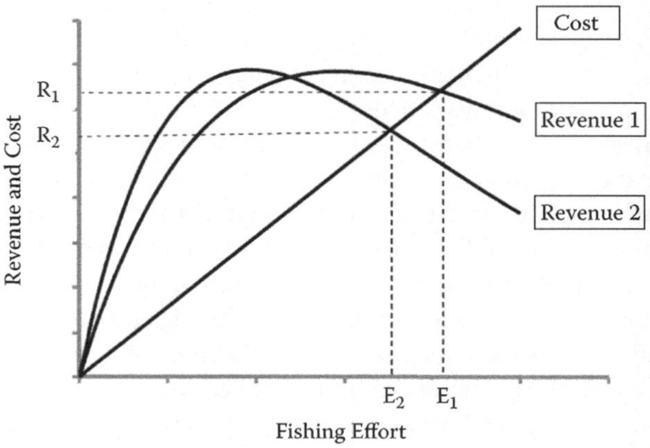

Figure 2.2 Effects on a commercial fishery of an increase in the efficiency of capture (aggregation scenario). The effect of an aggregation reef is to shift the revenue curve to the left, altering the open-access equilibrium such that effort falls to E2 and revenue to R2.

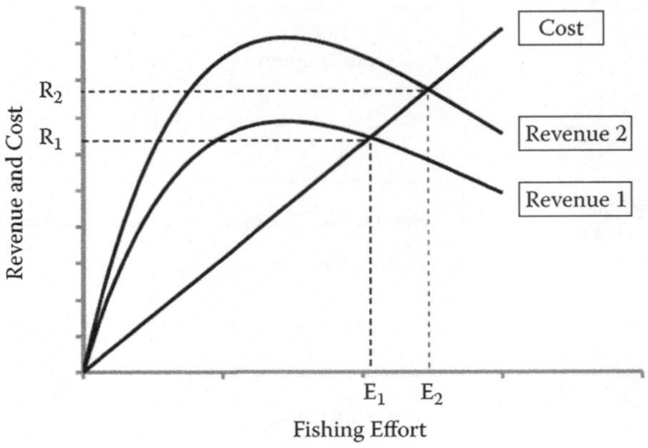

Figure 2.3 Effects on a commercial fishery of an increase in the environmental carrying capacity of the stock (enhancement scenario). Here, the starting equilibrium is the same as in Figure 2.2 (aggregation scenario), but the effect of an enhancement reef is to shift the revenue curve outward, and as a result effort and revenue increase.

scenario 2 (enhancement) the curve is shifted upwards and to the right. If the costs of artificial reef deployment are not passed on to the fishermen, which might be the case if this were a publicly funded program intended to benefit the community at large, the expenses directly incurred by fishermen will be unaffected and, hence, the cost function for the fishery will remain unchanged.

In Figure 2.2, the initial open-access equilibrium is such that the fishery is operating beyond MSY (maximum sustainable yield), at effort level E1 and revenue level R1. The effect of an aggregation reef is to shift the revenue curve to the left, altering the open-access equilibrium such that effort falls to E2 and revenue to R2. In Figure 2.3, the starting equilibrium is the same but, here, the effect of an enhancement reef is to shift the revenue curve outwards and as a result effort and revenue increase. On the basis of this simple analysis, therefore, it is possible to show that artificial reef deployment can affect the operating performance of an open-access fishery in a way which is dependent on the biological mechanics (i.e., aggregation or enhancement) of the reef system. The two scenarios produce contrasting results. In the first case, the establishment of an artificial reef eventually leads to lower yield (and hence revenue) and requires less fishing effort; in the second case, we get the opposite result. In each case, the changes in yield and effort will have knock-on consequences for employment and livelihoods.

There are some further lessons that can be learned from this. In both scenarios, the open-access assumption implies that any economic surplus (technically, resource rent) from the fishery will be dissipated as a result of competition. Artificial reefs, whether through aggregation or enhancement, increase the rent-generating potential of a fishery but, for society to benefit, there needs to be a way of solving the open-access problem and for the resource rent to be captured. Ownership rights represent one solution, though in practice this is complicated by the variety of legal regimes that apply to fisheries where artificial reefs have been established (Whitmarsh and Pickering, 1997). However, our concern in this chapter has less to do with the detailed institutional arrangements of fisheries management than with making the basic point that some form of regulatory control is likely to be imperative if the real economic value of these structures is to be optimized. As such, research on artificial reefs cannot be sensibly conducted outside of the mainstream of fisheries science (or, indeed, fisheries economics) since it is from these related disciplines that our understanding of how best to manage reef-based fisheries will be derived.

REPRESENTATIVE ARTIFICIAL REEF PRACTICES IN FISHERY SCIENCE AND ECOSYSTEM MANAGEMENT

Sustaining Artisanal Fisheries

Situation

Portuguese citizens rank third in fish consumption (after Japan and Iceland) with an annual rate of around 60 kg per inhabitant. In the Algarve region in the South, this figure is even higher, as a consequence of its gastronomic heritage and because it is Portugal's most important tourism region. Such strong demand is the basis of an important fishing industry. In 2007, the Algarve fishing fleet had around 1500 vessels, most of which (>90%) belong to the artisanal segment (<12 m length of boat). The fish fauna of Algarve coastal waters reflects the high species richness of this region, biogeographically between the Mediterranean Sea and the Northeast Atlantic Ocean. A strong demand for fish supply, easy access to inshore waters, and high commercial value of catch led to intensive exploitation of most fish populations, causing significant disruptions to local fisheries and ecosystems. Also, the multispecies characteristic and the great dynamics of these coastal resources caused management problems, as traditional technical management tools (e.g., minimum sizes, closed seasons, catch limits, closed areas, effort or gear restrictions) have been insufficient to guarantee fisheries sustainability.

Goal

The long and diverse Japanese experience in this field of science and technology inspired Portugal's IPIMAR (National Fisheries and Marine Science Institute) researchers to propose and develop a pilot project in the Algarve. The aim of this project was to evaluate, within the framework of local geoecological conditions, the effects of man-made structures at the ecological and fishing levels. Results of the pilot project (1990–1996) encouraged the national authorities to deploy a large artificial reef complex in the Algarve coastal waters (1998–2011). By the time it is concluded (2015) the Algarve artificial reef complex will represent an investment of 10 million euros, including around 20% of this allocated to scientific studies and monitoring.

Reef Description

Deployment of artificial reefs in Algarve coastal waters was initiated in 1990, when two identical small artificial reefs were set off the Ria Formosa lagoon (a highly productive nursery habitat) as a pilot project supported by the Integrated Plan for Regional Development. IPIMAR took responsibility for the entire process from site selection to reef design and monitoring. The following criteria were used for siting and designing the artificial reefs:

- Reef placement, on clean sandy/muddy bottom nearby in-shore nursery areas, to increase habitat complexity and nursery grounds
- Reef placement on deteriorated bottoms to mitigate low biodiversity
- Reef placement at least at a 15 m depth to guarantee module stability and avoid conflicts with an important local fishery (bivalve dredge)
- Reef modules with rough surfaces to favor settlement of sessile fauna, and large holes to avoid siltation
- Reef sets organized in a chaotic way to provide a heterogeneous habitat and shelters to fishes, crustaceans, and cephalopods

Figure 2.4 Modules deployed at artificial reef sites off Algarve, Portugal. A, small cubic concrete reef units; B, large rectangular concrete module; C, large octagonal concrete modules.

- Reef sets extending enough along the water column and the seabed to aggregate a wide range of species (benthic, nekton-benthic, and pelagic) and enhance natural populations
- Materials free of contamination and remain unchanged for many years

The Faro and Olhão reefs consist of sets of small-size (protection reef) and large-size (exploitation reef) concrete modules. The aim of the protection reef was to create a shelter for the juvenile fishes that cyclically migrate between the lagoon and continental shelf. It consisted of 735 small concrete cubic units (Figure 2.4A), each weighing 3 t with a unitary external volume of 2.7 m^3 (for a total volume of 2017 m^3), distributed in 21 groups, at depths of 15 to 22 m, occupying a total area of 39 ha. The exploitation reef consisted of 20 large reinforced concrete modules, with a total volume of 3036 m^3, comprising two different shapes: rectangular with a unitary external volume of 130 m^3 and weighing 30 t (Figure 2.4B), and octagonal with a unitary external volume of 174 m^3 and weighing 40 t (Figure 2.4C). These large modules were distributed in five groups, 25 to 40 m deep, and occupy a total area of 21 ha. The distance between the protection and the exploitation reefs varied from 0.9 km in the Olhão reef system to 2.7 km in the Faro reef system. Between 1998 and 2003, six new larger artificial reefs systems were deployed off the Algarve, with another scheduled for 2010. Each of these reefs has at least 2940 small-size concrete modules and 36 octagonal large-size modules. The reef complex will contain more than 21,500 modules, occupying in a discontinuous way a total area of 4 km^2, with an estimated area of influence of 70 km^2. To the authors' best knowledge, the Faro/Olhão artificial reef is the largest structure of this type in Europe (8.2 km in length by 1.5 km wide).

Objectives

The program had the following objectives:

1. Provide protection to the juvenile populations of fish, particularly to those of commercial interest
2. Promote biodiversity and fishing resources
3. Increase fishing yields on deteriorated grounds
4. Test alternative (and innovative) tools for marine environmental and fisheries management

Research Findings

1. The two pilot artificial reefs showed similar patterns in terms of residence status and spatial occupation of fish assemblages, as determined by visual censuses (Santos et al., 2005). Most species recorded were considered resident (45%), with 30% transient and the remaining 25% occasional. There was a strong affinity between the fish assemblages from the reefs and those of the nearby Ria Formosa lagoon (Algarve's most important nursery ground), as described by Santos et al. (1996). This reflects seasonal migration from the lagoon, namely, for juveniles of the Sparidae family (Santos et al., 2005; Leitão et al., 2009). High densities of young-of-the-year fishes have been systematically and primarily observed during summer and late fall, corresponding to periods of major migratory events.

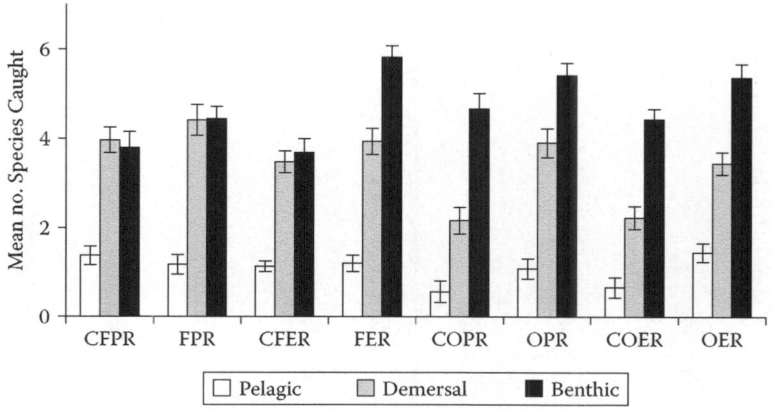

Figure 2.5 Mean number of species caught (±standard error [SE]) per group of fish at Faro and Olhão artificial reefs, Portugal, at protection reefs (FPR = Faro Protection Reef [PR], OPR = Olhão PR), at exploitation reefs (FER = Faro Exploitation Reef [ER], OER = Olhão ER), and at control sites (CFPR = Control, at Faro Protection Reef, CFER = Control, at Faro Exploitation Reef, COPR = Control, at Olhão Protection Reef, and COER = Control, at Olhão Exploitation Reef. (From Santos M.N. and C.C. Monteiro. 2007. *Hydrobiologia* 580: 225–231. With permission.)

2. With regard to reef ecology, Leitão et al. (2007) showed that the diet of *Diplodus sargus* was strongly associated with prey availability on the artificial reefs, highlighting the importance of these man-made habitats for increasing biomass of this species. Leitão et al. (2009) classified the artificial reefs as essential fish habitat, describing their role for three *Diplodus* species as recruitment, growth, and nursery areas for juveniles, and spawning/mating areas for adults.
3. Meanwhile, data on fishing yields were derived from direct site observations, records of commercial fleet landings, and bottom gillnets deployed following standard local fishing practices (Santos and Monteiro, 1997). The mean number of species caught per standard catch was continuously higher at the artificial reefs than at the control sites (Figure 2.5). Deployment of the man-made structures did not, however, change the composition of the fish assemblages caught by the gillnets, nor the equilibrium of the community, since the relative proportion of the different functional groups of fishes remained stable. Increased abundance of *Dicentrarchus labrax* (a highly valuable species) has been registered since the enlargement of the Faro artificial reef (from 0.6 to 12.2 km^2) in 2003 (IPIMAR, unpublished data). Such positive impact was noticed by the local fishing community, which again started using bottom longlines (formerly the most common fishing gear) instead of gillnets (Leitão et al., 2009).
4. Fishing yields from the artificial reefs continually exceeded those from the control sites (1.3–2.6 times) in terms of the mean CPUE (catch per unit of effort) in weight (see Figure 2.6 and Santos and Monteiro 1997, 1998, and 2007). Whitmarsh et al. (2008) showed that catches at the artificial reefs would be expected to earn approximately €13 per unit of effort more than they would at the control sites, other things being equal, when they were first installed. However, data collected in recent years show that as a consequence of the increasing differences between the catch rates from the reefs and control sites, as well as a slight increase in market prices, this figure is now between €25–35 per unit of effort more at the reefs than at the control sites.

Conclusion

The goal of the pilot project was realized; it provided sufficient objective information that led to the decision to "scale up" to a larger reef complex. Based on the positive outcome from the Algarve reefs (e.g., new reproductive and nursery sites, new food resources, production of new biomass, fishing income), it is reasonable to foresee that the use of artificial reefs will increase worldwide, but it should not necessarily be employed as the key element to guarantee sustainability of coastal

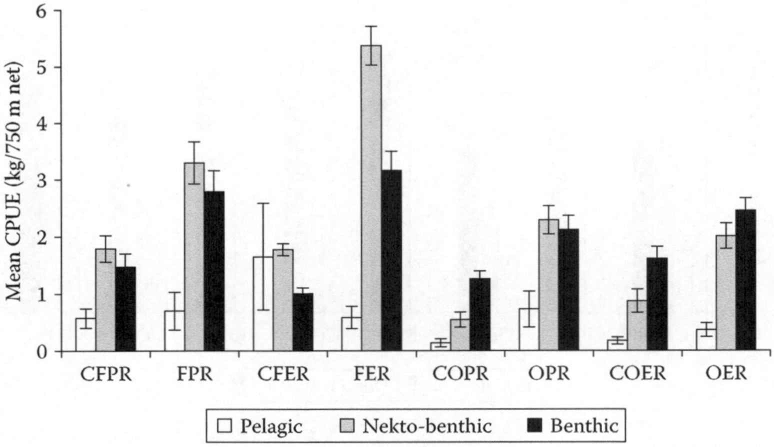

Figure 2.6 Mean CPUE in terms of weight (±SE) per group of fish at Faro and Olhão artificial reefs, Portugal. (Abbreviations for sites are the same as in Figure 2.5. (From Santos, M.N. and C.C. Monteiro. 2007. *Hydrobiologia* 580: 225–231. With permission.)

marine environments and fisheries, but rather as a complementary management action. An example of this is the identification of artificial reefs as essential fish habitats, which can be an important step toward the creation of Marine Protected Areas based on artificial reef deployment. In Portugal, there are not specific regulations or plans for artificial reefs. Multispecies fisheries make it difficult to promulgate species-specific regulations. Thus, deployment of a reef to meet a certain quota for fishery production or income has not been achieved yet (see the chapter by Santos et al. in this volume for related information).

Improving Small-Scale Fisheries and Reducing User Conflict

Situation

The western side of the central Adriatic Sea bordering Italy, is characterized by a flat, sandy mud seabed with some outcrops of limestone. Heavy river runoff carries a significant nutrient load, which causes eutrophication. The consequent high aquatic primary production provides abundant food for planktivorous fishes and filter feeders, mainly bivalves such as baby clam (*Chamelea gallina*), mussels (*Mytilus galloprovincialis*), and oysters (*Ostrea edulis* and *Crassostrea gigas*). The area is intensively exploited by several fishing activities. Hydraulic dredges for baby clam and small-scale fisheries using set nets and traps occur from the coast up to 5.5 km offshore, where trawling is forbidden by law. Pelagic trawling, otter trawling, and beam trawling should take place outside, but the vessels also fish illegally inside the forbidden coastal zone, often damaging the fixed gears. These fishing activities compete for space and resources; for example, small-scale fishing competes with both the hydraulic dredges for space and the illegal trawling for space and resources.

Goal

The need is to improve local small-scale fisheries through the enhancement and diversification of the fish population as well as development of extensive mussel culture, and reduce user conflict through reduction of illegal trawling. The first experimental artificial reef was constructed with financial assistance of the National Council of Research (CNR), Institute of Marine Sciences (ISMAR) of Ancona, and the Italian Merchant Marine Ministry (which regulated Italian fisheries up to 1993).

More recent initiatives were carried out by CNR–ISMAR, fishermen's associations, or the Marche Regional Authority with the financial assistance of the European Community. Reefs were planned and designed by CNR–ISMAR in cooperation with the fishermen's associations and, in some cases, the Marche Regional Authority. Monitoring of reefs was carried out by CNR–ISMAR.

Reef Description

Construction of small- and large-scale reefs was from 1974 to 2006. (Various configurations of blocks, cages, poles, and plates, described below, are depicted in Figure 2.7.) The following criteria were used for siting and designing the artificial reefs:

- Distance from the coastline to be ~5.5 km
- Reef units heavy enough to stop illegal trawling
- Reef units/structures placed sufficiently close to each other to block trawler entry inside the reef
- Antitrawling artificial reefs extend as much as possible, to protect a wide coastal zone
- Reef units/structures to extend enough along the water column to aggregate/enhance fishes with different habits (benthic, nekton-benthic, and pelagic) for diversifying the local fish population and developing natural populations of mussels and oysters
- Reef units with rough surfaces and holes to favor settlement of sessile fauna and provide shelter to fishes, crustaceans, and cephalopods
- Materials free of contamination and remaining unchanged for many years

Two small-scale reefs were experimental, while the third was constructed on behalf of a fishermen's association. They were placed at 10–12 m depth and 0.8–2 km from the shoreline, with two close-to-natural hard outcrops and the other in the open sea far from natural rocky habitats. Reefs were either concrete cubic block (2 × 2 × 2 m; weight 13 t) and provided with rough surfaces to favor settlement of benthic organisms and holes to provide shelter for fishes and macroinvertebrates, or else concrete cages for mussel culture (base: 4 × 5 m, height: 5 m). Blocks were arranged in 2-layer pyramids placed 15–20 m apart on gravel "mattresses" to avoid subsidence. The concrete cages were placed among the pyramids. Large-scale reefs were placed on the external border of the coastal zone, from 13 to 15 m depth 2.8 to 5.5 km from the shoreline. All the reefs were placed far from natural rocky habitats. Three arrangements of varied materials were employed: Concrete cubic blocks were arranged in 3-layer pyramids and deployed on gravel "mattresses." Stone piles were placed between the pyramids to make the reef continuous, and two wooden vessels were sunk at the center of area delimited by the pyramids. Small and big cubic blocks were dispersed in the overall area around the oasis. Concrete cubic blocks were arranged in 2-layer pyramids and deployed on concrete bases

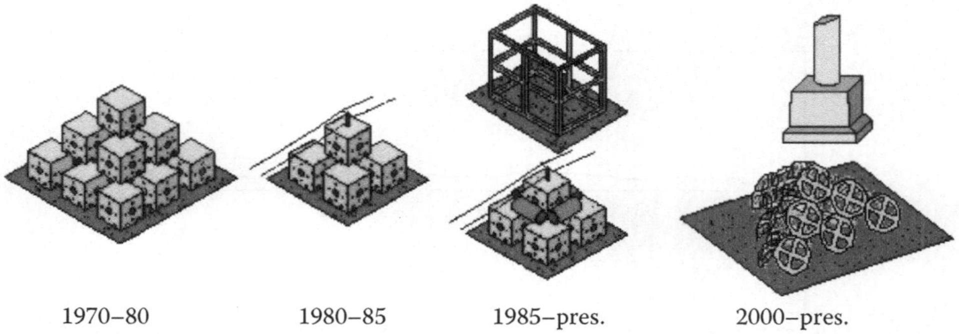

Figure 2.7 Appearance of artificial reef structures used during 1970 to 2010 in the coastal Adriatic Sea, Marche region of Italy.

to avoid subsidence. Antitrawling concrete poles were placed among the pyramids, both on the reef perimeter, at 20 m from each other, and inside the area, at 40 m from the pyramids. Finally, concrete plates were assembled in pyramidal structures (base: ~12 m^2, height: 3 m), which were deployed inside the reef area at 50–60 m from each other. Antitrawling concrete poles were placed 40 m apart along the reef perimeter.

Objectives

1. Prevent illegal trawling in the coastal area
2. Reduce conflicts among fishermen
3. Improve local reef-dwelling fish and hard-substrate bivalve populations
4. Enhance small-scale fisheries by creating suitable zones for fixed gears

Research Findings

Monitoring programs usually started 1 or 2 years before the reef construction and finished from 3 to 5 years after. (1) The deployment of the artificial reefs induced an increase of species richness (Figure 2.8) and diversity in the fish assemblage of the natural habitat. This was mainly due to the constant occurrence of reef-dwelling and partially reef-dwelling species, which are rare in the natural sandy mud habitat (Bombace et al., 2000; Fabi and Scarcella, 2007; Fabi et al., 2007, 2008). This effect was also evident by comparing the composition of the professional fishermen's catches obtained at the artificial reefs and in the open-sea soft bottoms (Figure 2.9; Fabi and Grati, 2005). (2) The use of different typologies of units favored the diversification of the reef fish assemblage. For example, the four-block pyramids are especially efficient in attracting benthic and nekto-benthic reef-dwelling fish, some of which settle permanently at the reef. On the other side, the concrete cages for extensive mussel culture are more efficient for gregarious, nekto-benthic, and pelagic

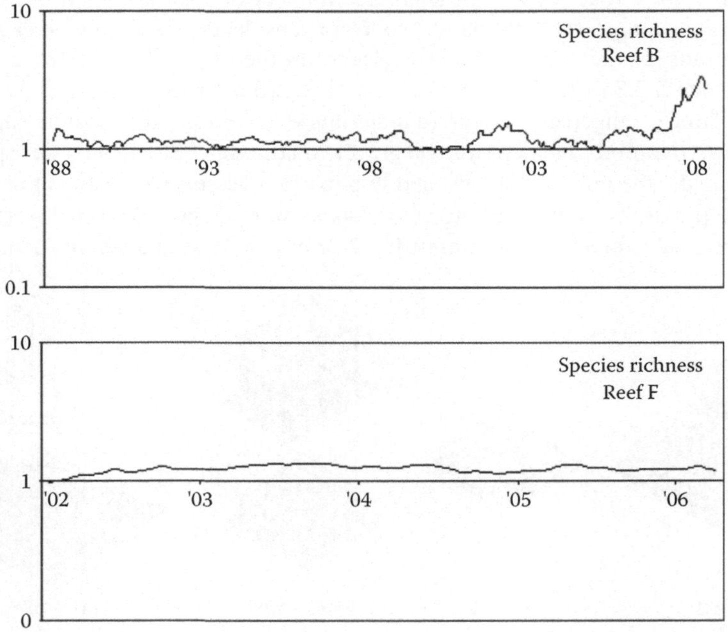

Figure 2.8 Species richness expressed as logarithmic trend of the ratio, reef/control (12 months running average) at Marche region reef sites, Italy.

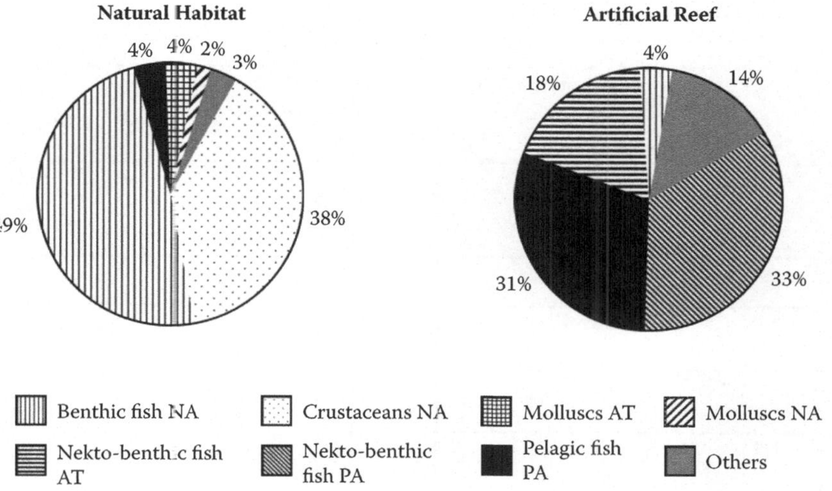

Figure 2.9 Species composition of the catches obtained with set nets by the professional fishermen at artificial reefs in Marche region, Italy. (NA = species nonattracted by hard substrates; PA = species partially attracted by hard substrates; AT = species attracted by hard substrates.)

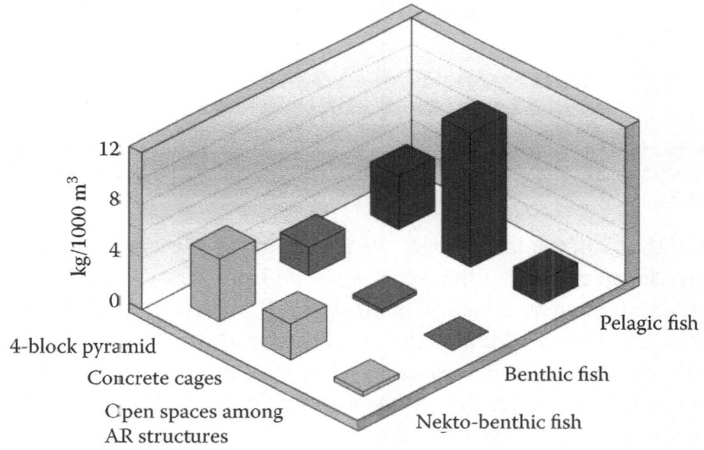

Figure 2.10 Biomass of benthic, nekto-benthic, and pelagic fish recorded at the 4-block pyramids, the concrete cages for mussel culture, and in the open spaces among the man-made structures at artificial reef B, Marche region, Adriatic Sea, Italy.

fish, which tend to concentrate close to structures that extend along the water column (Figure 2.10; Bombace et al., 1997). (3) Higher catch rates of reef-dwelling benthic and nekto-benthic fish as well as of partially reef-dwelling fish have been usually recorded at the artificial reefs in respect to the open-sea control sites (Figure 2.11) (Bombace et al., 2000; Fabi and Scarcella, 2007; Fabi et al., 2007, 2008). (4) A few benthic and nekto-benthic fish, such as the black scorpion fish, *Scorpaena porcus;* the brown meagre, *Sciaena umbra;* and the shi-drum, *Umbrina cirrosa,* established resident populations at the reefs. For example, at reef B, the catch ratio between the reef and control site of *S. porcus* ranged from 0.9 to 1.8 in the 16 years after the reef deployment (1988–2003) and

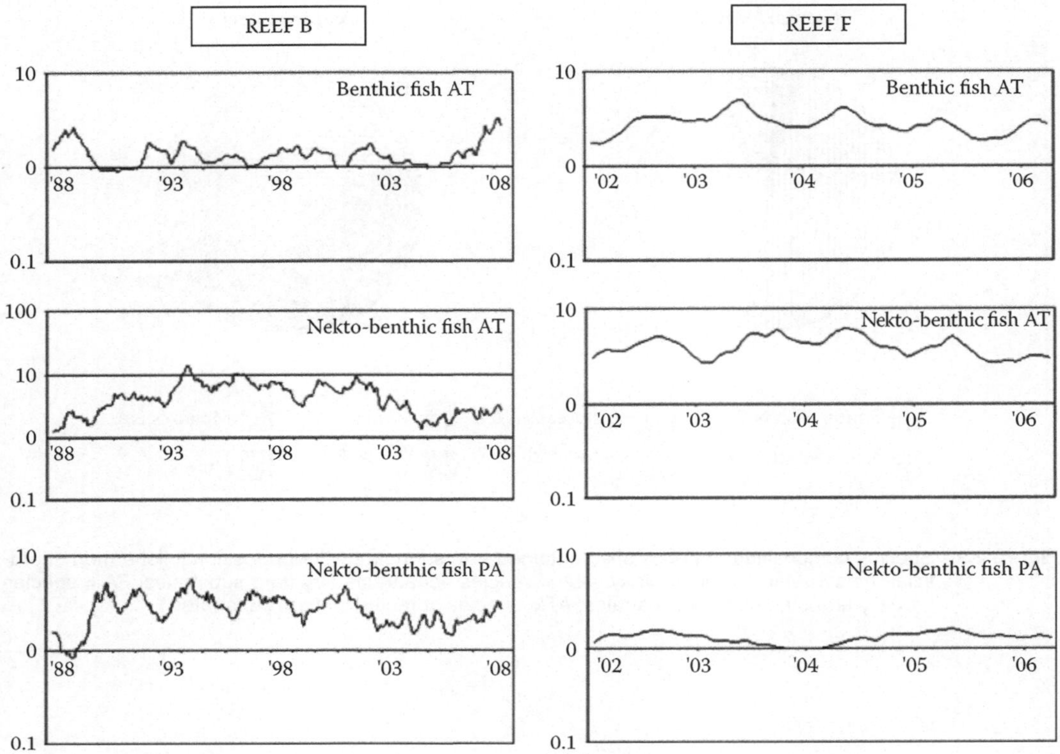

Figure 2.11 Mean catch of AT benthic, AT nekto-benthic, and PA nekto-benthic fish (kg/500 m/12 h), as logarithmic trend of the ratio, reef/control (12 months running average). (Abbreviations as in Figure 2.9.)

gradually increased to 3.3 in the last 5 years. At reef F, in the 6 years after the deployment of the artificial structures, the catch ratio of this species varied from 1.7 to 4.7, with a mean value of 2.8. The catch ratio of *S. umbra* went from 0.9 to 8.8 at reef B, with a mean value of 3.2, and from 1.7 to 4.2 (mean value = 2.8) at reef F. Finally, the mean catch ratio of *U. cirrosa* at reef B amounted to 2.6 in the period 1990–2004, but it decreased to 1.1 in the subsequent years. At reef F, it remained rather stable during the 6 years after the reef deployment, ranging from 1.4 to 3.0 with an average of 2.0. (5) The brown meagre, *S. umbra,* catches around 70% of its prey items on the artificial substrate, while the annular seabream, *Diplodus annularis,* and the striped seabream, *Lithognathus mormyrus,* gain 50%–57% of their nutrition from the artificial structures, also feeding on the soft seabed outside the reef and in the open sea (Figure 2.12). The two-banded seabream, *Diplodus vulgaris,* was seen eating on the artificial substrates (Fabi et al., 2006). (6) The presence of artificial structures seems to benefit *S. porcus* populations in terms of maximum size, growth performance, and longevity (Figure 2.13) (La Mesa et al., 2010). (7) The yearly mean biomass of commercial mussels (shell height ≥5 cm) settled on the units of the artificial reefs ranges from 17 to 55 kg m^{-2}.

Conclusion

Deployment of artificial reefs along the coast of the Marche Region of eastern Italy has produced positive effects from both the ecological and the socioeconomic point of view. Artificial reefs induce enrichment and diversification of the natural habitat fish assemblage, concentrating reef-dwelling fishes. They also contribute to increase the stock of some of these species, providing

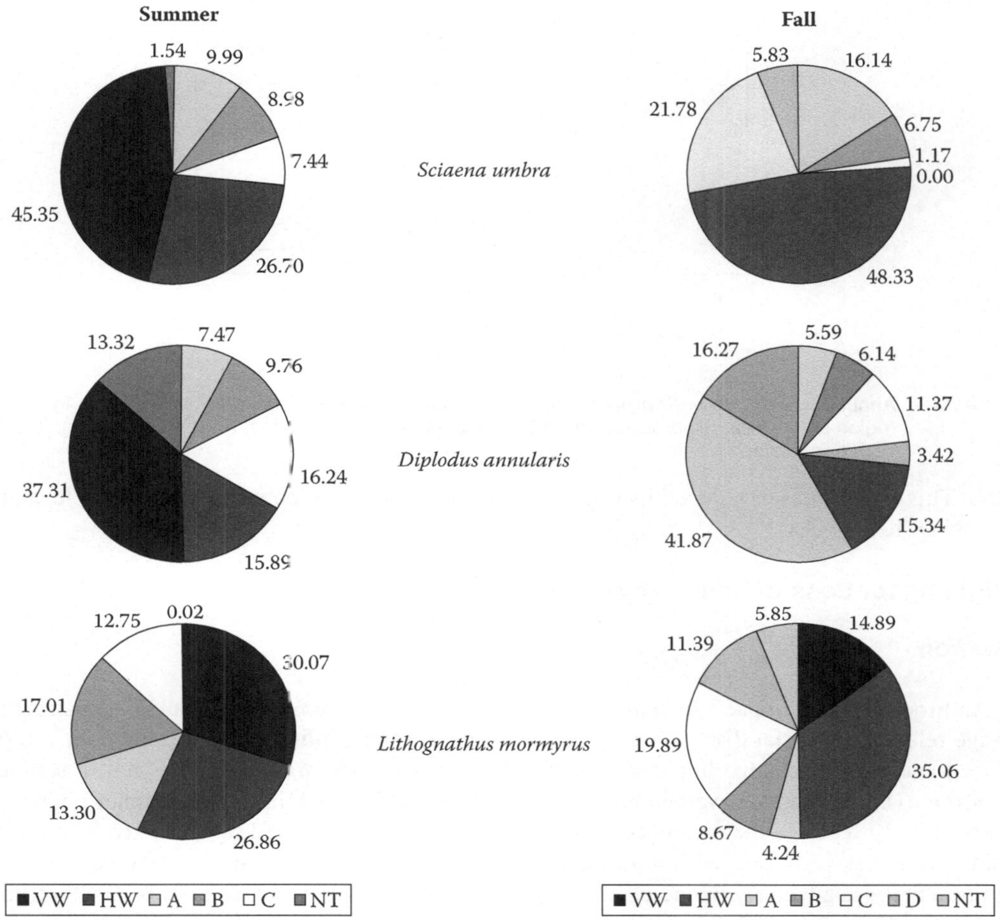

Figure 2.12 Contribution of different environments to the diet of three fish species at Marche region reefs, Italy. (NT = prey items not found in the investigated areas; VW = vertical walls of the reef structures; HW = horizontal walls of the reef structures. Summer: A = inner sites within the reef at 2, 5, and 10 m from the structures and outer sites at 2 m from the structures; B = outer sites at 5, 10, 20, and 50 m; C = open-sea control sites. Fall: A = inner sites within the reef at 2, 5, and 10 m from the structures; B = outer sites at 2, 5, 10, and 20 m; C = open-sea control sites; D = outer site at 50 m from the structures. The different environments were identified in each season on the basis of a cluster analysis.) (Reproduced from Fabi, G. et al. *Bull. Mar. Sci.* 78(1): 39–56, 2006. With permission.)

additional food and favoring the growth rate. The artificial reefs also contribute to increase the natural populations and exploitable biomass of filtering bivalves, whose larvae would be lost for the lack of suitable substrates. Where large-scale reefs were deployed, the conflicts between trawling and inshore fisheries (small-scale fisheries with set gears and hydraulic dredges) are noticeably reduced, as the former cannot operate in the coastal areas. Although the existing artificial reefs are not large enough to guarantee a satisfactory income to all the local small-scale fishermen year around, they represent a means for shifting part of the fishing effort to different resources rather than those traditionally exploited. The diversification of fishing practices and target species assumes a noticeable economic importance for the fleets, which operate on flat, sandy-mud bottoms and, hence, usually employ a low number of specialized gears aimed to catch a limited pool of species (Scarcella et al.,

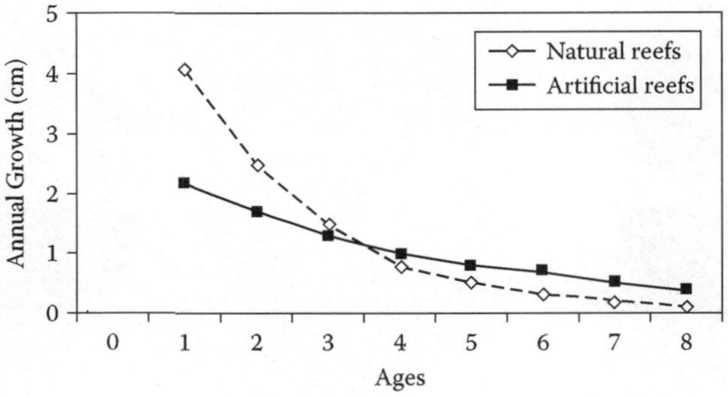

Figure 2.13 Annual growth rate of *Scorpaena porcus* at artificial reefs and natural rocky habitats in Marche region study area. (Reproduced from La Mesa, M. et al. *Sci. Marina*, 74(4): 677–685, fig. 8, 2010. With permission.)

2004). This contributes to reduced fishing effort on some species, such as the common sole, and to preserving their stocks.

Mitigating for Loss of Kelp Habitat

Situation

Anthropomorphic impacts, including nearshore seafloor pipe construction and deconstruction, sewage release, industrial discharges, and once-through water cooling systems, along with storm-induced and climate-change impacts on kelp (*Macrocystis pyrifera*) and kelp habitat in southern California, United States, have become more evident with the doubling of the human population over the last 40 years. In the region, commercial fishing has been reduced significantly, recreational fishing effort has increased with reduced catch, hard-substrate habitat has become covered with sand or otherwise adversely impacted, and natural cyclical rebounds in kelp have been slower than expected and/or measurably adversely impacted. Of concern to this case study is a 25.4 t reduction in the standing stock of kelp bed fishes from operation of the San Onofre Nuclear Generating Station. Hard-substrate habitat restoration and management is a logical goal of responsible fisheries management for the region since hard substrate with kelp forests provides about 19 times the ecological diversity and biomass compared to sand-bottom habitats (W. North, California Institute of Technology, personal communication) and only 20% of the natural nearshore area is rocky or hard and only some of this area is in the optimum depth range for kelp (10–16 m).

Goal

A state resource agency, the California Coastal Commission, has required mitigation in the form of a constructed artificial reef that will provide in-kind replacement for the loss of kelp forest habitat at San Onofre. Thus, the fisheries and ecosystems management goal of this in-kind mitigation project is to compensate for the loss of 61 ha of kelp habitat resources, including giant kelp understory algae, invertebrates, and fishes. The Wheeler North Reef at San Clemente, California, is the largest man-made kelp reef constructed in the United States.

Reef Description

Phase 1, the experimental reef (a 9-ha seafloor footprint, 18,500 tons of quarry rock, and 15,400 tons of broken concrete) was built in August–September 1999 (Figure 2.14). A 1999–2004 five-year

Figure 2.14 Constructing the Wheeler North Reef in California, United States: Skip-loader moving quarry rock into position to drop over the side of the barge.

biological feasibility performance study was conducted to verify the project's intended objectives could be met (Reed et al., 2005). The Phase 2 build-out reef (69-ha seafloor footprint, approximately 126,000 tons of boulder-size quarry rock) was built in June–September 2008. The project construction costs of the two phases are approximately U.S. $20 million. (An additional U.S. $20 million has been allocated for performance studies through 2025.) Criteria used for design and siting of reefs were

- Water depth between 11.5 and 15 m, which is optimal for local kelp
- Sand depth of 0.5 m so that reef substrate would not become buried
- Avoid hard-substrate areas that could support natural kelp beds
- Construction in areas with no kelp presence greater than 1 year in the historical database from 1967 to 2004
- Rock not to be deposited within 50 m of areas of special interest (e.g., existing fishing sites)
- Construction-barge anchor sites not located in a way that would impact areas of special interest, nor shall anchors be placed in areas that will impact existing hard substrate
- Located close to the natural San Mateo Kelp Bed so that no kelp seeding will be necessary
- The reef shall be constructed of single-layer (<1 m) quarry rock distributed on the seafloor at a percent coverage of at least 42% bottom coverage (as determined by point-of-contact method), so that the entire reef will cover at least 68 ha, and that there be a minimum overlap of rock material

Objectives

Specific, measurable performance standards were established to judge the success of the reef (and to determine whether remediation or adaptive management on the reef is necessary):

1. At least 90% of the hard-substrate area must remain available for attachment of reef biota
2. Sustained giant kelp density of at least four adult plants per 100 m^2
3. Resident fish assemblage similar in density and species number to natural reefs within region
4. Young-of-year fish assemblage similar in density and species number to natural reefs within region
5. Standing stock of fishes on the mitigation reef shall be at least 28 U.S. tons (25.4 t)

6. Benthic community (both algae and macroinvertebrates) coverage or density and number of species similar to natural reefs within region
7. Important functions of the reef shall not be impaired by undesirable or invasive benthic species

Research Findings

The ensuing 1999–2004 Phase 1 study produced promising results (Reed et al., 2005, 2006): (1) All the artificial reef Phase 1 designs and rock densities typically supported more species of resident fishes (mean = 10) than the natural reference reefs (mean = 5), in standard transects of 160 m^2. (2) Fish species richness was greatest near the bottom, where roughly twice as many species (mean = 9, artificial reefs; 4.5, natural) were observed compared to the midwater and surface regions. (3) The overall assemblages of resident fishes on the artificial reefs showed a high degree of similarity to that of the natural reefs, except for the appearance of kelp perch (*Brachyistius frenatus*) only on artificial reefs. (4) Statistical analysis showed that fish standing stock (20–75 t at rock artificial reefs versus 10–40 t at reference sites), density (300 versus 33 adult fish per transect), species richness (10 versus 5 per transect), and recruitment on all the artificial reef designs were either similar to or greater than that observed at the nearby natural reefs. (5) Rock density within the reef footprint had a substantial influence on the fish assemblage, with higher densities and numbers of species occurring on artificial reef modules with greater bottom coverage: for example, at "low" density and 17%–40% coverage, there were 320 fish per transect; at "medium" density 400; and at "high" density 550 fish. (6) Fish production was not evaluated on the 9-ha Phase 1 reef. (7) Giant kelp was found to be abundant on the Phase 1 reef modules, with 30 plants per 100 m^2 in 2003 and 20 in 2004 versus 20 and 4, respectively, at the reference reefs.

Conclusion

The definition of quantitative success criteria for this reef system made evaluation of performance objective, whereby the goal of mitigation, on a pilot basis, was judged to be attained. This led to the decision to invest substantial funds to build the large-scale Wheeler North Reef.

Enhancement of Fisheries Resources and Revenues in Marine Ranching

Situation

Coastal fisheries production in Korea has decreased since the 1970s, due to causes such as overfishing, pollution, and land reclamation. The proclamation of 200-mi Exclusive Economic Zones by many coastal nations has made deep-sea fishery access more difficult. In response, the Korean government is promoting a national marine ranching effort. This is similar to terrestrial farming where young livestock are put out to graze on pastureland, later to be harvested. In this project, hatchery-raised fish species are released into a managed coastal area where the environment has been modified through the deployment of artificial reefs, to build up fishing productivity sustainably (Kim et al., 2009). Marine ranching projects in Korea began in 1998, and will be promoted for 8–11 years with a total funding commitment of approximately $128 million (U.S.). One of them, at Tongyong, began in 1998 with projected funding of $24 million (U.S.) and ended in 2006. Approximately 99,000 m^3 of artificial reef volume was installed at 139 sites in an area of 20 km^2 (KMMAF 2006). The investment for artificial reefs was approximately $5.1 million (U.S.). Surveys (KMMAF 2006; Kim et al., 2007, 2008) have indicated that many artificial reefs in Korea have proven to be useful tools for enhancing the resources of fisheries.

Goal

Creation of artificial reefs aims to enhance fisheries resources in the Tongyong marine ranching area by increasing food supply, spawning sites, and nursery areas for designated fish species. Two strategies are used: One is through enhancing the resources potential by increasing primary productivity or through habitat enhancement using artificial reefs, so as to promote the regeneration of resident fish populations. The other is the "resource-added" type using hatchery-produced seed of the resident fish species and releasing them.

Reef Description

Prior to developing artificial reefs in Tongyong marine ranching waters, field surveys were conducted and data from fishing grounds and mariculture areas were evaluated. While numerous natural reefs exist in this area, sediments chiefly are muddy sand. Primary and secondary productivities are relatively rich, with scattered spawning and nursery grounds. The number of fish species in Tongyong coastal waters was 89, with 59 associated with artificial reefs (Kim et al., 1999). Two target fish species were selected for enhancement: Rockfish (*Sebastes schlegeli*) and porgy (*Pagrus major*). As rockfish tend to gather in dark and small spaces whereas porgy prefer a complex reef structure with a large volume (Kakimoto, 1998), the design of reef structures was made based upon their biological requirements. Reefs are classified by two types, on the basis of void space: Frame structural reefs such as the high-storied reef (for porgy) contain over 80% void space, while face structural reefs such as the box reef (for rockfish) have less than 80% void space (Figure 2.15). Reefs were installed at depths between 5 and 50 m.

Objectives

1. Newly constructed reefs must provide spawning and nursery grounds and refuge for target fish species.
2. Effectiveness of the new reef design should be equal to or higher than that of existing designs deployed in the same conditions.
3. The monetized value of a reef set benefit should exceed the cost of reef construction.
4. Amount of fish stock after placement of artificial reefs in the marine ranching area should be increased more than two times the level before deployment of reefs.
5. Income of fishermen engaged in the marine ranching area should be higher than that of neighboring regional fishermen.

Research Findings

Kim et al. (2007) surveyed 17 fish species and estimated annual catch volume of 1452 kg by trammel gillnet at a multifunctional high-storied steel reef (15 × 15 × 15 m), 43 m deep in the marine ranching area, with rockfish and porgy constituting 52% of total fish weight. When this value of catch is expressed in terms of "benefit" of the reef, the ratio of benefit value to construction cost of the reef is 1.41. This means that the economical efficiency of the reef is relatively high. Meanwhile, Kim et al. (2008), using trammel gillnets, found no great difference in the number of fish species at a box reef and a natural reef. However, the fish catch at the box reef was 4.4 times greater than that obtained at a natural reef. Rockfish were caught at the box reef only, and twice as many porgy were fished there as on the natural reef. Artificial reefs in the marine ranching area may affect the amount of fish stock. Fish stock in 1998 was estimated at 118 tons overall and at 40 tons for rockfish and, in 2006, when the project ended, it was estimated at 749 tons and at 180 tons for

Figure 2.15 Artificial reef structures used in Tongyong marine ranching project, Korea. A, Box steel reef (9.0 × 9.0 × 10.2 m); B, two-storied box-type steel reef (14 × 14 × 9 m); C, M300 steel reef (8 × 10 × 6 m); and D, Box reef (3 × 3 × 3 m), with more open upper areas, and more covered lower interior space.

rockfish (KMMAF, 2006). Income of fishermen in the marine ranching area increased about 80% in 2006 over that of 2002 (KMMAF, 2006).

Conclusion

The increased fish stock in the marine ranching area may come from various factors such as designation of a marine protected area and administrative measures that limit fishing activities within it. However, the observation of increased fish biomass at artificial reefs leads to the conclusion that reefs can be one of the useful tools for enhancing the resources of fisheries.

DISCUSSION

An emerging knowledge base of long-term ecological and socioeconomic performance of artificial reefs indicates tangible contributions of this technology both to varied, on-site stakeholders and to the broader scientific and management sector. In the four preceding case studies, ecological functional equivalency between artificial and natural reefs seems to be present for at least attributes of the system that were studied (e.g., production of biomass), and stakeholders seem to express satisfaction with management aims (e.g., fishing enhancement). The concerns of all five sets of stakeholders identified at the beginning of this chapter seem to have been engaged. For example, reef planning was done by resource managers and scientists; pilot studies established positive biological and investment expectations for full-scale reefs, and ultimately, the user received a satisfactory product (e.g., fish harvest, restored habitat).

Individually, and in concert, these four situations embody broader attributes that may unify, guide, and foster future investigations and management in various branches of aquatic sciences: First, sustainability of resource utilization is a universal concern that sets reefs in a broader social and environmental context. Secondly, maintaining or enhancing biodiversity is a common goal that extends reef objectives beyond exploitation. Next, engaging a broad variety of community stakeholders in reef development, funding, and assessment efforts in partnership with scientists provides benefits to education. Also, using artificial reefs as sites to investigate ecological principles or for manipulative experiments affords study designs that could not be conducted at natural habitats. Meanwhile, promoting multidisciplinary studies that include teams of ecologists, engineers, and social scientists offers a powerful approach for addressing complex environmental issues, as well as a means for assembling datasets for modeling of reef systems. And, adaptation of fundamental knowledge of species life history requirements to the design of physical structure of reefs combines basic research with the interests of society for applications. These and other considerations may be useful in planning new research and management related to both artificial reefs and fisheries and ecosystems in general.

We suggest that a global search of artificial reef applications would reveal other situations that emulate the success of the programs in Portugal, Italy, Korea, and the United States. As a guide to assessing or "grading" the attributes of other situations, we propose the list of attributes presented in Table 2.2. (We note that more established arms of fishery science, such as aquaculture, stock enhancement, and, of course, policy and resource allocation, might benefit from a similar review.)

Table 2.2 Criteria Proposed for Evaluating the Degree to Which An Artificial Reef Application Contributes to Fishery Science and Management of Aquatic Ecosystems

- Is the reef program nested in a larger natural resource management context?
- Has the potential impact of the reef upon the broader ecosystem been determined?
- Are complementary stakeholders engaged in appropriate aspects of the reef program (e.g., planning, funding, monitoring)?
- Are best available engineering data and practices for reef construction, durability, and maintenance followed?
- Are best available biological data and practices available (or being acquired) and being used in design of habitat structure (e.g., profile, complexity) consistent with species life histories?
- Are there reasonable forecasts of economic and social impacts, benefits, and costs for the project?
- Are goals of the reef program clearly stated?
- Are objectives of the reef program specific, and defined quantitatively, so that attainment can be measured according to "success criteria?"
- As necessary, are pilot studies used to test innovative or expensive concepts prior to full-scale reef deployment?
- If the reefs are to be used to establish or enhance biological assemblages, are quantitative measures of principal community attributes defined, such as for production (e.g., expected growth and addition of biomass), recruitment, or other life history factors?
- If the reefs are to be used in restoration, what is the baseline or other level of condition which is sought, and is it quantified?
- Is rigorous, interdisciplinary monitoring of reef performance to be conducted before, during, and after construction?
- Is an accompanying research plan, including hypothesis-driven studies, to be implemented to augment performance monitoring?
- Are the scientific results of reef studies being published in scientific journals, and data being incorporated into long-term formal databases, as well as Geographic Information Systems and ecological models as available?
- Is there opportunity for training of students as a means of enhancing research and management skills for developing professionals?

REFERENCES

Bombace, G., G. Fabi, L. Fiorentini, and A. Spagnolo. 1997. Assessment of the ichthyofauna of an artificial reef through visual census and trammel net: Comparison between the two sampling techniques. Pages 291–305 in Hawkins, L.E. and S. Hutchinson, with A.C. Jensen, M. Sheaderand, and J.A. Williams, Eds. The Responses of Marine Organisms to Their Environments. *Proceedings of the 30th European Marine Biological Symposium*. University of Southampton, Southampton, U.K.

Bombace, G., G. Fabi, and L. Fiorentini. 2000. Artificial reefs in the Adriatic Sea. Pages 31–63 in Jensen A., K. Collins, and A. Lockwood, Eds. *Artificial Reefs in European Seas*. Kluwer Academic Publ. Dordrecht, The Netherlands.

Fabi, G. and F. Grati. 2005. Small-scale fisheries in the maritime Department of Ancona (Central Northern Adriatic Sea). *AdriaMed Technical Documents*. 15: 64–84.

Fabi, G., S. Manoukian and A. Spagnolo A. 2006. Feeding behaviour of three common fishes at an artificial reef in the northern Adriatic Sea. *Bull. Mar. Sci.* 78(1): 39–56.

Fabi, G. and G. Scarcella. 2007. Monitoraggio volto alla valutazione degli effetti indotti dalla realizzazione della barriera artificiale a fini multipli P.to Recanati—P.to Potenza Picena. Report for the Marche Region—Service for Fishing Activities, Trade, and Protection of the Consumers, Hunting, and Sport Fishing. 29 pp.

Fabi, G., G. Scarcella, and A. Spagnolo. 2007. Monitoraggio volto alla valutazione degli effetti indotti dalla realizzazione della barriera artificiale a fini multipli Pedaso—Cupra Marittima. II anno di indagine (2006)—I anno dopo la posa in opera dei substrati artificiali. Report for the Marche Region—Service for Fishing Activities, Trade, and Protection of the Consumers, Hunting, and Sport Fishing. 47 pp.

Fabi, G., G. Scarcella, and A. Spagnolo. 2008. Monitoraggio volto alla valutazione degli effetti indotti dalla realizzazione della barriera artificiale a fini multipli Pedaso—Cupra Marittima. IV anno di indagine (2008)—III anno dopo la posa in opera dei substrati artificiali. Report for the Marche Region—Service for Fishing Activities, Trade, and Protection of the Consumers, Hunting, and Sport Fishing. 81 pp.

Garcia, S.M. 1990. Artificial reefs and fish aggregating devices in Southeast Asian fisheries: Management issues. Pages 153–163 in *Symposium on Artificial Reefs and Fish Aggregating Devices as Tools for the Management and Enhancement of Marine Fishery Resources*. Indo-Pacific Fishery Commission. Colombo, Sri Lanka.

Grossman, G.D., G.P. Jones, and W. Seaman. 1997. Do artificial reefs increase regional fish production? A review of existing data. *Fisheries* 22(4): 17–23.

Kakimoto, H. 1998. Creation of habitats for fishes by the artificial reef technique. *J. Nat. Coast. Fish. Dev. Assoc*. 1986: 1–23. (In Japanese.)

Kim, C.G., H.S. Kim, T.H. Kim, and B.S. Bae. 1999. Report on the Development of Marine Ranching Program in Tongyong, Korea. Korean Ministry of Maritime Affairs and Fisheries. 16 pages. (In Korean.)

Kim, C.G., H.S. Kim, T.H. Kim, and B.C. Baik. 2007. Report on the Development of Marine Ranching Program in Tongyong, Korea. Korean Ministry of Maritime Affairs and Fisheries. 17 pages. (In Korean.)

Kim, C.G., H.S. Kim, C.I. Baik, H. Kakimoto, and W. Seaman. 2008. Design of artificial reefs and their effectiveness in the fisheries of eastern Asia. *American Fish. Soc. Symp*. 49: 933–942.

Kim, C.G., B.G. Kim, Y.S. Choi and K.C. Cho. 2009. Development of sea ranching in the Yellow Sea. Pages 23–25 in *The Second Regional Mariculture Conference*. United Nations Development Programme/Global Environment Facility, Cheju, Korea.

KMMAF (Korean Ministry of Maritime Affairs and Fisheries). 2006. Pages 321–337 in Report on Tongyeong Marine Ranching Project. (In Korean.)

La Mesa, M., G. Scarcella, F. Grati, and G. Fabi. 2010. Age and growth of the black scorpionfish, *Scorpaena porcus* (Pisces, Scorpaenidae), from artificial structures and natural reef in the Adriatic Sea. *Sci. Marina*. 74(4): 677–685.

Leitão F., M.N. Santos, and C.C. Monteiro. 2007. Contribution of artificial reefs to the diet of the white sea bream, *Diplodus sargus* (Linnaeus, 1758). *ICES J. Mar. Sci*. 64 (3): 473–478.

Leitão, F., M.N. Santos, K. Erzini, and C.C. Monteiro. 2009. Artificial reef use by *Diplodus* spp. species: Importance for artisanal near shore fisheries. *Fish. Management Ecol*. 16 (2): 88–99.

Milon, J.W. 1989. Economic evaluation of artificial habitat for fisheries: Progress and challenges. *Bull. Mar. Sci*. 44(2): 831–843.

Milon, J.W. 1991. Social and economic evaluation of artificial aquatic habitats. Pages 237–270 in W. Seaman and L.M. Sprague, Eds. *Artificial Habitats for Marine and Freshwater Fisheries*. Academic Press, San Diego.

Morton, B. 1996. Artificial reefs, fish, and cyanide. *Mar. Poll. Bull.* 32(7): 522–523.

Pitcher, T.J., E.A. Buchary, and T. Hutton. 2002. Forecasting the benefits of no-take human-made reefs using spatial ecosystem simulation. *ICES J. Mar. Sci.* 59: S17–S26.

Pondella, D.J., J.S. Stephens, and M.T. Craig. 2002. Fish production of a temperate artificial reef based on the density of embiotocids (Teleostei: Perciformes). *ICES J. Mar. Sci.* 59: S88–S93.

Reed, D.C., S.C. Schroeter, and D. Huang. 2005. Final Report on the Findings and Recommendations of the Experimental Phase of the SONGS Artificial Reef Mitigation Project. Prepared for the California Coastal Commission. Marine Science Institute, University of California, Santa Barbara. 136 pp.

Reed, D.C., S.C. Schroeter, D. Huang, T.W. Anderson, and R.F. Ambrose. 2006. Quantitative assessment of different artificial reef designs in mitigating losses to kelp forest fishes. *Bull. Mar. Sci.* 78(1): 133–150.

Relini, G., M. Relini, G. Palandri, S. Merello, and E. Beccornia. 2007. History, ecology, and trends for artificial reefs of the Ligurian Sea, Italy. *Hydrobiologia* 580: 193–217.

Santos, M.N., C.C. Monteiro, and G. Lassèrre. 1996. Faune ichthyologique comparée de deux récifs artificiels du littoral de la Ria Formosa (lagune Portugal): Résultats préliminaires. *Oceanol. Acta* 19: 89–97.

Santos, M.N. and C.C. Monteiro. 1997. Olhão artificial reef system (south Portugal): Fish assemblages and fishing yield. *Fish. Res.* 30: 33–41.

Santos M.N. and C.C. Monteiro. 1998. Comparison of the catch and fishing yield from an artificial reef system and neighboring areas off Faro (south Portugal). *Fish. Res.* 39: 55–65.

Santos M.N., C.C. Monteiro, and G. Lasserre. 2005. Observations and trends on the intra-annual variation of the fish assemblages on two artificial reefs in Algarve coastal waters (southern Portugal). *Sci. Marina* 3: 415–426.

Santos M.N. and C.C. Monteiro. 2007. A fourteen-year overview of the fish assemblages and yield of the two oldest Algarve artificial reefs (southern Portugal). *Hydrobiologia* 580: 225–231.

Sayer, M.D.J. and T.A. Wilding. 2002. Planning, licensing, and stakeholder consultation in an artificial reef development: The Loch Linhe reef, a case study. *ICES J. Mar. Sci.* 59: S178–S185.

Scarcella, G., G. Bombace, G. Fabi, and F. Grati. 2004. Aumento controllato dello sforzo di pesca in una barriera artificiale dell'Adriatico settentrionale: Effetti sul popolamento ittico. *Biol. Mar. Medit.* 11(2): 21–32.

Ungson, J.R., Y. Matsuda, H. Hirata, and H. Shihara. 1995. An economic benefit-cost estimation of the red sea bream ranching in Kagoshima Bay, Japan. Pages 189–201 in D.S. Liao, Ed. *Proceedings of the Seventh Biennial Conference of the International Institute of Fisheries Economics and Trade*, July 18–21, 1994. Taipei, Taiwan.

Watanuki, N. and B.J. Gonzales. 2006. The potential of artificial reefs as fisheries management tools in developing countries. *Bull. Mar. Sci.* 78(1): 9–19.

Whitmarsh, D., J. Ramos, M.N. Santos, and C.C. Monteiro. 2008. Marine habitat modification off the Algarve (southern Portugal): An economic analysis of the fisheries and the prospects for management. *Ocean Coast Manage.* 51: 463–468.

Whitmarsh, D. and H. Pickering. 1997. Commercial exploitation of artificial reefs: Economic opportunities and management imperatives. CEMARE Research Paper 115. University of Portsmouth, Portsmouth, U.K.

Willmann, R. 1990. Economic and social aspects of artificial reefs and fish aggregating devices. Pages 384–391 in *Symposium on Artificial Reefs and Fish Aggregating Devices as Tools for the Management and Enhancement of Marine Fishery Resources*, May 14–17, 1990. Indo-Pacific Fishery Commission. Colombo, Sri Lanka.

CHAPTER 3

The Artificial Habitat as an Accessory for Improving Estimates of Juvenile Reef Fish Abundance in Fishery Management

Stephen T. Szedlmayer

CONTENTS

ABSTRACT

The juvenile stages of most reef fishes are difficult to sample due to cryptic habitat associations with complex reef habitats. Traditional methods such as trawls are usually inefficient even over reef habitats with relatively small (<1 m) relief. Artificial reefs offer an alternative method for the study of juvenile reef fishes as this information is essential in the effective management of benthic fisheries. If juvenile reef fishes can be attracted to artificial reefs, it is possible to use artificial reefs as part of the sampling method to study these important early stages. For example, the juvenile life stage of the recreationally and commercially important red snapper, *Lutjanus campechanus* (Poey 1860), from the northern Gulf of Mexico has received considerable attention because high mortalities during these stages are associated with later year classes. However, most abundance estimates of age 0–1 red snapper are based on trawl surveys, but as shown here, juvenile red snapper quickly become associated with reef structure during their first month or two of life, suggesting that juvenile abundance estimates using trawls may have both low precision and accuracy. Consistently high densities over several years of new recruits to artificial reefs (even in years of low adult red snapper abundance) and the lack of any increase in new recruit densities (even after significant shrimp trawl bycatch reduction, in recent years) indicates that early postsettlement processes influence subsequent red snapper year class strength. These conclusions are based on diver surveys of artificial reefs as opposed to traditional trawl sampling. The increased use of sampling juvenile reef fishes with artificial reefs as the "sampling tool" could provide new insight into the ecology and management of important reef fish species.

The various limitations and ecological bottlenecks of marine reef fishes have been extensively studied, with the greatest effort directed at coral reef fishes (e.g., Sale, 1991, 2002). Despite this effort, it is still unclear as to the relative contributions of environmental limitations on early life larval (Doherty, 1991, 2002) or postsettlement juvenile mortality limitations on subsequent adult reef-fish populations (Jones, 1991; Hixon, 1991; Hixon and Webster, 2002). This discussion continues (Armsworth, 2002; Anderson et al., 2007; Caddy, 2008; Oeberst et al., 2009) and is applicable to population regulation questions concerning red snapper stocks in the northern Gulf of Mexico. What has received less attention is the use of artificial reefs in the study of red snapper and their possible utility in defining the functioning mechanisms that affect red snapper stocks (Gallaway et al., 2009; Shipp and Bortone, 2009). Here, the early life history of red snapper is examined with inferences concerning the population dynamics based on artificial reef studies.

INTRODUCTION

Red snapper is an extremely important fish species in the northern Gulf of Mexico. They are harvested by both commercial and sport fishers and impact the economies of coastal communities tremendously. However, fishery assessment models indicate that the stock is considerably overfished (SEDAR7, 2005; SEDAR, 2009). A difficulty with red snapper is that management efforts have continually increased restrictions through bag limits (i.e., number of fish retained per trip), size limits, and total allowable catch over the last 19 years, yet the population models still show a severely overfished stock (SEDAR, 2009). These restrictions have led to extreme reductions in fishing effort, with the 2009 fishing season lasting only 2 months with a total catch near 2.3 million kg.

In contrast, there are several aspects of the red snapper life history and population dynamics that indicate the stock may be in better condition than predicted from previous assessments. First, there is an apparent expansion of the species to new habitats where no records of its previous occupation exist. For example, age-1 red snapper were recently collected 48-km inland within the city limits of Mobile, Alabama, a low-salinity location that has usually been considered outside of the normal red snapper habitat (Dute, 2008). Also, there have been several recent reports of red snapper from the southwest Florida coast, an area with few red snapper in the last 20 years (*Tampa Tribune*, 2009). Another difficulty with the assessment process is the lack of inclusion of larger and older red snapper collected in long-line catches on the outer portion of the continental shelf, which are typically not used in the management of this species (Mitchell et al., 2004). These populations of larger, older fish may represent a reserve population that could buffer extreme effect of fishing mortality. Also, some stock reports are more optimistic than others. Szedlmayer (2007) suggested higher spawning potential ratios compared to past estimates and, over the last 50 years, the construction of extensive artificial reefs along with oil–gas platforms has resulted in harvest levels that have exceeded what past stocks could be expected to sustain (Gallaway et al., 2009; Shipp and Bortone, 2009). One difficulty with the red snapper population assessment is that all models are based on a valid stock–recruit relation. The present fishery is regulated on the premise that sets maximum fishing mortality at some level that would maintain minimum levels of spawning potential ratio (SEDAR, 2009). The implicit assumption is that larger stocks produce larger number of eggs, which then are reflected in larger number of recruits. This stock–recruit relation has not been demonstrated thus far. In fact, the relation shows considerable variation and suggests that fishing mortality may have little effect on new, age-0 recruits (SEDAR, 2009). Without a valid stock–recruit relation, it is probable that other life history stages may be experiencing limiting bottlenecks that may lead to fluctuations in year class strength.

Another aspect that has probably led to the difficulty in managing red snapper has been the use of indices of abundance for young fishes. These indices have been essentially based on trawl surveys from SEAMAP surveys (Nichols et al., 2005). However, these trawl surveys were typically conducted during the summer (e.g., in June before new recruitment) or fall (e.g., in October when most new recruits have moved to structured habitat). For example, Szedlmayer and Lee (2004) showed that age-0 fish moved to nontrawlable, structured habitat about 4 months after hatching. Thus, fish that were collected in trawl surveys may only be a small percentage of the actual number of new recruits.

It is well established that small changes in mortality rates during early life history stages can have significant effects on subsequent year classes (Houde, 1987; Myers and Cadigan, 1993; Sogard, 1997). In red snapper stock assessments, particular importance is contingent on age-0 and age-1 total (Z) mortality and the partitioning of Z into natural (M) and bycatch (F) mortality. For example, in a 2009 assessment of red snapper (SEDAR, 2009), M values doubled from 0.98 to 2.0 when compared to an earlier assessment (SEADR7, 2005), resulting in a 20% increase in the projected yield from the fishery. Thus, accurate estimates of early life history mortalities are essential and can have significant effects on model predictions and fishery management decisions.

This chapter examines the early life history of red snapper. Specifically, it compares estimates of age-0 to age-1 abundance based on visual SCUBA surveys of red snapper on artificial reefs. From these data, seasonal specific mortality rates for juvenile red snapper from initial settlement (recruitment) through age-1 are then calculated. Seasonally specific mortality rates can subsequently be used to predict possible early life history bottlenecks for red snapper populations.

METHODS

The study sites were located approximately 15–30 km south of Dauphin Island, Alabama (Figure 3.1), in the Gulf of Mexico. The area lacked natural reef structure, was uniform in depth (20 m), and dominated by sand–mud–shell substrates (Schroeder et al., 1995; Dufrene, 2005). Small artificial reefs (1–4 m^2) were placed in the northeast area of the Hugh Swingle reef-building zone. Underwater observers, using SCUBA, counted all fish and assigned them to 25-mm size class intervals based on estimated length.

In July 1998 and 1999, reefs were deployed in 4-m^2 plots (0.1 m height) and made of oyster shell or shell and 10 concrete blocks (20 × 20 × 41 cm, each with two 12 × 14 cm oval holes). All reefs ($N = 60$ each year) were placed at 20 m intervals and alternated from shell to block type. In August 1998, a visual survey was completed on all reefs deployed in 1998, after which these reefs were destroyed by a tropical storm. Three visual surveys were completed on the 1999 reefs in September, October, and November.

In July 2000, 3 × 3 m oyster shell reefs ($N = 10$), and 2 × 2 m shell reefs ($N = 10$) were deployed in the same area. These reefs were placed in a grid pattern, 20 m between reefs, alternating between 2 × 2 and 3 × 3 m sizes. Each reef was visually surveyed three times by divers using SCUBA in early August, late August, and October 2000.

In July and August 2001, artificial reefs (1 m^2 plots) were built that consisted of oyster shell ($N = 20$), shell cage ($N = 20$), shell block (4 blocks; $N = 20$), and shell-block cage ($N = 20$). Reefs were placed in 8 transects with 10 reefs per transect and habitat types alternated at 20 m intervals. A PVC stake (1 m × 12 mm diameter) was placed vertically near each habitat to indicate shrimp trawl effects; that is, the stake would be displaced if the habitat was trawled. Reefs were surveyed in August, September, October, and November 2001 (Piko and Szedlmayer, 2007).

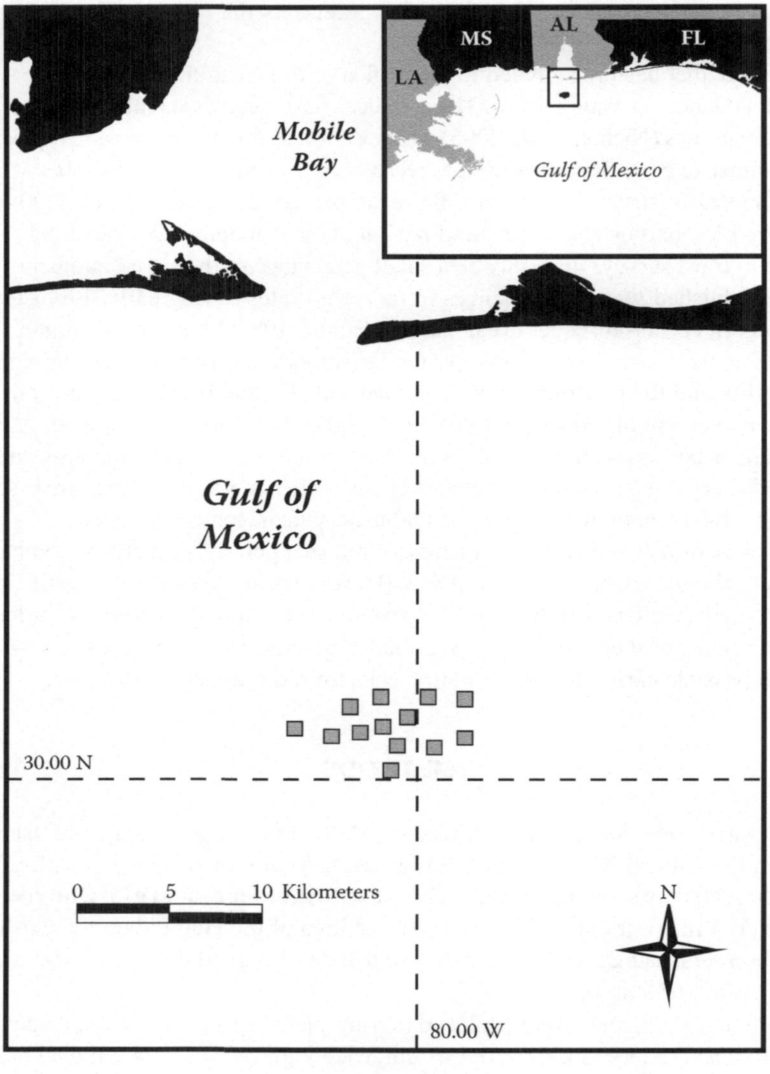

Figure 3.1 Study site locations in the northeast Gulf of Mexico. All artificial reefs over all years were deployed in the Hugh Swingle reef building zone, within 3 km of these sites from 2001.

In June and July 2002, a new set of reefs were deployed (N = 80) using the same design as employed in 2001. One visual survey was completed on all reefs in September 2002, after which all reefs were destroyed by a tropical storm. After the storm, the visual surveys confirmed that the PVC stakes were still intact, yet the reefs had become silted over (Lingo and Szedlmayer, 2006).

In September 2003, the artificial habitats deployed were constructed from plastic-coated wire (1.2 × 1.2 × 0.6 m; 3.6-cm mesh; N = 16). Each habitat contained two concrete blocks (20 × 20 × 41 cm) and ten sections of PVC pipe (31 cm long, 10 cm diameter) for added structure. Reefs were placed 24 m apart, at a depth of 23 m. A second reef design was also deployed at a depth of 20 m in October 2003 (steel cages 2.5 × 1.3 × 1.2 m; N = 16). These steel cages were placed 57 m apart. All reefs were surveyed in October, November 2003, April, and June 2004 (Chapin et al., 2009).

A third artificial reef design was deployed in June 2003 (2 × 2 × 0.2 m; N = 40). Each reef consisted of ten concrete blocks (each block = 20 × 20 × 41 cm) placed on a polyethylene mat

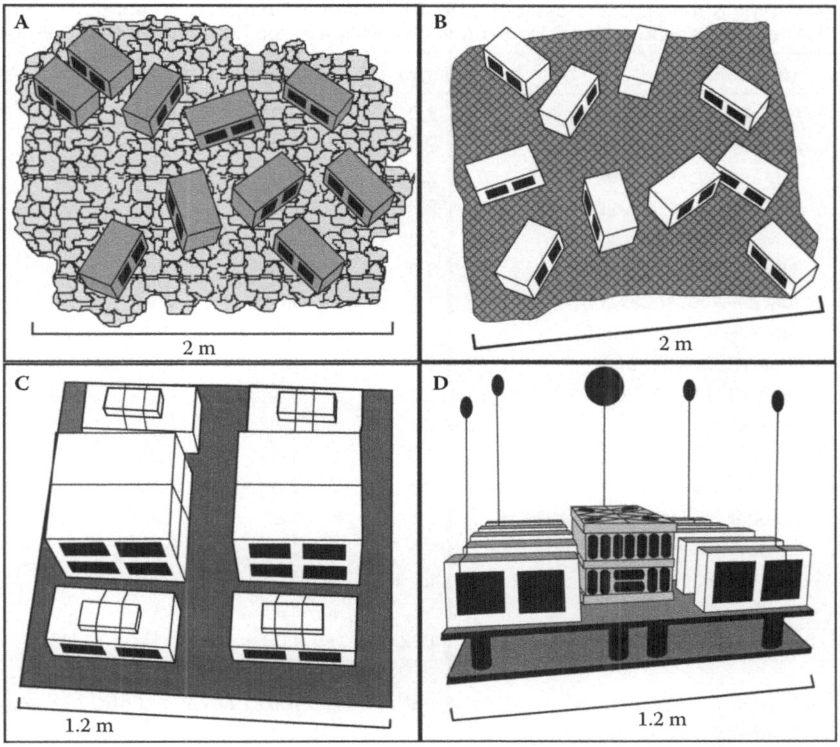

Figure 3.2 Artificial reef design of (A) 1998–1999, (B) 2003, (C) 2005 and 2006 habitats, and (D) 2007.

(4 m^2, 0.64-cm mesh; Figure 3.2A). Reefs were placed at 30 m intervals within each transect, with transects approximately 0.5 km apart. These reefs were visually surveyed in October, November 2003, and May 2004 (Redman and Szedlmayer, 2009).

In October and November 2004, wire reefs (1.2 × 1.2 × 1.2 m; N = 22) made of plastic-coated galvanized wire (5.1-cm mesh) containing four concrete blocks, each were deployed in 20 m of water. Reefs were surveyed in May and June 2005, before a tropical storm destroyed all reefs. A visual survey on 13 June 2005 indicated that the remaining wire reefs were displaced approximately 24 m from their original position and were in poor (i.e., broken and displaced) condition.

In October 2005, artificial reefs (N = 20) were deployed that consisted of 12 concrete blocks (20 × 20 × 41 cm) attached with cable ties to a plywood base (0.005 × 1.2 × 1.2 m). Two sets of four concrete blocks each were stacked and attached in the center of the board with the four remaining concrete blocks secured on either side (Figure 3.2B). Each reef was surveyed in October, December 2005, May, August, and December 2006. In July 2006, reefs (N = 40) with the 2005 design were deployed and surveyed in June 2006.

An additional reef design (1.22 × 1.22 × 0.42 m) was deployed (N = 30) in August 2007. Each of the reefs deployed in 2007 were comprised of concrete half-blocks (41 × 10 × 20 cm). Each half-block was attached to a polypropylene double pallet (1.22 × 1.02 × 0.14 m) with four cable ties (123 cm, 79 kg breaking strength). Five half-blocks were arranged in rows on each side of the pallet, and a plastic crate (61 × 30.5 × 28 cm) was placed in the middle. The plastic crate had various-sized holes (12.1 × 3.2, 14 × 3.8, 7.6 × 3.2 cm) on the top and the sides. One float (13 × 5 cm) was attached with 0.64 cm diameter line to each corner half-block of the pallet (N = 4), and floated 1 m above the habitat. A large circular float (15 cm diameter) was also tied in the center of the plastic pallet 1 m above the half-blocks (Figure 3.2C). All reefs were attached to a 1.8 m ground anchor with a 1.3 cm

Table 3.1 Reef-type Surveyed, Deployment Date, Dimensions, Number of Reefs Deployed, and Total Volume Per Individual Reef for Artificial Reefs Surveyed for Juvenile Red Snapper

Reef Type	Time Built	Length (m)	Width (m)	Height (m)	*N*	Volume (m^3)
Shell/block	April 24–July 8, 1998	2.0	2.0	0.3	30	1.20
Shell	April 24–July 8, 1998	2.0	2.0	0.1	29	0.40
Shell/block	July 7–August 3, 1999	2.0	2.0	0.3	30	1.20
Shell	July 7–August 3, 1999	2.0	2.0	0.1	30	0.40
Shell small	July 11–26, 2000	2.0	2.0	0.1	10	0.40
Shell large	July 11–26, 2000	3.0	3.0	0.1	10	0.90
Shell/block	July 24–August 28, 2001	1.0	1.0	0.3	60	0.30
Shell	July 24–August 28, 2001	1.0	1.0	0.1	20	0.10
Shell/block	June 12–July 31, 2002	1.0	1.0	0.3	60	0.30
Shell	June 12–July 31, 2002	1.0	1.0	0.1	20	0.10
Small cage	September 10, 2003	1.2	1.2	0.6	16	0.86
Large cage	October 7, 2003	2.5	1.3	1.2	16	3.90
Block/mat	June 15–25, 2003	2.0	2.0	0.2	40	0.80
Wire cage	October 28–November 22, 2004	1.2	1.2	1.2	22	1.73
Block/wood	October 10–12, 2005	1.2	1.2	0.4	20	0.58
Block/wood	July 17–26, 2006	1.2	1.2	0.4	40	0.58
Block/pallet	August 1–9, 2007	1.2	1.0	1.1	30	1.42
Block/pallet	July 9–10, 2009	1.22	1.02	1.14	20	1.42

diameter rope. Habitats were placed in six rows with five habitats per row, at 500 m intervals. Reefs were surveyed in May and June 2008.

Artificial reefs with the same design as those deployed in 2007 reefs were deployed in July 2009 ($N = 20$). All reefs were secured to the substrate with a 1.8 m ground anchor. Reefs were surveyed in August and September 2009.

The reef type used, time built, dimensions, number of reefs, and total volume are shown in Table 3.1. For each year, the mean number of age-0 and age-1 red snapper was estimated per reef volume for each survey. The highest mean number of individuals from each year was used to make comparisons across years. Also, after fish attained maximum abundance, the subsequent decline in age-0 and age-1 red snapper was used to estimate a mortality rate over all years.

RESULTS

All reefs over all years attracted high numbers of age-0 red snapper (Figure 3.3). The highest mean number was observed in 2000 with age-0 red snapper attaining a density of 451 fish per m^3 (Figure 3.4). Mean numbers of new recruits varied in other years, ranging from 58 in 1998 to 202 per m^3 in 1999. More recent estimates of red snapper recruitment in 2009 also indicated similar abundance at 77 per m^3, compared to 1998 and 2002 (Figure 3.4). The mean abundance of age-1 red snapper was an order of magnitude lower compared to age-0 fish, and ranged from 4 to 40 per m^3 over different years. There was an increase in abundance of age-1 red snapper in more recent years at 10 to 40 per m^3 from 2006 to 2009, compared to <6 fish per m^3 from 1998 to 2000 (Figure 3.4).

Over several years, it was possible to compare the initial abundance of age-0 to age-1 red snapper the following year to estimate instantaneous annual mortality (Z = natural log of survival). Mortality ranged from 2.2 (2001–2002) to 2.7 (2002–2003). A higher number of initial recruits of age-0 red snapper appeared to result in more age-1 fish the following year. This pattern was apparent for 1998, 2000, 2001, and 2000 cohorts (Figure 3.4).

Figure 3.3 ***See color insert.*** The "pallet" reef type with age-0 red snapper September 9, 2009.

The maximum densities of new recruits occurred in August (Figure 3.5). Mean abundance of age-0 and age-1 red snapper from each survey period were plotted for all years, which enabled a seasonal estimate of mortality over all years for age-0 and age-1 red snapper. There was a substantial change in the mortality rate over the first year. Almost all of the total annual mortality occurred during the first 3 months (August, September, and October) after the initial recruitment of red snapper to reef structure. Separated into time periods, mortality was 2.7 from August to October, after which mortality decreased to 0.7 over the next 9 months from November to July the following year. Thus, once fish survived the first 3 months after recruitment to reef structure, they most likely would survive to the following year. Then mortality increased for these age-1 fish to 2.0 from August to December (Figures 3.5 and 3.6).

DISCUSSION

Early Recruitment to Structured Habitat

Age-0 red snapper moved to reef structure early in their life history at the smallest-size classes (approximately 30 mm TL) just after metamorphoses. Here, the first juvenile settlers to benthic substrate are defined as new recruits (Armsworth, 2002). Previous studies indicated a longer occupancy of open-habitat types by new red snapper recruits before movement to more structured reef habitat. These previous studies also reported substantial differences in the length of time over open habitat; for example, a delay of approximately 4 months was shown by Szedlmayer and Lee (2004), while much longer delays were suggested with young fish staying over open habitat over their first year (Rooker et al., 2004; Wells et al., 2008). The longer occupancy of red snapper over open habitat through age-1 was considered part of the "normal" early life history of red snapper in stock assessments as estimates M for age-1 fish were based on the ability to capture age-1 red snapper with trawl gear (Nichols et al., 2005; SEDAR, 2009). The multiple year survey of small artificial reefs of the present study clearly showed peak recruitment of age-0 red snapper to structured habitat in August of their first year, at much earlier stages compared to all past studies. If densities are compared between the present artificial reefs surveys and past trawl surveys, it is apparent that trawl

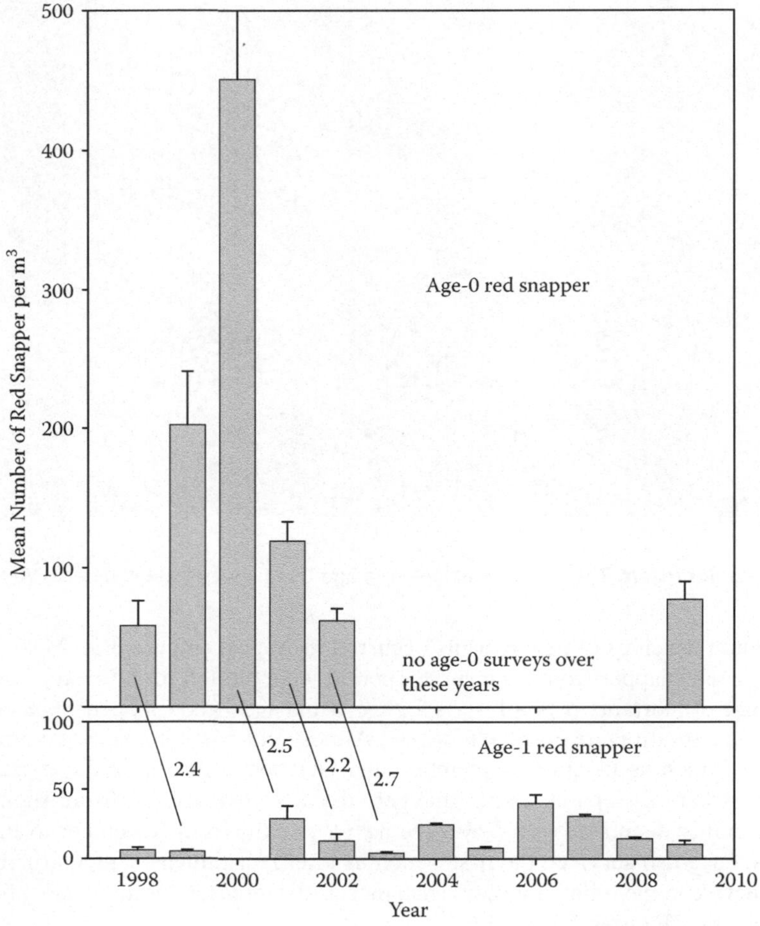

Figure 3.4 Age-0 and age-1 red snapper peak mean abundance per m^3 for each year. Diagonal lines show mortality (*Z*) estimates for years with surveys of age-0 fish in the first year, followed by estimates of age-1 fish in the following year.

surveys did not sample most of the new red snapper recruits; that is, they were simply sampling in the wrong place and time period. For example, the maximum mean number of age-0 red snapper from trawl surveys were approximately 10 fish ha^{-1} (Wells et al., 2008), 100 fish ha^{-1} (Rooker et al. 2004), and 1557 fish ha^{-1} (Szedlmayer and Conti, 1999). When the lowest density estimate over all years from the present study (1998) is converted to number ha^{-1} for comparisons to trawl densities the artificial reef estimated densities are three orders of magnitude greater, that is, 500,000 fish ha^{-1}. Thus, using visual sampling techniques on artificial reefs to address management questions probably results in a different depiction of red snapper early life history when compared to trawl surveys. Here, the pattern of early recruitment to structured habitat was consistent over many years (1998–2009), indicating that previous estimates of abundance, movements, mortalities, and habitat value based on fish sampled using trawl surveys are inadequate. This study suggests that previous trawl studies were only comparing the leftover or remaining surplus production of fish that were unable to compete for the structured habitat and would not ordinarily survive due to increased predation and reduced prey resources (Piko and Szedlmayer, 2007; Gazey et al., 2008; Caddy, 2008; Redman and Szedlmayer 2009).

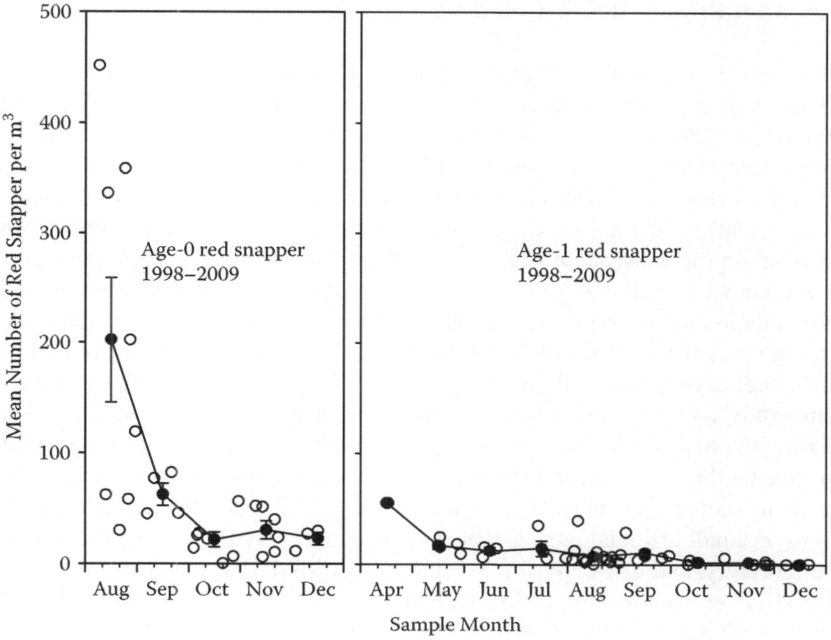

Figure 3.5 Age-0 and age-1 red snapper mean abundance per m^3 pooled over all years by survey period.

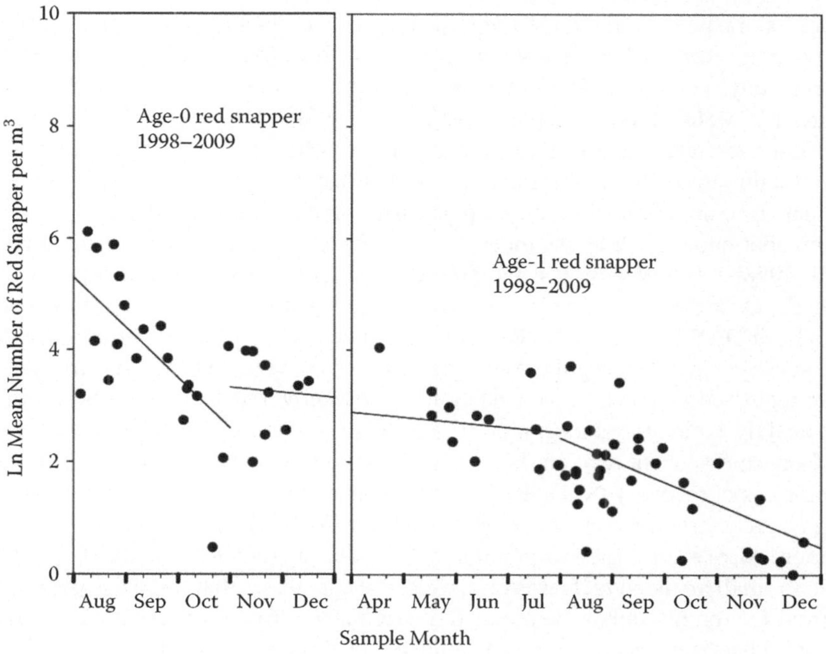

Figure 3.6 Natural logarithm of age-0 and age-1 red snapper mean abundance per m^3 pooled over all years by survey periods. Different lines show mortality estimates for different time periods. Not shown are January, February, and March, because surveys were not taken.

Postrecruitment Population Regulation

There has been a long-standing debate concerning the relative contributions of pelagic larval stage population regulation and postrecruitment juvenile-stage population regulation (Doherty, 1991; Jones, 1991; Hixon, 1991; Doherty, 2002; Armsworth, 2002; Hixon and Webster, 2002; Anderson et al., 2007; Oeberst et al., 2009). In a traditional view, many researchers have considered the early life history in marine fishes of little consequence and that stock–recruit relations were the driving force behind population fluctuations and all that matters are the sizes of the adult spawning stocks, which then regulate the subsequent year class strength (see a discussion of paradox: Rothschild, 1986). The present study indicates that, in fact, early life history stages are important and support postrecruit population regulation for red snapper. All data supporting this contention were based on the study of red snapper with artificial reefs. Probably, the most important factor that points in this direction were the very high natural mortalities ($M = 2.7$) of age-0 red snapper in the first 3 months after recruitment to reef structure. Over all years combined, 79% of the total first year mortality occurred during the first 3 months. Thus, it appears that if fish survive the first 3 months, they are likely to survive the first year. Most, if not all, mortality was considered M, because there were no indications of any reefs being subject to impacts by trawl gear. One criticism of estimating M from visual surveys on small artificial reefs is that estimates in M may be the result of emigration rather than natural mortality, and fish were simply leaving structured habitat and suffered bycatch or fishing mortality (F) over open habitat. If this were the case, it would be expected that age-0 and age-1 red snapper bycatch and SEAMAP trawl survey estimates would be correlated to present visual surveys; that is, higher reef densities would lead to more spill over to open habitat, thus leaving more fish susceptible to trawl gear. However, a comparison of densities from the artificial reefs in the present study to bycatch and CPUE from SEAMAP trawl studies over the same years (SEDAR, 2009) showed little correlation for either age-0 or age-1 red snapper (Figures 3.7 and 3.8). Also, there has been a drastic reduction in shrimp trawl effort over the last several years (SEDAR, 2009), and this reduction seems to have had no effect on densities of fish of age-0 over these same years in the present study. There was also a third period of differential mortality that showed substantial increase after fish were >1 year old (i.e., 13–18 months old). However, the decline in red snapper after age-1 from the "small" artificial reefs in the present study was probably owing to movement to larger reef structure rather than mortality. Fish of this size (i.e., >13 months) have been observed to move to larger structured habitat as red snapper grow (Gallaway et al., 2009).

Also, the continuous high abundances of age-0 recruits on the small artificial reefs over the time period (1998–2009) observed in the present study provide support for a postrecruit population bottleneck. Over this same time period, stock assessments suggested a severely overfished red snapper stock (SEDAR7, 2005; SEDAR, 2009). Lorenzen (2008) indicated that many fish populations show density-dependent regulation in the postrecruit stage, but that heavily exploited populations were regulated by stock–recruit relations. Apparently, red snapper do not fit this model as the stock is heavily exploited yet most mortality occurs just after settlement and there has been little evidence of any stock-recruit relation. Presently, according to the most recent assessment, the stock is displaying signs of recovery (SEDAR, 2009), but this recovery is not reflected in increased age-0 fish densities on reef structures in more recent years (2007–2009), again indicating a lack of a stock–recruit relation. Also, even at the low population estimates of 1998 to 2002, the highest recruitment of age-0 fish to small artificial reefs was observed. The contention that some fish populations are not regulated from the top but rather the population size results from postrecruitment processes in the early life history has been described for several fisheries (Caddy, 2008; Gazey et al., 2008; Lorenzen 2008). In the present study, red snapper stocks in the northern Gulf of Mexico appear to follow this type of population regulation. Moreover, two factors (i.e., prey resources and shelter from predation) are probably both enhanced from structured habitat. Additionally, the high recruitment and

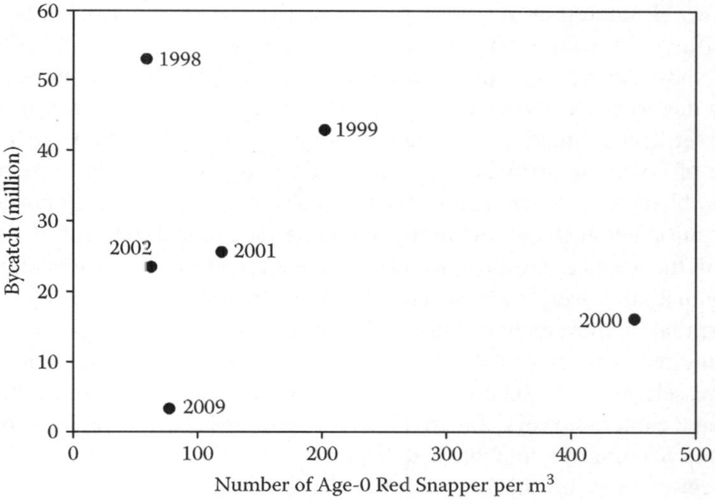

Figure 3.7 Age-0 red snapper mean abundance per m³ from visual surveys compared to bycatch mortality estimates from SEDAR (2009).

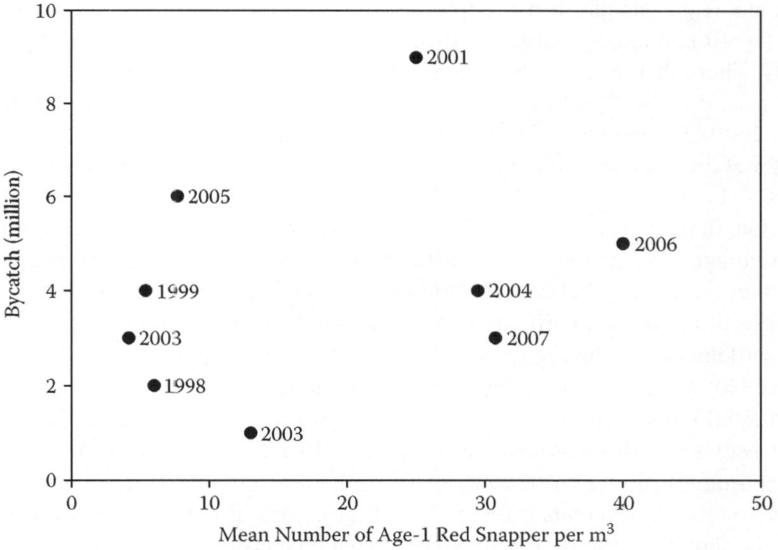

Figure 3.8 Age-1 red snapper mean abundance per m³ from visual surveys compared to bycatch mortality estimates from SEDAR (2009).

subsequent densities on artificial reefs suggests that such habitat is limited (Szedlmayer and Lee, 2004; Piko and Szedlmayer, 2007; Redman and Szedlmayer, 2009).

Future Use of Artificial Reefs in Fishery Management Questions

Future studies on early life history of red snapper will be advanced with the use of artificial reefs. Clearly, artificial reefs do not function in a vacuum, and adjacent habitats probably have a substantial effect (Lindberg et al., 1990; Frazer and Lindberg, 1994; Strelcheck et al., 2005). Presently, it is difficult to make inferences about such adjacent habitats on the differential habitat value of various substrate types for juvenile stages of red snapper. For the most part, this lack of understanding is due to cryptic habitat use patterns of red snapper where fish move to structured habitat just after settlement, making quantitative sampling with typical trawl gear difficult at best. A new approach of involving artificial reefs as the sampling "tool" could overcome many of the past sampling problems encountered in evaluating habitat value for this important species. The latest version of this small artificial reef design could be used to facilitate comparisons of different benthic habitats in the northern Gulf of Mexico. This reef type has several advantages: (1) it has the appropriate combination of weight and surface area to overcome many of the physical problems of past reef designs, that is, movement or burial; (2) the proper dimensions and structure to promote attraction of young red snapper and other reef fishes; (3) can be constructed and deployed from a relatively small vessel; (4) can be anchored to reduce storm effects; (5) it is small enough to allow for complete visual and camera surveys; and (6) it has concrete surfaces that increase benthic epifaunal growth (Miller and Szedlmayer, unpublished data).

Another aspect of early life history features of red snapper that could be enhanced through the use of artificial reef deployments is the relative effects of predation. If predation is a major ecological forcing function for populations of reef fishes, it would be expected that proximity to potential predators would also affect habitat selection by young new recruits. Thus, a prediction would be that small reefs (e.g., pallet reefs in the present study) that are near (approximately 15 m) large reefs (e.g., gas platforms, approximately 2000 m^3) would have reduced numbers of new recruits (age-0 and age-1) compared to small reefs that are further away (e.g., 500 m) from the large reefs. These distances are based on previous studies of movements and residency of red snapper on artificial reefs (Szedlmayer and Schroepfer, 2005; Schroepfer and Szedlmayer, 2006; Chapin et al., 2009), but may also function for other marine predators and prey. Reefs could then be surveyed for abundance and size estimates of associated reef fish communities with visual and video surveys.

In conclusion, in the northern Gulf of Mexico, young red snapper are present at higher densities 3 to 4 orders of magnitude greater at small artificial reefs than estimates of abundance of these fish from trawl surveys over open habitat. The high densities of the earliest settlers suggest that "optimal" habitat is limiting for their early life stages. These high densities quickly decline in the first few months after settlement, but then remain relatively stable for the remainder of the first year. There is little evidence for a stock–recruit relation, based on the consistent high recruitment of young red snapper to artificial reefs even in years that had low adult stock abundance estimates. Also, these high recruit densities on artificial reefs are not affected by a reduction in shrimp trawling, suggesting that an important, limiting bottleneck occurs in the postsettlement stages of red snapper. This lack of a stock–recruit relation has little to do with questions of red snapper adult stock condition (i.e., overfished or not overfished), only that the population has not attained such a severely depleted condition that the stock–recruit relations becomes important. Significant regulation at present stock levels is occurring at the juvenile stage and is, for the most part, natural mortality.

ACKNOWLEDGMENTS

I thank S. Beyer, A. Chapin, D. Miller, C. Furman, C. Gurshin, G. Lipkey, D. Moss, P. Mudrak D. Nadeau, A. Piko, M. Piraino, A. Purcell, R. Redman, C. Simmons, T. Syc, M. Topolski, D. Topping B. Wildberger, and R. Wingate for help building and surveying artificial reefs. This project was funded by the Mississippi-Alabama Sea Grant Consortium (MASGC R/SP-14) NOAA award number NA06OAR4170078 and Marine Resources Division, Alabama Department of Conservation and

Natural Resources. This study is a contribution of the Alabama Agricultural Extension Station and Department of Fisheries and Allied Aquacultures, Auburn University.

REFERENCES

Anderson, T. W., M. H. Carr, and M. A. Hixon. 2007. Patterns and mechanisms of variable settlement and recruitment of a coral reef damselfish, *Chromis cyanea. Mar. Ecol. Prog. Ser.* 350:109–116.

Armsworth, P. R. 2002. Recruitment limitation, population regulation, and larval connectivity in reef fish metapopulations. *Ecology* 83: 1092–1104.

Caddy, J. F. 2008. The Importance of "cover" in the life histories of demersal and benthic marine resources: A neglected issue in fisheries assessment and management. *Bull. Mar. Sci.* 83: 7–52.

Chapin, A. M., S. T. Szedlmayer, and R. P. Phelps. 2009. Survival and movement of hatchery-reared red snapper on artificial habitats in the northern Gulf of Mexico. *Fish. Manage. Ecol.* 16: 28–36.

Doherty, P. J. 1991. Spatial and temporal patterns in recruitment. Pages 261–293 in P. F. Sale, Ed. *The Ecology of Fishes on Coral Reefs*. Academic Press, San Diego, CA.

Doherty, P. J. 2002. Variable replenishment and the dynamics of reef fish populations. Pages 327–355 in P. F. Sale, Ed. *Coral Reef Fishes: Dynamics and Diversity in a Complex Ecosystem*. Academic Press, San Diego, CA.

Dufrene, T. A. 2005. Geological Variability and Holocene Sedimentary Record on the Northern Gulf of Mexico Inner to Mid-Continental Shelf. M.S. thesis, Louisiana State University, Baton Rouge, LA. 100 pp.

Dute, J. 2008. Yankee Red Snapper of Mobile Bay. July 6, 2008. Press-register, Mobile, AL.

Frazer, T. K. and W. J. Lindberg. 1994. Refuge spacing similarly affects reef-associated species from three phyla. *Bull. Mar. Sci.* 55: 388–400.

Gallaway, B. J., S. T. Szedlmayer, and W. J. Gazey. 2009. A life history for red snapper in the Gulf of Mexico with an evaluation of the importance of petroleum platforms and other artificial reefs. *Rev. Fish. Sci.* 17: 48–67.

Gazey, W. J, B. J. Gallaway, J. G. Cole, and D. A. Fournier. 2008. Age composition, growth, and density dependent mortality in juvenile red snapper estimated from observer data from the Gulf of Mexico penaeid shrimp fishery. *N. Am. J. Fish. Manage.* 28: 1828–1842.

Hixon, M. A. 1991. Predation as a process structuring coral reef fish communities. Pages 476–508 in P. F. Sale, Ed. *The Ecology of Fishes on Coral Reefs*, Academic Press, San Diego, CA.

Hixon, M. A. and M. S. Webster. 2002. Density dependence in reef fish populations. Pages 303–325 in P. F. Sale, Ed. *The Ecology of Fishes on Coral Reefs*, Academic Press, San Diego, CA.

Houde, E. D. 1987. Fish early life dynamics and recruitment variability. *Amer. Fish. Soc. Symp.* 2: 17–29.

Jones, G. P. 1991. Postrecruitment processes in the ecology of coral reef fish populations: A multifactorial perspective. Pages 294–328 in P. F. Sale, Ed. *The Ecology of Fishes on Coral Reefs*, Academic Press, San Diego. CA.

Lindberg, W. J., T. K. Frazer, and G. R. Stanton. 1990. Population effects of refuge dispersion for adult stone crabs (Xanthidae, *Menippe*). *Mar. Ecol. Prog. Ser.* 66: 239–249.

Lingo, M. E. and S. T. Szedlmayer. 2006. The influence of habitat complexity on reef fish communities in the northeastern Gulf of Mexico. *Environ. Biol. Fishes* 76: 71–80.

Lorenzen, K. 2008. Fish population regulation beyond "stock and recruitment": The role of density-dependent growth in the recruited stock. *Bull. Mar. Sci.* 83: 181–196.

Mitchell, K. M., T. Henwood, G. R. Fitzhugh, and R. J. Allman. 2004. Distribution, abundance, and age structure of red snapper (*Lutjanus campechanus*) caught on research longlines in U.S. Gulf of Mexico. *Gulf Mex. Sci.* 22: 164–172.

Myers, R. A. and N. G. Cadigan. 1993. Density-dependent juvenile mortality in marine demersal fish. *Can. J. Fish. Aquat. Sci.* 50: 1576–1590.

Nichols, S., G. Pellegrin, Jr., and G. W. Ingram, Jr. 2005. Estimation of juvenile M for red snapper based on SEAMAP survey data. Nat. Mar. Fish. Serv., SEDAR7-AW-15, Miami, FL.

Oeberst, R., B. Klenz, T. Gröhsler, M. Dickey-Collas, R. D. M. Nash, and C. Zimmermann. 2009. When is year-class strength determined in western Baltic herring? *ICES J. Mar. Sci.* 66: 1667–1672.

Piko, A. A. and S. T. Szedlmayer. 2007. Effects of habitat complexity and predator exclusion on the abundance of juvenile red snapper. *J. Fish Biol.* 70: 758–769.

Redman, R. A. and S. T. Szedlmayer. 2009. The effects of epibenthic communities on reef fishes in the northern Gulf of Mexico. *Fish. Manage. Ecol.* 16: 360–367.

Rooker, J. R., A. M. Landry, Jr., B. W. Geary, and J. A. Harper. 2004. Assessment of a shell bank and associated substrates as nursery habitat of post-settlement red snapper. *Estuar. Coast. Shelf Sci.* 59: 653–661.

Rothschild, B. J. 1986. *Dynamics of Marine Fish Populations*. Harvard University Press, Cambridge, MA. 277 pp.

Sale, P. F., Ed. 1991. *The Ecology of Fishes on Coral Reefs*, Academic Press, San Diego, CA. 754 pp.

Sale, P. F., Ed. 2002. Coral reef fishes: Dynamics and diversity in a complex ecosystem. Academic Press, San Diego, CA. 549 pp.

Schroeder, W. W., A. W. Shultz, and O. H. Pilkey. 1995. Late Quaternary oyster shells and sea-level history, inner shelf, northeast Gulf of Mexico. *J. Coast. Res.* 11: 241–258.

Schroepfer, R. L. and S. T. Szedlmayer. 2006. Estimates of residence and site fidelity for red snapper *Lutjanus campechanus* on artificial reefs in the northeastern Gulf of Mexico. *Bull. Mar. Sci.* 78: 93–101.

SEDAR7. 2005. Stock assessment report of SEDAR 7, Gulf of Mexico Red Snapper. NMFS, SEFSC, NOAA, Charleston, SC, 480 pp. Available from http://www.sefsc.noaa.gov via the Internet. Accessed January 28, 2009.

SEDAR. 2009. Stock Assessment of Red Snapper in the Gulf of Mexico: SEDAR Update Assessment. NMFS, SEFSC, NOAA, Miami, FL, 111 pp. Available from http://www.sefsc.noaa.gov via the Internet. Accessed January 28, 2009.

Shipp, R. L. and S. A. Bortone. 2009. A perspective of the importance of artificial habitat on the management of red snapper in the Gulf of Mexico. *Rev. Fish. Sci.* 17: 41–47.

Sogard, S. M. 1997. Size-selective mortality in the juvenile stage of teleost fishes: A review. *Bull. Mar. Sci.* 60: 1129–1157.

Strelcheck, A., J. H. Cowan, Jr., and A. Shah. 2005. Influence of reef location on artificial-reef fish assemblages in the north central Gulf of Mexico. *Bull. Mar. Sci.* 77: 225–440.

Szedlmayer, S. T. 2007. An evaluation of the benefits of artificial habitats for red snapper, *Lutjanus campechanus*, in the northeast Gulf of Mexico. *Proc. Gulf Caribb. Fish. Inst.* 59: 223–230.

Szedlmayer, S. T. and J. J. Conti. 1999. Nursery habitats, growth rates, and seasonality of age-0 red snapper, *Lutjanus campechanus*, in the northeast Gulf of Mexico. *Fish. Bull.* 97: 626–635.

Szedlmayer, S. T. and J. D. Lee. 2004. Diet shifts of red snapper, *Lutjanus campechanus*, with changes in habitat and fish size. *Fish. Bull.* 102: 366–375.

Szedlmayer, S. T. and R. L. Schroepfer. 2005. Long-term residence of red snapper on artificial reefs in the northeastern Gulf of Mexico. *Trans. Am. Fish. Soc.* 134: 315–325.

Tampa Tribune. 2009. Red Snapper Bite in the Gulf of Mexico is Red Hot. July 15, 2009, *Tampa Tribune*, Tampa FL.

Wells, R. J., J. H. Cowan, Jr., W. F. Patterson, III, and C. J. Walters. 2008. Effect of trawling on juvenile red snapper (*Lutjanus campechanus*) habitat selection and life history parameters. *Can. J. Fish. Aquat. Sci.* 65: 2399–2411.

CHAPTER 4

How Artificial Reefs Could Reduce the Impacts of Bottlenecks in Reef Fish Productivity within Natural Fractal Habitats

John F. Caddy

CONTENTS

ABSTRACT

Theoretical considerations suggest that a shortage of suitable cover for reef-dependent motile organisms may lead to high predation mortality and/or migration, constrain adult biomass and spawning potential, and result in a bottleneck in fisheries production. For free-living fish and motile invertebrates, the natural mortality rate often descends steeply with size and age toward maturity, but from simple simulations, the fractal properties of structurally complex marine habitats suggest that mortality rates may rise again for older, crevice-dependent fishes. That natural cover for commercial-sized organisms is limited can be inferred from the rapid colonization of artificial reefs. Estimating hole-size availability in the local environment could be important in optimizing crevice sizes prior to installing artificial reefs.

For shelter-dependent animals, competition for the progressively fewer large crevices predicted by fractal theory will inevitably increase, given longer residence times for older fishes. This may lead to migration or a rise in predation mortality, and eventually lead to a production bottleneck for larger fishes. This could explain the effectiveness of large-scale deployment of artificial reefs with large crevice sizes, but also the potential they have for overexploitation if access to artificial reefs by fishers is not controlled. Nonetheless, installing artificial reefs within reserves, with crevice sizes adjusted for the dimensions of mature fishes, could help compensate for natural habitat deficiencies and create spawning refugia supplying recruits to adjacent areas.

The possibility of production bottlenecks was investigated, reanalyzing size-frequency data from an unfished population, and applying a new method of checking size frequencies based on fractal theory that calculates the scaling coefficient *Q*, which is the rate of decline in log numbers with size. This procedure helps identify bottlenecks in habitat availability, judging from the intersection of quasi-linear segments of log-size frequency plots. Size-frequency data for an unexploited reef area suggest that, for some smaller reef fishes, trends in log numbers at size may reflect fractal expectation (Mandelbrot, 1983). For larger specimens, steep declines in numbers at size suggest that cover was limited, and predation or migration were occurring well before the maximum size for the species was reached.

INTRODUCTION

A common assumption in population dynamics is that, following the juvenile stages, death rates due to predation remain constant. However, for reef dwellers, predation rate likely reflects the availability of suitable cover for a given size. The migration between different habitat types during the life history could also be a response to the availability of suitable cover for different-sized animals. The absence of suitable "cover units" (hiding places) may render antipredator behavior ineffective and could increase natural mortality rate at size. For those demersal reef species that are dependent on cover, this shortage of structurally complex elements in the natural environment exposes fish to predation (see Walters and Juanes, 1993); a conclusion stemming from numerous field observations (e.g., Sale, 1978; Shulman, 1985; Eggleston et al., 1990; Holbrook and Schmidt, 2002). Field work also shows that the absence of cover may lead to bottlenecks in production, due to a high probability of predation in those large areas of sea floor that are without cover. Observations underwater by many investigators, summarized in Caddy (2007), suggest that both feeding and predation usually focus close to the boundary between structurally complex habitats providing shelter and the flat, sedimentary habitats where foraging often occurs. A method is proposed for identifying bottlenecks from plots of log-size frequencies that take into account the fact that the structured habitats used as cover can display fractal characteristics (Burrough, 1981; Bradbury et al., 1984; Bell et al., 1991; Li, 2000), and the practical implications of this for natural as well as artificial habitats are discussed in this chapter.

The possibility that the structure of natural habitats can explain observations on finfish demography remains to be widely accepted, but apparent "coincidences" are noteworthy. Tupper and Boutilier (1995) showed that cod early postrecruits experience good survival in structurally complex habitats of cobbles and seaweed beds, respectively. Nordeide (1993) found that stocking fjords with cod juveniles was ineffective in increasing adult populations if suitable habitat was limiting (see, e.g., Keats et al., 1987) and, for rockfish, O'Connell and Carlisle (1993) demonstrated that fish densities are habitat specific. Many similar observations can be found in the literature, summarized in Caddy (2007).

VISUALIZING FRACTAL REEF ARRAYS

The ideal hole frequency at size of artificial reefs designed to optimize throughput of fish biomass should reflect the biology and growth of the species sheltering there, and reduce the size-specific impact of predation on individuals using them for shelter. Hole sizes used in many artificial reefs tend to be large and relatively uniform; often containing a significant proportion of sexually mature fish (Munro, 1983). This may be logical, as suggested in the following text, if spawning populations are to be conserved. It may, however, render these population components vulnerable to overexploitation on artificial reefs if there is open access for fishers. It seems worth considering first how abundant shelter units of different sizes are in nature. Structurally complex habitats in nature have often been shown to have fractal characteristics, and the fractal coefficient of a habitat are often measured along a transect (e.g., Burrough, 1981; Bradbury et al., 1984) and range in value between $1 < D < 2$. When considering the dimensions of a surface with embedded crevices, however, the effective dimension must be converted to $2 < D' < 3$.

The effect of a fractal array of hole sizes on survival is illustrated graphically using the "Holes for Windows" software of Stamatopoulos and Caddy (1991). This application randomly disperses a prespecified number of holes at size over a surface, either as predicted by fractal theory or by other criteria. These fractal arrays help visualize how fractal dimension determines the numbers of different-sized crevices in a natural surface, and lead to simulation of how a habitat's limiting holding capacity at size changes with the individual growth rate of its inhabitants. The number of holes falls off more sharply with size as the fractal coefficient increases, and an array with a fractal coefficient approaching 3 (visualized as a convoluted surface with crevices leading into interior volumes, as in a coral reef) is dominated by small holes (Figure 4.1). As shown by the tables in Figures 4.1a and b, small fish confined to such a surface would rapidly exceed the numbers of suitable-sized vacant crevices as they grow in size.

The first array, Figure 4.1a, shows a surface created with $D' = 2.15$. For the same initial number of small holes, this allows survival of close to 30 times the number of organisms eventually occupying 60 cm holes than where the surface was created with a fractal coefficient approaching 3. Thus, a low fractal coefficient, or a natural surface supplemented artificially with larger holes, will facilitate a much less restrictive regime for growth and survival of larger crevice-dependent organisms than a natural fractal habitat.

THE ARTIFICIAL REEF CONTROVERSY

An argument against installing artificial reefs in areas of open access to fishers was made by Polovina (1991a,b) who argued that their high efficiency in aggregating reef fish made resident fishes vulnerable to overexploitation, but the reason for their high efficiency remains unexplained. Arguments have been made (e.g., Caddy, 2007) that structurally complex shelter is scarce in most flat sedimentary environments, and rendered more so by the impact of swept-bottom fishing gear in destroying cover of epifauna or epiflora. Accepting that artificial reefs are attractive to reef-dwelling fishes is another way of saying that suitable shelter from predators in the benthic environment is patchy and relatively uncommon in most seascapes (e.g., Schmid, 2000). This chapter suggests why increased vulnerability may occur later in life if habitats are fractal in nature.

ABUNDANCE AT SIZE RELATIONSHIPS ON FRACTAL SURFACES

For coral reefs and other structurally complex environments, an inevitable consequence of fractal theory (Caddy, 1986; Caddy and Stamatopoulos, 1990) is that large crevices are much rarer than small ones, and hence, reefs can accommodate many more small than large organisms. The last-cited authors proposed measuring the relative attractiveness of different hole diameters at a proposed reef site with a "crevice sampler" prior to installing artificial reefs with appropriate crevice sizes. Following a similar research strategy for stone crab populations, Beck (1997) demonstrated

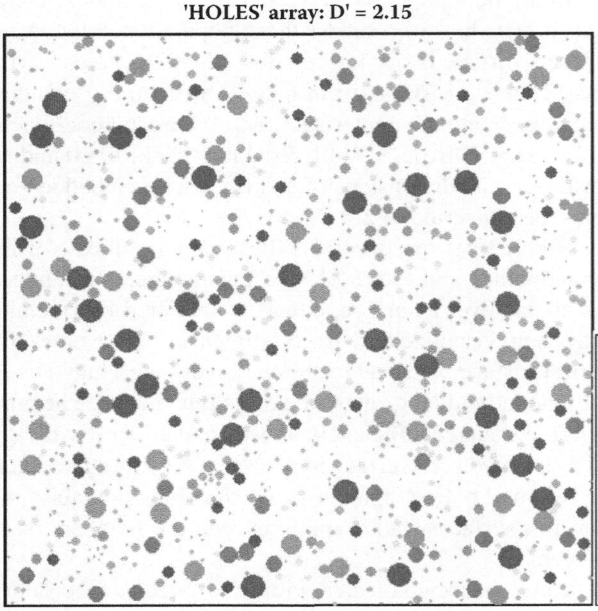

Entry diam. (cm)	# crevices
5	500
10	225
15	141
20	102
25	79
30	64
40	46
50	35
60	29

(a)

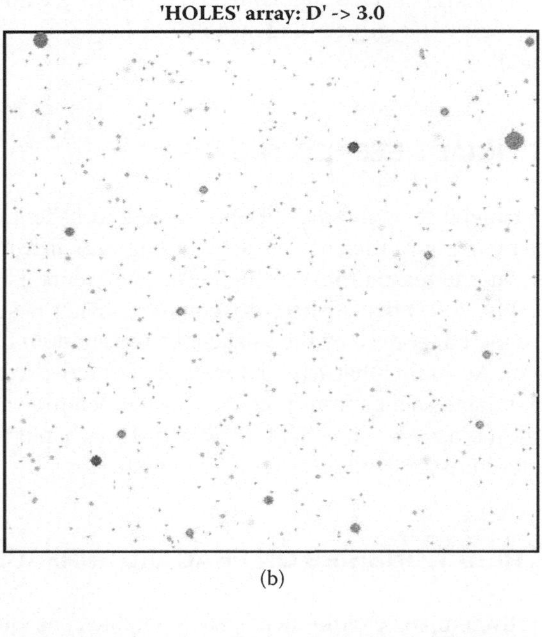

Entry diam. (cm)	# crevices
5	500
10	63
15	19
20	8
25	4
30	2
40	1
50	1
60	0

(b)

Figure 4.1 ***See color insert.*** Visualizing the relative frequency of small and large holes in two arrays corresponding to (a) $D' = 2.15$, and (b) $D' = 3$. The predicted size distributions of holes are shown in the accompanying tables. Arrays created by "Holes for Windows" (Stamatopolous and Caddy, 1991).

that the rarity of larger boulders in some Florida bays may stunt the growth of stone crabs hiding under them, and Smith and Tyler (1975) showed how the collapse of patch reefs reduced shelter availability under them for larger reef organisms.

A number of investigators (e.g., Shulman, 1985; Gorham and Alevizon, 1989; Gotceitas and Colgan, 1989; Gunnarsson, 1992; Gee and Warwick, 1994a,b) have described the relationship between the abundance of marine fauna at size and the fractal nature of their habitat. For terrestrial vegetation, Morse et al. (1985) suggested approximating the number of cover units on a fractal surface, hence the limiting number of organisms at size (N_L) inhabiting it, by:

$$N_L = k/L^{D+1} \tag{4.1}$$

where k is a constant, L is the size of the crevice (and at the limit, of the organism inhabiting it), and D is the fractal coefficient.

Consider that the size composition of a population of crevice-dependent organisms may also be affected by factors (e.g., fishing or migration) as well as habitat holding capacity, such that the relative abundance at size of larger animals is reduced at a consistent rate over a given size range. To describe the combined effect of all such causes, the term *scaling quotient* (Q) is substituted for $D + 1$ in Equation 4.1.

It is postulated that the size-frequency data of the fish themselves may be used to determine the value of a scaling quotient, Q, that measures stock depletion over linear segments of a logarithmic size frequency. From Equation 4.1, consider two crevice sizes that can just accommodate fish of lengths L_1 and L_2. The limiting number of individuals (following Equation 4.1) are postulated to be related to the size of individuals by:

$$N_{L2} = N_{L1}*(L_1{}^Q/L_2{}^Q) \tag{4.2}$$

This relationship can be fitted to successive pair of numbers at size (N_{L2},N_{L1}) over apparently linear segments of a species log-size frequency, to find values of Q such that the predicted line passes through the observed log numbers for that segment.

$$\text{Expressed as: } \ln (N_{L2}/N_{L1}) = \ln (L_1{}^Q/L_2{}^Q) \tag{4.3}$$

Equation 4.2 is equivalent to an instantaneous depletion rate per centimeter increase in fish size. Simulations with this relationship show that Q reflects the instantaneous rate of total mortality Z' for fished stocks, or M for unfished stocks, between two sizes L_1 and L_2 along a growth trajectory, and could provide a rough measure of the fractal coefficient of the surface. While for unfished stocks the value of Q may reflect habitat availability, if fishing mortality or migration also occurs in addition to habitat constraints, size-frequency analyses will yield Q values potentially much higher than the maximum spatial dimension, 3, and hence, Q will no longer have a dimensional significance. Where fractal processes are the main constraint on abundance-at-size, however, values of $Q \leq 3$ could reflect the fractal dimension of the habitat surface that is available for cover over the size interval.

A historical data set of size frequencies of research catches from Pedro Bank, Jamaica, then unfished, was analyzed by Munro (1983) as proportions at size of the total research catch. As Munro did when using length-converted catch curves (LCCCs), I took natural logarithms of numbers at size, and inspected the plots for linear segments. Q was determined empirically over those parts of the observed size range where a consistent log-linear relationship coincided visually with the log-size frequency. The value of Q giving the best fit to that section of the log-frequency curve was confirmed using the "Solver" routine of Microsoft EXCEL® software. Since there is inevitably an element of subjectivity in delimiting quasi-linear sections of a log-size frequency, values of Q were only

Table 4.1 Summary of Fits to Log-Linear Components of Size Frequencies from Munro (1983) for Species Showing Discontinuities in the Log-Size Frequency

Species	Estimate of 1st Linear Component of *Q*	Estimated Bottleneck Fork Length (cm)	Ratio Total Length/ Body Height	Estimated Bottleneck Body Height (cm)	Estimate of 2nd Linear Component of *Q*
Bar jack *(Caranx ruber)*	3	**30**	0.304	**9.1**	7
Blackfin snapper (*Lutjanus buccanella)*	1–2	**41**	0.328	**13.4**	15
Black snapper *(Apsilus dentatus)*	3	**51**	0.328	**16.7**	30
Yellowtail snapper *(Ocyurus chrysurus)*	3	**52**	0.304	**15.8**	>30
White grunt *(Haemulon plumieri)*	1–2	**31**	0.452	**14.0**	>30
Cottonwick grunt *(Haemulon melanurum)*	6	**25**	0.308	**15.4**	>30
Angelfish *(Pomacanthus arcuatus)*	3	**27**	0.618	**16.7**	15
Ocean surgeon *(Acanthurus bahianus)*	1–2	**19**	0.411	**7.8**	24
Coney *(Cephalopholis fulva)*	5	**28**	0.292	**8.2**	>30

Note: Estimates of body length and derived body heights at the estimated bottleneck size are shown in bold, with estimates of the scaling quotient for linear components. (For the first linear component, these frequently are close to the fractal coefficient ≥3 typical of a very broken reef surface.)

reported to one decimal place. For steeper slopes where few size intervals were present, values provided in Table 4.1 are indicative only, but taken together imply a high rate of depletion with size.

Log-size frequencies have been used to measure total mortality rate *Z*, and a straight line slope in a log-size frequency plotted against predicted age is one way of estimating depletion rates (e.g., Ziegler, 1979; Jones and van Zalinge, 1981; Pauly, 1982; Sparre et al., 1989). Consider, however, the slope of a log-size frequency fitted directly to size without converting the *x* axis of the plot to age. If the slope of the right-hand side of such a log-size frequency changes abruptly at a given size, a bottleneck in habitat holding capacity, or the onset of migration, may be explanations. For an exploited population, the onset of fishing above a given size at recruitment to the grounds would also increase the slope. One situation, where an increase in the slope of the line seems likely with increasing size, is when an organism "grows out of" the cover locally available, (e.g., a low algal furze is suitable only as cover for small fishes), if this habitat no longer offers the protection from predation it did at a smaller size.

FRACTAL THEORY AND FISH HABITAT

Various sources (e.g., Mark, 1984; Li, 2000; Purkis and Kohler, 2008) support the idea that many complex benthic habitats have fractal characteristics and that this determines changes in the holding capacity for motile organisms with size. Evidence from satellite imagery (Purkis and Kohler, 2008) showed that patch boundaries of coral reef areas are fractal and self-similar over scales ranging from <10 m^2 to >1000 m^2, with a fractal coefficient (or dimension along a linear transect) across the Puerto Rican shelf of $D = 1.2–1.3$, and higher values of 1.4–1.65 in the Arabian Gulf. Mark (1984) also found fractal coefficients approximating $D = 1.2$ for transects across a coral reef. A large number of structurally complex marine habitats are also fractal, as summarized in Caddy (2007, 2008) and are used as cover by motile demersal and benthic marine organisms at various life history stages. Reef-dependent species in their natural habitats often occupy broken

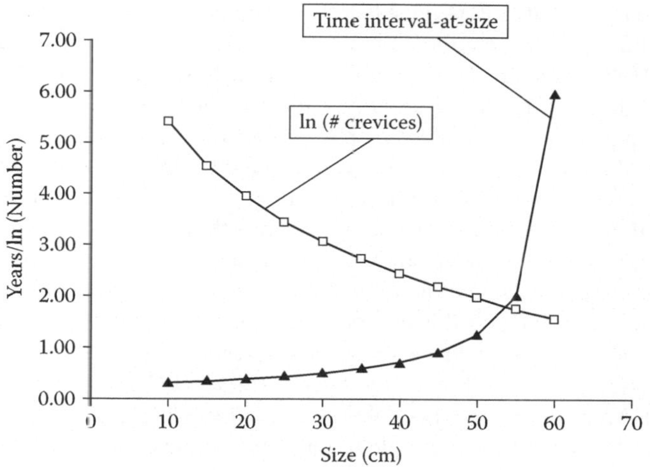

Figure 4.2 Decline in log numbers of crevices predicted on a fractal surface ($D' = 2.15$) as predicted by a von Bertalanffy growth function ($K = 0.3$; $L_\infty = 61$ cm) and the time spent in crevices by fishes of different lengths.

surfaces, ranging dimensionally between a surface (2) and a volume (3). An example comes from Purkis and Kohler (2008) who measured the fractal coefficient of coral reef surfaces (D' implies here the effective fractal coefficient of a surface) lay in the range $2 < D' < 3$. In their words, a coefficient for a surface approaching $D' \geq 3$ corresponds to a surface "so rough as to fill three-dimensional space," and they suggested that we should expect reef fish demography to reflect this underlying habitat scaling. (It is relevant to note that for cave systems on land, Tercafs (1997) found fractal coefficients approaching 3.)

Fish behavior in relation to cover is complex and is not dealt with in detail here, nor do I enter into detail on species behavior and predator avoidance strategies. However, the "musical chairs" analogy of Smith (1975) may apply to some species, and he suggested that the "rules" a fish follows in choosing a crevice might be summarized as follows:

Large fish cannot fit into small crevices.
Though small fish fit in large crevices, they are eaten or outcompeted for these by larger fish.
Hence fish tend to choose crevice sizes that correspond closely to some limiting body dimension.

Assuming these rules are approximately valid, and that both the von Bertalanffy growth model and a fractal decline in crevice numbers apply, output from a simple simulation of growth and mortality on a fractal surface is shown in Figure 4.2. Ages-at-size are given in terms of the von Bertalanffy parameters, by $t = -(1/K)*\ln((L_\infty - l_t)/L_\infty)) + t_0$; and the time to grow from L_1 (equivalent to the maximum body depth of a fish just fitting in crevice 1) to L_2 (the maximum body depth just fitting into the next larger crevice 2) is given by $\Delta t_{1\text{->}2} = t_2 - t_1$, where maximum body depth = a* (body length) is assumed to be the limiting body dimension for fitting into a crevice of the same diameter. For two successive crevice diameters in an array of crevices that increase in size by a constant increment, the predicted decrease in number of crevices at size is given by Equation 4.2.

What Information on Habitat Availability at Size Is Provided by Size-Frequency Data?

Since unexploited reef fish communities no longer exist in most world areas, Munro's data set has grown in value over time. Without the confounding effect of fishing, the size compositions described by Munro and collaborators must largely reflect predation as the main cause of declining numbers-at-size. Although immigration/emigration processes cannot be excluded, the distance to the Jamaican mainland makes this less probable for reef fishes on Pedro Bank.

Once the von Bertalanffy parameters were estimated, Munro and colleagues used the Beverton and Holt (1956) estimator to obtain values of Z/K ($\approx M/K$) from $(L_\infty - l_t)/(l_t - l_c)$ (l_t is the mean fish length at age t; L_∞ and K are parameters of the von Bertalanffy growth equation; and l_c is the size at first capture for the population). His natural mortality estimates ($Z \approx M$) were high: $M = 1.7 - 2.2$. (It is worth mentioning that the size compositions suggest that for reef fishes, a sharp change in the mortality rate per size increment occurs above a certain size shown for the species in Table 4.1.)

The slope of the right-hand side of a plot of logarithm of numbers of individuals against age is a "catch curve" (Ricker, 1975), and this can be used to estimate overall mortality rate. The slope of a catch curve of log numbers against effective age as deduced from the von Bertalanffy growth parameters can be used in a "length-converted catch curve" (LCCC) (Ziegler, 1979; Pauly, 1982; Sparre et al., 1989). Munro (1983) and collaborators used this method to obtain estimates of Z or M from the slope of log-size frequencies plotted against estimated age, as simulated from the fitted growth equation. A new approach to analyzing a plot of logarithm-transformed numbers of individuals at size is suggested here, which does not require conversion to ages. This approach was based initially on the assumption that as the number of appropriate-sized "cover units" (or crevices) declines in numbers with size as predicted by fractal theory, and so will the abundance at size of cover-dependent organisms. Individuals "out of cover" are assumed to be eliminated by higher predation mortality as suggested by Caddy and Stamatopoulos (1990) and Holbrook and Schmitt (2002), or they may migrate elsewhere.

That fractal theory provides a reasonable approximation to the vector of natural mortality at age for crevice-dependent species was suggested by Caddy (1986) and Caddy and Stamatopoulos (1990):

> Thus, isn't it logical to ask if habitat limitations, rather than the potential maximum size of a species, is what shows up in the size frequencies of reef fishes? If a limit to the number of large crevice sizes is the problem, and can be diagnosed, this should be relevant to artificial reef design.

When a consistent rate of decline in numbers at size can be fitted to a segment of a log-length frequency, the assumption is that one or more depletion processes are occurring in the equivalent time/size interval. Data from an unfished population are a useful place to start estimating the scaling quotient, Q. If no migration occurs, the apparent overall mortality rate: $Z = M + (E - \mathrm{I})$ then converges on M (terms defined as follows—M, E, and I are annual rates of natural mortality, emigration, and immigration, respectively). It will be necessary to judge which of these processes are involved from independent information. For an isolated population without immigration or emigration (presumably a valid assumption for an isolated shelf area like Pedro Bank), I and E can be set to zero. If fractal habitat constraints are progressively more severe for larger organisms, then, as suitable cover diminishes with size (a mechanism postulated by Lipcius et al., 1998 for Caribbean spiny lobsters), this will lead to increased predation or displacement elsewhere.

RESULTS

How Can a Bottleneck in Hole Size Lead to Higher Mortality Rates for Large Reef Fish?

A simulation of the demography of crevice-dwelling organisms was conducted with two values of the von Bertalanffy growth function and two fractal coefficients. Two different growth parameter values for von Bertalanffy K; (0.3; 0.6) and a common $L_\infty = 70$ cm were used, and two values of $D + 1$ (2.15; 3.0) were considered to describe the habitat of a cohort during its life history. Individuals displaced from a crevice were either supposed to be predated or migrated out of the area. In the simulation, fish lengths were adjusted to maximum body depth and the equivalent crevice height by a fixed ratio (maximum body depth/fork length = 0.4). The number of crevices or holes suitable for a given body depth was divided by the time interval to grow from the size fitting into the current crevice, to a body depth just able to enter the next larger crevice, as estimated from the von Bertalanffy equation. Evidently, the growth of an organism through an array of progressively larger crevices could lead to a population bottleneck if the number of fish of maximum body depth, (h) > number of crevices of diameter, (d).

For a surface of crevices with constant size increments, whose number declines with size by fractal expectation (e.g., Figure 4.1), the time interval spent in larger crevices inevitably increases as L_∞ is approached. Thus, the effective availability of crevices not only declines rapidly according to fractal expectation, but older fish occupy larger crevice sizes for longer periods. Thus, if a limiting mortality applies (all crevices are occupied), then a mortality index can be defined by

$$Z' = \ln(N_1/\Delta t_{1-2}) - \ln(N_2/\Delta t_{2-3}),$$

where N_1 and N_2 are crevice numbers (and presumably also fish numbers if all are occupied) at sizes 1 and 2, divided by the occupancy time at these sizes. Interval $\Delta_{1\text{->}2}$ is the time it takes a fish to grow through a given size interval, which for constant size increments increases with age (Figure 4.2).

This phenomenon may itself lead to a demographic bottleneck without any dramatic change in habitat configuration. Simulations (Figure 4.3) show a higher mortality Z' with size for smaller free-living, demersal finfishes (e.g., Sparholt, 1990; Caddy, 1991), but for older, reef-dependent species, there may be an increasing rate of exclusion and presumably a higher natural mortality rate per size increment. Although the same trend was seen for all input values used in the simulation (Figure 4.3), a fish with a "high von Bertalanffy K" showed a lower crevice exclusion rate than "a low von Bertalanffy K" fish. The predicted "crevice exclusion rate" was higher for high fractal coefficient surfaces than for lower ones. This conclusion seems compatible with the relatively high values of K reported for many reef fishes. Thus, although it is unlikely that the availability of smaller crevices will form a bottleneck constricting population size, this is not necessarily the case for

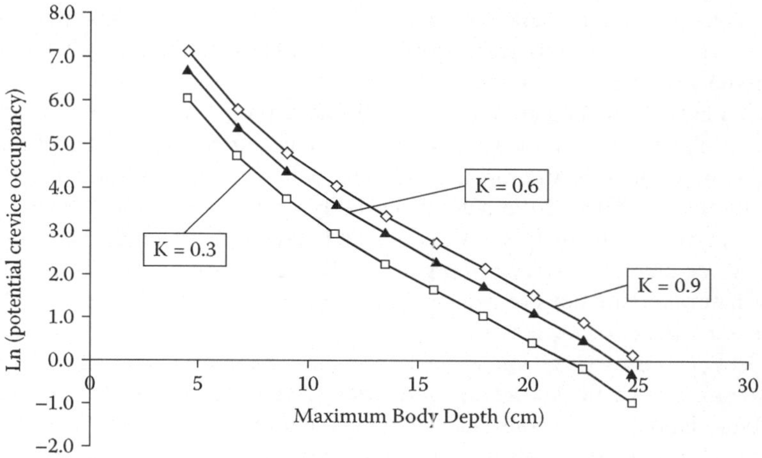

Figure 4.3 Predicted occupancy at size in a fractal environment, where the slope of the number of individuals (logarithm transformed) becomes steeper as the maximum body depth of a fish (equivalent to a length L_∞) is approached.

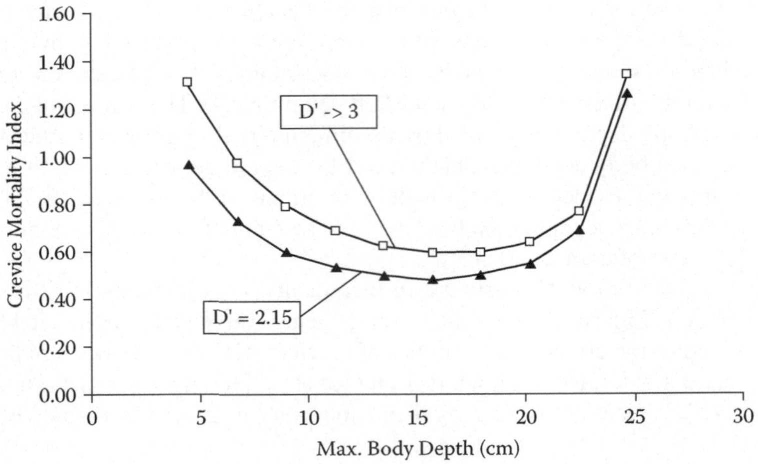

Figure 4.4 Simulating reduced availability of shelter on infauna survival in a fractal array of crevices (two values of the spatial coefficient D' are considered). Excluded individuals are assumed to be predated. The critical body dimension was the maximum body depth, and the index was expressed as the ratio: (log number of crevices/time interval to grow to the next crevice size). Time intervals were calculated from the Bertalanffy growth equation ($K = 0.6$; $L_{\infty} = 61$ cm; $t_0 = 0$).

larger individuals. A higher mortality for large fish might thus be expected, reflecting their slower growth, and hence the longer time interval Δt they take to realize a 1 cm increase in size than for smaller, younger fishes. While a small crevice might, in theory, be occupied in sequence by several juveniles in a given year, an older fish may occupy a larger crevice for more than a year.

This provides one explanation of why bottlenecks might be expected for larger reef-dwelling fish without postulating a change in habitat characteristics: A bottleneck for large mature reef fish in fractal habitats would be diagnosed by a steeper slope of the log catch curve for larger sizes (Figure 4.3). For species where the von Bertalanffy K parameter ranged from 0.3 to 0.9, Figure 4.4 shows the vectors of natural mortality to be expected with size. An advantage to growing fast is that high-K animals only occupy small crevices for short periods —hence, in a fractal environment, they experience less competition for smaller crevices, and they should be more competitive than slower-growing species. These conclusions seem broadly compatible with the analysis for several species from the Munro data set described in the following sections.

As it grows, a crevice-dependent organism will find it progressively more difficult to locate a crevice size close to its limiting body dimension on a fractal surface, and if it does not, will be vulnerable to predation. Using this argument, Caddy and Stamatopoulos (1990) created fractal arrays and simulated the likely shape of the natural mortality vector at size for species living in them, assuming different von Bertalanffy growth rates and fractal coefficients. They showed that the natural mortality vectors at size or age of organisms living in crevices on a fractal array resemble those found for natural populations of demersal fish (e.g., Sparholt, 1990; Caddy, 1991).

Two causes of bottlenecks are shown in Figure 4.5. One possible mechanism postulates that the scaling coefficient Q rises to a high value—not necessarily from a change in habitat characteristics (such as the absence of a particular crevice dimension, though this is possible), but from increased vulnerability to predation, or as a consequence of mass emigration from an increasingly unsuitable habitat. These mechanisms would result in a steeper slope in log numbers, an increase in Q, and a sharp rise in the apparent mortality index for older fishes. Interestingly, moderate fractal changes alone did not produce dramatic changes in the slopes of simulated size frequencies. Another potential cause of a bottleneck is postulated; a significant nonfractal drop in crevice numbers at size occurs over 35 cm, before the same fractal expectation resumes as before.

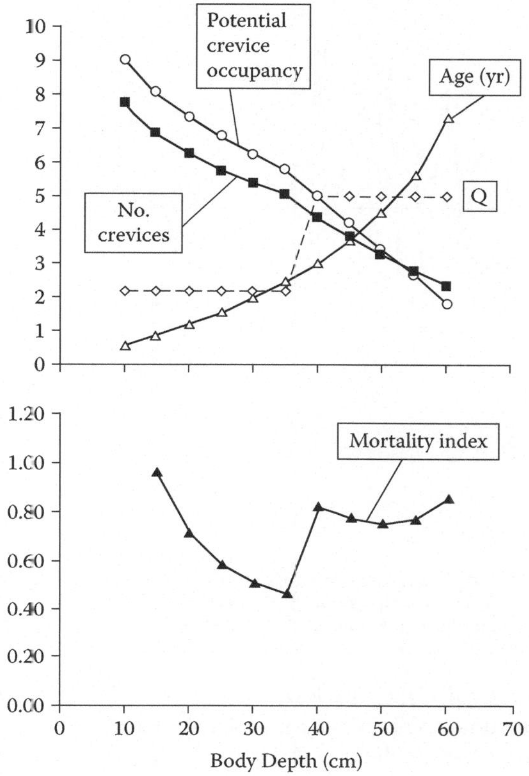

Figure 4.5 Possible causes of population bottlenecks for a fish with von Bertalanffy $K = 0.3$, and initial fractal coefficient $D' = 2.15$ for the habitat. The rate of decline in numbers increases as Q rises suddenly to 5 (implying perhaps the onset of migration to override fractal expectation?), or a drop in number of crevices occurs at crevice diameter 30 cm before fractal expectation resumes as before.

The Pedro Bank Data Set

With his collaborators, Munro (1983) published a detailed analysis of reef fish dynamics which is close to unique in the literature for multispecies reef fish populations. He stated that Pedro Bank, Jamaica, was unexploited in the 1960s–1970s. This precious data set helps elucidate the dynamics of reef fish populations, and is worth reinterpreting to facilitate understanding of the theoretical ideas mentioned earlier. The size frequencies of reef-dependent species in this locality should provide information on habitat preference.

Pedro Bank is an isolated offshore platform at 30–40 m depth south of Jamaica, which is sparsely covered with seagrass, algae, and low coral outcrops, and seems a useful site for investigating the preceding ideas. Especially, since coral reef surfaces are reported to have fractal characteristics, the extensive multispecies size-frequency data set collected by Munro (1983) and colleagues from Pedro Bank allows investigation of this question. A definite sill around Pedro Bank leads to a steep slope that drops almost vertically. A high abundance of sharks and limited cover must mean that reef fish populations there are likely to be vulnerable to predation and have few possibilities for migration elsewhere. The fish stocks of the bank at the time of study by Munro and collaborators were then unexploited, so the potential impact of fishing mortality on size frequencies can be discounted.

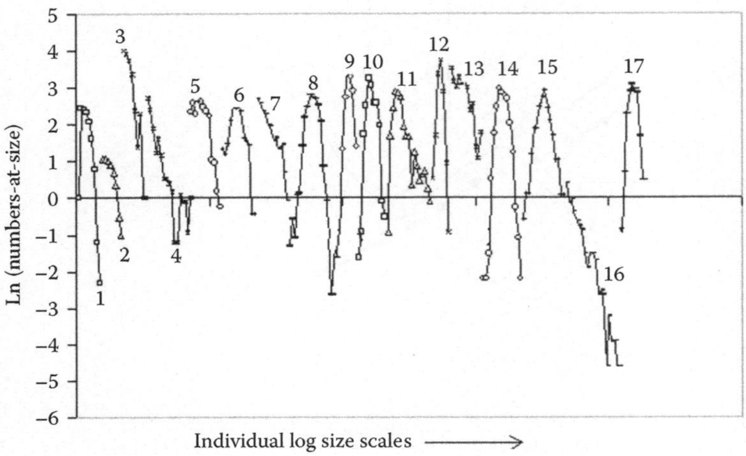

Figure 4.6 Group plot of percent individuals at size (numbers transformed logarithmically; separate scales in centimeters for each species along the *x*-axis are not shown) for 17 reef organisms taken from an unexploited population on Pedro Bank, Jamaica. (Data from Munro, J.L. 1983. Caribbean coral reef fishery resources. *ICLARM Stud. Rev. No* 7, p. 276.)

Due to gear selectivity, small fishes are less often retained by the main fishing gear used—the fish trap. This explains the initial rise in numbers with fish length in the size frequencies (Figure 4.6), while the declining slope of log numbers with length for larger fish is conventionally interpreted for an unfished stock as measuring the natural mortality rate (although migration to another habitat could cause the same effect). Since Caribbean Z traps have a pear-shaped entrance funnel about 30-cm high (Munro et al., 1971), declines in smaller coral reef fishes seem unlikely to be due to gear selection, which Munro (1983) felt did not pose a restriction on entry by the fish sizes sampled (see later). Reef fishes presumably react similarly to trap entrances as they do to natural crevices, since even unbaited traps catch some fish and may provide cover from predators. By comparing trends in log numbers at size of fish against the frequency of holes at size predicted by Equation 4.3, some regularities emerge. Figure 4.6 shows plots for 17 common species from the Munro study. Common features that emerge are close to straight-line segments in the decline of log abundance of most species once they are fully recruited to the gear.

For some species, a steep decline in log numbers at size was seen (Figure 4.7), perhaps corresponding to the high natural mortality rates derived by Munro and colleagues. However, for at least nine species in the Munro data set, more complex size frequencies with two quite different rates of decline in log number of fully recruited fishes were evident (Figure 4.8): Such a transition often occurred from a lower to a higher rate of decline in numbers at size (Figure 4.8). Rough estimates were made of the size of transition from a generally lower rate of depletion to a steep decline in numbers at size. Although the few large fish caught in large size intervals was limiting, making the estimates of higher slopes less reliable, these estimates are consistent and imply either a high rate of predation or the onset of migration—few fishes approached the maximum sizes (L_{∞}) estimated by Munro (shown by a black dot on the *x* axis of Figures 4.7 and 4.8). For species showing two rates of decline in numbers, the estimated fish length at the discontinuity between them shown in Table 4.1 was converted to the body dimension (maximum body depth), more likely than length to restrict entry to a fish trap or crevice. This calibration came from accurate scale drawings of each species shown in Fischer (1978). The maximum body height/fork length ratio varied from 0.33 for *Ocyurus*

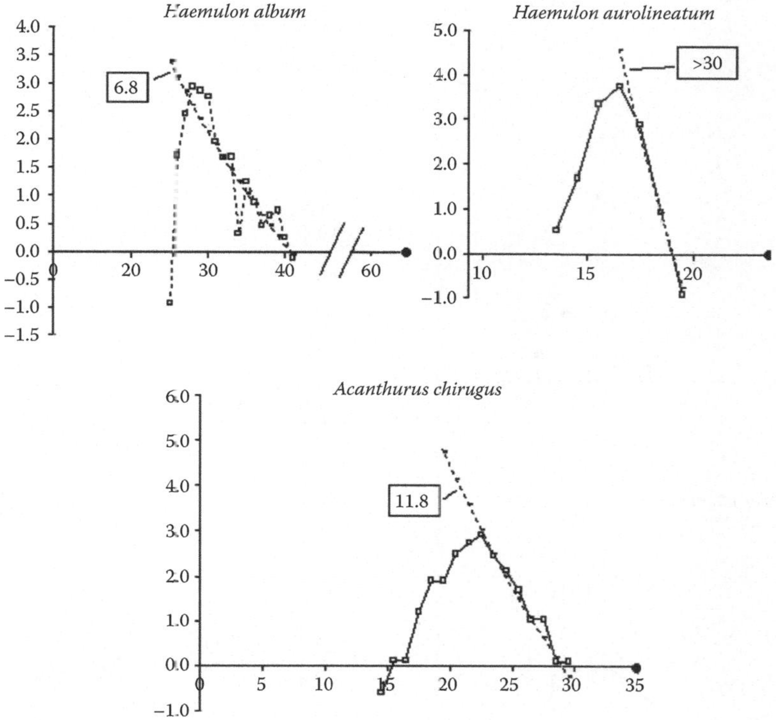

Figure 4.7 Log plots of proportions at size in research catches by Munro (1983) for three species taken on unexploited Pedro Bank that show single-steep declines in abundance once recruited. Superimposed on the log numbers at size are best-fitting lines for Equation 4.2 corresponding to values of Q.

chrysurus (yellowtail snapper) to 0.7 for *Pomacanthus arcuatus* (gray angelfish). These values were used to convert lengths to equivalent body heights in Table 4.1 and Figure 4.9.

For fully recruited smaller fish, a scaling quotient of $Q \leq 3$ cropped up fairly frequently (Table 4.1). One explanation for this estimate might be that it is tracking the fractal coefficient of the surface, D', and the relative availability of suitable volumes (dimension L^{-3}) of cover units, which is the critical constraint on survival, and hence limits the population to smaller fishes. In this sense, small values of Q may be measuring the fractal coefficient of the surface. The abrupt onset of a steep decline in catches for larger fishes above a given size must reflect a sharp change in their relative abundance on the bank, and this may correspond to a habitat bottleneck. It resembles Figure 4.5, except that the slope is more abrupt, suggesting that, at above this size range, large crevices may be rare. Massive shark predation, or migration elsewhere, seems to reasonably explain this transition. Munro (1983) says that, on maturity, larger *Balistes* spp. may migrate to deeper water on the sides of the bank. This may apply to other species also, but the availability of suitable habitat on the almost vertical slopes of Pedro Bank must be limited. Whether 8–17 cm is an effective limit on crevice dimensions on Pedro Bank seems worth further field investigation, however.

ESTIMATING THE FISH SIZE AT WHICH A BOTTLENECK TAKES EFFECT

As described by Munro et al. (1971), the "horse neck" funnel of Antillean fish traps is pear shaped and roughly 30 cm high. To investigate a possible effect of trap funnel size on size selectivity,

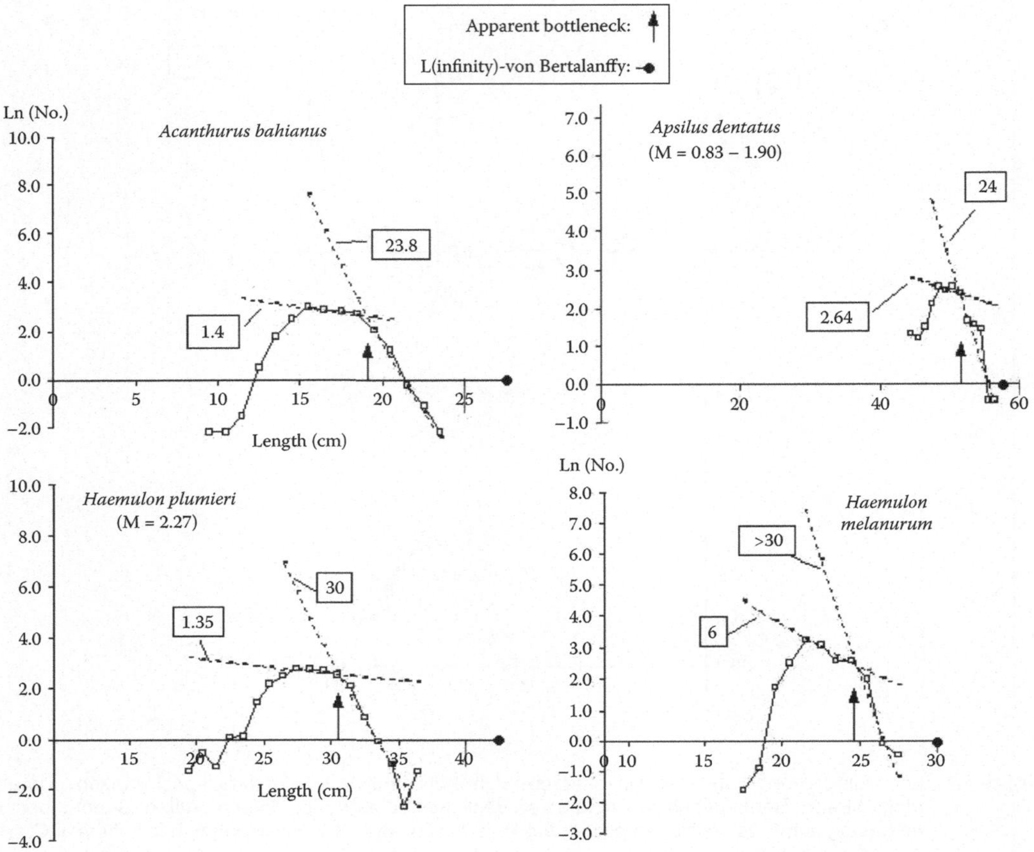

Figure 4.8 As per Figure 4.7, log plots of proportions at size in research catches by Munro (1983) for species showing a discontinuity between an earlier low rate of decline in numbers at size, and a later high rate. The filled circle is the von Bertalanffy L_∞, and the vertical arrow is the suggested fish length at which a bottleneck seems to occur. Values for natural mortality rates given by Munro (1983) are shown for comparison, where available.

scale drawings of the Pedro Bank species from Fischer (1978) were used to convert the estimated length of a species at a hypothesized bottleneck, to an equivalent maximum body depth that would be closer to the physical dimension of crevice occupied as shelter from predation. Table 4.1 provides estimates of the Q values obtained for nine species sampled by Munro (1983), which appeared to show a discontinuity of slope at a fork length ranging from 19–41 cm. This corresponds to a much tighter mean for the maximum body depth of 13.0 ± (s.d. = 3.5) cm for the nine species whose log-size frequency showed a discontinuity. For larger animals, very high values of Q and numbers-at-size in steep decline apply for animals in the range 20–50 cm fish length, or 8–17 cm (3–7 in.) maximum body depth. This is below the limiting dimension of the trap funnel (oval; 60 cm high—Munro et al., 1971) and well below the maximum size of most species (as indicated by the L_∞ parameter values shown as dots on Figures 4.7 and 4.8). An apparent bottleneck size below 30 cm maximum body depth seems unlikely to reflect a gear selection process—given that reef fishes are adept at entering small crevices in response to the presence of predators. Figure 4.9 shows that a potential funnel size limitation would be unlikely to limit ingress of most of the large individuals of each species shown in Table 4.1 to the traps. As Munro assumed, what is being registered by the change in numbers at size is an effective change in local abundance-at-size available to the traps,

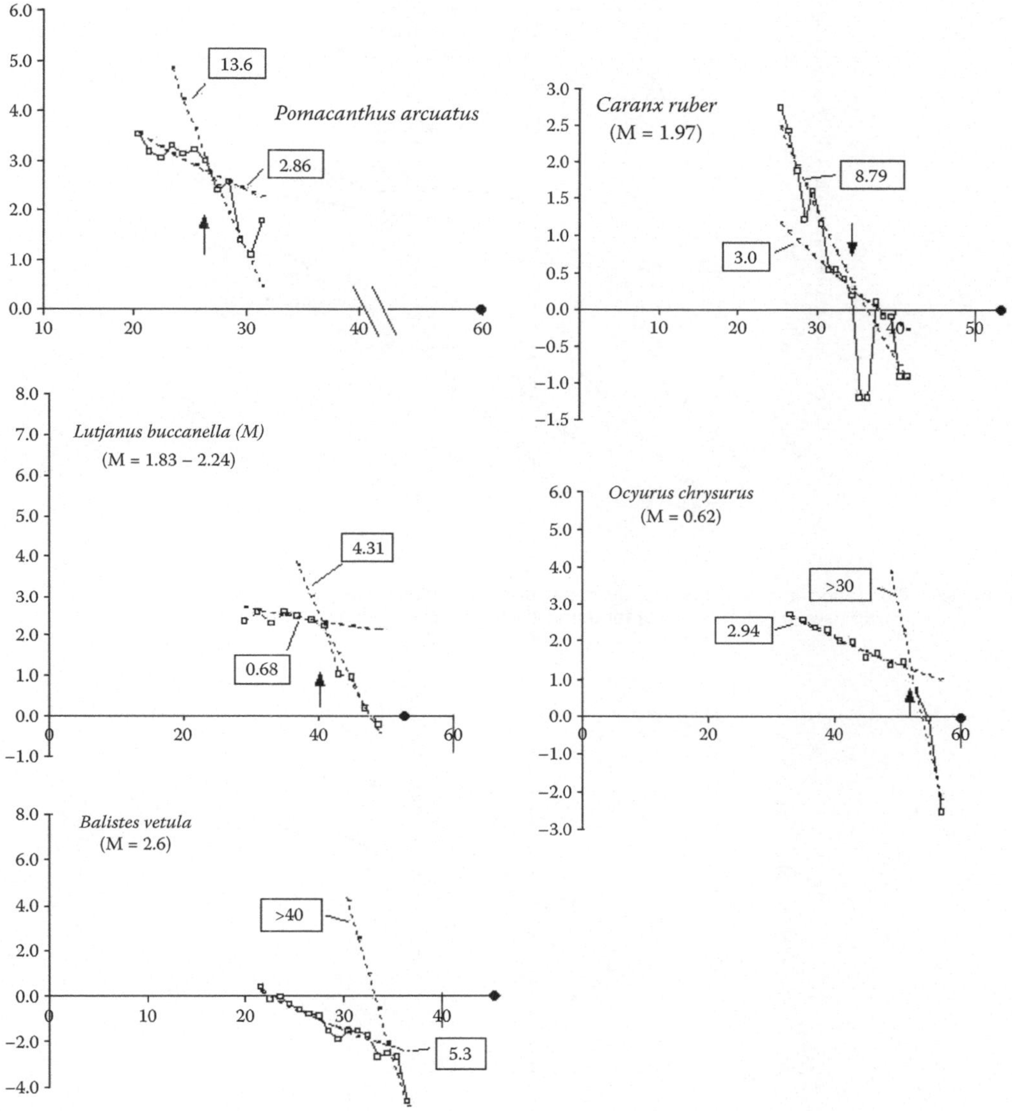

Figure 4.8 (continued).

not an effect of gear selectivity. If this is the case, it seems reasonable to hypothesize that the size range (8–17 cm body depth) by which larger fish become rare may represent an effective upper limit for common crevice sizes on the bank. This deduction seems worth pursuing by underwater field observations.

A COMPARISON WITH OTHER METHODOLOGIES

Munro (1983) estimated both growth rates and natural mortalities, and found a wide range of values for both parameters in the 20+ species studied. The von Bertalanffy coefficient, K, was correlated with the natural mortality rates (Figure 4.10), and there was a wide variation in both

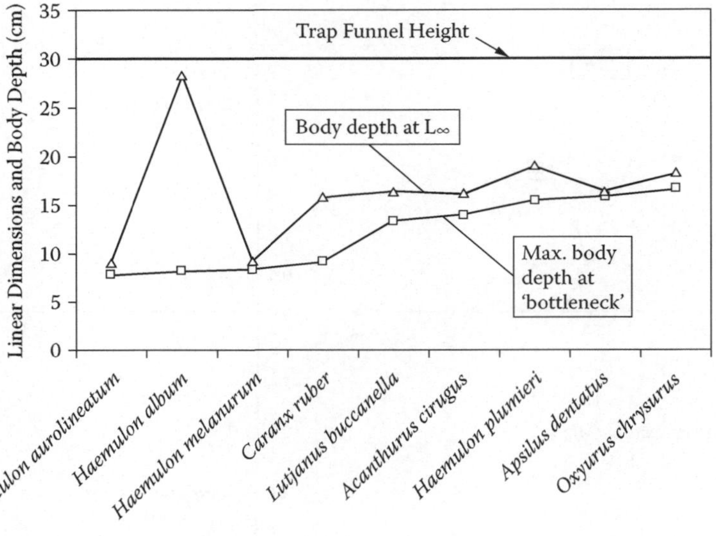

Figure 4.9 Illustrating that maximum body depth does not seem to have been a constraint for entry to the Z-trap used by Munro (1983) for sampling Pedro Bank reef fishes.

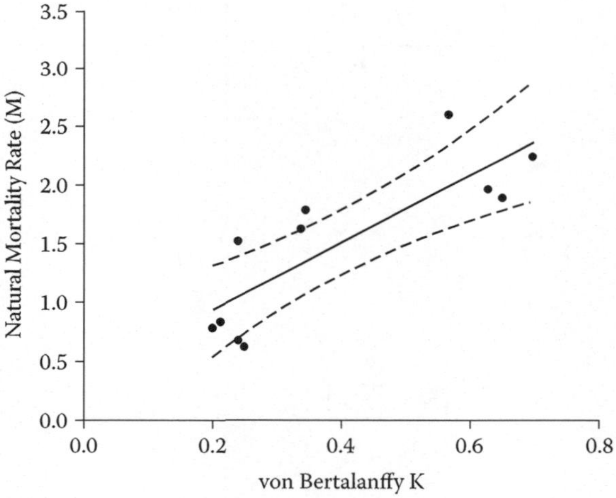

Figure 4.10 Relationship between natural mortality rate and growth rate for a sample of 11 species from Pedro Bank, using values for *M* and *K* given in Munro (1983).

parameters between fast- and slow-growing fish species. If habitat limitations were age specific, the slopes of the log numbers with size would have varied between fast-growing and slow-growing fish, but if the effect was size dependent, both slow- and fast-growing species would show a common rate of decline in numbers with size. This appears substantially the case and suggests a common habitat constraint applies to mortalities of unfished stocks. The length-converted catch curve (LCCC) used to estimate mortalities by Munro (1983) makes a number of assumptions which also apply to the present method, namely, that catchability should not be size-related, growth and natural mortality do not change markedly between years, and that the species size frequencies should represent the whole stock. Because species growth and mortality rates are correlated (Figure 4.10), the LCCC method is dependent on accurate estimates of the von Bertalanffy parameters in a situation where few fish survived as a basis for estimating maximum potential size.

DISCUSSION

This paper has described an approach to size-frequency analysis starting with fractal theory, and showed how this may be relevant for preassessment of the potential for establishing artificial reefs in an area. Samples of size-frequency data from the literature on reef-associated species were analyzed and showed a range of rates of decline in log numbers at size, with larger reef fishes showing much higher rates of decline per size increment. The widespread occurrence of fractal environments and their association with reef fishes may be the reason why the abundance of younger reef fish in an unexploited population decline at rates $Q \leq 3$, suggestive of shelter volume or crevice dependency. Given that coral reefs show fractal coefficients typical of convoluted surfaces, this is circumstantial evidence that both mortality rate and the scaling coefficient Q are tracking the impact of a fractal environment on younger fish. The study also suggests at least one mechanism underlying a bottleneck in survival of larger reef fishes and shows that the predicted change in log frequency of larger fish is substantiated for a number of species.

PRACTICAL IMPLICATIONS

If the fractal crevice theory applies, its practical significance is to suggest that increasing the number of large crevices in a protected area would enhance the reproductive capacity of reef fishes, and this might be achieved by installing large-holed artificial reefs within MPAs. It also suggests an explanation for the attractiveness of artificial reefs with large hole diameters that takes the conclusions of Polovina (1991a,b) into account: namely, the rarity in most areas of large crevice sizes attracts large fishes to artificial reefs. At the same time, it would logically suggest protecting mature fish by enclosing at least some artificial reefs with large crevices within Marine Protected Areas (MPAs) to avoid depletion of the broodstock by fishing.

Many Pedro Bank reef fish showed a discontinuity in the slope of the log-size frequency that I have called a "bottleneck," at a body depth of between 8 and 17 cm. Above this size discontinuity, the rate of decline Q increased dramatically with loss of a species from the catch well before reaching the maximum size (both L_∞ and a size limited by trap entrance height); that is, it represents a fast decline in abundance with size. A possible explanation for this abrupt change might be migration but, given the isolated nature of Pedro Bank, increased predation due to a decline in crevice numbers with size seems the most probable explanation. A simulation of changes in survival at size in fractal environments (demonstrated by the literature to apply to coral reefs) suggests that, in addition to declining availability of large crevices, the longer residence time for slower-growing older fish also plays a role in narrowing habitat availability with size. This hypothesis may explain the high attractiveness of reef crevices noted by Polovina (1991a,b). The existence of an abnormally high abundance of large crevices in artificial reefs, contrasting with the general rarity of large crevices in many marine biotopes, may explain why artificial reefs act as aggregation devices. This can either lead to overexploitation or, if artificial reefs are protected within MPAs, may provide a local nucleus of large spawners for outwelling of gametes to adjacent fished areas.

Where the value of Q changes markedly above a critical size, this may be tentatively identified as a "habitat bottleneck," and the mechanism should be investigated further by in situ habitat measurements, or by the use of a "fractal sampler."

If competition for the limited structural elements used in predation avoidance is critical, life history migrations of larger fish (e.g., to offshore banks or canyons) may be initiated in response to

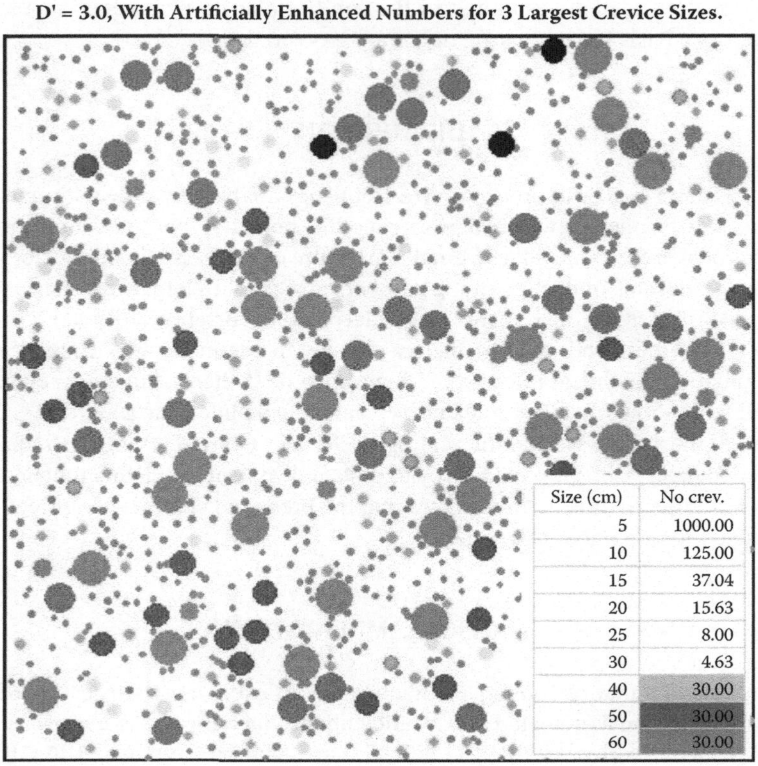

Figure 4.11 ***See color insert.*** Illustrating a reef surface with enhanced numbers of crevices in the size range 40–60 cm superimposed on a natural reef with fractal coefficient $D' = 3.0$.

declines in adequate cover. The use of Equation 4.3 to iteratively fit Q to linear segments of log catch curves at size provides an indicator of the possible mechanisms underlying linear runs of log numbers at size, and may help decide on supplementing the crevice sizes needed to protect spawners from predation within MPAs. Adding more large holes to natural surfaces (or artificial reefs) as in Figure 4.11 could markedly enhance the holding capacity of a reef for fish of reproductive ages.

REFERENCES

Beck, M.W. (1997). A test of the generality of the effects of shelter bottlenecks in four stone crab populations. *Ecology*, Vol. 78, No. 8, pp. 2487–2503.

Bradbury, R.H., R.E. Reichelt, and D.G. Green. (1984). Fractals in ecology: Methods and interpretation. *Mar. Ecol. Prog. Ser.,* Vol. 14, pp. 295–296.

Burrough, P.A. (1981). Fractal dimensions of landscapes and other environmental data. *Nature,* Vol. 294, pp 240–242.

Caddy, J.F. (1986). Modeling stock-recruitment processes in Crustacea: Some practical and theoretical perspectives. *Can. J. Fish. Aquat. Sci.,* Vol. 43, pp. 2330–2344.

Caddy, J.F. (1991). Death rates and time intervals: Is there an alternative to the constant natural mortality axiom? *Rev. Fish. Biol. Fisheries.,* Vol. 1, pp. 109–138.

Caddy, J.F. (2007). *Marine Habitat and Cover: Their Importance for Productive Coastal Fishery Resources.* Oceanographic Methodology Series, UNESCO Publishing, 253 pp.

Caddy, J.F. (2008). The importance of "cover" in the life histories of demersal and benthic marine resources: A neglected issue in fisheries assessment and management. *Bull. Mar. Sci.* Vol. 83, No. 1, pp. 7–52.

Caddy, J.F. and C. Stamatopoulos (1990). Mapping growth and mortality rates of crevice-dwelling organisms onto a perforated surface: The relevance of cover to the carrying capacity of natural and artificial habitats. *Estuar. Coast. Shelf Sci.* Vol. 31, pp. 87–106.

Eggleston, D.B., R.N. Lipcius, D.L. Miller, and L. Coba-Cetina. (1990). Shelter scaling regulates survival of juvenile Caribbean spiny lobster *Panulirus argus*. *Mar. Ecol. Prog. Ser.*, Vol. 62, pp. 79–88.

Fischer, W. (Ed.). (1978). FAO species identification sheets for fishery purposes. *Western Central Atlantic (Fishing Area 31).* (7 volumes) FAO/UNDP.

Gee, J.M. and R.M. Warwick. (1994a). Body size distribution in a marine metazoan community and the fractal dimensions of macroalgae. *J. Exp. Mar. Bio. Ecol.*, Vol. 178, No. 2, pp. 247–249.

Gee, J.M. and R.M. Warwick. (1994b). Metazoan community structure in relation to the fractal dimension of marine algae. *Mar. Ecol. Prog. Ser.*, Vol. 103, pp. 141–150.

Gorham, J.C. and W.S. Alevizon. (1989). Habitat complexity and the abundance of juvenile fishes residing on small scale artificial reefs. *Bull. Mar. Sci.*, Vol. 44, pp. 662–665.

Gotceitas, V. and P. Colgan. (1989). Predator foraging success and habitat complexity: Quantitative test of the threshold hypothesis. *Oecologia*, Vol. 80, pp. 158–166.

Gunnarsson, B. (1992). Fractal dimension of plants and body size distribution in spiders. *Funct. Ecol.*, Vol. 6, pp. 636–641.

Holbrook, S.J. and R.J. Schmitt. (2002). Competition for shelter space causes density-dependent predation mortality in damselfishes. *Ecology*, Vol. 83, No. 10, pp. 2855–2868.

Jones, R. and N.P. van Zalinge. (1981). Estimates of mortality rate and population size for shrimp in Kuwait waters. Kuwait. *Bull. Mar. Sci.*, Vol. 2, pp. 273–288.

Keats, D.W., D.H. Steele, and G.R. South. (1987). The role of fleshy algae in the ecology of juvenile cod (*Gadus morhua* L.) in inshore waters off eastern Newfoundland. *Can. J. Zool.*, Vol. 65, pp. 49–53.

Li, B-L. (2000). Fractal geometry applications in description and analysis of patch patterns and patch dynamics. *Ecol. Modelling*, Vol. 132, pp. 33–50.

Lipcius, R.N., D.B. Eggleston, and T. Luhrs. (1998). The habitat-survival function for Caribbean spiny lobsters: An inverted size effect and non-linearity in mixed algal and sea grass habitats. *Mar. Freshwat. Res.*, Vol. 49, No. 8, pp. 807–816.

Mandelbrot, B.B. (1983). *The Fractal Geometry of Nature.* W.H. Freeman and Co., San Francisco.

Mark, D.M. (1984). Fractal dimension of a coral reef at ecological scales: A discussion. *Mar. Ecol. Prog. Ser.*, Vol. 14, pp. 293–294.

Morse, D.R., J.H. Lawton, M.M. Dodson, and M.H. Williamson. (1985). Fractal distribution of vegetation and the distribution of arthropod body lengths. *Nature*, Vol. 314, pp. 731–733.

Munro, J.L. (1983). Caribbean coral reef fishery resources. *ICLARM Stud. Rev. No.* 7, p. 276.

Munro, J.L., P.H. Reeson, and V.C. Gaut. (1971). Dynamic factors affecting the performance of the Antillean fish trap. *Proc. Gulf Caribb. Fish. Inst.* Vol. 23, pp. 184–194.

Nordeide, J.T. (1993). Experiments with Reared Juvenile Coastal Cod (*Gadus morhua* L.) in order to Enhance the Stock in a Fjord. Ph.D. thesis, University of Bergen, Norway, 102 pp.

O'Connell, V.M. and D.W. Carlisle. (1993). Habitat-specific density of adult yelloweye rockfish *Sebastes ruberrimus* in the eastern Gulf of Alaska. *Fish. Bull.*, Vol. 91, pp. 304–309.

Pauly, D. (1982). Studying single species dynamics in a tropical multispecies context. Pages 33–70 in *ICLARM Conf. Proc.* 9; 360 pp.

Polovina, J.J. (1991a). Ecological considerations on the applications of artificial reefs in the management of artisanal fisheries. *Trop. Coastal Area Manage.*, Vol. 6, Nos. 1/2, pp. 1–4.

Polovina, J.J. (1991b). Fisheries applications and biological impacts of artificial reefs. Pages 153–176 in Seaman, W. Jr. and L.M. Sprague (Eds.). *Artificial Habitats for Marine and Freshwater Fisheries.* Academic press, San Diego, p. 285.

Purkis, S.J. and K.E. Kohler. (2008). The role of topography in promoting fractal patchiness in a carbonate shelf landscape. *Coral Reefs*, Vol. 27, pp. 977–989.

Ricker, W.E. (1975). Computation and interpretation of biological statistics of fish populations. *Bull. Fish. Res. Bd. Canada*, Vol. 119, 382 pp.

Sale, P.F. (1978). Coexistence of coral reef fishes—a lottery for living space. *Environ. Biol. Fish.*, Vol. 3, pp. 85–102.

Schmid, P.E. (2000). Fractal properties of habitat and patch structure in benthic ecosystems. *Adv. Ecol. Res.*, Vol. 30, pp. 339–400.

Shulman, M.J. (1985). Recruitment of coral reef fishes: Effects of distribution of predators and shelter. *Ecology*, Vol. 66, pp. 1056–1066.

Smith, C.L. (1975). Analysis of a coral-reef fish community: Size and relative abundance. *Hydro-Lab J.*, Vol. 3, pp. 31–38.

Smith, C.L. and J.C. Tyler. (1975). Succession and stability in fish communities of dome-shaped patch reefs in the West Indies. *Am. Museum Novitates*, No. 2572, pp. 1–18.

Sparre, P., E. Ursin, and S.C. Venema. (1989). Introduction to tropical fish stock assessment. Part 1—Manual. FAO Fish. Tech. Pap. 306/1, 337 pp.

Sparholt, H. (1990). Improved estimates of the natural mortality rates of nine commercially important fish species included in the North Sea multispecies VPA model. *J. Cons. Perm. Explor. Mer.*, Vol. 46, No. 2, pp. 211–223.

Stamatopoulos, C. and J.F. Caddy. (1991). "Holes for Windows"—*(Informal software available from the senior author)*.

Tercafs, R. (1997). A fractal model of cave ecological environment. *Mémoires de Biospéologie*, 24: 1–8.

Tupper, M. and R.G. Boutilier. (1995). Effects of habitat on settlement, growth, and post-settlement survival of Atlantic cod (*Gadus morhua*). *Can. J. Fish. Aquat. Sci.*, Vol. 52, pp. 183–1841.

Walters, C.J. and F. Juanes. (1993). Recruitment limitation as a consequence of natural selection for use of restricted feeding habitats and predation risk taking by juvenile fishes. *Can. J. Fish. Aquat. Sci.*, Vol. 50, pp. 2058–2070.

Ziegler, B. (1979). Growth and mortality rates of some fishes of Manila Bay, Philippines, as estimated from length frequencies. Thesis, Kiel University, 115 pp.

CHAPTER 5

An Integrated Coastal Area Management Strategy to Deploy Artificial Reefs

Sylvain Pioch, J.-C. Raynal, G. Lasserre, and C. Aliaume

CONTENTS

ABSTRACT

The growing world demand for marine fisheries resources (95 million tons in 2006, scheduled to increase by 2% to 3% globally each year—UNESCO, 2001) is compounded by the prospect of a growing population and is a real threat to world fish stocks. This threat particularly affects offshore or coastal fish stocks in both the short and long term (Worm et al., 2006; Cury and Miserey, 2008). When facing sociocultural food issues in a crisis situation, fisheries resources managers generally focus on two fundamental principles: preserving stocks and developing marine ranching.

In this chapter, methods and planning proposals are examined to implement an Integrated Coastal Area Management (ICAM) strategy, including the most successful fishery facilities (Denis et al., 2001; Henocque, 2006) with the example of the Japanese artificial reefs and a feasibility study conducted in the town of Saint-Leu (Reunion Island, West Indian Ocean). A geo-socio-systemic approach is used to create artificial reefs in coastal areas. It is assumed that an artificial reef should be part of a concerted strategy using an eco-socio-systemic approach in the framework of sustainable territory enhancement, so that all the long-term objectives of restoring, maintaining, and/or increasing fish populations can be achieved.

INTRODUCTION

Coastal areas are subject to multiple, human-induced pressures that seriously affect natural marine habitats and species (Denis et al., 2001). This narrow strip between land and sea hosts almost two-thirds of the world's population, with occupancy rates that are fivefold higher than the average density of inhabited inland areas. By 2050, an estimated 75% to 80% of the world population will live near the coast (Saunier and Laffitte, 2007). The growing demand for marine fisheries resources (95 million tons landed in 2006, with a global increase of 2% to 3% per year forecast by UNESCO, 2001) is compounded by the increase in the world population, which is expected to reach 8.9 billion in 2050 (UNFPA, 2004) implying serious pressure on fish stocks (Worm et al., 2006; Cury and Miserey, 2008). Human needs for water and energy, as well as for leisure activities, will increase with a resulting increase in pollution and waste that will also affect marine ecosystems. Coastal areas are also threatened by the removal of marine resources through fishing. Currently, the world's oceans are the only source of protein for one-quarter to one-third of the world population. The FAO outlook on world fishery resources states that "As stated before in "The State of World Fisheries and Aquaculture," the maximum wild capture fisheries potential from the world's oceans has probably been reached, and a more closely controlled approach to fisheries management is required ..." (FAO, 2008, p. 7–8). The pressure exerted by people on the marine environment is not without consequences. The increase in world demand for aquatic products is about 2% per year, suggesting increased pressure on the environment (Saunier and Laffitte, 2007). Seventy-seven percent of marine fisheries are already fully exploited, overexploited, or depleted (FAO 2004–2006). According to Worm et al. (2009), 63% of assessed fish stocks worldwide require rebuilding, and lower exploitation rates are needed to reverse the collapse of vulnerable species. There is thus an urgent need for policies aimed at the recovery of stocks.

In practice, when resources managers are faced with these food and sociocultural issues, they generally focus on two fundamental principles:

1. Changing the pressure on fishing mainly by catch restrictions, gear modification, and closed areas. An example is a fishery management approach such as TAC (total annual catch) to assign an annual quota that, if exceeded, will terminate the fishery for that year; the total allowable catch is set at a level to prevent a catch so large that the stock will be overfished.
2. Affecting the environment by increasing resources through extensive "mariculture." For example, increasing the number of juveniles by restocking or by enhancing or restoring habitats by creating artificial reefs.

In conjunction with such actions for the management of environmental and fishery resources, other more technical actions are also implemented, including prohibiting illegal, unreported, and unregulated (IUU) fishing; designing control instruments; and creating management plans (European Commission, 2007b; UNESCO, 2001). To enable coastal management agencies to play an effective role in restoring, maintaining, or even increasing fish populations, it is crucial that the policies they implement are part of a concerted approach aimed at goals that balance regional and ecosystem requirements.

In the following, we examine methods and planning proposals to implement an ICAM strategy, including the most successful fishery facilities (Denis et al., 2001; Henocque, 2006) and using as examples the Japanese experience with artificial reefs and a feasibility study conducted in the town of Saint-Leu (Reunion Island, West Indian Ocean).

THE GEO-SYSTEM: A WAY TO INTEGRATE MANAGEMENT IN COASTAL ZONES

For lay individuals, "marine exploitation" often conjures up images of searching for sunken treasure with nonlimited resources given by the ocean (currents theories of biologist such as Jean-Baptiste Lamarck [1744–1829] and Thomas Huxley [1825–1895]). This vision is now chronologically out of place, even though some developers continue to justify the concept with varying success (Miossec, 1998). Learning to manage global maritime space (e.g., by restoring fishing grounds in coastal areas) to reestablish its biodiversity and increase its fishery productivity is part of our heritage. Our heritage also includes the concept of "ocean space" by coastal inhabitants. Social and cultural aspects of our coastal culture are strongly linked with "good management" practices or the long-term use of natural resources, especially if the aim is self-regulation or educational (i.e., assimilated by the citizen and not only by the authorities; Hardin 1968). As coastal communities struggle to respond positively to the demand for seafood products and recreational fishing, there is an opportunity to promote the development of artificial reefs. We use a geosystem approach, that is, the study of the relationship between societies and nature (Scheibling, 1994). A geosystem is a spatial unit formed by balancing the climate, soil, fauna, flora, and structure, and is used by geographers to describe the "natural environment" (Brunet et al., 1995). The approach is different from that used by many ecologists because it takes human actions into account without necessarily perceiving humans as disruptive to the ecosystem. Thus, nature is influenced by humans because they are a part of it. A geosystem also includes all past and present human interventions with regard to the environment (Veyret, 2004). It is important to consider the functional and structural reality of a geosystem as an organized population that consumes resources extracted from the biophysical environment. For convenience sake, here the term *geosystem* is used when considering the links between human actions and different ecosystems.

With regard to the problems mentioned earlier, we are currently responsible for the dynamic stewardship along our coasts. Our stewardship envisages coastal areas supported by fishing, fish farming, agriculture, urban development, and industrial, cultural, and recreational activities. It is essential that we, as a community of environmental stewards, have the foresight to understand and influence public policies from their conception to their intended outcome (see Figure 5.1).

When the geosystemic approach focuses on the management of coastal areas, it resembles the widely known ICZ(A)M strategy (Integrated Coastal Zone [or Area] Management; Cicin-Sain and Knecht, 1998). This is a dynamic process that brings together government and local communities, scientists, policy makers, and public and private interests to prepare and implement a plan to protect and develop coastal resources. The approach is designed to optimize choices among management options to achieve long-term objectives. ICAM cannot replace sectoral planning (i.e., planning in different activity sectors) but focuses on the relationship between the activities in different sectors to achieve broader objectives. Policies concerning maritime transport, fisheries, energy, and tourism have evolved separately. These policies sometimes result in failures, inconsistencies, and conflicts in the use of resources (European Commission, 1999; Doumenge, 2004; Cadoret, 2006). Effective

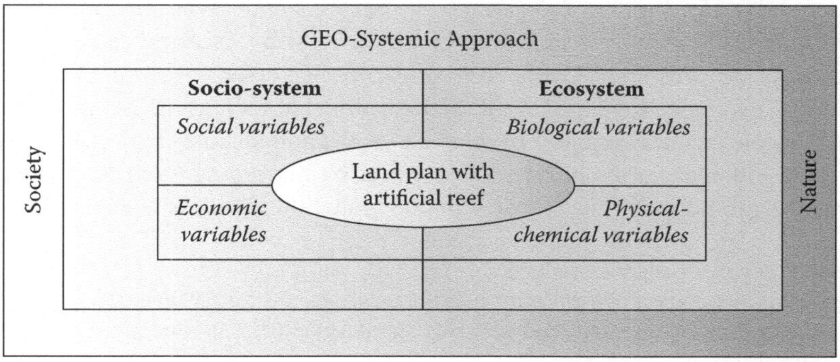

Figure 5.1 Geosystemic approach and the relationship between society and nature. (From Pioch, S. 2008. Artificial habitat: A strategic tool for integrated coastal area management? Proposal guideline for their development in coastal fisheries. Ph.D. dissertation, Geography and Coastal Land Planning, University Montpellier and Tokyo University of Marine Science. 295 pp.)

coastal development is a way to reduce these conflicts by introducing rules for the use of a delimited space. More importantly, effective coastal development can facilitate the overall economic development of coastal communities while ensuring the relative stability of ecosystems even if the natural resources (including fish) are impacted by humans (Meur-Ferec, 2007). The European Parliament (European Commission, 2007a) currently promotes the implementation of ICAM strategies in Europe, and these have been integrated in the national legislation of all member countries of the EU (decision of Interministerial Planning Committee, July 2001, September 2004). This integrative approach should ensure long-term benefits toward making marine resources exploitation more sustainable. This management strategy could include land planning in conjunction with an artificial reef deployment. Indeed, artificial reefs could be used as a transactional benefit for fishermen to encourage them to experiment with a new fishing strategy (e.g., new gear, new fishing area). It would also encourage fishing in areas previously thought to be unproductive for fishing (i.e., areas of naturally low productivity or those areas destroyed by other human activities).

THE JAPANESE MANAGEMENT OF ARTIFICIAL REEFS THROUGH IMPLEMENTATION OF AN ICAM-TYPE STRATEGY

The Japanese experience in deploying artificial reefs is an example of the successful implementation of ICAM facilitated by a "mono-actor" approach in the use of natural resources and coastal access. The Japanese strategy was to be close to the actors (fishermen) while heeding the demand for seafood, which required the implementation of several policies designed by a technical–scientific commission, and aimed at conserving and managing resources and activities. Japan is regarded as the world leader in the successful development of the seabed for the sustainability of coastal fisheries (Mottet, 1985; Bailly, 1989; Baine, 2001). Extensive coastal mariculture (Latin *mar* meaning "sea") began with the first Japanese artificial reefs. The 1652–1655 chronicles of Emperor Joo's reign described these artificial reefs as being made of boulders and designed to expand exploitable fishing zones near coastal villages. The first scientific studies were undertaken in 1903 by the Fukuoka prefecture. However, the most significant use of artificial reefs began in 1952 with a government plan for the "development of fisheries areas in deep waters" (Simard, 1985; Mariojouls, 2004). What is considered to be the most original contribution to the creation of artificial reefs to enhance Japanese fisheries was included fisheries and aquaculture in the overall management of the coastal zone to increase marine production. For the past 70 years, the enhancement of coastal resources has continued in several areas off the coast of Japan. In addition to the development of the fisheries through an organized research network and highly efficient professional administration, the large-scale release of juveniles of more than 90 species is under way (Hogan, 2010). In addition, these artificial reefs help support fisheries stocks by increasing carrying capacity and productivity in fishing zones. According to Denis et al. (2001: p. 21), extensive mariculture (used to mean "marine ranching") is understood as " … all actions aimed at increasing the production of marine plant or animal species without inputs." In some cases, the human contribution to the success of the action is limited to increasing the number of juveniles by restocking them. The Japanese plan for reorganization and promotion of the sector is divided into "three pillars" (MAFF, 2007):

1. Ensuring the supply of aquatic products
2. Facilitating the development of the fishing industry and the quality and health safety of aquatic products
3. Encouraging advanced research, and ensuring the development and monitoring of research and development

This policy relies on a network of important players, such as the fisheries research agency, university research centers, prefectural centers, and private companies. The fisheries agency comprises the nine Japanese fisheries research institutes. University research in Japan is very dynamic and includes 34 research departments specializing in fishing or aquaculture and is distributed among Japan's universities. The prefectural centers are the most efficient interfaces between the Japanese administration, fishermen, and the general public. They are responsible for local ecological monitoring of the marine environment and conduct research in collaboration with national environmental and fisheries programs. Private companies are also developing innovative models of artificial reefs and generally work with their own research units or in collaboration with other agencies in the network.

In Japan, the principles that guide the morphological design of artificial reefs are adapted to the objectives and parameters of eco-social systems, based on relationships (including ethological relationships) between species and their habitat (Nakamura, 1980; Kakimoto, 1985; Kakimoto et al., 1995). Two major factors are taken into account during the design phase:

1. Artificial reefs have to be adapted to the vital biophysical parameters of the target species.
2. The three main ecological functions of a habitat—nursery, breeding ground, and shelter—have to be promoted (Nakano, 2007).

Other aspects considered in the design and developments are

- Physical resistance of materials and mechanical aspects (Katoh, 1985).
- Social aspects that include the settlement of territorial disputes, self-management, appropriate fishing equipment, and deterrence of poaching and overfishing (Kakimoto, 2004).

The artificial reefs in Japan have three main goals: to restore, maintain, and develop the ecosystem and "fishing areas" to sustain coastal fisheries (improving working conditions, attracting young people to the sector, etc.). The system and process of governance implemented in fisheries management depend on the size of the project, its objectives, and the species targeted. The effectiveness of Japanese artificial reefs is ecologically evaluated in terms of the performance of the reef relative to target species. Monitoring was conducted, in part, by local fishermen. This demonstrates the involvement of local stakeholders in the project. The focus for artificial reefs is on the catch level and/or socioeconomic balance in Japan. The three main criteria to evaluate the reef are functionality (i.e., providing habitat for spawning, protection, and increasing the food base for target species), the abundance of target species, and the level of satisfaction of the fishermen relative to fishing time, travel distance, etc.

The Japanese policy of proactive resources management through extensive mariculture is achieved partly through the use of artificial reefs and partly through a deliberate policy of sea fertilization called *saibai gyogyo* that involves parallel restocking.

The Japanese management of fisheries resources is considered a shared social venture and fisheries resources are a national priority. There is only one main objective for all fishery resource projects: They must be self sufficient (the success rate of these projects should be around 65% in 2012). This is an ambitious objective since Japan is the world's largest importer of fish and produces only about 10% of world production (10 million tons out a total of 100 million tons fish production in the world). This translates into 65.7 kg/year per capita consumption of fish in Japan.

Fishing cooperatives and the central government play a central role in the plan. Fishing cooperatives start projects and are assisted by their local fisheries agency and the Japanese Institute of Technology on Fishing Ports, Grounds, and Communities (JIFIC). Together, these entities ensure the coordination and development of coherent programs for the improvement of fishing grounds.

Significant funding is available for these kinds of projects. From 2007 to 2008, 602 million euros (close to $1 billion) were provided by the government (National Statistics of Ministry of Internal Affairs, Japan, 2008). The exclusive use of coastal areas by fishermen has enabled the implementation of this form of management, while elsewhere in the world the fact many different stakeholders (each with their own agenda) are involved hampers this type of development. The Japanese experience shows us which management rules and options to implement but, outside Japan, multistakeholder collaboration and a geosystemic approach is required. While Japanese institutions set up a process to create rights to a territory where the resource (considered as a national patrimony) is exclusively managed by fishermen, in France (and in many other countries), the situation is different. It is the role of Maritime institutions to provide solutions taking into account the many actors and multipurpose uses.

THE ARTIFICIAL REEF, AN APPROPRIATION TOOL FOR THE MANAGEMENT FOR COASTAL AREAS: THE EXAMPLE OF SAINT-LEU, REUNION ISLAND, FRANCE, WEST INDIAN OCEAN

The range of different actors and of different possible uses led many managers to adopt a systemic approach in the geo-context of integrated coastal zone (ICAM strategy). In our case study here, such an approach was implemented. The study was conducted under the authority of the municipality of Saint-Leu (Reunion, France, Department in the West Indian Ocean) and was similar to that of other development schemes for artificial reefs conducted in France (in Marseille, Agde, etc.). The case study was conducted in the coastal village of Saint-Leu (Reunion Island, France), where the demand for artificial reefs resulted from a spatial conflict. The choice of the area was justified as it was a "patrimonial," inherited estate, coral zone, rich in fish, and of great ecological value (high biodiversity). The site is visited by a variety of users including scientists, nature associations, tourists, and professional and recreational fishermen. Management goals are paradoxically shared between the exploitation of natural resources and species protection. Fishing (by both professional and recreational fishermen) is both a traditional activity and an important local source of food. The legal status of this natural site changed from local (of limited value but historical/traditional economic importance) to a supranational world heritage (coral conservation at the world scale through programs like "CRISP," which started in 2002 (French Development Agency: CRISP, 2008) The coastal artisanal activities, which were a source of supplementary income and of food, were faced with the creation of a marine park (a legitimate requirement for the conservation of an ecological world heritage site). The preelectoral context added force to the voices of local users who were faced with the creation of a "no-access" reserve. Although the creation of the reserve was ecologically justified, local users believed the reserved was being imposed by the State and by Europe, despite the public debate and conciliation work undertaken before the beginning of the project (European Commission, 2005).This underlines the need to involve all stakeholders in identifying and validating targets before any project involving artificial reefs is undertaken (Ducloy, 2006).

Both institutional agents and users came under the jurisdiction of Saint-Leu authorities, who, in addition to being the spokesperson of the fishermen, also answered to several administration levels, that is, the Department of Reunion, the French government, and Europe (France has pledged to respect biodiversity conservation policies to promote protected marine areas covering 10% of the 11 million km² of the French economic area by 2012—Commission of the European Communities, 2006). The implementation of artificial reefs seemed to be an innovative approach to attain divergent goals that actually converged, but were expressed differently. A three-pronged approach was instituted:

- Preserve coastal fauna and flora by offering a new zone of laying, food, and shelter for local species (Bolopion et al., 2000)
- Propose new areas for commercial fishing and recreation for local users and tourists
- Increase the fishing resource in the long term

The deployment of artificial reefs seems to be a complementary measure (rather than a mitigation measure to completely protect marine areas; Lasserre and Monteiro, 2002). Artificial reefs help create habitats for targeted species that are of major socioeconomic interest to fisheries. The specific area of deployment was composed of sand-bottom areas with the least apparent biological marine production (Gabrié and Montaggioni, 1985). Such grounds are located between a depth of 0 to 300 m, and the only exploitable resource is the sandy "crab giraffe" (Spanner crab, *Ranina ranina*). The decrease in catch by recreational or professional activities is a widespread phenomenon that results from concentrating coastal exploitation and allowing habitat loss to occur. Both features impede species replenishment (i.e., settlement and reproduction; Tessier, 2005). Artificial reefs could potentially permit a balance between the creation of new fishing areas and the management of target species. To achieve this goal, suitable management strategies and the strict distribution of the activities in space and time are indispensable; otherwise, the attraction versus production dilemma could reduce the benefits of such a project (Bortone, 1998).

In practice, the process of implementation of artificial reefs involves five major stages:

- Phase 1: Development of the charter of use
- Phase 2: Site selection, volume, and reef design
- Phase 3: Selection of concession jurisdiction
- Phase 4: Financing plan
- Phase 5: Evaluation policy and scientific monitoring

Phase 1: Development of the Charter of Use

In Reunion Island, all five stages were formalized by the implementation of a "charter of use" (Pioch and Pary, unpublished). Setting up a charter early in the process seemed to answer a need and was the wish of the different users (the concept and commitment being the subject of a social consensus). The principal of the charter was accepted by the local government agencies, stakeholders, and by the French authorities, because of its simplicity and the advantages offered by self-management. The implementation of the charter was based on creation of an organization comprising local government agencies and the local fishermen's association.

The charter sets out the objectives of the organization, along with rules for responsible fishing and sustainable development. It points out the need to function as in a cooperative.

Generally, the acceptance of the content of a charter by the end users is a prerequisite for the creation of an artificial reef in a developed area. In Reunion Island, this commitment was formalized by individual signatures in exchange for a license allowing the signatory to use the reef. Noncompliance would involve financial penalties and/or a ban on access to the sites (Figure 5.2).

In Saint-Leu, the charter in the following was proposed to users.

The charter included three levels of control:

- Internal control by all the signatories during their activities
- Control by a sworn signatory
- External control by the coast guard or the municipal police

The charter represented an additional guarantee in the appropriation of a common space. It also started a process toward a form of "privatization" of the marine area through an access control system, giving a feel of ownership to the user. This approach was of course inspired by the Japanese fisheries, but also by the system of the Mediterranean *prud'hommes*. This is an exception to other French marine fisheries in which the signatories have rights to self-managed uses within a delimited territorial area, and the fisheries can create and implement regulations and solve conflicts in the *prud'homme* court, as well as impose sanctions (Cazalet, 2007).

Charter Framework		
	A Professional Angler	B Recreational Activities
Signatories	• Professional organization • Coast Guard • Local Municipality • Environmental Agency	• Association of recreational activities (divers, anglers) • Coast Guard • Local Municipality • Environmental Agency
Impacted Group(s)	Professional anglers	Divers, recreational anglers
Goals	Only signatories are allowed on the reef area	
Activity Calendar	Week	Weekend
Area Under Charter Validity	The charter applies to the territory of the commune of Saint Leu. Fishermen charter signatories recognize having read the attached map, which accurately shows details of the artificial reefs areas.	
License	Personal permit is delivered.	
Assessment	The signatories to the Charter undertake to transmit an estimate of their catch in developed areas, according to specific forms of fishing and anonymous (to be determined in future follow-up).	
Public Debate	One meeting with all signatories each year (local municipality organization).	
Validity	One year	
Optional Measures		
	A Professional Angler	B Recreational Activities
Biological	• Increased minimum weight • No catch during spawning season	• No catch during breeding and spawning season • Minimum weight emphasize • No-kill will be enhance ("no-kill for good-picture" advertising)
Technical	• Rules regarding the size of boat or power	• Forbid fishing on the reef • Forbid swimming inside the reef
Logo for "sustainable reef" for signatories"	≅ *Signatories manager* ≅ ≅ *artificial reef of St Leu* ≅	
	Could be fired from the area after 3 no-respect chart and activities.	

Figure 5.2 Management measures implemented with the "charter of use."

Phase 2: Site Selection, Volume, and Reef Design

Once the charter was approved, the extent of the area to be developed had to be delimited. Site selection is mainly concerned with biotic environmental conditions (including the presence and behavior of target species) but must also take socioeconomic development and seabed characteristics (including topography, sedimentology, and hydrodynamics) into account.

All these parameters from these variables can be entered into a geographic information system (GIS). GIS is an efficient data management system to help identify suitable locations for artificial reefs (Raynal and Pioch, 2007). The system is based on a conceptual data model built on a hypergraph-based data structure (HBDS) that revolves around objects, classes, attributes, and relationships.

In Saint-Leu, the results of data treatment using GIS helped identify a potential artificial reef deployment area located south of the village that would be able to accommodate 2000 m^2 of artificial reefs. The use of GIS combined with management guidelines (choice of target species, sharing space, etc.) led to designation of three reefs:

- Areas up to a depth of 20 m, coral bottom representing complex shelter for species that do not move far (lobsters, snappers, groupers, etc.).
- Areas at a depth of 30 m, coral resources providing little complex shelter.
- Finally, areas at a depth of more than 50 m, reefs for homeless and highly migratory pelagic fish such as kingfish, dogtooth tuna, barracuda, etc. Because of their depth, these reefs are inaccessible to undersea fishermen.

Phase 3: Selection of Concession Jurisdiction

In France, the appropriation of a public marine space by a process of "privatization" is legally impossible (Pioch, 2004). Three regimes are appropriate for the occupation and/or immersion of artificial reefs. The legal procedures or "regimes" differ according to whether:

- A permit for marine farming was awarded (Decree No. 83-228 of March 22, 1983 as amended, establishing rules for the authorization of mariculture farms).
- A temporary occupation permit was awarded (Article 28 of the state domain).
- A concession was awarded for the use of the maritime public domain outside ports (CUDPM) (Decree No. 2004-308 of March 29, 2004).

Given the local situation and the needs of Saint-Leu with respect to the opportunities offered by these different legal "regimes," the permit for the use of a public maritime domain outside a port was of the greatest interest. This scheme allows professional and amateur fishermen, and divers to continue to operate.

Phase 4: Financing Plan

A project of this magnitude could have benefited from a financing plan involving the local community and Europe through the European Fund for Fisheries Region (Regional Board), as well as of the French Departmental Council (*Conseil General*). Finally, other partners may also contribute (associations, NGOs, fund for nature, private partners, marine park, etc.).

In addition to covering the investment and operating costs, the financing plan must also include a scientific monitoring plan to evaluate the effectiveness of the reefs and their integration in the economic and ecological process.

Phase 5: Evaluation Policy and Scientific Monitoring

At this stage, the support of all stakeholders is essential (Monteiro and Santos, 2000). Close collaboration with fishermen and divers should provide information on landings, fishing effort, and on the sustainability of operations by determining an acceptable level of fishing. The supervision of these operations can be assigned to scientists or research organizations to ensure coherent analysis and interpretation of results. These agencies will also be able to perform surveys, monitor populations, etc. Their findings can then be used to adjust the management plan and to complete the equipment if necessary.

Building on the recent creation in Reunion Island of the marine park and of the protection of coral zones in the NMR (Natural Marine Reserves), a network of artificial reefs created on sand substratum should provide suitable habitats for species "exported" from the reserve (owing to the spill-over effect). The expected effect of the reserve should have a direct and positive effect for fishermen, and will also benefit adult-introduced species in authorized zones with adjacent artificial reefs, as—if the latter are well managed—they will supplement the fishing reserves of the marine park. This would complete existing installations in the Regional Underwater Land Plan. An initial experiment concerning artificial reefs was conducted by Tessier in 2003 in Saint-Paul Bay (northwestern Reunion Island), and the results indicated that large concentrations of fish were accumulating around and within these boulder reefs, particularly, juveniles of targeted species (Tessier, 2005).

In conclusion, artificial reef deployments can improve the protection and sustainable management of coral reef ecosystems. The direction of the marine park, the French government, and the public were able to find converging objectives in terms of management and, finally, to reach a consensus concerning the artificial reefs. We believe that correctly analyzed and well-managed artificial reefs are able to protect and/or to produce resources. These reefs should lead to the emergence of "conscious management" because they involve (and engage) the end user in an appropriation process, which is the first step in management, through the creation of a new "known" territory delimited by its physical installation. Ultimately, the development of an artificial reef could trigger a new reflection process concerning resources management. It is important to involve fishermen in the decision-making process right from the beginning and, at the same time, using a broad but long-term approach, such as the geosystemic method ICAM. The ecosystem step of ecosystem approach of fisheries (FAO, 1995) is often focused on underwater ecological ecosystems and fails to pay sufficient attention to other crucial factors, such as the social dimension of fishing activities.

REFERENCES

Bailly, D. 1989. Pêche et aquaculture au Japon. Nantes, IFREMER. 24 pp.

Baine, M. 2001. Artificial reefs: A review of their design, application, management, and performance. *Ocean Coastal Manage.* 44: 241–259.

Bolopion, J., A. Forest, and L.J. Sourd. 2000. Rapport sur l'exercice de la pêche côtière dans la zone côtière de la France. *Ifremer. Ifremer.* 137 pp.

Bortone, S.A. 1998. Resolving the Attraction-production Dilemma in Artificial Reef research: Some Yeas and Nays. *Fisheries* 23(3): 6–10.

Brunet, R., R. Ferras, and H. Théry. 1995. *Les mots de la géographie*. Montpellier/Paris. 4e éd. R.-L. d. française: 518 pp.

Cadoret, A. 2006. Conflits d'usage liés à l'environnement et réseaux sociaux: enjeux d'une gestion intégrée? Géographie. Montpellier, Université Paul Valéry, Montpellier 3. 591 pp.

Cazalet, B. 2007. Les droits d'usage territoriaux, de la reconnaissance formelle à la garantie juridique. Le cas des aires marines protégées ouest-africaines. *Mondes en développement* 138: 61–76.

Cicin-Sain, B. and R.W. Knecht. 1998. *Integrated Coastal and Ocean Management, Concepts, and Practises.* Island Press. 517 pp.

CRISP. 2008. Available on http://www.crisponline.

Cury, P. and Y. Miserey. 2008. *Une mer sans poissons*. Calmann-Lévy. Paris. 283 pp.

Denis, J., Y. Henocque, M. Antona. et al. 2001. Des outils et des hommes pour une gestion intégrée des zones côtières. Guide méthodologique. Vol. II, Paris. Manuels et guides de la Commission Océanographique Intergouvernementale UNESCO. n°42: 64 pp.

Ducloy, P. 2006. Bilan halieutique des immersions de récifs artificiels en Languedoc-Roussillon, perspectives et proposition d'action, Agro campus Rennes. 92 pp.

Doumenge, J.-P. 2004. Passé, présent et solutions d'avenir d'une politique appliquée à la gestion durable d'un espace sur-convoité, le littoral. UPV—Montpellier 3. DESS Activités et aménagements littoraux et maritime. Cours 2° année. 32 pp.

European Commission. 2005. Stratégie thématique pour la protection et la conservation du milieu marin. Bruxelles. Communication de la commission au conseil et au Parlement Européen. 10 pp.

European Commission. 2006. Fonds européen pour la pêche, Règlement (CE) n°1198/2006 du Conseil, du 27 juillet 2006. Bruxelles. Conseil de la Communauté Européenne. 44 pp.

European Commission. 2007a. Documentation et politique générale de l'Europe. Available on http://europa.eu

European Commission. 2007b. PCP, Politique Communautaire de la Pêche. Available on http://ec.europa.eu/fisheries/cfp_fr.htm.

European Commission. 1999. Towards a European Integrated Coastal Zone Management ICZM Strategy: The General Principles and Policy Options, a reflection paper, Directorates-General Environment, Nuclear Safety and Civil Protection Fisheries Regional Policies and Cohesion.

FAO 2008. The State of World Fisheries and Aquaculture 2008. ftp://ftp.fao.org/docrep/fao/011/i0250e/i0250e.pdf.

FAO 1995. Code of Conduct for Responsible Fisheries. Rome. Food and Agriculture Organization of the United Nations.

Gabrié, C. and L. Montaggioni. 1985. Sediments from fringing reefs of Reunion Island. *Indian Ocean. Sedim. Geol.* 31: 281–301.

Hardin, G. 1968, The tragedy of the commons, *Science*, 162, pp.1243–1248.

Henocque, Y. 2006. Leçons et futur de la gestion intégrée des zones côtières dans le monde. VertigO—La revue en sciences de l'environnement. 7: 11 p.

Kakimoto, 1985. Distribution and behavior of fish at the artificial reef. Actes du colloque pluridisciplinaire Franco-Japonais océanographie: Les aménagements côtiers et la gestion du littoral. Edition Société Franco-Japonnaise d'Océanographie, Ceccaldi H., Champalbert J.; Marseille, 16–21 Septembre 1985: Fascicule 6, Récifs artificiels. P 123–133.

Katoh, J. 1985. Study on artificial reefs installation program from the viewpoint of hydraulic environmental engineering. Actes du colloque pluridisciplinaire franco-japonais océanographie: Les aménagements côtiers et la gestion du littoral. Edition Société Franco-Japonnaise d'Océanographie, Ceccaldi H., Champalbert J., Marseille, 16–21 Septembre 1985: Fascicule 6, Récifs artificiels: pp. 65–83.

Lasserre, G. and C.C. Monteiro. 2002. Les récifs artificiels en Europe: Application à l'aménagement intégré. Colloque international sur la promotion de méthodes et de pratiques de pêche durables en Afrique de l'Ouest. Dakar. Senegal 18–20 mars 2002. World Wide Fund for nature (WWF).

MAFF. 2007. Ministère de l'Agriculture de la Fôret et des Pêches. Disponible sur le site http://www.maff.go.jp/eindex.html.

Mariojouls, C. 2004. Le projet d'aménagement de la bande côtière japonaise. Bulletin de la Société Franco Japonaise d'Océnaographie. 2 pp.

Meur-Ferec, C. 2007. La GIZC à l'épreuve du terrain: premiers enseignements d'une expérience française. Développement durable et territoire. Varia. Disponible sur http://developpementdurable.revues.org/document4471.html.

Miossec, A. 1998. Les littoraux, entre nature et aménagement, Collection Campus Géographie, SEDES: 192 pp.

Monteiro, C. and M. Santos. 2000. Portuguese artificial reefs. *Artificial reefs in European Seas*. Kluwer Academics pub. Dordrecht, the Netherlands: 249–261.

Mottet, M.G. 1985. Enhancement of the marine environment for fisheries and aquaculture. In *Japan. Artificial Reefs: Marine and Freshwater Applications*. 2nd ed. F.M. D'itri ed. Lewis Publishers, Chelsea, Michigan. pp. 13–112.

National Statistics of Japan Ministry of Internal Affairs. 2008. Available on http://www.stat.go.jp/english/index.htm.

Pioch, S. 2004. Bilan des immersions en récifs artificiels en Languedoc Roussillon. Montpellier, Université Montpellier 2/CEPRALMAR/Créocean. 120 pp.

Pioch, S. 2008. Artificial habitat: A strategic tool for integrated coastal area management? Proposal guideline for their development in coastal fisheries. Ph.D. Geography and Coastal Land Planning, University Montpellier 3, Montpellier, France - Tokyo University of Marine Science, Tokyo, Japan. 295 pp.

Pioch, S., and B. Pary. 2007. Etude de faisabilité pour la création d'un site de récifs artificiels pour la pêche côtière et la chasse sous marine en apnée à Saint Leu. Sainte Marie, Ile de la Réunion, Egis Eau/commune de Saint Leu. (unpublished) 219 pp.

Raynal, J.-C. and S. Pioch. 2007. Analyse multicritère du site de Saint-Leu pour déterminer la faisabilité d'implantation d'habitats artificiels grâce à un SIG. Egis Eau. 25 pp.

Saunier, C. and P. Laffitte. 2007. L'apport de la science et de la technologie au développement durable. La biodiversité: L'autre choc ? L'autre chance? Paris. Office parlementaire d'évaluation des choix scientifiques et technologiques. Assemblée nationale. 192 pp.

Scheibling, J. 1994. Qu'est ce que la géographie? Paris, Hachette. 199 pp.

Simard, F. 1985. La technologie de la pêche miraculeuse. *Sciences et Avenir*. 461: 40–47.

Tessier, E. 2005. Dynamique des peuplements ichtyologiques associés aux récifs artificiels à l'île de la Réunion Ouest de l'océan Indien—Implication dans la gestion des pêcheries côtières. Ecologie marine. Université de la Réunion. U.F.R. sciences et technologies. 240 pp.

UNESCO. 2001. Disponible sur le site www.unesco.org.

UNFPA. 2004. Etat de la population Mondiale 2004—Le consensus du Caire 10 ans après: la population, la santé en matière de reproduction et l'effort mondial pour éliminer la pauvreté. Fond des nations Unies pour la Population. 124 pp.

Veyret, Y. 2004. *Geo-environnment*. Paris. Armand Collin Coll. Campus. Geographie. 186 pp.

Worm, B., R. Hilborn, J.K Baum, T.A. Branch, J.S. Collie, C. Costello, M.J. Fogarty et al. 2009. Rebuilding global fisheries. *Science* (New York) 325, no. 5940 (Juillet 31): 578–585. doi:10.1126/science.1173146.

Worm, B., E.B. Barbier, N. Beaumont et al. 2006. Impacts of biodiversity loss on ocean ecosystem services. *Science* 314: 787–790.

CHAPTER 6

Artificial Reefs for Lobsters

An Overview of Their Application for Fisheries Enhancement, Management, and Conservation

Ehud Spanier, Kari L. Lavalli, and Dor Edelist

CONTENTS

ABSTRACT

Adult lobsters of the families Palinuridae, Nephropidae, and Scyllaridae are important fisheries resources in all world oceans. All benthic stages are nocturnally active and shelter during the day when residing in structured environments, presumably to avoid predation. Recent concerns about habitat destruction and/or degradation through anthropogenic and natural occurrences, as well as previous interest in enhancement of local populations for exploitation purposes, has led to the investigation of providing additional sheltering opportunities for lobsters in shelter-limited habitats. Although various man-made structures (e.g., ship wrecks, sunken aircraft) are known to attract adult lobsters, modern development of artificial reefs currently focuses on imitating natural shelters (particularly those in hard substrates) and are designed according to the spatial preferences of a given species and its ontogenetic phases.

A review of five decades of literature indicates an increase in the number of studies on artificial reefs for lobsters stemming from improved knowledge about recruitment processes of important commercial species, as well as remediation attempts due to the decline of many commercial populations from overfishing, disease, man-made environmental disasters, and other natural causes. Collectors developed for postlarvae of several species of spiny and clawed lobsters have been very successful in recruiting this life history stage for a variety of uses, including substitution for loss of natural sheltering habitat, to project future subadult or adult populations, or for future enhancement efforts.

Literature on artificial reefs is dominated by the considerable efforts associated with the construction of *casitas* and *pesqueros* used for fisheries of the spiny lobster, *Palinurus argus*, in the Caribbean and Gulf of Mexico, and construction of concrete structures for spiny lobster enhancement in Japanese fisheries. However, due to successes in spiny lobster enhancement, new efforts have focused on artificial reefs for *Homarus* and *Scyllarides* spp. Even so, while local density of lobsters may increase at such reefs, there is insufficient evidence that speaks to their effectiveness of increasing population levels of the target species.

Keywords: Artificial reefs, man-made reefs, casitas, collectors, benthic collectors, pelagic collectors, lobsters, enhancement.

INTRODUCTION

Adult and juvenile lobsters are relatively large decapods that increase activity levels at dusk to forage and then decrease activity around dawn (Kanciruk and Herrnkind, 1973; Atema and Cobb, 1980; Ennis, 1984; Herrnkind, 1980; Jones, 1988; Spanier and Almog-Shtayer, 1992; Childress and Herrnkind, 1994; Smith et al., 1999; Martinez et al., 2002; Lavalli et al., 2007). Within the photic zones of the continental shelf, benthic lobsters of all ontogenetic stages prefer to shelter in complex substrates (Cobb, 1971; Marx and Herrnkind, 1985; Jernakoff, 1990; Sharp et al., 1997; Ratchford and Eggleston, 1998; Robertson and Butler, 2003) or bury in soft sediments during daytime hours (Jones, 1988; Faulkes, 2006); little is known about the activity levels of deep-water lobsters. These diverse sheltering behaviors are assumed to be predator avoidance adaptations (Barshaw and Lavalli, 1988; Eggleston et al., 1990, 1992; Smith and Herrnkind, 1992; Wahle, 1992a,b; Wahle and Steneck, 1992; Barshaw et al., 1994) and, when predators are absent (at least in laboratory settings), lobsters become active during the day (Wahle, 1992b; Barshaw and Spanier, 1994a). As such, human activities that impact the presence of predators (via removal of species or alteration of species by fishing practices) or the presence of suitable shelter-providing habitat (via coastal development projects, certain fishing techniques, or pollution events) can have profound effects on the behavior and survival of lobsters comprising local populations (Caddy, 2008).

Similarly, natural events, such as algal blooms, excessive river run-off during heavy rains, hurricanes, and/or thermal changes via climate change, can also negatively impact established lobster populations that may take years to recover. Where such perturbations to the environment have occurred, the use of man-made habitats may help to ameliorate such effects by attracting lobsters back into the area and therefore may aid in maintaining local lobster populations.

This chapter examines scientific and technical publications dealing with lobsters and man-made habitats by focusing mainly on the two (out of five) lobster families that have considerable commercial importance. It summarizes our current state of knowledge of what types of structures are most successful in attracting various ontogenetic stages of lobsters and provides a series of guidelines that can be used for future work on artificial reefs specifically aimed at improving lobster populations.

Trends in Commercially Exploited Lobster Species

The species that support the greater part of the total world lobster fishery come from the following groupings: clawed lobsters (Nephropidae), spiny lobsters (Palinuridae) and, to a much lesser degree, slipper lobsters (Scyllaridae) (Spanier and Lavalli, 2007; Herrnkind and Cobb, 2008). All fished species of lobsters are considered luxurious delicacies and are among the most costly seafood products worldwide (e.g., Wallace, 2004). Reported global fisheries production of lobsters in 2007 was 226,805 metric tons, of which clawed lobsters made up 158,933 tons or 70%, spiny lobsters made up 63,972 tons or 28%, and slipper lobsters contributed only 3,900 tons or <2% (FAOSTAT, 2010). This is only a partial estimate since catches of some minor and artisanal lobster fisheries, especially in developing countries, are not recorded and reported. Demand for lobsters has increased as expressed in an increase in prices for live whole bodies, tails, and frozen meat (FAOSTATS, 2010; Figure 6.1). This demand drives enhanced worldwide production of lobsters (FAOSTATS, 2010; Figure 6.2) that results in increased fishing effort. Between 1950 and 2008, there has been a threefold increase in lobsters production, and a 2.4-fold increase in worldwide exports was reported between 1986 and 2006 (FAOSTATS, 2006). As a result, local stocks have been overfished, and certain fisheries have even experienced collapses. For example, the sharp declines seen in stocks of *Homarus gammarus* in Norway (van der Meeren, 2003), *H. americanus* along the Atlantic coast of Canada (Garnick, 1989), the Tasmanian rock lobster *Jasus edwardsii* (Bradshaw, 2004), the Caribbean spiny lobster (*Panulirus argus*) in the southern Florida fishery (but not in the sanctuary of Dry Tortugas National Park) (Bertelsen and Matthews, 2001), the Mediterranean slipper lobster (*Scyllarides latus*) along the Mediterranean coasts of Europe (Spanier, 1991; Pessani and Mura, 2007) and in the Azores Islands (Martins, 1985), and *Thenus orientalis* and the spiny lobster *Palinurus polyphagus* in India

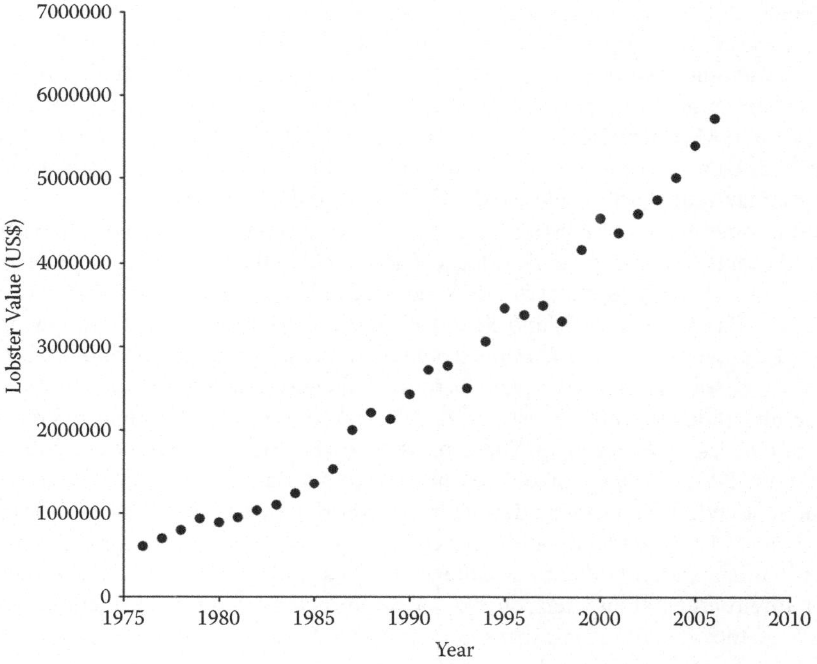

Figure 6.1 FAO commodity trade and production values in U.S. dollars for lobsters in all forms (live, frozen, tails). (From FAO—Fisheries and Aquaculture Information and Statistics Service, July 31, 2010.)

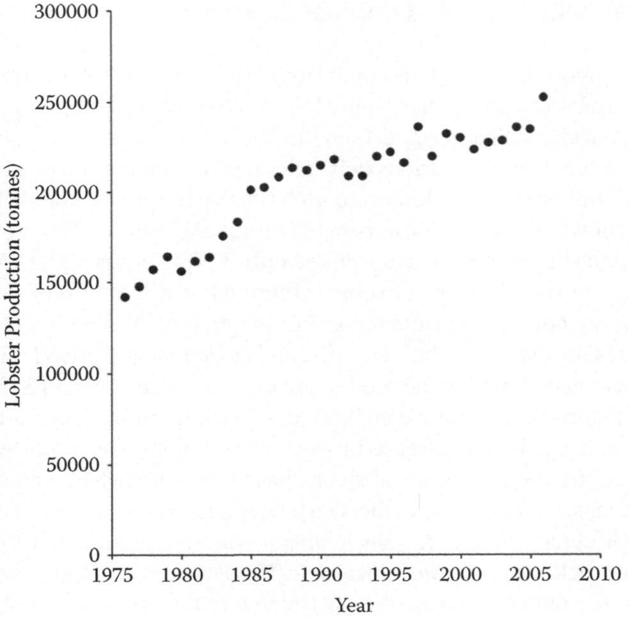

Figure 6.2 FAO commodity trade and production quantity (in tons) for lobsters in all forms (live, frozen, tails). (From FAO—Fisheries and Aquaculture Information and Statistics Service, July 31, 2010.)

(Radhakrishnan et al., 2007) all illustrate the unsustainability of current fishing and management practices. In some cases, overfishing of one or more species in one family of lobster increases fishing pressure on another family, as has been seen in the Galapágos and Hawaiian Islands where slipper lobsters are targeted as spiny lobsters become more rare (Polovina et al., 1995; Hearn, 2006; Hearn et al., 2007). Most fisheries scientists and managers believe that the natural lobster fisheries have been harvested to a worldwide maximum and that most lobster fisheries are currently operating at or above maximum sustainable yields (Herrnkind and Cobb, 2008).

In addition, some species of lobsters have experienced dramatic declines due to harmful natural environmental events or anthropogenic effects. These include the mass mortalities of *Jasus lallandii* along the west and south coasts of South Africa due to low oxygen conditions (Cockcroft, 2001); the massive die-offs in the Long Island Sound *Homarus americanus* fishery following 2 years of smaller, more localized, die-offs that were attributed to warmer water temperatures, hypoxia, heavy metal poisoning, pesticides, and alkylphenols (potential endocrine disruptors) (Biggers and Laufer, 2004; Pearce and Balcom, 2005; Tlusty et al., 2007; Vogan et al., 2008 and references therein); the decline in catches of *P. argus* in Yucatan, Mexico (Briones-Fourzán et al., 2000), as well as in Florida (Hunt, 2000) because of severe hurricanes that damaged habitat; the loss of *P. argus* juveniles due to a widespread sponge die-off in Florida Bay that provided shelter for early benthic stages (Butler et al., 1995); and the ~45% decline in recent landings of *P. argus* compared to historical averages that is apparently due to a pathogenic virus (PaV1) (Shields and Behringer, 2004; Li et al., 2008). Environmental disasters, such as oil spills, also can cause considerable loss of lobster populations (e.g., the oil spill in Narragansett Bay, Rhode Island, USA, in 1989; Castro et al., 2001). Such events have negative effects on local populations and recovery—which takes years—is further slowed when commercial exploitation of the remaining population continues.

Artificial Reefs as a Solution to Population Declines

One possible solution for the population declines is the construction of artificial reefs for lobsters. Various man-made structures such as shipwrecks, not originally designed or deployed to attract lobsters, have been known for decades to attract these large crustaceans (e.g., Howard, 1980; Werz, 2007). Other artificial structures such as breakwaters, jetties, and canal walls have attracted considerable numbers and wide size ranges of lobsters (Relini, 2000; Barnabé et al., 2000). Fishermen, knowing of this tendency of lobsters to be attracted to these unintentional artificial reefs, have set their traps and lobster pots there to increase their lobsters catch. Divers have also known about the concentration of lobsters in shipwrecks and preferred these man-made sunken structures when diving for lobsters (Berg, 2009).

Artificial reefs that were constructed for sessile organisms and designed to attract other taxa have also attracted lobsters, such as the Japanese "tsukiiso," (Sahoo and Ohno, 2000). Net fisheries or "ghost" nets entangled around other man-made structures are known to catch considerable numbers of lobsters (Matsuoka et al., 2005). Despite the attractive nature of these structures, they are far from ideal man-made structures for lobsters and seldom imitate natural lobster dens, or provide appropriate shelters for all benthic stages of lobsters (juvenile to adult). Hence, an efficient and effective design of artificial reefs for lobsters needs to account for particular ontogenetic stages and their needs.

The desired effect of artificial reefs is to increase long-term abundance and productivity of the target species. Therefore, when assessing the efficacy of artificial reefs, experimental resolution has to be fine enough to enable detection of differences between a localized attractive effect that merely aggregates and concentrates specimens and a true and permanent increase in abundance of specimens in the local area. Assessment has to also last long enough to observe the limits of artificial reef effects. Brickhill et al. (2005) stressed that improved study designs are imperative for temporal attraction versus production studies.

Even if artificial reefs result in attraction and concentration of members of a species, this does not necessarily equate to increased production of local populations, which is one of the major criticisms of artificial reefs (see discussion in Seaman, 2000). Proposals presented by entities from more than 30 countries deploying artificial reefs state that the main purpose of these reefs is fishery related (Jensen, 2002), and this is also reflected in scientific literature on such reefs (Bortone, 2008). Benefits for conservation of species and habitat restoration were occasionally mentioned in both scientific literature and proposals, but tended to be overstated and secondary to fishing enhancement. In the long term, such biases may shift our thinking about the roles of artificial reefs as a way to restore depleted populations.

When artificial habitats are deployed, three main types of effects on local fauna may take place: (1) biomass redistribution, (2) aggregation, which increases only the exploitable biomass, and (3) an increase in total biomass via production (Polovina, 1991). Biomass redistribution assumes that a mixed effect of species immigrating to and emigrating from the artificial reefs will take place with no significant one-sided trend. The attraction hypothesis predicts that the aggregation of organisms to the artificial reefs will not be augmented by new production or recruitment to the natural habitat from which species emigrated. The production hypothesis, however, predicts mitigation for the fauna attracted to the artificial reefs by new arrivals to the natural habitat and also suggests positive effects of faunal export from the artificial reefs, which may eventually serve as an enhanced gene pool for the local population. Which effect will take place depends on the ecosystem components present and how humans affect those components (e.g., fishing pressure). Many surveys of artificial reefs have reported the postdeployment presence of local fish aggregations, but little direct evidence indicates permanent increases in total population size or fish stock (e.g., Bohnsack et al.,

1997; Osenberg et al., 2002). Hence, some believe that artificial reefs simply represent another location where species could suffer overexploitation by fishery practices (Brickhill et al., 2005).

In general, it may be concluded that a site-specific and species-specific approach is necessary, because a wide range of factors affect the mechanisms underlying artificial reefs function. For lobster (as well as other crevice-dwelling organisms), this means that elements of monitoring the enhanced sheltering the artificial reef offers, as well as the spatial and temporal cues of key life stage events, must be incorporated into the study design in a way that allows separation of production effects from those of the natural habitat.

LIFE HISTORY FEATURES OF LOBSTERS RELEVANT FOR ARTIFICIAL REEFS

Lobsters are found in all oceans along the continental shelf and upper continental slope (for details, see Holthuis, 1991, 2002; Webber and Booth, 2007, and references therein). Information on adult habitats is readily available for commercially important species of clawed, spiny, and slipper lobsters, but is less available for species that are captured as a by-product of other fisheries, caught only in recreational fisheries, or are unexploited. From what we know, adult lobsters use a variety of habitats, ranging from those that provide complex structure via rocks, boulders, ledge, and coral outcroppings to those that provide no structure (mud, sand). Some species are well adapted for digging and burrowing and can actively manipulate the substrate to suit their needs; other species simply find crevices in which to shelter.

Clawed lobsters are found in wide variety of habitats largely because of their ability to burrow into substrates or to fit into crevices. Inshore populations of all benthic phases are found on mud, cobble, bedrock, peat reefs, eelgrass beds, and within sandy depressions (Cooper et al., 1975; Hudon, 1987; Able et al., 1988; Heck et al., 1989; Wahle and Steneck, 1991; Lawton and Robichaud, 1992). Offshore populations are found in mud, bedrock, within sandy depressions, or in clay (Cooper and Uzmann, 1980). Within their geographical range, clawed lobsters have a wide temperature tolerance (–1°C to 30.5°C) and, although generally considered stenohaline organisms, they are broadly tolerant of salinities ranging from those of coastal and offshore habitats (>25 ppt) to estuarine areas (Thomas, 1968; Harding, 1992). The earliest benthic phases of clawed lobsters seem to preferentially settle into crevice-providing habitats (Lawton and Lavalli, 1995), but they are found throughout the range of the species, where such habitats might not be readily available at time of settlement.

Adult spiny lobsters are widely distributed in tropical, subtropical, and temperate zones of all oceans, and occur from the intertidal to depths approaching 1000 m (Holthuis, 1991). Habitats of spiny lobsters are very diverse and vary according to life history stage and behavior (solitary versus social species)—lifestyles include shallow, semisocial, residential dwellers on coral reefs to gregarious, migratory species that live on open soft substrates at depths greater than 300 m (Butler et al., 2006). Usually, different genera do not co-occur, having distributions that are distinctive both in latitude and depth (Butler et al., 2006). However, species within a genus often do co-occur in a particular region (i.e., *Panulirus argus* and *P. guttatus* in the northern Caribbean), but are generally segregated by habitat and behavior (Berry, 1971; Lozano-Álvarez and Briones-Fourzán, 2001; Briones-Fourzán et al., 2006). This would be the case in the Indian Ocean where ten species are present (Phillips and Melville-Smith, 2006) or in southern Africa where seven genera co-occur (Berry, 1971).

Adult and subadult slipper lobsters are distributed in a variety of geographical regions and can be found in temperate, subtropical, and tropical parts of all oceans and adjacent seas with latitudinal ranges from 4°S-45°N and depths of 0 to at least 800 m (Holthuis, 1991, 2002; Brown and Holthuis, 1998; Webber and Booth, 2007). Latitudinal and depth variations are associated with differences in several environmental factors such as temperature, light, salinity, and pressure. These life history phases are also found in a variety of habitats, from featureless flat soft substrates, such as mud, sand, and shell-sand to rubble, macroalgae, sea weed, and sedentary invertebrates (sponges and branching corals), to harder and complex substrates of rocky outcrops and coral reefs (Webber and Booth, 2007). One can divide the substrate habitats of slipper lobsters into two groups: Those that are complex, such as rocks, coral reefs, and rocky caves, and are attractive to species of *Acantharctus, Arctides, Scyllarides,* and *Scyllarus.* The second substrate group is noncomplex and featureless, such as sand or mud, and these are attractive to species of *Thenus, Ibacus,* and *Evibacus princeps* (Haddy et al., 2007; Jones, 2007; Radhakrishnan et al., 2007). *Parribacus* spp. seems to dwell in both complex (coral, stone, or shore reefs) and plain substrates (Sharp et al., 2007). Despite these generalities, very little is known about the actual preferences of slipper lobsters for most of their life history stages; hence, design of artificial reefs targeting these species is greatly hampered.

ARTIFICIAL STRUCTURES FOR POSTLARVAE OF LOBSTERS

Man-made structures for postlarval stages (passive collectors) have been useful for evaluating patterns of settlement and recruitment of a variety of benthic decapod crustaceans including lobsters (e.g., Phillips and Booth, 1994; Phillips et al., 2001; Wahle, 2003; Wahle et al., 2009). Materials that mimic the natural nursery habitat of the target species have been critical to the successful development of postlarval collectors (Wahle et al., 2009). Therefore, most of the research on artificial reefs for postlarvae has been conducted on commercial species of clawed and spiny lobsters whose natural nursing grounds are known and, for spiny lobsters, this represents a very small portion of the total species within the family. In the case of slipper lobsters, settlement habitats are unknown (Spanier and Lavalli, 1998).

Ultimately, the goal of such collectors is to supplement natural habitats so that postlarvae that might not normally survive because of failure to find appropriate settlement sites could use the artificial structures and recruit into the local populations. Such a result would enhance production of specific species that are commercially important. Currently, however, the artificial collectors have been used primarily to understand postlarval settlement patterns, determine carrying capacity of settlement habitats, and provide a predictive tool in fisheries models for projecting recruitment into adult populations (e.g., *Panulirus cygnus*; Pearce and Phillips, 1988).

Clawed Lobsters

Some postlarvae of the American lobsters, *Homarus americanus*, have recruited to benthic concrete artificial shelters designed primarily for larger lobsters (Sheehy, 1976; Herrnkind and Cobb, 2008). Similarly, some small juvenile European lobsters, *H. gammarus*, were found in a pulverized fuel ash artificial reef designed primarily for adults, suggestive of settlement there (Jensen et al., 1994, 2000a, b). Nevertheless, these types of habitats are not ideal for settlement and, based on laboratory and field studies of ideal settlement substrates for postlarval survival (Lavalli and Barshaw, 1986; Barshaw and Lavalli, 1988; Wahle and Steneck, 1991), later studies have focused on mimicking natural habitats that are attractive to postlarvae and subsequent benthic stages. Of course, quantitative behavioral and physiological information on the earliest benthic life stages is necessary to understand processes influencing larval and postlarval supply that occurs prior to settlement to a new site offered by an artificial reef—a "supply-side" context for postsettlement dynamics (Wahle et al., 2009).

Such information is largely available for *H. americanus.* Development from larvae to postlarvae takes 4–8 weeks and is largely dependent on water temperature. During the latter part of the postlarval stage, individuals dive repetitively to seek suitable shelter-providing nursery habitat, but remain in the plankton until such habitat is found or the molt to the first juvenile stage is imminent.

Ideal habitat consists of rocky crevices amongst cobbles and boulders (Lavalli and Barshaw, 1986; Barshaw and Lavalli, 1988; Wahle and Steneck, 1991, 1992; Wahle and Incze, 1997), but postlarvae also successfully survive in salt marsh peat (Able et al., 1988; Barshaw et al., 1994) and lower intertidal habitats (Cowan et al., 2001). Following settlement, the postlarva molts to the first benthic stage and remains closely associated with the settlement habitat during its first year of life (Lawton and Lavalli, 1995). Thus, it is likely that any artificial reefs designed to attract postlarvae will remain important for young-of-the-year and possibly second-year juvenile lobsters.

Research on clawed lobsters in the 1980s focused heavily on understanding postlarval recruitment processes and, in an attempt to use previously successful techniques for postlarval spiny lobsters, passive collectors were deployed in coastal regions where postlarval supply was known to occur. These collectors failed to retain postlarvae, most likely because clawed lobsters do not use floating algae as an intermediate settlement site between planktonic and benthic phases; instead, they settled directly into the benthos (Wahle et al., 2009). Once preferred settlement habitat was identified in the early 1990s, other types of passive collectors were designed to more closely mimic such habitats (Wahle et al., 2009). These collectors consisted of standard trap wire mesh lined with a finer mesh and filled with cobble stones (Incze et al., 1997) or stacked, cross-wise layers of PVC tubing fastened to a wire mesh tray lined with a finer mesh (Palma et al., 1998, 1999; Wilson, 1999). For these designs, divers were required to both deploy and retrieve the collectors. In addition, divers either covered the collectors in fine mesh before bringing them to the surface or sampled them in situ with suction samplers—labor-intensive practices that limit the ability to assess the success of such artificial structures in enhancing lobster populations over the long term.

Despite the intense labor involved, suction sampling has been used to monitor long-term recruitment onto artificial reefs and has recorded settlement of wild *H. americanus* postlarvae and presence of young-of-the-year, subadult, and adult lobsters in densities similar to that in natural areas (Castro et al., 2001), suggesting that the artificial reefs enhanced recruitment and settlement locally. However, attempts to seed with hatchery-reared, microwire-tagged juveniles released to the reefs have not been successful since only a few of more than 6000 tagged lobsters were recovered in the reefs. These failures may have been the result of flaws in the delivery of the lobsters to the reef, the attraction of predators to the site at the time of lobster delivery, the failure to seed the reefs with appropriately aged juveniles that would immediately settle into the cobble, the emigration of the settlers, or the failure to appropriately condition the hatchery-reared animals to predator smells, as has been done with *Homarus gammarus* restocking efforts in Norway (Svåsand, 2007). More work needs to be done to demonstrate that such reefs could be used to mitigate population decreases from natural or anthropogenic effects via seeding with hatchery-reared lobsters.

Thus far, artificial reefs and other structures designed to attract postlarvae have been deployed in shallow, near-shore nurseries, primarily because of limitations in the ability of divers to safely sample the structures. To reveal settlement patterns in zones beyond the practical limits of diving, Wahle et al. (2009) constructed collectors comprised of wire mesh trays lined with fine screening on the floor and walls and filled with cobble to simulate natural nursery habitat (Figure 6.3). These collectors were deployed from a vessel equipped with a standard pot hauler and then sampled in a comparative fashion by diver-operated suction sampling and by haul up to the surface. No newly settled lobsters were lost during the haul-up, retrieval process (as compared to what could be sampled directly in situ), and densities of young-of-year lobsters found in the hauled collectors were similar to those in directly adjacent natural cobble habitat sampled by divers employing suction samplers. Plans exist to use these collectors to determine if postlarvae settle at greater depths. Experiments using these same collectors are also ongoing in Norway to determine if they can successfully recruit *H. gammarus* postlarvae (van der Meeren, Institute of Marine Research, Bergen, Norway, personal communication). At this point, these structures have been used mainly to determine their level of attractiveness to settling lobsters rather than enhancing population numbers.

Figure 6.3 A passive collector for postlarvae of American clawed lobsters, *Homarus americanus*. Dimensions: 61.0 × 91.5 × 15.0 cm. (From Wahle, R. A. et al., 2009. *N.Z. J. Mar. Freshwat. Res.* 43: 465–474. Used with permission.)

Spiny Lobsters

Compared to all other taxa of lobsters, great progress has been made in the development of spiny lobster postlarval (puerulus) collectors, and a number of different types of collectors now exist that are fine-tuned for a particular species (Butler and Herrnkind, 2000). Originally, collectors were designed to capture pueruli for further study of growth in the laboratory (Witham et al., 1968; Phillips, 1972). However, once the likely locations of pueruli became known, collectors were deployed to investigate settlement patterns and timing (see review by Butler and Herrnkind, 2000 and references therein) or to collect large numbers of postlarvae for mariculture or grow-out purposes (Hirata et al., 1988). Additionally, collectors are used to develop settlement indices to predict commercial catches a number of years in advance so that management of the fishery can be finely tuned to environmentally induced or anthropogenic-induced changes in larval production (Phillips, 1986; Caputi et al., 1995).

Design of collectors consists of either crevice-providing collectors for pueruli that would be most likely to settle directly into the benthic environment (e.g., *Jasus edwardsii*, Kensler, 1966; Booth, 1979; and some *Panulirus* spp., Tholasilingam and Rangarajan, 1986; Hirata et al., 1988) or seaweed-like collectors for pueruli that have an algal-dwelling phase prior to benthic life (e.g., *Panulirus argus*, Marx, 1986). Sampling using these collectors involves washing lobsters from the crevice-providing collectors or shaking and tapping the seaweed-like material so that the lobsters fall off. Timing of settlement is critical, as is knowledge of the time of molt into the first juvenile stage, so that appropriate deployment/retrieval intervals can be determined (Phillips and Booth, 1994; Butler and Herrnkind, 2000).

Crevice-like collectors include the "pallet collector" and the "Booth crevice collector." The pallet collector is constructed from timber and consists of two panels with ~15 slats separated from each other by 1 cm and 2 internal baffles. The panels are hinged, so the collector can be opened

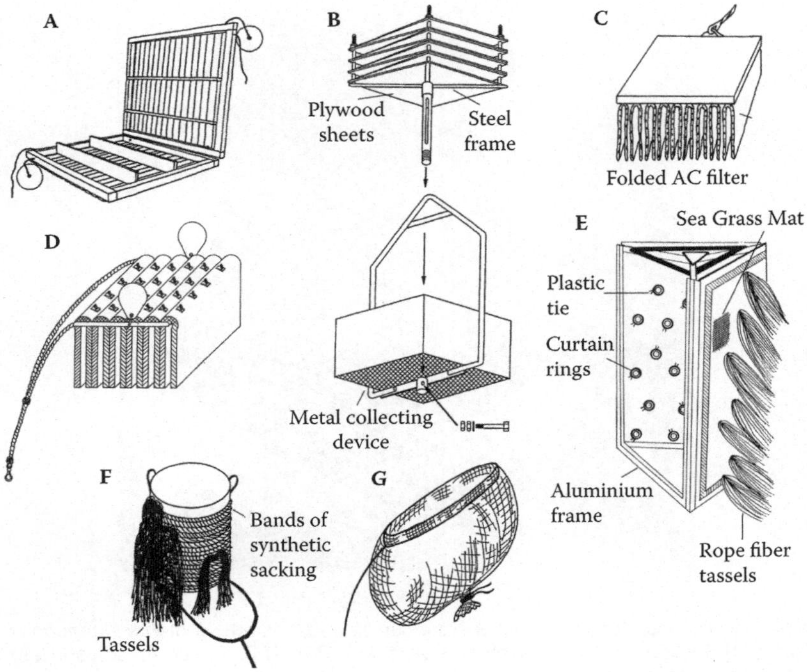

Figure 6.4 Passive collectors for postlarvae of spiny (and occasionally also for slipper) lobsters: (A) hinged pallet collector; (B) Booth crevice device and mesh box used to enclose it prior to recovery; (C) Witham collector; (D) PVC-based hogs-hair collector; (E) Phillips artificial seaweed collector; (F) GuSi collector; (G) Serfling and Ford collector. (Modified from Witham, B. et al. 1968, *St. Flor Bd. Conserv. Tech. Ser.* 53: 1–31; Phillips, B.F. 1972, *Crustaceana* 22(2): 147–154; Phillips, B. F. and J. D. Booth. 1994. *Rev. Fish. Sci.* 2(3): 255–289. Used with permission.)

to check for pueruli (Figure 6.4A). It is buoyed along its length at the surface and attached to a mooring chain and bottom weight with a swivel clip so that it can spin in the water according to currents. The collector hangs broadside to the current (Phillips and Booth, 1994) and was designed to capture pueruli of *Jasus edwardsii* (Lewis, 1977). The Booth crevice device was also developed to capture pueruli of *Jasus edwardsii* and consists of eight treated plywood sheets in a galvanized steel frame that creates seven wedge-shaped crevices (Figure 6.4B). These collectors can be set on the substrate in waters 1 to 10 m deep or suspended in the water column and are easily recovered by divers. Typically, the Booth crevice device is moored in place by attachment to car tires filled with concrete. Prior to retrieval, the collector is placed into a mesh-based metal box to prevent escape by the pueruli (Booth et al., 1991). Depending on how the collector is used, it not only captures pueruli but also juveniles. The Booth crevice collector has also captured nistos of the scyllarid *Antipodarctus aoteanus* (Phillips and Booth, 1994).

Collectors that mimic a floating algal environment are based on an original design by Witham et al. (1968)—the "Witham collector." Originally, the Witham collector consisted of a polyurethane float 30 cm^2 and ~2.5 cm thick, coated with fiberglass resin, onto which sheets of nylon webbing similar to air-conditioning filters, were attached on one side (Figure 6.4C). This design was modified numerous times when different materials for the sheets of webbing became available and were then discontinued; most collectors today use the readily available air-condition filters for the sheets and these generally hang 15 to 30 cm down from the float (Phillips and Booth, 1994). Float material has also changed, and today, it primarily consists of a rectangular framework of PVC pipe with cross supports to hang the filter material (Figure 6.4D). The number of sheets range from 8 to 12 and pueruli collect on both sides of each sheet. This collector is still known as the Witham collector, but is also called the "Hunt hogs-hair collector" or simply the "hogs-hair collector" and is primarily used to collect *Panulirus argus*, although modifications of the design have been used to capture *Panulirus japonicus* (Phillips and Booth, 1994). A major modification of this design was made by Phillips (1972) who constructed a three-sided aluminum frame into which were inserted sheets of gray PVC, 61 cm high × 35 cm wide × 0.6 cm thick. Woven polypropylene material was glued to the PVC sheet, and tassels of synthetic rope (Tanikalon fiber) were then attached to this material. The tassels were ~1.5 m long and divided into three equal lengths from which four tassels split off. Each tassel was tied to the PVC panel, and 25 tassels were evenly spread across the surface of the PVC panel (Figure 6.4E). The top of the collector floats at the water's surface, and it is moored in place with a chain and weight. This collector is known as the "Phillips collector" and is used primarily to capture pueruli of *Panulirus cygnus* in Western Australia (Phillips and Booth, 1994), although it has also been successful in capturing pueruli of *P. argus* in Cuba (Cruz et al., 1991) and nistos of the scyllarid *Ibacus peronii* in Western Australia (Phillips, 1975). Prior to retrieval of either the Witham or Phillips collectors, a bag is placed around the panels/sheets to prevent escape of pueruli.

A low-cost, low tech, algal mimic collector called the "GuSi collector" was developed in Mexico and consists of a 19 L bucket covered with bands of synthetic sacking to which are tied 110 tassels made of strips of artificial seaweed. The strips are typically shredded to produce smaller-width tassels (Figure 6.4F). A circular polyurethane plate is placed inside the base of the upside-down bucket to keep the collector at the water's surface, and the bucket is moored in place by attaching rope to the holes for the bucket's original handle (Gutierrez-Carbonell et al., 1992). This collector is successful in collecting pueruli of *P. argus* and is generally more effective than Witham or hogs-hair collectors (Briones-Fourzán and Gutierrez-Carbonell, 1992; Guiterrez-Carbonell et al., 1992). Some researchers have designed collectors using live surfgrass and red algae by enclosing these into nylon bags that are then tied to a float to remain near the water surface where illumination is high (Figure 6.4G). These collectors, known as "Serfling and Ford collectors," were designed specifically to capture *Panulirus interruptus* pueruli (Serfling and Ford, 1975), but have also been successful in capturing *P. argus* (Phillips and Booth, 1994).

All collectors have limited lifespans: Wood, even when marine grade and treated, will eventually succumb to ship worms; tassels last less than a year and have to be replaced; and nylon meshing will also degrade over time. Many of the materials used in these collectors require continual checking and replacement, and few last longer than 5 years (Phillips and Booth, 1994). In addition, many of the collectors require conditioning in situ (either to leach chemicals or to allow fouling organisms to coat surfaces) before they become attractive to pueruli, and this too can affect lifespan of the collector. Consequently, financial considerations become important when assessing the deployment of spiny lobster puerulus collectors, and often, these are based on the purpose behind the deployment (e.g., aquaculture versus monitoring).

Special artificial reefs have been used to capture postpueruli for grow out ("seed lobsters") in the aquaculture industry in Vietnam (primarily *P. ornatus*) (Hung and Tuan, 2009). The seed lobsters are collected from their settling grounds and kept in cages in the coastal waters where they are supplied with artificial shelters (wooden poles, pieces of corals, and other materials drilled with 5 to 10 mm holes) and fed with trash fish (Long and Hoc, 2009). This same technique of grow out is also used in Indonesia (Williams, 2009).

Slipper Lobsters

Scyllarid postlarvae (nistos) have only occasionally been reported on collectors designed and deployed for spiny lobsters postlarvae (pueruli) (e.g., *Ibacus peronii* and *Antipodarctus aotenatus*; Phillips and Booth, 1994). Sekiguchi et al. (2007) suggested that scyllarid settlement requirements differ from those of palinurids and that nistos might not have the same function as pueruli. They argued that whereas palinurids appeared to prefer holes and crevices in hard substrates or structurally complex algal or coral habitats in shallow waters, the flattened form of scyllarids, including the nisto, suggested adaptation to burrowing into softer, more homogenous substrates. Until areas of postlarval supply are identified for many scyllarid species, or at least those that are commercially important, developing appropriate collectors will not be possible.

ARTIFICIAL STRUCTURES FOR JUVENILE AND ADULT LOBSTERS

All artificial reefs for juveniles, subadults, and adults of the three commercially important families of lobsters are benthic and shelter providing. Studies of sheltering preferences and requirements are dominated by spiny lobsters, followed by clawed and then slipper lobsters.

Clawed Lobsters

Throughout the range of clawed lobsters, appropriate shelter-providing, near shore habitat is thought to be a limiting factor that creates a population bottleneck that affects the distribution and abundance of fishable lobsters (Fogarty and Iodine, 1986; Richards and Cobb, 1986; Steneck, 2006). American clawed lobsters are most common in the Gulf of Maine in the United States, in the Gulf of St. Lawrence, and close to Nova Scotia in Canada despite occurring all the way to North Carolina. This biogeographic pattern is likely the result of glacial deposits left from Pleistocene glacier advances and retreats that left heavy concentrations of gravel in a broad arc around the periphery of the Gulf of Maine and the inner rocky shelf near Nova Scotia, as well as in and between isolated banks in the gulf (Pratt and Schlee, 1969). Artificial reefs, therefore, represent a means by which local abundance might be increased on featureless terrain, particularly in the southern regions of distribution where glacial deposits are absent. As opposed to studies with spiny and slipper lobsters, much of the early research on artificial reefs has focused on temporarily altering the local distribution pattern of lobsters by providing shelter-providing structure on barren substrates rather than attempting to understand the components that make a reef site location successful and features that make the reef particularly appealing to lobsters.

One of the earliest attempts to construct artificial reefs for *Homarus americanus* was done in the mid-1960s by Scarratt (1968, 1973) who deployed a naturalistic reef covering nearly 3000 m^2 that was made of sandstone rocks up to 1 m in size, and which was assembled on a sandy bottom mixed with small cobble in Northumberland Strait, Canada, 2.5-km away from the nearest known preferred lobster habitat. Colonization by lobsters was slow throughout the first two years but, within 7 years, biomass of immigrant lobsters exceeded that of nearby natural areas, with a similar, broad-size distribution of all life history phases (Scarratt, 1968, 1973).

Nonnatural materials have also been used to create reefs. One artificial reef (Kismet Reef), 457-m long × 46-m wide, consisting of two submerged barges and bundled tires, was deployed on a sand and gravel bottom in 6 to 7 m of water in Great South Bay, New York, and another (Fire Island reef), 1.6-km long × 0.2 km wide, consisting of rock and building rubble, was deployed in deeper waters (21 m) of the Atlantic (Briggs and Zawacki, 1974). These reefs were originally designed to attract finfish and increase their biomass, but also attracted lobsters. However, there were differences in the sizes and sex ratios of the lobsters recruited into the two different sites, primarily due to the oceanic reef attracting the larger, offshore individuals and the Great South Bay reef primarily attracting subadults that had not yet recruited to the fishery. In addition, the Great South Bay reef was male dominated, while the oceanic reef had a 1:1 sex ratio (Briggs and Zawacki, 1974).

In what was one of the first attempts to match reef characteristics with behavioral preferences of lobsters, Sheehy (1976, 1977) designed shelter units for *H. americanus* based on work by

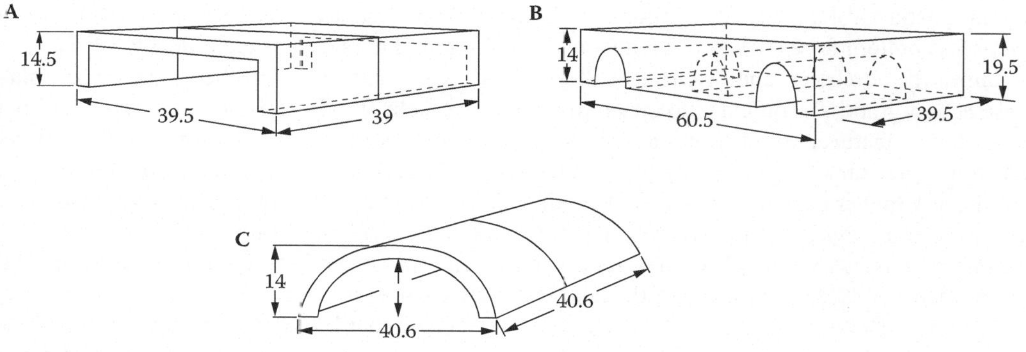

Figure 6.5 Artificial reef shelters for American clawed lobsters: (A) two-piece single-unit shelter; (B) triple-unit shelter; (C) high stable, half cylinder single-unit shelter. (Modified from Sheehy, D. J. 1976. *J. Fish. Res. Board Can.* 33: 1615–1622; Sheehy, D. J. 1977. A study of artificial reefs from unit shelters designed for the American lobster (*Homarus americanus*). Ph.D. dissertation, University of Rhode Island. Used with permission.)

Cobb (1971). His pumice concrete shelters consisted either of two-piece, single-chamber crevices (39.5 cm wide × 14 cm high × 39 cm deep) or a single smaller crevice (39.5 cm wide × 14 cm high × 19.5 cm deep) (Figure 6.5A). They were deployed off Rhode Island on a featureless sand substrate about 0.6 km from the nearest lobster supporting habitat. Lobsters preferred shelters that were oriented with the openings perpendicular to the current rather than those parallel to the current, most likely because the latter configuration was unstable in storms. A year later, lobsters ranging from postlarvae to egg-bearing females resided on the reef, and multiple occupancy of shelters rose to 35%. Lobster biomass was higher on the artificial reefs than on nearby natural lobster ground. Triple units (60.5 cm wide × 19.5 cm high × 39.5 cm deep with three 11 cm wide openings spaced 6.5 cm apart, Figure 6.5B) were occupied at a higher rate than similar volume single units; however, these units were difficult for divers to handle and space.

A more stable, half-cylinder, single-chamber unit was subsequently developed with a curved roof design (40.6 cm wide × 14 cm high × 40.6 cm deep, Figure 6.5C), and these were deployed at six different sites with bimonthly monitoring over 2 years (Sheehy, 1977). Overall occupancy of shelters was highest during winter months, and multiple occupancy also peaked in winter with smaller, wounded individuals comprising the majority of multiple occupants. While an earlier study (Sheehy, 1976) suggested that spacing interval of the units was important and affected the size of lobsters recruiting, this latter study (Sheehy, 1977) demonstrated no such interaction between spacing and lobster size. However, large lobsters occupied the periphery of all reefs, while smaller lobsters occupied the units within the reef and 1 m was suggested as the minimum spacing between units (Sheehy, 1977). A more recent study by Steneck (2006), using artificial shelters composed of hemicylindrical PVC pipe (20.3 cm wide × 47.7 cm long), demonstrated effects more similar to Sheehy's (1976) work. Steneck (2006) suggested that while 1 m spacing of shelters led to significantly increased population densities, it also resulted in a greater proportion of empty shelters, a higher incidence of aggressive interactions, and primarily attracted lobsters of smaller sizes. Differences in the results between Sheehy (1977) and Steneck (2006) may have been due to differences in shelter material—one used concrete structures more similar to natural substances and one used a smooth, nonnaturalistic substance that was not conditioned prior to deployment—that could have impacted the behavior of the lobsters. Hence, further investigation is needed to better understand the dynamics involved in the spacing of individuals with different materials before appropriate reefs can be deployed for subadult and adult clawed lobsters. Nevertheless, analysis of communities pre- and postdeployment demonstrated that the artificial reefs increased local productivity by attracting settling postlarval lobsters. In addition, the artificial reefs increased the carrying capacity of the featureless bottom for a variety of organisms, including predators of lobsters (Sheehy, 1977).

Similarly, Bologna and Steneck (1993) found that artificial kelp beds (made of black construction-grade plastic cut into strips to mimic live kelp fronds and mounted onto steel bars embedded in featureless substrate) attracted similar densities of subadult lobsters as did live kelp beds that were transplanted into featureless terrain, and that both live and artificial kelp beds had significantly higher densities of lobsters than did adjacent featureless terrain. Both live and artificial beds were immediately attractive with lobsters colonizing the bed within 24 h. However, as the size of the beds increased, lobster density decreased and was strongly and positively influenced by the perimeter-to-area relationship of the kelp bed, with perimeter length being the important factor (i.e., an edge effect). Bologna and Steneck (1993) concluded that kelp beds, or even artificial beds, had the capacity to affect local lobster population densities by concentrating individuals along the edge of the bed and could, therefore, increase local carrying capacities of featureless habitats. Given that the Japanese have developed artificial reefs for kelp, deployment of such reefs in the western Atlantic could also have a positive effect on lobster populations. Other studies that show positive effects on enhancing lobster density include the artificial reef constructed to mitigate for large-scale lobster loss following the 1989 oil spill in Narragansett Bay, Rhode Island (Castro et al., 2001). It consisted of six 10 m × 20 m replicate artificial reefs made of rock rubble deployed on featureless shell-mud bottoms. The artificial reefs were monitored for 5 years by diver surveys, tagging, and trapping. Adults and large juveniles of *H. americanus* immigrated into the reef within 3 months and, from the second year forward, their densities in the man-made reefs were equal to or larger than nearby natural site. Despite these successes in increasing lobster density and species biomass, Sheehy (1976, 1977) cautioned that reef location should be carefully selected as grain size, water depth, wave activity, and current conditions all have important ramifications for long-term stability of the artificial reef components.

Miller et al. (2006) examined such locality effects—shelter type, substrate on which the shelter resides, and area effects of a shelter pile—to determine how these factors influence the ability to shelter by subadult and adult lobsters in two sizes ranges (50–59 mm CL and 70–79 mm CL or 82–89 mm carapace length). Comparisons between low-entrance concrete bricks (37 mm high × 110 m wide) and high-entrance bricks (57 mm × 110 mm wide) on a sand–gravel substrate demonstrated that lobsters of any of the size groups tested required the high-entrance bricks to be able to occupy shelters without having to excavate substrate (a time-consuming task), but could occupy the low-entrance bricks after excavation. However, smaller lobsters had more difficulty than larger lobsters in the excavation process. When presented with rock piles on a sand–gravel bottom versus a hard bottom, the size of the rocks impacted ability to shelter with smaller lobsters (50–59 mm and 70–79 mm carapace length), occupying piles on the soft bottom and excavating into the sand–gravel under the rocks. Coarseness of the sand–gravel affected time to excavate and influenced shelter occupancy, such that smaller-grain (1 to 2 cm) substrates were more easily excavated by small lobsters than larger-grain substrate (3 to 5 or 6 to 8 cm). Finally, larger diameter piles of rocks, with fewer rock layers, resulted in higher densities of lobsters.

Additional studies have attempted to enhance American lobster populations using artificial shelters in the field (e.g., Hruby, 2009), but none has had any significant impact on fisheries. As a result of the failure of prior attempts to significantly enhance production for fisheries or to mitigate effects of habitat loss or degradation, Barber et al. (2009) developed a systematic plan for artificial reef site selection that specifically targeted *H. americanus* prior to deploying cobble/boulder reefs as part of a mitigation project for habitat loss due to a gas pipeline. Their model included substrate mapping of potential sites; determination of grain size, depth, and slope; sampling of mobile benthic macrofauna for reference points; determination of postlarval supply via collectors; and ranking of sites. The results of this stepwise process allowed the selection of a site that had low sedimentation rates, suitable slope and depth, appropriate bottom substrates to support the weight of an artificial reef, natural larval supply, and low species diversity before reef deployment. Thus far, the artificial reef has successfully recruited larvae of various invertebrate species, including lobster, and species diversity is approaching that of natural reefs nearby (Barber et al., 2009). This approach is consistent with Sheehy (1977) suggestions for site selection and holds great promise for actually increasing production and enhancing natural population levels at such reefs.

European clawed lobsters, *H. gammarus*, have also been the target of several artificial reef projects in the United Kingdom. The Poole Bay artificial reef off the central south coast of England, deployed in 1989 as a materials test experiment, originally consisted of units made from blocks of stabilized, pulverized fuel-ash (PFA), placed in 10 m × 30 m arrays of eight conical 4 m × 1 m piles on a 12 m deep sandy bottom 2 km away from the nearest natural lobster habitat. In 1998, tire modules were added. Lobsters were recorded in the reef 3 weeks after deployment (Jensen et al., 1994, 2000a, b; Jensen, 2002). Berried females recruited into the reef 2 years after its deployment, and small juveniles were found on the reef 3 years later (Jensen and Collins, 1995). Some lobsters were repetitively tagged and recaptured on the reef system for over 4 years (Smith et al., 1999). Electromagnetic telemetry of lobsters detected predominantly nocturnal movements between and among the eight reef units, with more frequent movements in spring and summer than in winter. Smaller lobsters moved more frequently than larger individuals in early and late autumn (Smith et al., 1999).

The Loch Linne artificial reef, deployed off the west coast of Scotland from 2001 to 2006 at a depth of 10 to 20 m, was designed to understand how reef construction and species interact (Sayer and Wilding, 2002; Wilding and Sayer, 2002). In this reef, 30 separate reef modules are clustered into eight groups with a single module containing 4000 blocks of two types (solid and ones with two voids for nesting spaces) constructed in a conical pile 3 to 4.5 m in height and 10 to 15 m in diameter. The different kinds of blocks were deployed in different hydrological conditions and different sediments (cobble, silty-sand, and mud) to study colonization and habitat utilization at different scales and habitat complexity. A monitoring program was initiated prior to development in 1998 and continues today. Fixed belt transect surveys conducted monthly over a year demonstrated that there were no differences in animal abundance and diversity among the groups of reef modules and natural reefs in summer, autumn, or winter, but in the spring, the simple reef modules (those with solid blocks) had less abundance and reduced diversity compared to the complex reef modules (blocks with voids) and the natural reef (Hunter and Sayer, 2009). Overall abundance of obvious fish and macro invertebrates was 2–3 meters higher on the complex block artificial reef modules than in either the simple block artificial reef modules or nearby natural reefs (Hunter and Sayer, 2009). However, lobsters were not found within the belt transects on either the artificial reef or the natural reef, even several years after deployment of the first six groups of blocks were in place (W.R. Hunter, Scottish Association for Marine Science, Dunstaffnage Marine Laboratory, Dunbeg, Oban, Argyll PA37 1QA, UK, personal communication). Hence, it is clear that lobster populations must be nearby for artificial reefs to attract them.

France has experimented with a number of artificial reef materials along both its Atlantic and Mediterranean coasts. Artificial reefs deployed along the Atlantic coast had serious problems of siltation and were difficult to survey by divers; thus, those projects were largely abandoned and replaced by more intensive efforts in the Mediterranean (Barnabé et al., 2000). Hydrological and geological differences divide France's Mediterranean coast into east and west sections, and these differences affect colonization at the deployed reefs. Six artificial reefs were deployed off France, representing 19,840 m^3 along its east coast (the Provence-Alpes-Cote d'Azur region), and seven reefs were deployed along the west coast, representing 19,226 m^3 (the Languedoc–Roussillon region). The artificial reefs along the east coast were deployed largely to mitigate habitat degradation due to coastal development, while the reefs along the west coast were deployed at the request of artisanal fishermen to protect their static fishing gear and longlines from illegal trawling (Barnabé et al., 2000). As in the Atlantic, the earliest Mediterranean reefs deployed in the late 1960s consisted of old cars; these were followed by tire reefs in the late 1970s and 1980s. By the mid-1980s, more preplanning

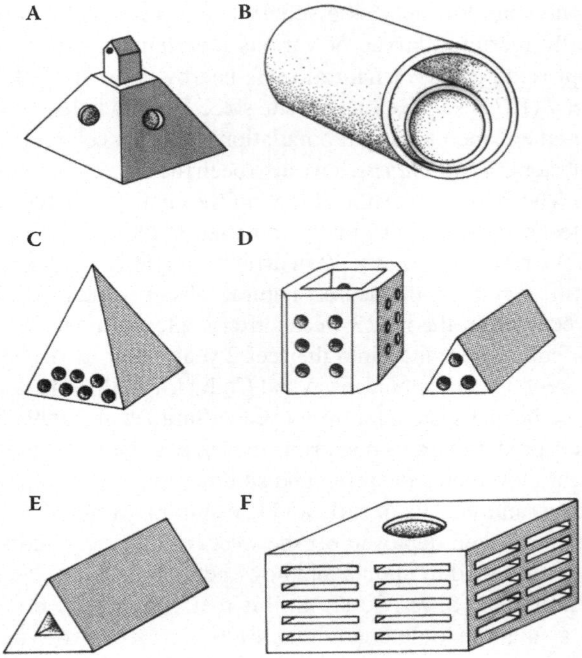

Figure 6.6 French and Japanese concrete artificial reefs for lobsters: (A) sea rock-type block used as a trawling obstacle; (B) pipe-within-pipe module used for protection of mussel beds and as a trawling obstacle; (C) pyramidal concrete block (70 cm) used for both spiny lobster reef and agar-agar cultivation in Shizuoka Prefecture; (D) blocks used by Shizuoka and Nagasaki Prefecture for lobster reefs; (E) triangular concrete block used in Wakayama Prefecture; (F) large rectangular block used by Shizuoka Prefecture. (From Barnabé, G., et al. 2000. Pages 167–184 in A. Jensen, K. J. Collins, and A. P. M. Lockwood, Eds. Artificial Reefs in European Seas. Kluwer Academic Publishers, Dordrecht, The Netherlands; and modified from Fishery Civil Engineering Study Association. 1982. Pages 16–22 in S. F. Vik, Ed. Japanese Artificial Reef Technology: Translations of Selected Recent Japanese Literature and An Evaluation of Potential Applications in the United States. Aquabio Technical Rep. 604, Aquabio, Inc., Arlington, MA. Used with permission.)

went into both deployment and material selection, such that most artificial reefs were composed of concrete and were designed as specific modules to predetermined configurations.

Deployments along the east coast ended in 1989, although in 1997 a new reef of concrete telegraph poles was deployed. Artificial reefs along the west coast were deployed in 1985, 1988, and 1995 (Barnabé et al., 2000). Two of the west coast Mediterranean reefs have attracted lobsters. The first was deployed in 1985 for the purpose of providing an obstacle to trawling. This reef consisted of 410 modules of a "sea-rock" type (flat topped, pyramidal concrete structure with voids on the pyramidal faces, Figure 6.6A) covering 640 m^3. Extensive colonization by oysters, mussels, fish, octopus, and lobsters were reported by Tocci (1996) for this reef. The second reef was deployed in 1995 to protect a molluskan culture zone, and this reef was constructed from two concrete pipes, one of 1 m diameter that fit into another of 1.9 m diameter, each of 2.5 m length, weighing 8.5 ton (Figure 6.6B). Units were spaced 200 m from each other; 60 units were placed off Marseillan in 1992 and 200 units were placed off Agde in 1995. The units in Marseillan became colonized by mussels, oysters, conger eel, sea bass, and numerous lobsters (Barnabé, 1995), while those in Agde mainly attracted mussels and conger eels (Barnabé et al., 2000). Differences among reefs in their attractiveness to lobsters indicate that even when the same materials are used in different locations, some nearby population of lobster must be present for emigration into the reef to occur. Thus, if lobsters become the target species for reef projects, basic information about their distribution must be present before appropriate reef sites can be chosen.

Spiny Lobsters

Compared to clawed and slipper lobsters, there is a wealth of information on artificial reefs for spiny lobsters that mainly arises from extensive field and laboratory research on the behavior and ecology of various life history stages of one commercial species—the Caribbean spiny lobster, *Panulirus argus*—although there is also much information on the postlarval stage of *P. cygnus*. Data from this research has provided a baseline for construction of appropriate artificial shelters for this species and, generally, these data are lacking for other species of spiny lobster.

Sheltering in natural structures has been studied for several species of spiny lobsters. Species in the genera *Palinurus*, *Panulirus*, and *Jasus* typically seek shelter in crevices in rocks, corals, sponges, or under ledges or vegetation (e.g., Kanciruk, 1980; Spanier and Zimmer-Faust, 1988; MacDiarmid, 1994; Childress and Jury, 2006). Many spiny lobster species have an ontogenetic habitat shift from the postlarval settlement habitat of algae, kelp, or seagrass to benthic crevices as larger benthic juveniles, subadults, and adults (e.g., Butler and Herrnkind, 2000; Butler et al., 2006; Childress and Jury, 2006). The species that demonstrate such ontogenetic shifts of habitats often display a more gregarious habit, inclusive of sheltering together, as they grow in size (e.g., Eggleston and Lipcius, 1992; Eggleston et al., 1992; Mintz et al., 1994; Zimmer-Faust and Spanier, 1987; MacDiarmid, 1994) and do so even when unoccupied natural dens are plentiful (see review by Nevitt et al., 2000). Other species, however, do not exhibit such ontogenetic habitat shifts and settle directly onto adult habitat; often, these are the obligate coral-reef dwelling species (i.e., *Panulirus guttatus*; Sharp et al., 1997; Robertson and Butler, 2003).

Spiny lobsters prefer dens that have shaded cover with multiple entrances and avenues of escape (e.g., Spanier and Zimmer-Faust, 1988; Eggleston et al., 1990). Predators can influence microhabitat preferences, such that lobsters become less selective in the presence of a predator (Gristina et al., 2009). For species with ontogenetic habitat shifts, the attraction of dens is further increased if conspecifics are present (Zimmer-Faust and Spanier, 1987; Ratchford and Eggleston, 1998), likely due to a "guide effect" from odors that emanate from such conspecifics (Zimmer-Faust and Spanier, 1987; Childress and Herrnkind, 1994, 1996, 2001). Hence, for social spiny lobsters, artificial reefs designers have to incorporate the ability of multiple individuals to co-den in crevices, something that is not necessary for clawed lobsters or solitary species of spiny lobsters.

Artificial structures for spiny lobsters have mainly been deployed to: (1) mitigate the loss of local shelters, (2) enhance depressed local populations due to fishing pressure or shelter limitation (Davis, 1985; Butler et al., 1995; Herrnkind et al., 1999), or (3) examine the ecological role of shelter on *P. argus* (Eggleston et al., 1990; Butler and Herrnkind, 1992, 1997; Lozano-Alverez et al., 1994; Mintz et al., 1994; Cruz et al., 2007). Less frequently, artificial reefs have been used as replicating collecting devices to facilitate population sampling (e.g., Cruz et al., 1986; Behringer and Butler, 2006) or to evaluate and compare the distribution and abundance of small juvenile lobsters in shallow waters (Arce et al., 1997) and in different habitats (Behringer et al., 2009). Cruz et al. (1986, 1995) used arrays of small artificial reefs constructed of 60 standard concrete blocks to successfully predict commercial catches in subsequent years and to monitor populations for fishery management models (Baisre, 2000).

Spiny lobster reef design has a long history. Investigators in Japan began experimenting with bamboo-framed structures as artificial reefs as early as the late 1700s, and based on successes in increasing local catches of fish, moved on to use old boats, sand bags, and cut stones (Tsumura et al., 1999). In the 1930s, the Japanese government began experimenting with concrete blocks and by 1950 these became the exclusive material for government-subsidized artificial reef projects

(Oshima, 1964), although other materials were tested (steel, old tires, ceramic products/earthen pipes, and synthetic resin products, old boats, old buses). Each prefecture in Japan now has its own preferred reef material (e.g., see Figure 6.6C–F). In some areas, simple stone beds and piers have been deployed in communities of agar-agar seaweed to create spiny lobster grounds (Nonaka et al., 2000). Despite decades of work to actively enhance productivity of lobster via artificial reefs, the catch today consists of lobsters attracted to artificial reefs located in previously poor fishing grounds, and does not represent an increase in stock size (Nonaka et al., 2000). Polovina (1989) argues that the real benefit of these reefs is not an increase in production, but an aggregative effect to concentrate fishing activities such that traditional small fishing vessels within Japanese fishing communities can remain economically viable.

Similarly, artificial reefs deployed for *Panulirus argus* have been used to concentrate lobsters and enhance fishery catches in Cuba and Mexico (Cruz et al., 1986; Cruz and Phillips, 2000; Briones-Fourzán et al., 2000, 2007). For more than 60 years, Cuban and Mexican fishermen have increased their catch of lobsters using simple, inexpensive, durable, and easily harvested artificial shelters called a *pesquero* in Cuba (e.g., Cruz and Phillips, 2000) or a *casita* in Mexico (e.g., Briones-Fourzán et al., 2000). These shelters are modified from the indigenous fishermen's earlier designs constructed of mangrove branches or similar palm trunks, 8 to 12 cm in diameter with parallel sticks creating a three to four layer 4 m^2 raft (Figure 6.7A). These were positioned on shallow substrates where natural shelters were scarce and currents were mild. A single *pesquero* could concentrate as many as 200 lobsters, which were then captured by divers using encircling nets. In this way, an estimated average of 16 tons of marketable lobsters could be acquired per diver per year (Cruz and Phillips, 2000). After it became illegal to cut mangroves in Cuba, other low-priced, durable building materials were used, including PVC pipes, all-cement structures, and

A

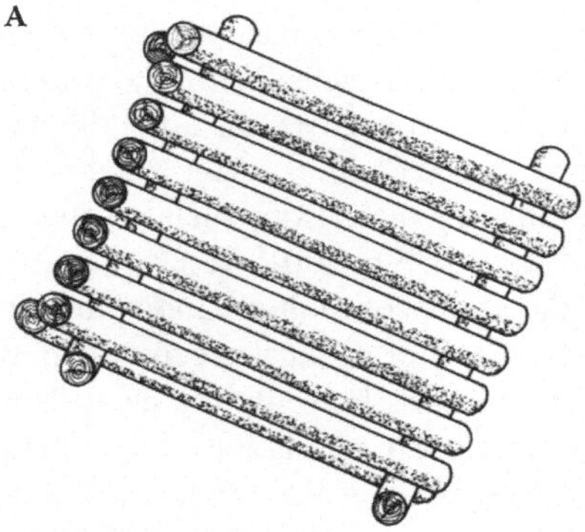

B

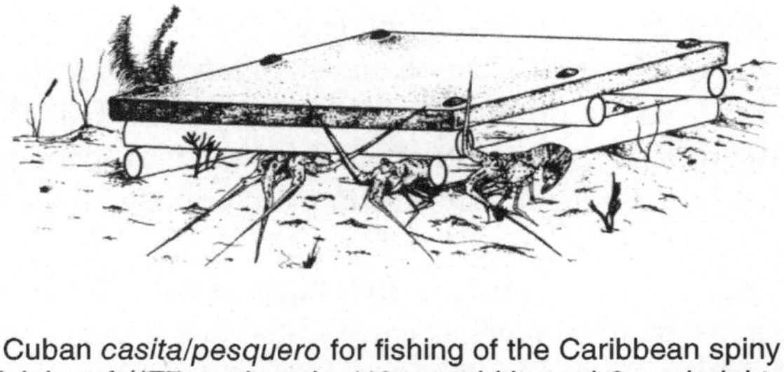

Figure 6.7 (A) A typical Cuban *casita/pesquero* for fishing of the Caribbean spiny lobsters, *Panulirus argus*. (B) This artificial reef (177 cm length, 118 cm width, and 6 cm height of opening) is constructed with a frame of PVC pipe and a roof of cement. (From National Research Council. 1988. Fisheries Technologies for Developing Countries. Office of International Affairs, National Research Council. National Academy Press, Washington, D.C. and Eggleston, D. B. and R. N. Lipcius. 1992. *Ecology* 73(3): 992–1011. Used with permission.)

ferro-cement shells mounted onto two or more wooden branches (Figure 6.7B). These devices were placed in coastal waters and have revolutionized lobster fishing, as well as fishery management in these regions (for details, see Cruz and Phillips, 2000; Briones-Fourzán et al., 2000). Hundreds of thousands of these artificial shelters have been successfully used for spiny lobster fishing, chiefly off Cuba and Mexico, but also in the Bahamas, United States, Virgin Islands, Florida, Africa, and elsewhere (Herrnkind and Cobb, 2008). Lobster fishermen believe that the casitas/pesqueros, besides enhancing fisheries, increase lobster populations by helping lobsters to codefend against natural predators (Briones-Fourzán et al., 2000) (e.g., via collective prey vigilance and collective defense as per Herrnkind et al., 2001). Casitas/pesqueros placed in habitat lacking in natural shelter also potentially allow lobsters more efficient and longer access to their prey and thus faster growth (Briones-Fourzán et al., 2000). This permits the exploitation of food resources over extended areas by providing lobsters with scattered shelters throughout a large area.

Briones-Fourzán et al. (2000) emphasized that casitas/pesqueros were most effective in shallow water habitat lacking natural crevices—habitats such as sea grass. These habitats are frequently proximate to nursery grounds where juveniles continually emerge and from which they must find new shelters and food (Herrnkind, 1980; Kanciruk, 1980; Lipcius and Eggleston, 2000; Herrnkind and Cobb, 2008). Used in this manner, casitas/pesqueros may increase production of lobster populations by decreasing the time juveniles are exposed to predators while locating shelter and/or food thereby increasing juvenile survivorship (Herrnkind et al., 2001). However, the benefits to local fisherman may be delayed as these structures act primarily as grow-out facilities. Settlement and juvenile habitats are usually in shallow water, while reproductive adult habitats are in deeper water (Kanciruk and Herrnkind, 1976; Lipcius and Herrnkind, 1987). Consequently, most lobsters in the casitas/pesqueros are below the minimum fishery size limit (Herrnkind and Cobb, 2008). Thus, stakeholders should be aware that the purpose of shallow-water casitas/pesqueros may differ greatly from the purpose of deeper-water casitas/pesqueros that would be more likely to attract legal-sized adults.

In addition to enhancing production and/or concentrating individuals for fishery purposes, artificial reef blocks and casitas/pesqueros have been used to mitigate habitat loss. Davis (1985) used hollow pyramids made of standard 2-hole concrete blocks to mitigate crevice loss for the more than one thousand juvenile *Panulirus argus* displaced by rock-fill during marina reconstruction. Lobsters moved into the pyramids and remained there over a 14-month period. Similarly, in 1991–1993 during a mass sponge die-off from a cyanobacteria bloom in Florida Bay, Herrnkind et al. (1997a,b, 1999) experimentally deployed a 1-hectare array of 240 double-stacked, three-hole concrete partition blocks (10 × 20 × 40 cm) as potential mitigation for loss of sponge crevices. Almost all large, crevice-bearing sponges supplying about 70% of dens for small juvenile lobsters (<50 mm carapace length) were destroyed over several hundred square kilometers (Butler et al., 1995), which dramatically increased the predation rate on these juveniles (Herrnkind and Butler, 1986; Smith and Herrnkind, 1992; Herrnkind et al., 1997a, 1997b). Blocks and their crevices were attractive and protective, recruiting densities of juveniles similar to those found in sponge-rich sites after only 3 months of deployment (Childress and Herrnkind, 1994; Herrnkind et al., 1997a, b). Analysis of microwire tag recapture data also supported the ideas that shelter was a key to survival of the small juveniles and that large numbers of benthic settlers were required to strongly affect the ultimate numbers of surviving juveniles (Herrnkind and Cobb, 2008). Similarly, Briones-Fourzán and Lozano-Álvarez (2001), demonstrated that small, scaled down, casita-like artificial shelters designed for postalgal juveniles were rapidly colonized by considerable numbers of late algal and early postalgal juveniles when placed in a crevice-poor, vegetated lagoon at Puerto Morelos, Yucatan, Mexico. Deployment of the artificial shelters resulted in a sixfold increase in juvenile density and a sevenfold increase in biomass compare to control sites lacking natural crevice shelters (Briones-Fourzán et al., 2006). Tag-recapture experiments revealed that this level of enhancement was achieved not by promoting individual growth, but by increasing survival, persistence, and foraging ranges of small and large juveniles. Briones-Fourzán et al. (2006) suggested that casitas mitigated for the lack of natural shelter and, simultaneously, increased sociality, such that smaller, more vulnerable juveniles cohabited with larger conspecifics with greater defensive abilities. Likewise, Cruz et al. (2007) has similarly suggested that introduction of artificial shelters might help reduce natural mortality of postpueruli and juveniles and increase recruitment to fishing areas.

This understanding of the mitigating effects of casitas/pesqueros comes about only by experiments designed to examine the effects of shelter in different habitat conditions with different life history stages. Earlier field studies showed that shelter selection by large juveniles and adults depended on lobster size, shelter dimensions, and lobster density (Eggleston and Lipcius, 1992; Ratchford and Eggleston, 1998). When large juveniles and adult *Panulirus argus* were experimentally tethered in place, they survived significantly better in a casita than just outside the artificial shelter or far away in open seagrass (Herrnkind and Cobb, 2008). The limited opening and height of the casita roof either prevented entry by predators of large lobsters or restricted an effective attack by the predator (triggerfish) within the shelter (Lozano-Alvarez and Spanier, 1997). Under experimentally high predation risk, lobsters sought larger shelters accommodating larger numbers of conspecifics, suggesting theoretical benefits from increasing collective defense and/or creating a dilution effect (e.g., Herrnkind et al., 2001). However, the same aggregative benefit does not necessarily hold for postalgal phase lobsters when tethered together (Butler et al., 1997; Childress and Herrnkind, 2001). Mintz et al. (1994) found that juvenile lobsters tethered in smaller, artificial sponge dens that could hold relatively few individuals had similar survival rates to those in casitas, suggesting that artificial shelters need to be appropriately scaled for the life history stage targeted (Eggleston et al., 1990).

Herrnkind and Cobb (2008) suggested that the most convincing argument for the protective role and enhanced survivorship of casitas would be a direct comparison showing higher, long-term (i.e., several months) survival by casita-resident lobsters versus same-aged lobsters roaming about large areas of sparse natural shelter. This requires sufficient knowledge of the abundance and distribution of lobsters in the absence of casitas. Current evidence strongly suggests that casitas may enhance populations by protecting shelter-seeking, postalgal juveniles when ready shelter is not otherwise available. However, at present, research results do not provide unequivocal evidence that casitas used for concentrating fishable adult lobsters also provide enhancement of lobster survival from natural predation at the population level.

A few artificial reefs in the Mediterranean have recruited limited numbers of European spiny lobsters, *Palinurus elephas* (e.g., in the Ligurian Sea, Italy—Relini et al., 2007; in the north Aegean Sea, Greece—Sinis et al., 2000). Additional laboratory experiments have demonstrated how shelter shape, size, and substrate slope affect the sheltering choices of *P. elephas* juveniles, and enhance their protection and survival rate (Gristina et al., 2009).

Despite the positive effects of enhancing the survival of benthic juveniles and concentrating legal-sized lobsters for fishing practices, Briones-Fourzán et al. (2000) and Herrnkind and Cobb (2008) indicate that there are some potential negative effects of artificial structures. These man-made shelters are large enough to attract predators (e.g., crabs, octopus, groupers, sharks, and triggerfish) that prey on juvenile lobsters, particularly smaller individuals (Mintz et al., 1994; Arce et al., 1997, but see Lavalli and Herrnkind, 2009 showing that smaller animals were not necessarily the most vulnerable). Some predators (e.g., octopus) may enter the artificial devices and prey on the lobsters there, while other predators (e.g., triggerfish) may pull lobsters from the shelters by grabbing onto their long antennae (Weiss et al., 2008). Some smaller predators can even compete with the lobsters for shelter (Butler and Lear, 2009). Artificial shelters that concentrate high numbers of lobsters may make otherwise scattered lobsters more vulnerable not just to predation by natural predators (see review by Briones-Fourzán et al., 2000), but also to overfishing. Additionally, crowding lobsters in casitas/pesqueros may facilitate the spread of diseases and parasites (e.g., Shields and Behringer, 2004; Behringer et al., 2008; Li et al., 2008), although evidence to date indicates that healthy lobsters are capable of detecting infected conspecifics and avoid contact with them (Behringer et al., 2008). Finally, Davis (1981) argued that where many inhabitants of the artificial shelter are undersized for the fishery, repeated handling during incidental capture in fishing gear might cause injury and reduce growth or delay maturity.

Slipper Lobsters

Development of artificial reefs for slipper lobsters lags behind that for spiny and clawed lobsters, mostly as a result of not understanding the sheltering preferences of the target species. Only the Mediterranean slipper lobsters (*Scyllarides latus*) has been studied for shelter preferences in both laboratory and field conditions (Spanier et al., 1988, 1990, 1991, 1993; Spanier, 1994; Spanier and Almog-Shtayer, 1992) and, based on these studies, several types of artificial reefs have been designed for deployment. Preferences included horizontally oriented shelters with small openings that supplied shade, reduced visual detection, and hampered entry of high body profile fishes (Spanier and Almog-Shtayer, 1992). Shelters with multiple openings were also preferred, presumably because these enable escape through a "back door" if a predator successfully penetrates the den.

Based on such preferences, an artificial reef constructed of used tires weighted with concrete in their lower parts was deployed on a 20 m flat substrate in the southeastern Mediterranean (Figure 6.8). Predation by the gray triggerfish was significantly less on lobsters tethered in the artificial reefs compared to those tethered in open areas (Barshaw and Spanier, 1994b). More recently, small experimental artificial reefs (1.2 m sided, cubical, steel reinforced, concrete structures fitted with 16 sections of 25 cm diameter polyethylene pipes open on both sides) were deployed in the same general area (Edelist and Spanier, 2009) and have successfully recruited *S. latus* (Figure 6.9). These initial successes suggest that artificial reefs might be useful to aggregate slipper lobster species in areas where shelter-providing habitat is lacking but lobsters are present. However, given the wide diversity of habitats exploited by slipper lobsters, researchers need to be cognizant of the habitat preferences of the targeted slipper lobster species, as not all species reside on complex substrates.

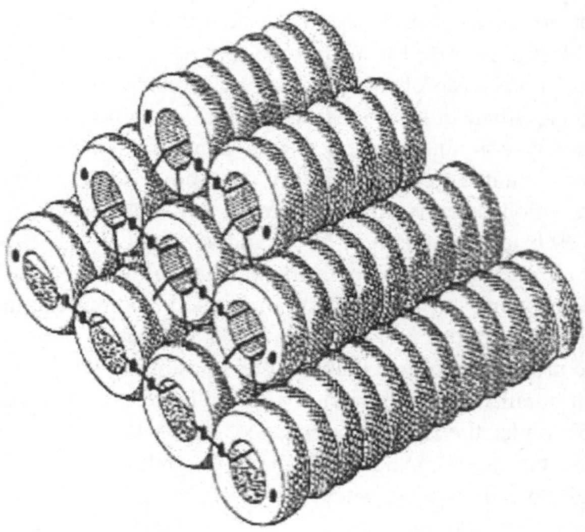

Figure 6.8 An artificial reef unit that successfully recruited Mediterranean slipper lobsters, *Scyllarides latus*. The man-made structure was made of used car tires (32 cm inner diameter, 65 cm outer diameter, and 17 cm tire width) connected with 18 mm steel bars and weighted with concrete poured into the lower part of the first row of tires. (From Spanier, E. et al.,1988. *Mar. Ecol. Prog. Ser.* 42: 247–255. Used with permission.)

Figure 6.9 Mediterranean slipper lobsters recruited to an experimental artificial reef unit (1.2 m-sided cubical steel-reinforced concrete structures, weighing 1500 kg in water and fitted with 16 sections of 25 cm diameter polyethylene pipes opened on both sides), designed and constructed according to the behavioral–ecological preferences of *Scyllarides latus* for shelter and deployed on a flat rocky substrate in the southeastern Mediterranean. (From Spanier, E. and K. L. Lavalli. 2007. Pages 377–391 in K. L. Lavalli and E. Spanier, Eds. The Biology and Fisheries of Slipper Lobsters. Crustacean Issues 17. CRC Press, New York. Used with permission.)

LOBSTER ENHANCEMENT, MARINE-PROTECTED ARTIFICIAL REEFS, AND OWNERSHIP ISSUES

In recent years, there have been considerable advances in hatchery rearing of several commercially important species of lobsters (see reviews by Aiken and Waddy, 1995, and Nicosia and Lavalli, 1999, for *Homarus*; Booth, 2006, for *Jasus*; Phillips and Melville-Smith, 2006, for *Panulirus*; Groeneveld et al., 2006, for *Palinurus*; and Mikami and Kuballa, 2007, for *Thenus*). Hatchery production of postlarvae of lobsters may be directed toward on-growing on shores and in sea cages or for stock enhancement. Sea-cage culture of spiny lobsters has been used in some parts of the world (mostly Asia and Mexico) for on-growing of juveniles with mixed results, due largely to reduced growth and increased mortality from poor water quality, infection, or increased aggressive encounters amongst individuals (Creswell, 1984; Assad et al., 1996; Lozano-Alvarez, 1996; Brown et al., 1999; Jeffs and James, 2001, and see reviews of grow-out attempts by Booth and Kittaka, 2000 and Williams, 2009). In Tasmania, Australia, fishermen can capture pueruli and young juveniles for outgrowing in lieu of fishing their quota and must return 50% of those animals to the sea the following year. However, collection costs have been high, aquaculture aspects have been difficult, and the production of legal-size lobsters has been low (Booth, 2006).

Enhancing natural populations with hatchery-reared lobsters that come from wild stock brooding females is possible under the right circumstances, but it is necessary to undertake significant biological research on the target species beforehand (Agnalt et al., 2007; Svåsand, 2007; Oliver et al., 2008). Although considerable advances have been made in rearing the full cycle of several species of lobsters in captivity, the main constraints for developing aquaculture programs to a commercial scale are economically based. Aquaculture, however, could provide stock for reseeding wild populations (and fisheries) (see review by Phillips and Melville-Smith, 2006). Such restocking can be successful if properly designed artificial structures are supplied to the predator-sensitive early stages and if these early stages are appropriately conditioned to predator and substrate odors so that they respond correctly when released. Some successes, particularly with hardy spiny lobster species, have occurred, but generally only in countries with low labor costs and fast-growing local species (Booth and Kittaka, 2000).

Recently, there has been increased interest in marine protected areas (MPAs) to protect and conserve populations of marine organisms. MPAs, combined with the use of artificial reefs, could be used in lobster management (e.g., Childress, 1997; Goñi et al., 2001; 2006; Follesa et al., 2008; Pettersen et al., 2009; and see review by Butler et al., 2006). Such marine reserves can be used to protect endangered or overfished populations and to create a sanctuary for reproductive populations (e.g. Bertelsen and Matthews, 2001). Despite the nomadic or migrating nature of some adult lobsters (e.g., Herrnkind and Cobb, 2008), MPAs can be effective for early benthic and juvenile lobsters (and even for subadults and adult lobsters). It is essential, however, that these structures are not too small in scale (see Eggleston and Dahlgeren, 2001; Stockhusen and Lipcius, 2001), are properly managed and protected, and contain the proper natural habitats for respective life history stages or have featureless habitat complemented by deployment of artificial habitats. In species that have limited movements for foraging and/or reproduction, MPAs, even if relatively small, may help enhance overfished stocks (Goñi et al., 2001). To be truly effective, the population structure of the protected species should be understood before establishment of such entities occurs (Cannas et al., 1998; Tuck and Possingham, 2000).

One of the most important aspects of using artificial reefs for lobster fisheries is the question of site ownership. Entities that invest in the construction and deployment of artificial reefs naturally desire to be the sole beneficiary of their investment. The identification of the stakeholder groups, allocation of rights, ownership (include possible lease/purchase agreements of the sea bottom), and the acceptance of potential liability (due to the effects of disconnected artificial reefs that can be swept away in severe weather conditions) can be controversial issues (e.g., Sayer and Wilding, 2002) with different solutions in different countries (e.g., Jensen, 2002). Generally speaking, most countries today have developed licensing procedures and protocols for development of artificial reefs to deal with such ownership and usage issues.

SUMMARY, CONCLUSIONS, AND RECOMMENDATIONS

The results of numerous studies during the last five decades on a limited number of commercial species of lobster indicate that artificial reefs designed especially for specific species, and life history stages seem effective in aggregating lobsters, facilitating lobster recruitment into the benthos, and mitigating for habitat or shelter loss. The most successful reefs have been designed and constructed according to ecological needs and behavioral preferences of the target species and its particular life history stage, and are deployed in the proper habitat and season. Nevertheless, the long-term effectiveness of these artificial habitats in enhancing production at the population level is still questionable. Only long-term, large-scale studies that compare populations of lobsters (of the same species, sex ratio, and size range) with and without man-made reefs may supply clearer answers. Such studies should, perhaps, be done in MPAs to control for harvesting effects by humans.

Few artificial reefs designed for lobsters were constructed directly for ecotourism, conservation of remaining portions of the populations, and production of future populations. Aesthetic artificial reefs for lobsters can be useful for such purposes and may alleviate harmful diving pressure on sensitive natural habitats (such as coral reefs), as well as conserve rarer species of lobster. In view of the recent decline of quite a few lobster populations, and the deterioration of their natural habitats, the use of artificial reefs for lobster conservation and lobster habitat mitigation is needed. In addition, since several lobster species are now reared in captivity, these steps can be supplemented by release of juvenile stages (in the right season and time of the day) at an MPA with the right artificial habitats for those particular stages. Restocking and enhancement of natural population by hatchery-reared lobsters from wild stock females should be done only after validating that it actually enhances production and does not simply displace natural stocks. As more information becomes available about the biology of additional species, future studies on artificial reefs and lobsters should be expanded to a variety of lobster taxa and geographical regions, and incorporate broad ecological theories such as habitat selection theory (e.g., Rosenzweig, 1981) and ideal free distribution (e.g., Kacelnik et al., 1992).

Several lobster scientists have indicated potential disadvantages of concentrating lobsters in artificial reefs. These suggestions should be tested in carefully designed field and mesocosm experiments (like those described in Lavalli and Herrnkind, 2009) so that the use of artificial reefs can be better understood. Finally, artificial reefs are deployed in areas that generally have low densities of the target species, but this does not mean that these areas are not productive grounds for other species. The impact that artificial reefs have on ecosystems within these areas needs to be better understood so that we do not target enhancement of one species to the detriment of many others.

REFERENCES

Able, K. W., K. L. Heck, M. P. Fahay, and C. T. Roman. 1988. Use of salt-marsh peat reefs by small juvenile lobsters on Cape Cod, Massachusetts. *Estuaries* 11: 83–86.

Agnalt, A. L., K. E. Jørstad, T. Kristiansen, E. Nøstvold, E. Farestveit, H. Næss, O. I. Paulsen, and T. Svåsand. 2007. Enhancing the European lobster (*Homarus gammarus*) stock at Kvitsøy Islands: Perspectives on rebuilding Norweigan stocks. Pages 413–443 in K. M. Leber, S. Kitada, H. L. Blankenship, and T. Svåsand, Eds. *Stock Enhancement and Sea Ranching,* 2nd edition. Fishing News Books, Blackwell Science, Oxford.

Aiken D. E. and S. L. Waddy, 1995. Aquaculture. Pages 153–175 in J. R. Factor, Ed. *Biology of the Lobster, Homarus americanus.* Academic Press, New York.

Arce, A. M., W. Aguilar-Davila, E. Sosa-Cordero, and J. F. Caddy. 1997. Artificial shelters (casitas) as habitats for juvenile spiny lobsters *Panulirus argus* in the Mexican Caribbean. *Mar. Ecol. Prog. Ser.* 158: 217–224.

Assad, L. T., D. S. Gondim, and M. Ogawa. 1996. The growout of spiny lobster juveniles in marine cages. *Arquivos Cienias do Mar, Fortaleza* 30: 13–19.

Atema, J. and J. S. Cobb. 1980. Social behavior. Pages 409–450 in J. S. Cobb, and B. F. Phillips, Eds. *The Biology and Management of Lobsters,* Vol. 2. Academic Press, New York.

Baisre, J. 2000. The Cuban spiny lobster fishery. Pages 135–152 in B. F. Phillips and J. Kittaka, Eds. *Spiny Lobsters: Fisheries and Culture,* 2nd edition. Fishing News Books, Blackwell Science, Oxford.

Barber, J. S., D. M. Chosid, R. P. Glenn, and K. A. Whitmore. 2009. A systematic model for artificial reef site selection. *N.Z. J. Mar. Freshwat. Res.* 43: 283–297

Barnabé, G. 1995. Suivi de la zone de pêche protégée de Marseillan. *1er rapport. Univ. Montpellier II, Station Méditerranéenne de l'Environmental Littoral.* 1–10.

Barnabé, G., E. Charbonnel, J.-Y. Marinaro, D. Ody, and P. Francour. 2000 Artificial reefs in France: Analysis, assessments and prospects. Pages 167–184 in A. Jensen, K. J. Collins, and A. P. M. Lockwood, Eds. *Artificial Reefs in European Seas.* Kluwer Academic Publishers, Dordrecht, The Netherlands.

Barshaw, D. E. and K. L. Lavalli. 1988. Predation upon postlarval lobsters *Homarus americanus* by cunners *Tautogolabrus adspersus* and mud crabs *Neopanope sayi* on three different substrates. Eelgrass, mud, and rocks. *Mar. Ecol. Prog. Ser.* 48: 119–123.

Barshaw, D. E. and E. Spanier. 1994a. The undiscovered lobster—a first look at the social behavior of the Mediterranean slipper lobster, *Scyllarides latus* Decapoda, Scyllaridae. *Crustaceana* 67(2): 187–197.

Barshaw, D. E. and E. Spanier, 1994b. Anti-predator behaviors of the Mediterranean slipper lobster *Scyllarides latus. Bull. Mar. Sci.* 55(2–3): 375–382.

Barshaw, D. E., K. Able, and K. L. Heck, Jr. 1994. Salt marsh peat reefs as protection for postlarval lobsters *Homarus americanus* from fish and crab predators: Comparisons with other substrates. *Mar. Ecol. Prog. Ser.* 106: 203–206.

Behringer D. C. and M. J. Butler. 2006. Density-dependent population dynamics in juvenile *Panulirus argus* (Latreille): The impact of artificial density enhancement. *J. Exp. Mar. Biol. Ecol.* 27: 84–95.

Behringer D. C., M. J. Butler, and J. D. Shields. 2008. Ecological and physiological effects of PaV1 infection on the Caribbean spiny lobster (*Panulirus argus* Latreille). *J. Exp. Mar. Biol. Ecol.* 359: 26–33.

Behringer, D. C., M. J. Butler IV, W. F. Herrnkind, J. H. Hunt, C. A. Acosta, and W. C. Sharp. 2009. Is seagrass an important nursery habitat for the Caribbean spiny lobster, *Panulirus argus,* in Florida? *N.Z. J. Mar. Freshwat. Res.* 43: 327–337.

Berg, D. 2009. Lobster Diving-How to Shipwreck Diving Guide. www.aquaexplorers.com.

Berry, P. F. 1971. The spiny lobsters (Palinuridae) of the east coast of southern Africa: Distribution and ecological notes. *Invest. Rep. Oceanogr. Res. Inst. S. Afr.* 27: 1–23.

Bertelsen, R. D. and T. R. Matthews. 2001. Fecundity dynamics of female spiny lobster (*Panulirus argus*) in a south Florida fishery and Dry Tortugas National Park lobster sanctuary. *Mar. Freshwat. Res.* 52(8): 1559–1565.

Biggers, W. J. and H. Laufer. 2004. Identification of juvenile hormone-active alkylphenols in the lobster *Homarus americanus* and in marine sediments. *Biol. Bull.* 206: 13–24.

Bohnsack, J. A., Ecklund, A. M., and A. M. Sjumant. 1997. Artificial reef research: Is there more than the attraction-production issue? *Fisheries* 22: 14–16.

Bologna, P. A. X. and R. S. Steneck. 1993. Kelp beds as habitat for American lobster *Homarus americanus. Mar. Ecol. Prog. Ser.* 100: 127–134.

Booth, J. D. 1979. Settlement of the rock lobster, *Jasus edwardsii* (Decapoda: Palinuridae), at Castlepoint, New Zealand. *N.Z. J. Mar. Freshwat. Res.* 13: 395–406.

Booth, J. D. 2006. *Jasus* species. Pages 340–358 in B. F. Phillips, Ed. *Lobsters—Biology, Management, Aquaculture, and Fisheries*, Blackwell Science, Oxford.

Booth, J. D. and J. Kittaka. 2000. Spiny lobster growout. Pages 556–585 in B. F. Phillips and J. Kittaka, Eds. *Spiny Lobsters: Fisheries and Culture,* 2nd edition. Fishing News Books, Blackwell Science, Oxford.

Booth, J. D., A. D. Carruthers, C. D. Bolt, and R. A. Stewart. 1991. Measuring depth of settlement in the red rock lobster, *Jasus edwardsii. N.Z. J. Mar. Freshwat. Res.* 25: 123–132.

Bortone, S. A. 2008. Coupling fisheries with ecology through marine artificial reef deployments. Pages 917–924 in J. Nielsen, J. Dodson, K. Friedland, T. Hamon, J. Musick, and E. Verspoor, Eds. *Reconciling Fisheries with Conservation. Proceedings of the Fourth World Fisheries Congress*, Vol. 49(1). American Fisheries Society.

Bradshaw, M. 2004. A combination of state and market through ITQs in the Tasmanian commercial rock lobster fishery: The tail wagging the dog? *Fish. Res.* 67: 99–109.

Brickhill, M. J., S. Y. Lee, and R. M. Connolly. 2005. Fishes associated with artificial reefs: Attributing changes to attraction or production using novel approaches, *J. Fish. Biol.*: 53–71.

Briggs, P. T. and C. S. Zawacki. 1974. American lobsters at artificial reefs in New York. *N.Y. Fish Game J.* 21: 73–77.

Briones-Fourzán, P. and D. Gutierrez-Carbonell. 1992. Postlarval recruitment of the spiny lobster, *Panulirus argus* (Latreille 1804), in Bahia de la Ascension, *Q.R. Proc. Gulf Caribb. Fish. Inst.* 41: 492–507.

Briones-Fourzán, P. and E. Lozano- Álvarez. 2001. Effects of artificial shelters (casitas) on the abundance and biomass of juvenile spiny lobsters *Panulirus argus* in a habitat-limited tropical reef lagoon. *Mar. Ecol. Prog. Ser.* 221: 221–232.

Briones-Fourzán, P., E. Lozano-Álvarez, and D. Eggleston. 2000. The use of artificial shelters (casitas) for research and harvesting of Caribbean spiny lobsters in Mexico. Pages 420–446 in B. F. Phillips and J. Kittaka, Eds. *Spiny Lobsters: Fisheries and Culture, 2nd edition*. Fishing News Books, Blackwell Science, Oxford.

Briones-Fourzán, P., M. Pérez-Ortiz, and E. Lozano-Álvarez. 2006. Defense mechanisms and antipredator behavior in two sympatric species of spiny lobsters, *Panulirus argus* and *P. guttatus. Mar. Biol.* 149: 227–239.

Briones-Fourzán, P., E. Lozano- Álvarez, F. Negrete-Soto, and C. Barradas-Ortiz. 2007. Enhancement of juvenile Caribbean spiny lobsters: An evaluation of changes in multiple response variables with the addition of large artificial shelters. *Oecologia* 151: 401–416.

Brown, D. E. and L. B. Holthuis. 1998. The Australian species of the genus *Ibacus* (Crustacea: Decapoda: Scyllaridae), with the description of a new species and addition of new records. *Zool. Med. Leiden* 72: 113–141.

Brown, P., R. Leader, S. Jones, and W. Key. 1999. Use of new water-stable feed for culture of the spiny lobster (*Panulirus argus*). *Purdue Exp. Stn. Tech. Pub.* 1: 1–8.

Butler, M. J. IV and W. F. Herrnkind. 1992. Spiny lobster recruitment in south Florida: Quantitative experiments and management implications. *Proc. Gulf Carib. Fish. Inst.* 41: 508–515.

Butler, M. J. IV and W. F. Herrnkind. 1997. A test of recruitment limitation and the potential for artificial enhancement of spiny lobster populations in Florida. *Can. J. Fish. Aquat. Sci.* 54: 452–463.

Butler, M. J. IV, and W. F. Herrnkind. 2000. Puerulus and juvenile ecology. Pages 276–301 in B. F. Phillips and J. Kittaka, Eds. *Spiny Lobsters: Fisheries and Culture,* 2nd edition. Fishing News Books, Blackwell Science, Oxford.

Butler, M. J. IV and J. A. Lear. 2009. Habitat-based intraguild predation by Caribbean reef octopus *Octopus briareus* on juvenile Caribbean spiny lobster *Panulirus argus*. *Mar. Ecol. Prog. Ser.* 386: 115–122.

Butler, M. J. IV, J. H. Hunt, W. F. Herrnkind, M. J. Childress, R. Bertelsen, W. Sharp, T. Matthews, J. M. Field, and H. G. Marshall. 1995. Cascading disturbances in Florida Bay, USA: Cyanobacteria blooms, sponge mortality, and implications for juvenile spiny lobsters, *Panulirus argus*. *Mar. Ecol. Prog. Ser.* 129: 119–125.

Butler, M. J. IV, W. F. Herrnkind, and J. H. Hunt. 1997. Factors affecting the recruitment of juvenile Caribbean spiny lobsters dwelling in macroalgae. *Bull. Mar. Sci.* 61: 3–19.

Butler, M. J., R. S. Steneck, and W. H. Herrnkind. 2006. Juvenile and adult ecology. Pages 263–309 in B. F. Phillips, Ed. *Lobsters—Biology Management, Aquaculture, and Fisheries*. Blackwell Science, Oxford.

Caddy J. F. 2008. The importance of "cover" in the life histories of demersal and benthic marine resources: A neglected issue in fisheries assessment and management. *Bull. Mar. Sci.* 83: 7–52.

Cannas, A., M. Pinna, R. Savarino, E. Secci, and C. Tidu. 1998. Plan for repopulation of the spiny lobster (*Palinurus elephas* Fabr.) in an area of the Gulf of Asinara: Biological, technical, and productive aspects. *Biologia Marina Mediterranea* 5: 1442–1449.

Caputi, N., R. S. Brown, and C. F. Chubb. 1995. Regional prediction of the western rock lobster, *Panulirus cygnus*, commercial catch in Western Australia. *Crustaceana* 68(2): 245–256.

Castro, K. M., J. S. Cobb, R. A. Wahle, and J. Catena. 2001. Habitat addition and stock enhancement for American lobsters, *Homarus americanus*. *Mar. Freshw. Res.* 52(8): 1253–1261.

Childress, M. J. 1997. Marine reserves and their effects on lobster populations: Report from a workshop. *Mar. Freshwat. Res.* 48: 1111–1114.

Childress M. J. and W. F. Herrnkind. 1994. The behavior of juvenile Caribbean spiny lobster in Florida Bay: Seasonality, ontogeny, and sociality. *Bull. Mar. Sci.* 54: 819–827.

Childress, M. J. and W. F. Herrnkind. 1996. The ontogeny of social behaviour among juvenile Caribbean spiny lobsters. *Anim. Behav*. 51: 675–685.

Childress, M. J. and W. F. Herrnkind. 2001. The guide effect influence on the gregariousness of juvenile Caribbean spiny lobsters. *Anim. Behav*. 62: 465–472.

Childress M. J. and S. H. Jury. 2006. Behaviour. Pages 78–112 in B. F. Phillips, Ed. *Lobsters—Biology Management, Aquaculture, and Fisheries.* Blackwell Science, Oxford.

Cobb, J. S. 1971. The shelter-related behavior of the lobster *Homarus americanus*. *Ecology* 52: 108–115.

Cockcroft, A. C. 2001. *Jasus lalandii* "walkouts" or mass strandings in South Africa during the 1990s: An overview. *Mar. Freshwat. Res.* 52(8): 1085–1093.

Cooper, R. A. and J. R. Uzmann. 1980. Ecology of juvenile and adult *Homarus*. Pages 97–142 in J. S. Cobb and B. F. Phillips, Eds. *The Biology and Management of Lobsters*. Academic Press, New York.

Cooper, R. A., R. A. Clifford, and C. D. Newell. 1975. Seasonal abundance of the American lobster, *Homarus americanus*, in the Boothbay region of Maine. *Trans. Am. Fish. Soc.* 104: 669–674.

Cowan, D. F., A. R. Solow, and A. Beet. 2001. Patterns in abundance and growth of juvenile lobster, *Homarus americanus*. *Mar. Freshwat. Res.* 52(8): 1095–1102.

Creswell, R. L. 1984. Increased production of spiny lobsters (*Panulirus argus*) through post harvest impoundment. *Proc. Gulf and Carib. Fish. Inst.* 43: 23–35.

Cruz, R. and B. Phillips. 2000. The artificial shelters (*Pesqueros*) used for the spiny lobster (*Panulirus argus*) fisheries in Cuba. Pages 400–419 in B. F. Phillips and J. Kittaka, Eds. *Spiny Lobsters: Fisheries and Culture, 2nd Edition*. Fishing News Books, Blackwell Science, Oxford.

Cruz, R., R. Brito, R. Diaz, and R. Lalana. 1986. Ecologia de la langosta (*Panulirus argus*) al se de la Isla de la Juventud. I. Colonizacion de arricefes artificiales. *Rev. Invest. Mar.* (Cuba) 7: 3–17.

Cruz, R., M. E. de Leon, E. Diaz, R. Brito, and R. Puga. 1991. Reclutamiento de puerulus de langosta (*Panulirus argus*) a la platforma Cubana. *Rev. Inv. Mar.* 12(1–3): 66–75.

Cruz, R., M. E. de León, and R. Puga. 1995. Prediction of commercial catches of the spiny lobster *Panulirus argus* in the Gulf of Batabanó, Cuba. *Crustaceana* 68 (2): 238–244.

Cruz, R., R. Lalana, M. Báez-Hidalgo, and R. Adriano. 2007. Gregarious behaviour of juveniles of the spiny lobster, *Panulirus argus* (Latreille, 1804) in artificial shelters. *Crustaceana* 80(5): 577–595.

Davis, G. E. 1981. Effects of injuries on spiny lobster, *Panulirus argus*, and implications for fishery management. *Fish. Bull.* 78: 979–984.

Davis, G. E. 1985. Artificial structures to mitigate marina construction impacts on spiny lobster, *Panulirus argus*. *Bull. Mar. Sci.* 37: 151–156.

Edelist, D. and E. Spanier. 2009. Influence of Levantine artificial reefs on the fish assemblage of the surrounding seabed. *Med. Mar. Sci.* 10(1): 35–54.

Eggleston, D. B. and C. P. Dahlgren. 2001. Distribution and abundance of Caribbean spiny lobsters in the Key West National Wildlife Refuge: Relationship to habitat features and impact of an intensive recreational fishery. *Mar. Freshwat. Res.* 52: 1567–1576.

Eggleston, D. B. and R. N. Lipcius. 1992. Shelter selection by spiny lobster under variable predation risk, social conditions, and shelter size. *Ecology* 73(3): 992–1011.

Eggleston, D., R. Lipcius, L. Coba-Centina, and D. Miller. 1990. Shelter scaling regulates survival of juvenile spiny lobster, *Panulirus argus*. *Mar. Ecol. Prog. Ser.* 62: 79–88.

Eggleston, D. B., R. N. Lipcius, and D. L. Miller. 1992. Artificial shelters and survival of juvenile Caribbean spiny lobster *Panulirus argus*—Spatial, habitat, and lobster size effects. *Fish. Bull. U.S.* 90: 691–702.

Ennis, G. P. 1984. Territorial behavior of the American Lobster *Homarus americanus*. *Trans. Am. Fish. Soc.* 113: 330–335.

FAOSTATS, 2006—Fisheries and Aquaculture Information and Statistics Service, *FAO Year Book—2006*. FAO, Rome, 81 pp.

FAOSTATS, 2010—FIGIS-Fisheries and Aquaculture Information and Statistics Service—2007. http://www.fao.org/fishery/statistics/global-production/query/en (July, 2010).

Faulkes, Z. 2006. Digging mechanisms and substrate preferences of shovel nosed lobsters, *Ibacus peronii* (Decapoda: Scyllaridae). *J. Crustac. Biol.* 26: 69–72.

Fishery Civil Engineering Study Association. 1982. Materials traditionally used in artificial reef design and construction by the national government, prefecture governments, and fishing industry. Pages 16–22 in S. F. Vik, Ed. *Japanese Artificial Reef Technology: Translations of Selected Recent Japanese Literature and An Evaluation of Potential Applications in the United States*. Aquabio Technical Rep. 604, Aquabio, Inc., Arlington, MA.

Fogarty, M. J. and J. S. Iodine. 1986. Recruitment dynamics in an American lobster (*Homarus americanus*) population. *Can. J. Fish. Aquat. Sci.* 43: 2368–2376.

Follesa, M. C., D. Cuccu, R. Cannas, S. Cabiddu, M. Murenu, A. Sabatinini, and A. Cau. 2008. Effects of marine reserve protection on spiny lobster (*Palinurus elephas* Fabr., 1787) in a central western Mediterranean area. *Hydrobiologia* 606: 63–68.

Garnick, E. 1989. Lobster *(Homarus americanus)* population declines, sea urchins, and "barren grounds": A space-mediated competition hypothesis. *Mar. Ecol. Prog. Ser.* 58: 23–28.

Goñi, R., O. Renones, and A. Quetglas. 2001. Dynamics of a protected Western Mediterranean population of the European spiny lobster *Palinurus elephas* (Fabricius, 1787) assessed by trap surveys. *Mar. Freshwat. Res.* 52: 1577–1587.

Goñi, R., A. Quetglas, and O. Renones. 2006. Spillover of spiny lobsters *Palinurus elephas* from a marine reserve to an adjoining fishery. *Mar. Ecol. Prog. Ser.* 306: 207–219.

Gristina M., F. Fiorentino, G. Garofalo, and F. Badalamenti. 2009. Shelter preference in captive juveniles of European spiny lobster *Palinurus elephas* (Fabricius, 1787). *Mar. Biol.* 156: 2097–2105.

Groeneveld, J. C., R. Goni, and D. Latrouite. 2006. *Palinurus* species. Pages 385–411 in B. F. Phillips, Ed. *Lobsters—Biology, Management, Aquaculture, and Fisheries*, Blackwell Science, Oxford.

Guiterrez-Carbonell, D., J. Simonin-Diaz, and P. Briones-Fourzán. 1992. A simple collector for postlarvae of the spiny lobster *Panulirus argus*. *Proc. Gulf Caribb. Fish. Inst.* 41: 516–527.

Haddy, J. A., J. Stewart, and S. K. J. Graham. 2007. Fishery and biology of commercially exploited Australian fan lobsters (*Ibacus* spp.). Pages 359–375 in K. L. Lavalli and E. Spanier, Eds. *The Biology and Fisheries of Slipper Lobsters*. Crustacean Issues 17. CRC Press (Taylor & Francis Group), New York.

Harding, G. C. 1992. American lobster (*Homarus americanus* Milne Edwards): A discussion paper on their environmental requirements and known anthropogenic effects on their populations. *Can. Tech. Rep. Fish. Aquat. Sci.* 1887.

Hearn A., 2006. Life history of the slipper lobster Scyllarides astori Holthuis 1960, in the Galapagos Islands, Ecuador. *J. Exp. Mar. Biol. Ecol.* 328: 87–97.

Hearn A., V. Toral-Granda, C. Martinez, and G. Reck. 2007. Biology and fishery of the Galăpagos slipper lobster. Pages 287–308 in K. L. Lavalli and E. Spanier, Eds. *The Biology and Fisheries of Slipper Lobsters.* Crustacean Issues 17. CRC Press (Taylor & Francis Group), New York.

Heck, K. L., Jr., K. W. Able, M. P. Fahay, and C. T. Roman. 1989. Fishes and decapod crustaceans of Cape Cod eelgrass meadows: Species composition, seasonal abundance patterns, and comparisons with unvegetated substrates. *Estuaries* 12: 59–65.

Herrnkind, W. F. 1980. Spiny lobsters: Patterns of movement. Pages 349–407 in J. S. Cobb and B. F. Phillips, Eds. *The Biology and Management of Lobsters*, Vol. I. Academic Press, New York.

Herrnkind, W. F. and M. J. Butler, IV. 1986. Factors regulating postlarval settlement and juvenile microhabitat use by spiny lobster *Panulirus argus*. *Mar. Ecol. Prog. Ser.* 34: 23–30.

Herrnkind, W. F. and J. S. Cobb. 2008. Artificial shelters for clawed and spiny lobsters: A critical review of enhancement efforts. *Amer. Fish. Soc. Symp.* 49: 925–932.

Herrnkind, W. F., M. J. Butler, IV, J. H. Hunt, and M. J. Childress. 1997a. Role of physical refugia: Implication from a mass sponge die-off in a lobster nursery in Florida. *Mar. Fresh. Res.* 48: 759–769.

Herrnkind, W. F., M. J. Butler, IV, and J. H. Hunt. 1997b. Can artificial habitats that mimic natural structures enhance recruitment of Caribbean spiny lobster? *Fisheries* 22: 24–27.

Herrnkind, W. F., M. J. Butler, IV, and J. H. Hunt. 1999. A case for shelter replacement in a disturbed spiny lobster nursery in Florida: Why basic research had to come first. *Amer. Fish. Soc. Symp.* 22: 421–437.

Herrnkind, W. F, M. K. Childress, and K. L. Lavalli. 2001. Cooperative defense and other benefits among exposed spiny lobsters: Inferences from group size and behaviour. *Mar. Freshwat. Res.* 52: 1113–1124.

Hirata, K., T. Shinoda, and Y. Nakayama. 1988. Communities developed on the floating artificial reef for the puerulus larvae of the Japanese spiny lobster in Nayaura Inlet facing the sea of Kumano. *Mar. Fouling* 7: 49–62. (in Japanese)

Holthuis, L. B. 1991. *Marine Lobsters of the World. An Annotated and Illustrated Catalogue of the Species of Interest to Fisheries Known to Date.* FAO Species Catalogue No. 125, Vol. 13: 1–292.

Holthuis, L. B. 2002. The Indo-Pacific scyllarinid lobsters (Crustacea, Decapoda, Scyllaridae). *Zoosystema* 24: 499–683.

Howard A. E. 1980. Substrate controls on the size composition of lobster (*Homarus gammarus*) populations. *ICES J. Mar. Sci.* 39(2): 130–133.

Hruby, T. 2009. Experimental lobster ranching in Massachusetts. *J. World Aquacult. Soc.* 10: 194–202.

Hudon, C. 1987. Ecology and growth of postlarval and juvenile lobster, *Homarus americanus*, off Îles de la Madeleine (Québec). *Can. J. Fish. Aquat. Sci.* 44: 1855–1869.

Hung, L. V. and L. A. Tuan. 2009. Lobster seacage culture in Vietnam. Pages 10–17 in K. C. Williams, Ed. *Spiny Lobster Aquaculture in the Asia-Pacific Region*. ACIAR, Canberra, Australia.

Hunt, J. H. 2000. Status of the fishery for *Panulirus argus* in Florida. Pages 189–199 in B. F. Phillips and J. Kittaka, Eds. *Spiny lobsters: Fisheries and culture. Second edition.* Fishing News Books, Blackwell Science, Oxford.

Hunter, W. R. and M. D. J. Sayer. 2009. The comparative effects of habitat complexity on faunal assemblages of northern temperate artificial and natural reefs. *ICES J. Mar. Sci.* 66: 691–698.

Incze L. S., R. A. Wahle, and J. S. Cobb. 1997. Quantitative relationships between postlarval production and benthic recruitment in lobsters, *Homarus americanus*. *Mar. Freshw. Res.* 48: 729–743.

Jeffs, A. G. and P. James. 2001. Sea-cage culture of spiny lobster *Jasus edwardsii* in New Zealand. *Mar. Freshwat. Res.* 52: 1419–1424.

Jensen, A. C. 2002. Artificial reefs of Europe: Perspective and future. *ICES J. Mar. Sci.* 59: S3–S13.

Jensen, A. C. and K. L. Collins. 1995. The Poole Bay artificial reef project 1989 to 1994. *Biol. Mar. Medit.* 2(1): 111–122.

Jensen, A. C., K. J. Collins, E. K. Free, and R. C. A. Bannister. 1994. Lobster (*Homarus gammarus*) movement on an artificial reef: The potential use of artificial reefs for stock enhancement. *Crustaceana* 67: 198–211.

Jensen A. C., J. A. Wickins, and R. C. A. Bannister. 2000a. The potential use of artificial reefs to enhance lobster habitat. Pages 379–401 in A. C. Jensen, Ed. *Artificial reefs in European Seas*. Kluwer Academic Publishers, Dordrecht, The Netherlands.

Jensen, A. C., K. J. Collins, and I. P. Smith. 2000b. The Poole Bay artificial reef project. Pages 263–287 in A. C. Jensen, K. J. Collins, and A. P. M. Lockwood, Eds. *Artificial reefs in European Seas*. Kluwer Academic Publishers, Dordrecht, The Netherlands.

Jernakoff, P. 1990. Distribution of newly settled western rock lobsters, *Panulirus cygnus*, *Mar. Ecol. Prog. Ser.* 66: 63–74.

Jones, C. 1988. The Biology and Behaviour of Bay Lobsters, *Thenus* spp. (Decapoda: Scyllaridae) in Northern Queensland, Australia. Ph.D. thesis, University of Queensland, Brisbane.

Jones, C. 2007. Biology and fishery of the Bay Lobster, *Thenus* spp. Pages 325–258 in K. L. Lavalli and E. Spanier, Eds. *The Biology and Fisheries of Slipper Lobsters*. Crustacean Issues 17. CRC Press (Taylor & Francis Group), New York.

Kacelnik, A., J. R Krebs, and C. Bernstein. 1992. The ideal free distribution and predator-prey populations. *TREE.* 7 (2): 50–55.

Kanciruk, P. 1980. Ecology of juvenile and adult Palinuridae (spiny lobsters). Pages 59–96 in J. S. Cobb and B. F. Phillips, Eds. *The Biology and Management of Lobsters*, Vol. II. Academic Press, New York.

Kanciruk, P. and W. F. Herrnkind. 1973. Daily and seasonal locomotor rhythms in spiny lobster, *Panulirus argus*. *Mar. Behav. Physiol.* 1: 351–359.

Kanciruk, P. and W. F. Herrnkind. 1976. Autumnal reproduction in *Panulirus argus* at Bimini, Bahamas. *Bull. Mar. Sci.* 26: 417–432.

Kensler, C. B. 1966. Ecological notes on the marine crayfish *Jasus edwardsii* (Hutton): Puerulus and post-puerulus stages. *N.Z. J. Mar. Sci. Newsl.* 8: 32–34.

Lavalli, K. L. and D. E. Barshaw. 1986. Burrows protect postlarval lobsters *Homarus americanus* from predation by the non-burrowing cunner *Tautogolabrus adspersus* but not from the burrowing mud crab *Neopanope texani*. *Mar. Ecol. Prog. Ser.* 32: 13–16.

Lavalli, K. L. and W. F. Herrnkind. 2009. Defensive strategies of Caribbean spiny lobsters: Effects of group size and predator group size. *N.Z. J. Mar. Freshwat. Res.* 43(1): 15–28.

Lavalli, K. L., E. Spanier, and F. Grasso. 2007. Behavior and sensory biology of slipper lobsters. Pages 133–181 in K. L. Lavalli and E. Spanier, Eds., *The Biology and Fisheries of Slipper Lobsters*. Crustacean Issues 17. CRC Press (Taylor & Francis Group), New York.

Lawton, P. and K. L. Lavalli. 1995. Postlarval, juvenile, and adult ecology. Pages 47–88 in J. R. Factor, Ed. *The Biology of the Lobster Homarus americanus*. Academic Press, New York.

Lawton, P. and D. A. Robichaud. 1992. Lobster habitat ecology research in the Bay of Funday. Pages 53–56 in *Science Review of the Bedford Institute of Oceanography, the Halifax Fisheries Research Laboratory, and the St. Andrews Biological Station, 1990 & 1991.* Dept. Fish. Oceans, Dartmouth, Nova Scotia, Canada.

Lewis, R. K. 1977. Studies on the puerulus and post-puerulus stages of southern rock lobster (*Jasus novaehollandiae* Holthuis) in the south eastern region of South Australia. Pages 36–37 in J. S. Cobb and B. F. Phillips, Eds. *Workshop on Lobster and Rock Lobster Ecology and Physiology. Div. Fish. Oceanogr. Circ.* 7.

Li, C., J. D. Shields, R. Ratzlaff, and M. J. Butler. 2008. Pathology and hematology of the Caribbean spiny lobster experimentally infected with *Panulirus argus* virus 1 (PaV1). *Virus Res.* 132: 104–113.

Lipcius, R. N. and W. F. Herrnkind. 1987. Control and coordination of reproduction and molting in the spiny lobster *Panulirus argus*. *Mar. Biol.* 96: 207–214.

Lipcius, R. N. and D. B. Eggleston. 2000. Introduction: Ecology and fishery biology of spiny lobsters. Pages 1–41 in B. F. Phillips, J. S. Cobb, and J. Kittaka, Eds. *Spiny Lobster Management.* Fishing News Books, Blackwell Science, Oxford.

Long, N. V. and D. T. Hoc. 2009. Census of lobster seed captured from the central coastal waters of Vietnam for aquaculture grow-out, 2005–2008. Pages 52–58 in K. C. Williams, Ed. *Spiny Lobster Aquaculture in the Asia-Pacific Region*. ACIAR, Canberra, Australia.

Lozano-Álvarez, E. 1996. Ongrowing of juvenile spiny lobsters, *Panulirus argus* (Latreille, 1804) (Decapoda, Paliunridae), in portable sea enclosures. *Crustaceana* 69: 958–972.

Lozano-Álvarez, E. and P. Briones-Fourzán. 2001. Den choice and occupation patterns of shelters by two sympatric lobster species, *Panulirus argus* and *Panulirus guttatus*, under experimental conditions. *Mar. Freshwat. Res.* 52: 1145–1155.

Lozano-Álvarez, E. and E. Spanier. 1997. Behavior and growth of captive sub-adults spiny lobsters, *Panulirus argus,* under the risk of predation. *Mar. Freshwat. Res.* 48(8): 707–713.

Lozano-Álvarez, E., P. Briones-Fourzan, and F. Negrete-Soto. 1994. An evaluation of concrete block structures as shelter for juvenile Carribean spiny lobsters, *Panulirus argus*. *Bull. Mar. Sci.* 55: 351–362.

MacDiarmid, A. B. 1994. Cohabitation in the spiny lobster *Jasus edwardsii* (Hutton, 1875). *Crustaceana* 66: 341–355.

Martínez, C. E., M. V. Toral-Granda, and G. J. Edgar. 2002. Langostino. Pages 216–232 in E. Danulat and G. J. Edgar, Eds. *Reserva Marina de Galápagos, Línea Base de la Biodiversidad.* Fundación Charles Darwin and Servicio Parque Nacional Galápagos, Santa Cruz, Galápagos, Ecuador.

Martins, H. R. 1985. Biological studies of the exploited stock of the Mediterranean locust lobster *Scyllarides latus* (Latreille, 1803) (Decapoda: Scyllaridae) in the Azores. *J. Crust. Biol.*, 5(2): 294–305.

Marx, J. 1986. Settlement of spiny lobster, *Panulirus argus*, pueruli in south Florida: An evaluation from two perspectives. *Can. J. Fish. Aquat. Sci.* 43: 2221–2227.

Marx, J. M. and W. F. Herrnkind. 1985. Macroalgae (Rhodophyta: Laurencia spp.) as habitat for young juvenile spiny lobsters, *Panulirus argus*. *Bull. Mar. Sci.* 36: 423–431.

Matsuoka, T., T. Nakashima, and N. Nagasawa. 2005. A review of ghost fishing: Scientific approaches to evaluation and solutions. *Fish. Sci.* 71: 691–702

Mikami, S. and A. V. Kuballa. 2007. Factors important in larval and postlarval molting, growth, and rearing. Pages 91–110 in K. L. Lavalli and E. Spanier, Eds., *The Biology and Fisheries of Slipper Lobsters.* Crustacean Issues 17. CRC Press (Taylor & Francis Group), New York.

Miller, R. J., G. J. Sharp, and E. M. O'Brien. 2006. Laboratory experiments on artificial reefs for American lobsters. *J. Crust. Biol.* 26(4): 621–627.

Mintz, J. D., R. N. Lipcius, D. B. Eggleston, and M. S. Seebo. 1994. Survival of juvenile Caribbean spiny lobster: Effects of shelter size, geographic location, and conspecific abundance. *Mar. Ecol. Prog. Ser.* 112: 255–266.

National Research Council. 1988. *Fisheries Technologies for Developing Countries.* Office of International Affairs, National Research Council. National Academy Press, Washington, D.C.

Nevitt, G., N. D. Pentcheff, K. J. Lohmann, and R. K. Zimmer. 2000. Den selection by the spiny lobster *Panulirus argus*: Testing attraction to conspecific odors in the field. *Mar. Ecol. Prog. Ser.* 203: 225–231

Nicosia, F. and K. Lavalli. 1999. Homarid lobster hatcheries: Their history in research, management, and aquaculture. *Mar. Fish. Rev.* 61(2): 1–57.

Nonaka, M., H. Fushimi, and T. Yamakawa. 2000. The spiny lobster fishery in Japan and restocking. Pages 221–242 in B. F. Phillips and J. Kittaka, Eds. *Spiny Lobsters: Fisheries and Culture,* 2nd edition. Fishing News Books, Blackwell Science, Oxford.

Oliver, M. D., A. B. MacDiarmid, R. A. Stewart, and C. Gardner. 2008. Anti-predator behavior of captive-reared and wild juvenile spiny lobster (*Jasus edwardsii*). *Rev. Fish. Sci.* 16(1): 186–194.

Osenberg, C. W., C. M. St. Mary, J. A. Wilson, and W. J. Lindberg. 2002. A quantitative framework to evaluate the attraction–production controversy. *ICES J. Mar. Sci.* 59: 214–221.

Oshima, Y. 1964. *Artificial Reefs*. Fishery Propag. Ser. 8. Japan Fishery Resources Protective Association.

Palma, A. T., R. A. Wahle, and R. S. Steneck. 1998. Different early post-settlement strategies between American lobsters *Homarus americanus* and rock crabs *Cancer irroratus* in the Gulf of Maine. *Mar. Ecol. Prog Ser.* 162: 215–225.

Palma, A. T., R. S. Steneck, and C. J. Wilson. 1999. Settlement-driven, multiscale demographic patterns of large benthic decapods in the Gulf of Maine. *J. Exp. Mar. Biol. Ecol.* 241: 107–136.

Pearce, J. and Balcom, N. 2005. The 1999 Long Island Sound lobster mortality event: Findings of the comprehensive research initiative. *J. Shellfish Res.* 24(3): 691–697.

Pearce, F. and B. F. Phillips. 1988. ENSO events, the Leeuwin Current, and larval recruitment of the western rock lobster, *Journal du Conseil international pour l'Exploration de la Mer.* 45: 13–21.

Pessani, D. and M. Mura. 2007. The biology of the Mediterranean scyllarids. Pages 263–286 in K. L. Lavalli and E. Spanier, Eds. *The Biology and Fisheries of Slipper Lobsters*. Crustacean Issues 17. CRC Press (Taylor & Francis Group), New York.

Pettersen, A. R., E. Moland, E. M. Olsen, and J. A. Knutsen. 2009. Lobster reserves in coastal Norway: The process towards establishment. Pages 178–188 in E. Dahl, E. Moksness, and J. Støttrup, Eds. *Integrated Coastal Zone Management.* Wiley-Blackwell, NJ.

Phillips, B. F. 1972. A semi-quantitative collector of the puerulus larvae of the western rock lobster *Panulirus longipes cygnus* George (Decapoda, Palinuridae). *Crustaceana* 22(2): 147–154.

Phillips, B. F. 1975. Effect of nocturnal illumination on catches of the puerulus larvae of the western rock lobster by collectors composed of artificial seaweed. *Aust. J. Mar. Freshwat. Res.* 26: 411–414.

Phillips, B. F. 1986. Prediction of commercial catches of the western rock lobster *Panulirus cygnus*. *Can. J. Fish. Aquat. Sci.* 43: 2126–2130.

Phillips, B. F. and J. D. Booth. 1994. Design, use, and effectiveness of collectors for catching the puerulus stage of spiny lobsters. *Rev. Fish. Sci.* 2(3): 255–289.

Phillips, B. F. and R. Melville-Smith. 2006. *Panulirus* species. Pages 359–384 in B. F. Phillips, Ed. *Lobsters—Biology, Management, Aquaculture, and Fisheries*. Blackwell Science, Oxford.

Phillips B. F., R. Melville-Smith, Y. W. Cheng, and M. Rossbach. 2001. Testing collector designs for commercial harvesting of western rock lobster (*Panulirus cygnus*) puerulus. *Mar. Freshwat. Res.* 52(8): 1465–1473.

Polovina, J. J. 1989. Artificial reefs: Nothing more than benthic fish aggregators. *CalCOFI Rep.* 30: 37–39

Polovina, J. J. 1991. Fisheries applications and biological impacts of artificial habitats. Pages 153–176 in W. Seaman, Jr. and L. M. Sprague, Ed. *Artificial Habitats for Marine and Freshwater Fisheries*. Academic Press, New York.

Polovina, J. J., W. R. Haight, R. B. Moffitt, and F. A. Parrish. 1995. The role of benthic habitat, oceanography, and fishing on the population dynamics of the spiny lobster, *Panulirus marginatus* (Decapoda, Palinuridae), in the Hawaiian Archipelago. *Crustaceana* 68(2): 203–212.

Pratt, R. M. and J. Schlee. 1969. Glaciation on the continental margin off New England. *Geol. Soc. Amer. Bull.* 80: 2335–2342.

Radhakrishnan, E. V., M. K. Manisser, and V. D. Deshmukh. 2007. Biology and fishery of the slipper lobster, *Thenus orientalis*, in India. Pages 309–324 in K. L. Lavalli and E. Spanier, Eds. *The Biology and Fisheries of Slipper Lobsters*. Crustacean Issues 17. CRC Press (Taylor & Francis Group), New York.

Ratchford, S. G. and D. B. Eggleston. 1998. Size- and scale-dependent chemical attraction contribute to an ontogenetic shift in sociality. *Anim. Behav.* 56: 1027–1034.

Relini, G. 2000. The Loano artificial reef. Pages 129–150 in A. Jensen, K. J. Collins, and A. P. M. Lockwood, Eds. *Artificial reefs in European Seas*. Kluwer Academic Publishers, Dordrecht, The Netherlands.

Relini, G., M. Relini, G. Palandri, S. Merello, E. Beccornia. 2007. History, ecology, and trends for artificial reefs of the Ligurian Sea, Italy. *Hydrobiologia* 580: 193–217.

Richards, R. A. and J. S. Cobb. 1986. Competition for shelter between lobsters (*Homarus americanus*) and Jonah crabs (*Cancer borealis*): Effects of relative size. *Can. J. Fish. Aquat. Sci.* 43: 2250–2255.

Robertson, D. N. and M. J. Butler, IV. 2003. Growth and size at maturity in the spotted spiny lobster, *Panulirus guttatus*. *J. Crust. Biol.* 23(2): 265–272.

Rosenzweig, M. L. 1981. A theory of habitat selection. *Ecology* 62(2): 327–335.

Sahoo, D. and M. Ohno. 2000. Rebuilding the ocean floor—Construction of artificial reefs around the Japanese coast. *Curr. Sci.* 78(3): 228–230.

Sayer, M. D. J. and T. A. Wilding. 2002. Planning, licensing, and stakeholder consultation in an artificial reef development: The Loch Linnhe reef, a case study. *ICES J. Mar. Sci.* 59: S178–S185.

Scarratt, D. J. 1968. An artificial reef for lobsters (*Homarus americanus*). J. *Fish. Res. Bd. Can.* 25: 2683–2690.

Scarratt, D. J. 1973. Lobster populations on a man made rocky reef. *ICES Res. Doc. C.M.* 1973/K:47 Copenhagen.

Seaman, W. C. 2000. *Artificial Reef Evaluation with Application to Natural Marine Habitats*. CRC Press, Boca Raton, 246 pp.

Sekiguchi, H., J. D. Booth, and W. R. Webber. 2007. Early life histories of slipper lobsters. Pages 69–90 in K. L. Lavalli and E. Spanier, Eds. *The Biology and Fisheries of Slipper Lobsters*. Crustacean Issues 17. CRC Press (Taylor & Francis Group), New York.

Serfling, S. A. and R. F. Ford. 1975. Ecological studies of the puerulus larval stage of the California spiny lobster, *Panulirus interruptus*. *Fish. Bull. (U.S.)* 73: 360–377.

Sharp, W. C., J. H. Hunt, and W. G. Lyons. 1997. Life history of the spotted spiny lobster, *Panulirus guttatus*, an obligate reef-dweller. *Mar. Freshwat. Res.* 48(8): 687–698.

Sharp, W. C., Hunt, J. H. and W. H. Teehan. 2007. Observations on the ecology of *Scyllarides aequinoctialis*, *Scyllarides nodifer*, and *Parribacus antarcticus* and a description of Florida scyllarid lobster fishery. Pages 231–242 in K. L. Lavalli and E. Spanier, Eds. *The Biology and Fisheries of Slipper Lobsters*. Crustacean Issues 17. CRC Press (Taylor & Francis Group), New York.

Sheehy, D. J. 1976. Utilization of artificial shelters by the American lobster (*Homarus americanus*). *J. Fish. Res. Board Can.* 33: 1615–1622.

Sheehy, D. J. 1977. A Study of Artificial Reefs From Unit Shelters Designed for the American lobster (*Homants americanus*). Ph.D. dissertation, University of Rhode Island.

Shields, J. D. and D. C. Behringer. 2004. A pathogenic herpes-like virus in the Caribbean spiny lobster, *Panulirus argus*, from the Florida Key. *Dis. Aquat. Org.* 59: 109–118.

Sinis, A. I., C. C. Chintiroglou, and K. I. Stergiou. 2000. Preliminary results from the establishment of experimental artificial reefs in the N. Aegean Sea (Chalkidiki, Greece). *Belg. J. Zool.* 130 (Supplement 1): 139–143.

Smith, K. N. and W. F. Herrnkind. 1992. Predation on juvenile spiny lobsters, *Panulirus argus*: Influence of size, shelter, and activity period. *J. Exp. Mar. Biol. Ecol.* 157: 3–18.

Smith, I. P., K. J. Collins, and A. C. Jensen. 1999. Seasonal changes in the level and diel pattern of activity in the European lobster, *Homarus gammarus* (L.). *Mar. Ecol. Prog. Ser.* 186: 255–264.

Spanier, E. 1991. Artificial reefs to insure protection of the adult Mediterranean slipper lobster, *Scyllarides latus* (Latreille, 1803). Pages 179–185 in C. F. Boudouresque, M. Avon, and V. Gravez, Eds. *Les Espéces Marines á Protéger en Méditerranée*. GIS Posidonia Publ., France.

Spanier, E. 1994. What are the characteristics of a good artificial reef for lobsters? *Crustaceana* 67(2): 173–186.

Spanier, E. and G. Almog-Shtayer. 1992. Shelter preferences in the Mediterranean slipper lobster: Effects of physical properties. *J. Exp. Mar. Biol. Ecol.* 164: 103–116.

Spanier, E. and K. L. Lavalli. 1998. Natural history of *Scyllarides latus* (Crustacea Decapoda): A review of the contemporary biological knowledge of the Mediterranean slipper lobster. *J. Nat. Hist.* 32(6): 1769–1786.

Spanier, E. and K. L. Lavalli. 2007. Slipper lobster fisheries—Present status and future perspectives. Pages 377–391 in K. L. Lavalli and E. Spanier, Eds. *The Biology and Fisheries of Slipper Lobsters*. Crustacean Issues 17. CRC Press (Taylor & Francis Group), New York.

Spanier, E. and R. K. Zimmer-Faust. 1988. Some physical properties of shelter that influence den preference in spiny lobsters. *J. Exp. Mar. Biol. Ecol.* 122: 137–149.

Spanier, E., M. Tom, S. Pisanty, and G. Almog. 1988. Seasonality and shelter selection by the slipper lobster *Scyllarides latus* in the southeastern Mediterranean. *Mar. Ecol. Prog. Ser.* 42: 247–255.

Spanier, E., M. Tom, S. Pisanty, and G. Almog-Shtayer. 1990. Artificial reefs in the low productive marine environment of the southeastern Mediterranean. *P.S.Z.N.I. Mar. Ecol.* 11(1): 61–75.

Spanier, E., D. Weihs, and G. Almog-Shtayer. 1991. Swimming of the Mediterranean slipper lobster. *J. Exp. Mar. Biol. Ecol.* 145: 15–31.

Spanier, E., G. Almog-Shtayer, and U. Fiedler. 1993. The Mediterranean slipper lobster *Scyllarides latus*, the known and the unknown. *BIOS* 1(1): 49–58.

Steneck, R. 2006. Possible demographic consequences of intraspecific shelter competition among American lobsters. *J. Crust. Biol.* 26(4): 628–638.

Stockhusen, W. T. and R. N. Lipcius. 2001. Single large or several small marine reserves for the Caribbean spiny lobster? *Mar. Freshwat. Res.* 52: 1605–1614.

Svåsand, T. 2007. Why juvenile quality and release strategies are important factors for success in stock enhancement and sea ranching. Pages 59–70 in K. M. Leber, S. Kitada, H. L. Blankenship, and T. Svåsand, Eds. *Stock Enhancement and Sea Ranching*, 2nd edition. Fishing News Books, Blackwell Science, Oxford.

Tholasilingam, T. and K. Rangarajan. 1986. Prospects on spiny lobster, *Panulirus* spp. culture in the east coast of India. *Proc. Symp. Coastal Aquacult.* 4: 1171–1175.

Thomas, M. L. H. 1968. Overwintering of American lobster, *Homarus americanus* in burrows in Bideford River, Prince Edward Island. J. *Fish. Res. Bd. Can.* 25: 2725–2727.

Tlusty, M. F., R. M. Smolowitz, H. O. Halvorson, and S. E. DeVito. 2007. Host susceptibility hypothesis for shell disease in American lobsters. *J. Aquat. Anim. Health* 19: 215–225.

Tocci, C. 1996. Artificial reefs fields standing an obstacle against fraudulent trawling. *J. Rech. Oceanogr. Paris* 21: 42–44.

Tsumura, K., H. Kakimoto, and M. Noda. 1999. The history and future of artificial reefs and related aquatic habitats in Japan. *Proceedings of the Seventh International Conference on Artificial Reefs (7th CARAH)*: 264–271.

Tuck, G. N. and H. P. Possingham. 2000. Marine protected areas for spatially structured exploited stocks. *Mar. Ecol. Prog. Ser.* 192: 89–101.

van der Meeren, G. I. 2003. The Potential of Ecological Studies to Improve on the Survival of Cultivated and Released Aquatic Organisms. Ph.D. dissertation, University of Bergen, Norway.

Vogan, C. L., A. Powell, and A. F. Rowley. 2008. Shell disease in crustaceans—just chitin recycling gone wrong? *Environ. Microbiol.* 10(4): 826–835.

Wahle, R. A. 1992a. Substratum constraints on body size and the behavioral scope of shelter use in the American lobster. *J. Exp. Mar. Biol. Ecol.* 159: 59–75.

Wahle, R. A. 1992b. Body-size dependent anti-predator mechanisms of the American lobster. *Oikos* 65: 52–60.

Wahle, R. A. 2003. Revealing the stock-recruitment relationship in lobsters and crabs: Is experimental ecology the key? *Fish. Res.* 65: 3–32.

Wahle R. A. and R. S. Steneck. 1991. Nursery grounds of the American lobster *Homarus americanus*: A demographic bottleneck? *Mar. Ecol. Prog. Ser.* 69: 231–243.

Wahle R. A. and R. S. Steneck. 1992. Habitat restrictions in early benthic life: Experiments on habitat selection and *in situ* predation with the American lobster. *J. Exp. Mar. Biol. Ecol.* 157: 91–114.

Wahle, R. A. and L. S. Incze. 1997. Pre- and post-settlement processes in recruitment of the American lobster. *J. Exp. Mar. Biol. Ecol.* 217: 179–207.

Wahle, R. A., C. Wilson, M. Parkhurst, and C. E. Bergeron. 2009. A vessel-deployed passive postlarval collector to assess settlement of the American lobster *Homarus americanus. N.Z. J. Mar. Freshwat. Res.* 43: 465–474.

Wallace D. F. 2004. Consider the lobster. *Gourmet*, August 2004, 7 pp.

Webber, W. R. and J. D. Booth. 2007. Taxonomy and evolution. Pages 3–21 in K. L. Lavalli and E. Spanier, Eds. *The Biology and Fisheries of Slipper Lobsters*. Crustacean Issues 17. CRC Press (Taylor & Francis Group), New York.

Weiss, H. M., E. Lozano-Álvarez, and P. Briones-Fourzán. 2008. Circadian shelter occupancy patterns and predator-prey interactions of juvenile Caribbean spiny lobsters in a reef lagoon. *Mar. Biol.* 153: 953–963.

Werz, B. E. J. S. 2007. A suggested blueprint for the development of maritime archaeological research in Namibia. *J. Namib. Stud.* 2 (2007): 103–121.

Wilding, T. A. and M. D. J. Sayer. 2002. Evaluating artificial reef performance: Approaches to pre- and post-deployment research. *ICES J. Mar. Sci.* 59: S222–S230.

Williams, K. C. (Ed.), 2009. Spiny lobster aquaculture in the Asia-Pacific region. *Proceedings of an International Symposium,* Nha Trang, Vietnam, December 9–10 2008. ACIAR Proceedings No. 132, Australian Centre for International Agriculture Research, Canberra, 162 pp.

Wilson, C. 1999. Bathymetric and Spatial Patterns of Settlement in American Lobster, *Homarus americanus*, in the Gulf of Maine: Insights into Processes Controlling Abundance. Unpublished MSc thesis, University of Maine, Orono, Maine, United States.

Witham, B., R. M. Ingle, and E. A. Joyce, Jr. 1968. Physiological and ecological studies of *Panulirus argus* from the St. Lucie estuary. *St. Flor. Bd. Conserv. Tech. Ser.* 53: 1–31.

Zimmer-Faust, R. K. and E. Spanier. 1987. Gregariousness and sociality in spiny lobsters: Implication for den habitation. *J. Exp. Mar. Biol. Ecol.*105: 51–71.

CHAPTER 7

A Case Study of Artificial Reefs in Fisheries Management

Enhancement of Sandfish, Arctoscopus japonicus, *by Artificial Reefs in the Eastern Waters of Korea*

Chang Gil Kim, Sung Il Lee, Hyung Kee Cha, Jae Hyeong Yang, and Yong Soo Son

CONTENTS

ABSTRACT

This chapter describes the enhancement of sandfish (*Arctoscopus japonicus*) stocks by development of artificial reef in the eastern waters of Korea. It is a prime example of the application of artificial reefs in the institutional management of a fishery. The chapter focuses on two areas: analysis of sandfish stocks and the subsequent development of artificial reefs for the enhancement of its resources. Approximately 25,000 tons of sandfish were caught in Korea in 1971. However, the catch decreased to less than 4,000 tons in 2008. Reckless fishing operations in shallow waters reduced the number of incoming mature fish recruited into the spawning grounds where sargassum (gulfweed) grows, and also damaged the spawning grounds. Artificial reef structures to enhance sandfish spawning were of two types: Egg collectors and weight of block type. Selection of egg collector material was made based upon the comparative survey of gulfweed and polypropylene (PP) rope and polyethylene (PE) fishing net. The rope egg collector was modeled after the shape of gulfweed. Rope was cut into 40 cm long pieces and separated into strands. Strands were put between two strand-twisted ropes in the shape of brush for a test tube. The size of each brush-type egg collector was D 0.4 × L 1.5 m. Results of a comparative survey show that egg lumps of sandfish were attached in greater numbers on gulfweed than on PP and PE egg collectors. Results also varied according to the species of gulfweed. More mature fish were found in *Sargassum fulvellum* than *Sargassum horneri*. Based on these results, artificial reefs for sandfish were built of concrete blocks (L 0.5 × W 0.3 × H 0.1 m) with transplanted *S. fulvellum* on the upper surface. Sixty artificial reefs were installed in a depth of about 9 m in November 2008. The reefs currently play an important role in the enhancement of sandfish stocks.

Keywords: Sandfish, artificial reef, spawning grounds, egg collector, gulfweed, brush-type egg collectors.

INTRODUCTION

Sandfish is the common name for members for the Trichodontidae: *Arctoscopus japonicus* is the scientific name for the species of sandfish that is the subject of this investigation. The fork length of a typical sandfish adult is about 20 cm. They are often found in the eastern waters of Korea and off the coast of Japan. Sandfish inhabit the sandy sea bottom at a depth of 200 to 400 m and migrate to shallow waters (2 to 10 m) during the winter season (November to January) to spawn. Female sandfish carrying eggs are commercially valuable in Korea.

In Korea, sandfish are chiefly caught in the eastern coastal waters from September to November, when they migrate to spawn in the shallow waters. Approximately 25,000 tons of sandfish were caught in 1971. However, that figure decreased to less than 4,000 tons in 2008 (Lee et al., 2009). To cope with the sharp reduction in the numbers of sandfish, the Korean government instituted a fish stock recovery plan that includes the restoration of the numbers of sandfish (KMAFM, 2005). The fish stock recovery plan consists of two parts: One part includes the restriction of fishing operations, such as reducing the number of fishing boats, limiting the fishing season, and regulating mesh size of fishing gear and imposing limits of the size of fish that can be caught. The second part of the plan encompasses the enhancement of resources and includes the development of fishing grounds using artificial reefs and release of hatchery-raised fish species into the sea. One strategy for the recovery of sandfish is through enhancing the resource potential by increasing the primary productivity in the designated waters. Another strategy is through habitat enhancement using artificial reefs to promote the regeneration of the resident fish population. Artificial reef technology can potentially enhance the reproduction rate of sandfish by artificially improving the current food chain system in the eastern coastal waters of Korea. The recovery of sandfish focused on creating spawning grounds by using artificial reefs. The spawning grounds were designed with consideration of the behavioral characteristics of sandfish.

Fish use specific spawning sites because the sites: facilitate spawning, lack of predators, or simply serve as areas in which fish aggregate (Yamahira, 1997). Some fish species may also use local vegetation as spawning beds (Tsukidate, 1992; Coston-Clements et al., 1991). Indeed, many countries have already installed artificial reefs specifically designed for spawning (man-made spawning bed) in order to enhance the stock of target species (Ueda et al., 2002). Although the technology for marine ecosystem restoration using artificial reefs is still in its infancy, there is a valid role for artificial reefs in marine ecosystem restoration (Seaman, 2007).

Sandfish tend to lay their eggs on seaweed, such as gulfweed, or nettings of stationary nets such as the gape net and the set net (JSRFRL, 1956; Tanaka, 1988; Minami et al., 1989; Huzida and Horita, 1997). Sandfish deposit a spherical egg mass tightly on *Sargassum.* spp in such a way that supporting stems pass through the axis of the mass (Oklyama, 1990). Maruyama et al. (2001) stated that mature sandfish are attracted to certain species of gulfweed (*Sargassum*) during their breeding season. This indicates that sandfish are selective for a specific substrate (structure) when they spawn. To investigate this further, we conducted a study to examine the effects of artificial reef structures on the spawning of sandfish.

MATERIALS AND METHODS

Estimation of Catch Trends and the Resources of Sandfish in Korea

The present study focused on analysis of sandfish (Figure 7.1) stocks and development of artificial reefs for the enhancement of sandfish resources. The stock survey was conducted using the annual fish catch data, the locations of the fishing grounds, and the fishing gears used. Analysis of fork length composition of sandfish, their age and growth, the maturation of their gonads, and their spawning season was also conducted. These data were used to determine the spawning time of sandfish.

Enhancement of Sandfish Stock by Artificial Reefs

Artificial reef structures for sandfish spawning consisted of two parts: The egg collector and a concrete block acting as a weight. For the egg collector, the materials used in this study consisted of both natural material and artificial material. The natural material was *Sargassum* (gulfweed), and the artificial material was polypropylene rope. A fishing net with a mesh size of 5 × 5 cm was also used. The shape of the artificial egg collector was modeled on the shape of gulfweed. Rope was cut into 40 cm long pieces and separated into strands. These strands were then placed between two strand-twisted axis ropes to form a brush-like shape for a test tube. The top of the axis rope was maintained vertically by a buoy. The base of the egg collector was attached via shackles to a block or support frame. The diameter of the axis rope was 2 cm. The size of each brush-type egg collector was D 0.4 × L 1.5 m. The natural (*Sargassum*) egg collector was similar fashion to the artificial egg collector. Two species of *Sargassum* were employed: *S. fulvellum* and *S. horneri*. The polyethylene fishing net egg collector was similar in shape to an elliptical cylinder. The diameter of the net strands was 4 mm: Similar to that of a *Sargassum* branch. A concrete block (D 1.2 × H 0.3 m), with loop on the upper surface, was used as a weight to support the egg collectors. Two egg collectors were attached per weight block (Figure 7.2). Twelve artificial reefs were installed near rocks, at a 4 m water depth, in October 2006 (Figure 7.3). Our survey included the number of sandfish egg lumps and the location on each egg collector of the attached eggs. Egg collectors were inspected weekly (except on occasions of inclement weather) from December 2006 to March 2007. Data was analyzed using ANOVA.

Figure 7.1 ***See color insert.*** Sandfish, *Arctoscopus japonicus.*

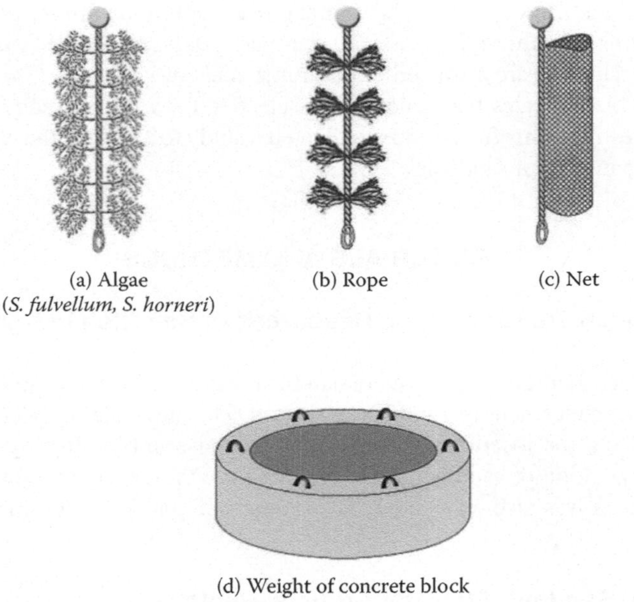

Figure 7.2 Experimental artificial reef structures for spawning of sandfish. Three different substrates for egg collector were used; *Sargassum* (gulfweed), polypropylene rope strands, and polyethylene fishing net. (From Yong, J.H. et al. 2009. *J. Korean Soc. Fisheries Technol.*, 45(4), 234–242. With permission.)

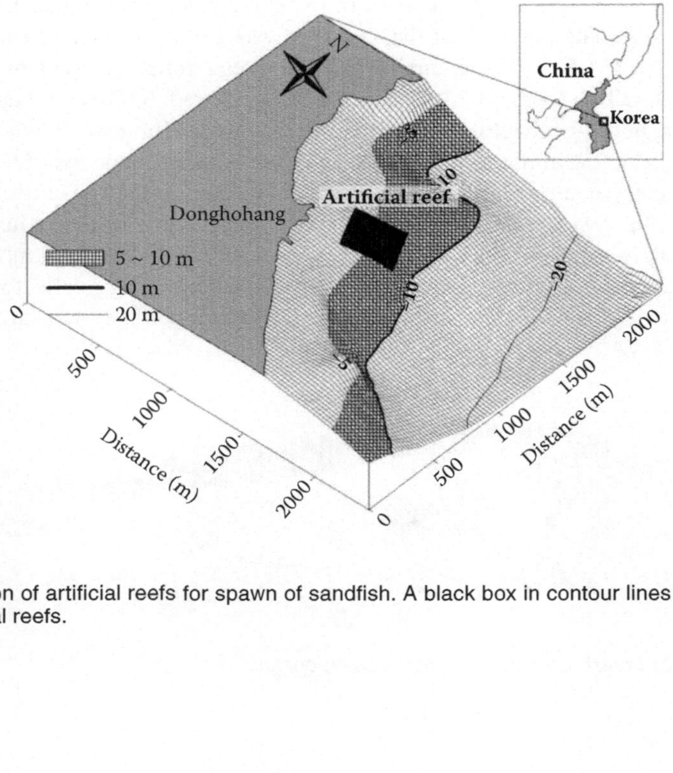

Figure 7.3 Location of artificial reefs for spawn of sandfish. A black box in contour lines indicates the site of artificial reefs.

RESULTS

Catch Trends for Sandfish and Their Resources in Korea

Catch Trends for Sandfish in Korea

Sandfish in Korea are chiefly caught in its eastern coastal waters from September to November, when fish migrate to spawn in shallow waters. Approximately 25,000 tons of sandfish were caught in Korea in 1971. However, that catch decreased to less than 4,000 tons in 2008 (Figure 7.4). Sandfish are mainly caught by Eastern sea Danish seine, West southern Danish seine, and coastal gillnet. Of these types of fishing gear, coastal gillnets are only gears used in shallow water, and the other types of gear are used in deep water. According to the data for 2008, coastal gillnets were used to catch 43% of the total sandfish caught, followed by 39% caught using Eastern Danish seine (Figure 7.5). Meanwhile, the catch per unit effort (CPUE) for Eastern sea Danish seine fluctuated widely from 1980 to 1995 (Figure 7.6). The fluctuations decreased from the end of 1990s, with the numbers of fish caught increasing slowly from 2000 onwards.

Resource Ecology of Sandfish

According to the size composition of sandfish-based on fork length that were caught using Eastern sea Danish seines for the period of 2005 to 2006, the fork length of female ranged from 11 to 25 cm and for males 10 to 20 cm (Figure 7.7). The mean fork length of females ranged from 17 to 18 cm and for males 14 to 16 cm. These data suggest that the fork length of females is longer than that of males. Concomitantly, the mean fork length of females was 18.2 cm in 2005 and 17.3 cm in 2006. That of males was 15.7 cm in 2005 and 15.2 cm in 2006. The mean fork length of sandfish was slightly smaller in 2006 than in 2005. The gonad somatic index of females begins to increase in September, peaks in November, and declines to a minimum in January (Figure 7.8). Females with ripening eggs are caught between August and September, with the peak yield occurring between October and November. Spent females are caught between November and January. Meanwhile, males with ripening sperm initially caught from July, with spent individuals being caught from September. From these results, the main spawning season of sandfish is generally

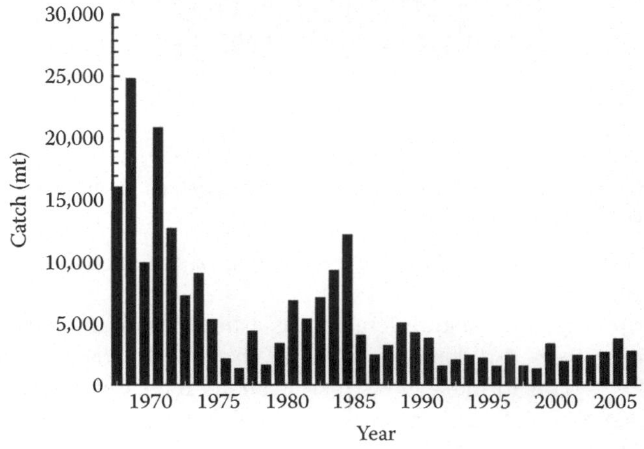

Figure 7.4 Variations in annual catches of *A. japonicus* in Korean waters from 1970 to 2008.

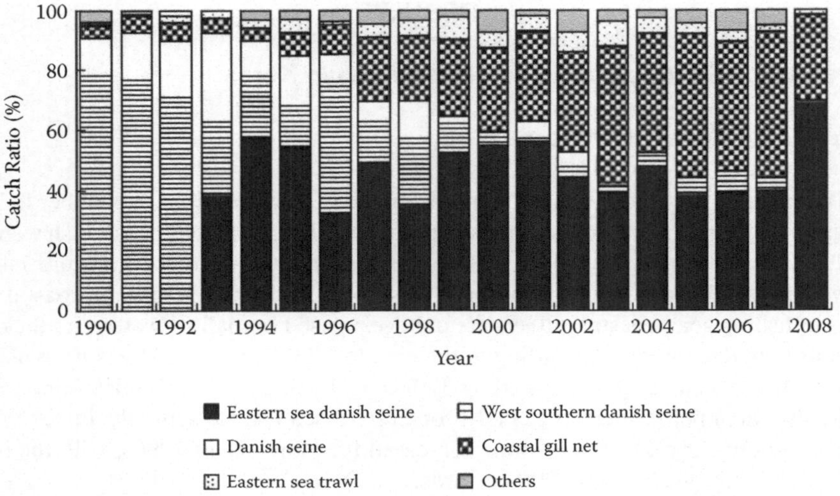

Figure 7.5 Variations in annual catch ratios of *A. japonicus* in Korean waters from 1990 to 2008.

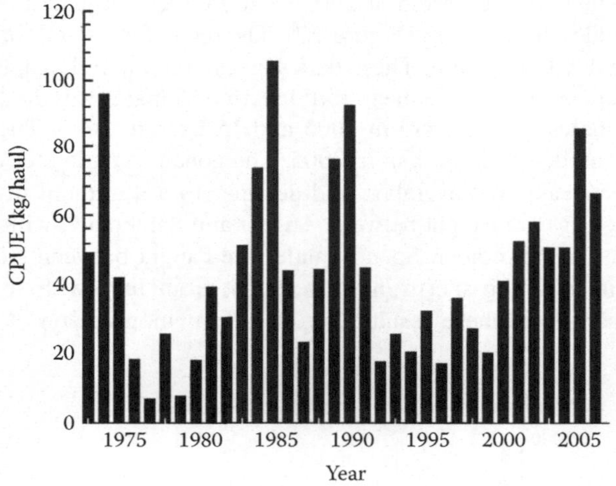

Figure 7.6 Variations in annual CPUE (kg/haul) of *A. japonicus* caught by the eastern sea Danish seine in Korean waters from 1975 to 2008.

between November and December. Thus, the installation of the artificial reef for spawning was planned for this period.

The number of ripe eggs in a mature female was estimated to range between 583 and 2,707. From the relationship between fork length and the number of eggs, the fecundity of sandfish was estimated at $0.541FL^{2.596}$. This formula shows that fecundity increases with increasing fork length (Figure 7.9). Female sandfish initiate reproduction when they attain a size of 16.1 cm FL according to the relationship between their fork length and fecundity (Figure 7.10). This determination was based upon appearance of mature individuals and the weight of females caught between October and November. The age of the sandfish recruited for reproduction was estimated at about 1.8 years using the van Bertalanffy growth formula.

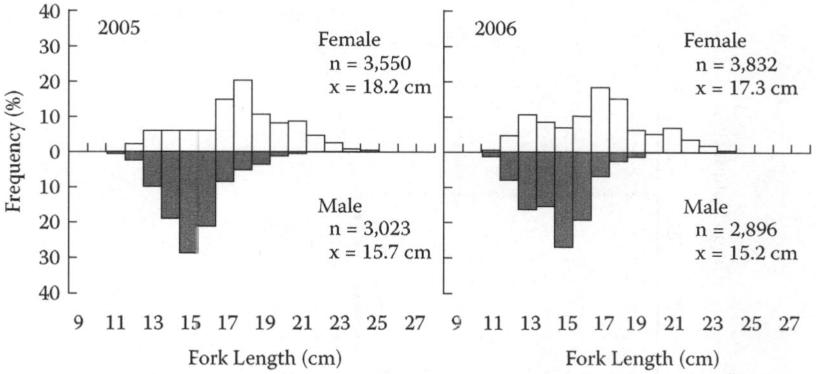

Figure 7.7 Length–frequency distributions of *A. japonicus* caught by the eastern sea Danish seine in Korean waters from 2005 to 2006. N and X represent the number of individuals and mean of FL, respectively.

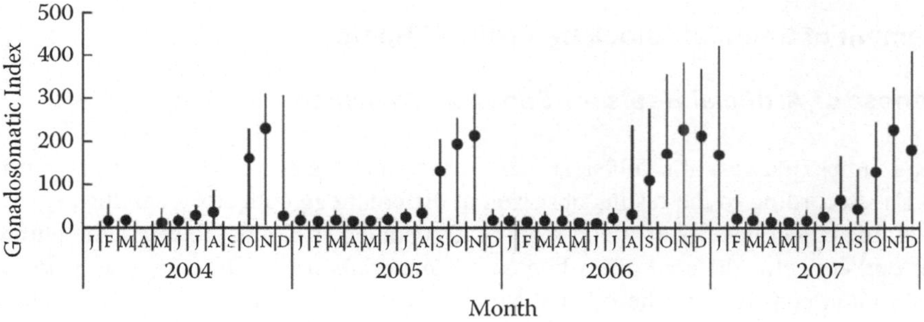

Figure 7.8 Monthly changes in gonadosomatic index of *A. japonicus* in Korean waters from January 2004 to December 2007.

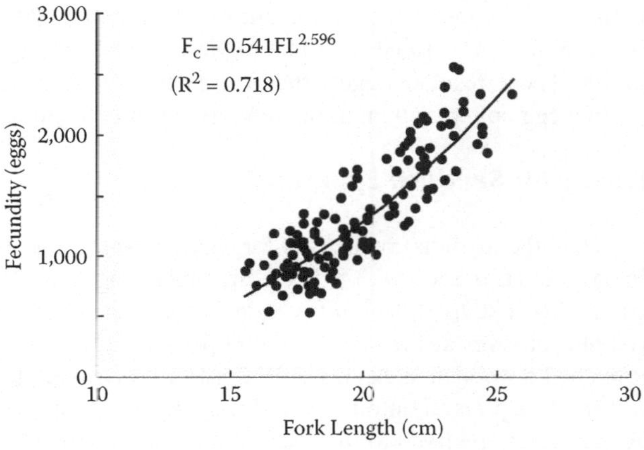

Figure 7.9 Relationship between fork length and fecundity of *A. japonicus* in Korean waters.

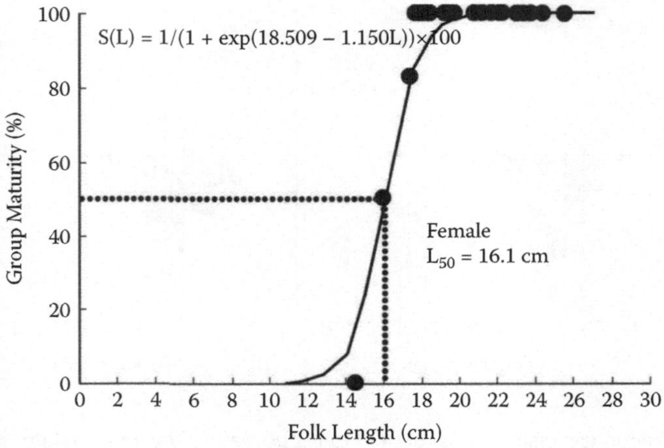

Figure 7.10 Relationship between fork length and group maturity of female *A. japonicus* in Korean waters.

Enhancement of Sandfish Stock by Artificial Reefs

Effectiveness of Artificial Reefs for Sandfish Spawning

The egg lumps (masses) of sandfish strongly adhered to the egg collectors of our artificial reefs (Figure 7.11). According to the results observed at different egg collectors, sandfish eggs were first observed on the egg collectors artificial reefs on December 11, 2006. However, the number of egg lumps per egg collector differed according to the materials used. They were attached in greater numbers on gulfweed than on the other materials, that is, PP brush-type egg collectors and the PE fishing net-type egg collectors (Tables 7.1 and 7.2). Also, the reproductive activity of sandfish varied according to the species of gulfweed used. More mature fish were found to a greater extent in *S. fulvellum* than *S. horneri.* Sandfish eggs were found attached to *S. fulvellum* on December 11, 2006 and peaked at 115 lumps on February 8, 2007. After March 17, 2007, no eggs were observed attached to Sargassum. Sandfish eggs were observed on December 27, 2006 attached to *S. horneri.* Here, the number of egg lumps was fewer. It is notable, however, that PP brush-type egg collectors on artificial reefs contained fewer eggs than on *S. fulvellum.* The number of sandfish egg lumps per egg collector ranged from 61.4 to 2.0 (Table 7.2). The highest number of egg lumps was recorded occurring on *S. fulvellum.* These results suggest that a greater number of mature sandfish were attracted to the *S. fulvellum* egg collector than to the other types of collectors.

Modified Artificial Reefs for Sandfish Spawning

Based upon these results, the artificial reefs used for sandfish spawning were modified to consist of a concrete artificial reef (L 0.5 × W 0.3 × H 0.1 m) containing transplanted *S. fulvellum* on the upper surface (Figure 7.12). The modified reef was designed with a focus on the egg collectors. The reef has two horizontal grooves at the edge of the upper side and a hole in the center (Yang et al., 2009). The grooves had a wooden stick (L 2 × W 2 × H 1 cm) affixed to them. About 8 to 10 shoots of *S. fulvellum* were transplanted onto the wooden stick. A hole in the upper section of the reef was made to allow the reef to be fixed to the rock on the sea bottom. This was achieved using an aluminum bolt (D 12 × L 20 mm) attached to a rock on the sea bottom (Figure 7.13). The fixation of the reef on a rock was intended to facilitate embryos propagation of *S. fulvellum* on nearby

(a)

(b)

(c)

Figure 7.11 ***See color insert.*** Egg lumps of sandfish attached to each egg collectors, *S. fulvellum* (a), polypropylene rope (b), and polyethylene fishing net (c).

rocks and to grow naturally there. Sixty artificial reefs were installed in at a depth of about 9 m in November 2008.

DISCUSSION

In this study, the egg masses of sandfish were found attached in greater numbers on gulfweed than on PP brush-type egg collectors. Results, however, varied according to the species of gulfweed used. More mature sandfish were found associated with *S. fulvellum* than *S. horneri.* A rope egg collector was also used that was modeled on the shape of gulfweed. The rope was cut into 40 cm long pieces and separated into strands. These strands were put between two strand-twisted ropes to form a brush-like shape for a test tube. The size of each brush-type egg collector was D 0.4 × L 1.5 m. Although the rope egg collector was shaped similar to gulfweed, the space between the

Table 7.1 The Number of Egg Mass of *A. japonicus* Attached by the Quality of Materials of Artificial Reefs for Spawning

		Total Number of Egg Mass				Number of Egg Mass/Line			
Date		*S. fulvellum*	*S. horneri*	Rope Strand	Fishing Net	*S. fulvellum*	*S. horneri*	Rope Strand	Fishing Net
2006	Dec. 11	19	—	1	1	3	—	<1	<1
	Dec .12	27	—	—	—	5	—	—	—
	Dec. 27	52	2	—	2	9	<1	—	—
2007	Jan. 1	52	2	—	2	9	<1	—	<1
	Jan. 12	620	24	23	13	103	4	4	2
	Jan. 17	620	24	23	13	103	4	4	2
	Jan. 23	618	24	21	22	103	4	4	4
	Jan. 30	618	24	21	22	103	4	4	4
	Feb. 8	690	24	13	16	115	4	2	3
	Feb. 22	540	19	11	16	90	3	2	3
	Mar. 17	134	—	—	—	32	—	—	—

Source: Yong, J.H. et al. 2009. *J. Korean Soc. Fisheries Technol.*, 45(4), 234–242. With permission.

Table 7.2 The Results of ANOVA-Test for the Number of Egg Mass of *A. japonicus* Attached by the Quality of Materials of Artificial Reefs for Spawning

Material Type	Mean ± SD
S. fulvellum	61.4 ± 48.7^b
S. horneri	3.0 ± 1.7^a
Rope strand	2.7 ± 1.4^a
Fishing net	2.0 ± 1.4^a

Source: Yong, J.H. et al. 2009. *J. Korean Soc. Fisheries Technol.*, 45(4), 234–242. With permission.

Note: Values (mean ± SD) with different letters are significantly different ($p < 0.05$) for each material type.

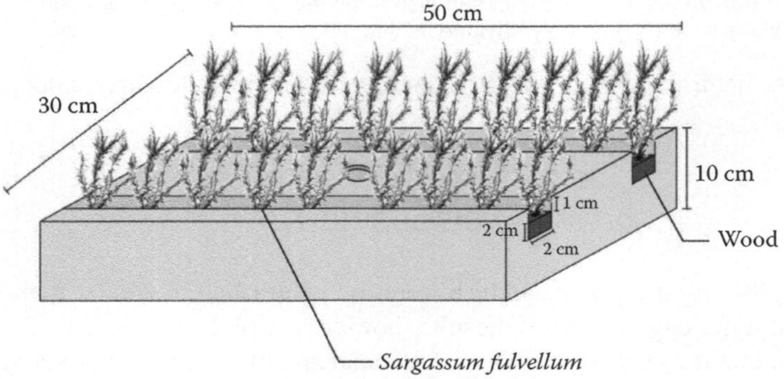

Figure 7.12 Modified artificial reefs for spawn of sandfish (L 0.5 × W 0.3 × H 0.1 m). (From Yong, J.H. et al. 2009. *J. Korean Soc. Fisheries Technol.*, 45(4), 234–242. With permission.)

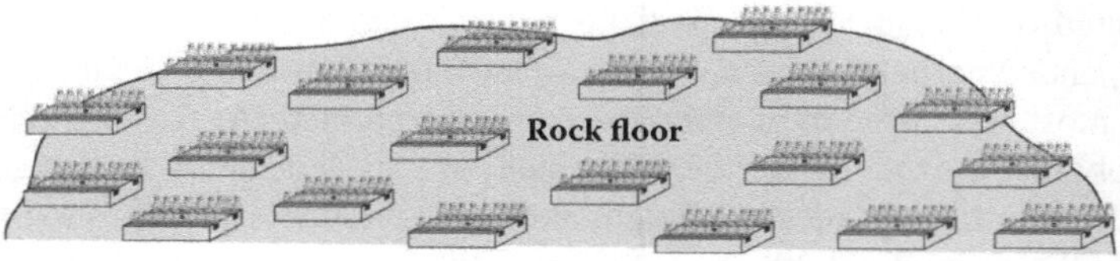

Figure 7.13 An example of installation of the modified artificial reefs on the rock.

strands may have been too narrow for sandfish to spawn there. Also, the rope egg collector may not have afforded suitable protection for the spawned eggs from both water current and illumination. As the diameter of a rope collector is small at about 40 cm long, the resistance against the current may be weak. Thus, the eggs may easily become damaged by heavy waves or strong currents. Also, high illumination may affect the hatching rate of the eggs due to attachment of algae (Nigata prefecture, 1985). Since egg survival in the environment constitutes the first step for a species' survival, egg adhesiveness plays an important role in allowing recolonization and preservation of biodiversity (Gaino and Rebora, 2003). Aquatic plants provide the removal of carbon dioxide and production of oxygen through photosynthesis (Cowx and Welcomme, 1998). They perform the important physical functions of filtering coastal waters and dissipating wave energy. Some fish may also use the vegetation as spawning beds for attachment of eggs (Tsukidate, 1992; Coston-Clements et al., 1991). It is known that sandfish deposit a spherical egg mass on *Sargassum* spp. in such a way that supporting stems pass through the axis of the mass (Oklyama, 1990). Maruyama et al. (2001) found that mature sandfish tended to attract a certain species of *Sargassum* during breeding times. Successful deposition of secure substrata is of prime importance for survival of eggs, because egg capsules that detach from the substrata become stranded ashore and perish (Oklyama, 1990).

As mentioned earlier, sandfish with matured eggs were found to a greater extent in *S. fulvellum* than *S. horneri*. Our results, thus, show that sandfish exhibit a selective preference for spawning substrates. *S. fulvellum* and *S. horneri* are generally brown or dark green in color. They have a holdfast that is perennial, stipes, leaves, lateral branches, and air bladders. Also, they have a rough sticky texture that, together with a robust but flexible body, helps it to withstand strong current. However, this species shows morphological differentiation. *S. fulvellum* is usually 1–2 m in length, and has a stipe and a holdfast and various branches irregularly stretching from the stipe. Leaves are oval or shaped like a rice scoop. *S. horneri* grow to a length of 3–5 m and have various stipes with prickles on their surface (KEM, 1968). The stipes are twisted together to form a bundle (ESFRI, 2007a). As the stipes form a bundle (tuft), their branches become entangled and resulting in little space remaining between them. The long stipes of *S. horneri* are apt to swing back and forth due to the water current or heavy waves. This movement result in a less protective environment from predators for the eggs, yet it does prevent eggs from drifting away, even in a strong current (Bohlen, 2003). The stipes with various prickles may also exert an influence on the courtship behaviors of the sandfish (Gibran et al., 2004). Thus, the current findings and supporting literature indicate that sandfish prefer *S. fulvellum* to *S. horneri* when spawning. A conclusive verification of this theory, however, requires further study.

In this study, the artificial reef used for our sandfish spawning studies consisted of two parts; the egg collectors and a concrete weight for their support. The latter was designed as a durable structure. Indeed, artificial reefs need to be durable. However, the former, *S. fulvellum*, was a vulnerable plant that, although being perennial, required transplantation at intervals. Thus, this method has the disadvantage of not being very durable. Therefore, any artificial reef utilizing seaweed should consider the recruitment of target species. The attachment of *S. fulvellum* was affected by various factors such as deposited mud and water current among others (JHFGA, 2003). Suspended particles in the waters impede the photosynthesis of seaweed and also hinder their attachment by becoming deposited on the surface of artificial reefs (Toda, 1989; Umar et al., 1998); however, deposition of suspended particles decreases with an increasing surface gradient of the artificial reef. This tendency is most noticeable in moderate stream environment (Arakawa and Morinaga, 1994). Our findings imply that a structure for the attachment of seaweed should be designed. This should incorporate steeply slanting surfaces at a height above the muddy layer deposited on the surface of the artificial reef (Terawaki et al., 2001). However, as the egg (embryo) of *S. fulvellum* is relatively large, surfaces that slanted too steeply may decrease the chances of their initial attachment due to slow precipitation (JHFGA, 2003). Current affects not only the transportation of the zoospores or embryos of seaweed but also their accumulation. Zoospores and embryos, subject to a moderate current, accumulate where the current converges or the wake zone is created (AJADPCI, 1994; JHFGA, 2003). When the waves pass over submerged structures (e.g., breakwater), a current is initiated in the spaces between the submerged structures, and a surface circulation results (AJADPCI, 1994). This current circulation may promote the dispersion of the zoospores and embryos of seaweed in the direction of onshore to offshore or vice versa (Toda, 1989). Also, in flow caused by vertical current circulation, a wake zone with eddies and vortices may be created around a reef on its downstream side (Sheng, 2000). This may result in higher accumulation of the zoospores or embryos of the seaweed.

The *S. fulvellum* transplanted to the reef should also be adequately managed for a certain period. This should include the protection of the reef from heavy waves and anthropogenic disturbances such as repeated fishing operations and the invasion of herbivore like sea urchins. Measures to restrict fishing operations include sandfish-size restrictions, aggregation period sales bans, gear restrictions, quotas, aggregation site-based marine reserves and total harvest bans (Sadovy, 1994; Domeier and Colin, 1997). Herbivory has an important influence on *Sargassum* distribution on these reefs (McCook, 1997). The main herbivore of *S. fulvellum*, in the current study, was the sea urchin (ESFRI, 2007b). The protection of *S. fulvellum* from sea urchins may be achieved by both suppressing sea urchin invasions of the reef and by removing them. The former may be conducted by luring the sea urchins away from the *S. fulvellum* beds to other seaweed beds, installing fences, or by structuring the reefs to be unsuitable to their inhabitation. For this last measure, the reef structure may be designed to be irregular with an extremely rough surface, contain a projected, sloped section, or incorporate spaces between the projected sections, amongst others (Choi et al., 2002; JFA, 2007). However, as the methods used to manage a reef differ according to condition of the sea, any method for management should be based upon the behavioral characteristics of the target species, as well as the sea conditions.

CONCLUSIONS

Sandfish in Korea are chiefly caught in the eastern coastal waters, from September to November, when migrate to spawn in shallow waters. Approximately 25,000 tons of sandfish were caught in Korea in 1971. However, that decreased to less than 4,000 tons in 2008. Females with ripening eggs are initially caught between August and September, with the peak yield occurring between October and November. Spent females were caught between November and January. Meanwhile, males with ripening sperm begin to get caught from July, and spent individuals from September. From these results, the main spawning season of sandfish is estimated to be between November and December. Thus, the installation of the artificial reef for spawning was planned for this period. The egg lumps (masses) of sandfish were strongly attached to the egg collectors of our artificial reefs. The number of egg lumps per egg collector differed according to the materials used. They were attached in greater numbers on gulfweed than on the other materials, that is, PP brush-type egg collectors and the PE fishing net-type egg collectors. Also, results varied according to the species of gulfweed used. More mature fish were found to a greater extent in *S. fulvellum* than *S. horneri*.

ACKNOWLEDGMENTS

We are grateful to Dr B. S. Bae for the assistance of the reef design and their installations. For their assistance, we also thank H. M. Cho and H. W. Loh.

REFERENCES

AJADPCI (All-Japan Association for the Development and Promotion of Coastal Industries).1994. *Coastal Fishery Projects, Construction, Maintenance, and Development*. Tokyo, A.A. Balkema Publication, Brookfield, 236–244.

Arakawa, H. and Morinaga, T. 1994. The rate of brown algal zoospores adhered seaweed substrate in relation to gradient of substrate. *Nippon Suisan Gakkaishi* 60(4): 461–466. (In Japanese)

Bohlen, J. 2003. Spawning habit in the spined loach. *Cobitis taenia* (Cypriniformes:Cobitidae). *Ichthyological Research* 50: 98–101.

Choi, C.G., Takeuchi, Y., Terawaki, T., Serisawa, Y., Ohno, M., and Sonhn, C.C. 2002. Ecology of seaweed beds on two types of artificial reef. *Journal of Applied Phycology* 14: 343–349.

Coston-Clements, L., Settle, L.R., Hoss, D.E., and Cross, F.A. 1991. Utilization of the *Sargassum* habitat by marine invertebrates and vertebrates, a review. NOAA Technical Memorandum NMFSSEFSC-296.

Cowx, I.G. and Welcomme, R.L. (Eds.) 1998. *Rehabilitation of Rivers for Fish*. Oxford, Fishing News Books, 160 pp.

Domeier, M.L. and Colin, P.L. 1997. Tropical reef fish spawning aggregations: Defined and reviewed, *Bulletin of Marine Science* 60: 698–726.

ESFRI (East Sea Fisheries Research Institute). 2007a. Pictorial Book of the Korean East coastal seaweed, Kangneung, 64–65. (In Korean)

ESFRI (East Sea Fisheries Research Institute). 2007b. Manual for the Construction of Seaweed Forest on the East Coast of Korea. Kangneung, 51 pp. (In Korean)

Gaino, E. and Rebora, M. 2003. Adhesiveness of the eggs of *Ecdyonurus venosus* to siliceous and calcareous substrate. *Physiology, Morphology and Ultrastructure*, 437–443.

Gibran, F.Z., Santos, F.B., Santos, H.F.D., and Sabino, J. 2004. Courtship behavior and spawning of the hairy blenny *Labrisomus nuchipinnis* (Labrisomidae) in southeastern Brazil. *Neotropical Ichthyology* 2(3): 163–166.

Huzida, D. and Horita, K. 1997. Recent decline of sandfish of Toyama prefecture of Japan. *Bulletin of Toyama Prefecture Fisheries Laboratory* 9: 19–34. (In Japanese)

JFA (Japanese Fisheries Agency). 2007. Guidelines for measure of seaweed weathering phenomenon, Tokyo, Japanese Fisheries Agency, 114–116. (In Japanese)

JHFGA (Japan Harbor and Fishing Ground Association). 2003. Guidelines of survey and design for seaweed bed creation type of harbor structure. Tokyo, Japanese Harbor and Fisheries Ground Association 8–12. (In Japanese)

JSRFRL (Japan Sea Regional Fisheries Research Laboratory). 1956. Benthic fishery and their resources in Japan Sea. Nigata, *Bulletin of the Japan Sea Regional Fisheries Agency* 4:197–215. (In Japanese)

Korean Education Ministry. 1968. Illustrated encyclopedia of fauna & flora of Korea. Seoul. *Korean Education Ministry* 8:154–171. (In Korean)

Korean Maritime Affairs and Fisheries Ministry. 2005. Report on fisheries resources recovery plan. *Korean Maritime Affairs and Fisheries Ministry*, Seoul, 451 pp. (In Korean)

Lee, S.I., Yang, J.H., Yoon, S.C., Chun, Y.Y., Kim, J.B., Cha, H.K., and Choi, Y.M. 2009. Biomass estimation of sailfin sandfish, *Arctoscopus japonicus* in Korean waters. *Journal of the Korean Fisheries Society*. 42: 239–334. (In Korean)

Maruyama, S., Takeda, Y., and Nakada, S. 2001. A study on coastal structures as fish spawning grounds light and flows around spawning grounds of sandfish. *Collection of Learned Papers of Oceanographic Development* 17: 147–151. (In Japanese)

McCook, L.J. 1997. Effects of herbivory on zonation of *Sargassum* spp. within fringing reefs of the Central Great Barrier Reef. *Marine Biology* 129: 713–722.

Minami, T., Narita, K., and Imamura, A. 1989. Spawning of sandfish in Toyama Bay of Japan. 3rd Sandfish Study Conference Report, Akita Fisheries Development Center, Akita, 42–44. (In Japanese)

Nigata prefecture. 1985. Comprehensive study and project of artificial reef in the Nigata prefecture coastal waters of Japan. Nigata, *Nigata Prefecture,* 167–168. (In Japanese)

Oklyama, M. 1990. Contrast in reproductive style between two species of sandfishes (Family Trichodontidae). *Fishery Bull., U.S.,* 88:543–549.

Sadovy, Y. 1994. Grouper stocks of the western central Atlantic: The need for management and management needs. *Proc. Gulf Carib. Fish. Inst.* 43:43–65.

Seaman, W. 2007. Artificial habitats and the restoration of degraded marine ecosystem and fisheries. *Hydrobiologia* 580:143–155.

Sheng, Y.P. 2000. Pages 5–64 in Seaman, W., Jr., Ed. *Artificial Reef Evaluation with Application to Natural Marine Habitats.* CRC Press, New York.

Tanaka, M. 1988. Information on sandfish, *Arctoscopus japonica,* In Japan Sea. 2nd Sandfish Study Conference Report, Akita Fisheries Promotion Center, Akita, 17–20. (In Japanese)

Terawaki, T., Hasegawa, H., Arai, S., and Ohno, M. 2001. Management-free techniques for restoration of *Eisenia* and *Ecklonia* beds along the central pacific coast of Japan. *Journal of Applied Phycology* 13:13–17.

Toda, S. 1989. Effects of water movement on dispersion of drifting seeds and larvae. Bulletin of National Research Institute, *Aquaculture* 16:1–82. (In Japanese)

Tsukidate, J. 1992. Ecology of *Sargassum* spp. and *Sargassum* forest formation. *NOAA Technical Report NMES* 106:63–73.

Ueda, Y., Tagaki, T., Hirosawa, A., Namimatsu, A., Goto, S., and Toshikawa, T. 2002. Preliminary observations on the mating and spawning behavior of Oval Squid (*Sepioteuthis lessoniana*) at a man-made spawning bed settled in the coastal waters of Tokushima prefecture in Japan. *Bulletin of Tokushima Prefecture Fisheries Research Institute* 1:33–36. (In Japanese)

Umar, M.J., McCook, L.J., and Price, I.R. 1998. Effects of sediment deposition on the seaweed *Sargassum* on a fringe coral reef. *Coral Reefs* 17:169–177.

Yamahira, K. 1997. Proximate factors influencing spawning site specificity of the puffer fish *Takifugu niphobles. Marine Ecology Progress Series* 147:11–19.

Yong, J.H., Lee, S.I., Bae, B.S., Cha, H.K., Yoon, S.C., Chun, Y.Y., Kim, J.B., and Chang, D.S. 2009. Development of artificial spawning seaweeds of the sandfish, *Arctoscopus japonicus*. *Journal of Korean Society of Fisheries Technology*. 45(4): 234–242. (In Korean).

CHAPTER 8

Assessing Artificial Reefs for Fisheries Management

A 10-Year Assessment off the Northern Coast of Rio de Janeiro

Luciano Neves dos Santos, Daniel Shimada Brotto, and Ilana Rosental Zalmon

CONTENTS

ABSTRACT

Artificial reefs were deployed in March 1996 on a flat bottom, 9 m deep and 5 km area offshore Southeast Brazil, and temporal changes of the associated fish assemblages were assessed over a period of 10 years. Fishes were collected through bottom gillnets at 1 month, 6 months, 1 year, 6 years, 7 years, and 10 years after artificial reef deployment. A total of 961 fishes belonging to 41 species and 18 families were caught, with *C. chrysurus* (Carangidae) and *C. nobilis* (Haemulidae) as the dominant species and Sciaenidae as the richest family (N = 11 species). No single species occurred throughout the surveys and fish species composition differed temporally among <1 year, 1–6 year, and 7–10 year time intervals. Fish richness, abundance, biomass, and size varied significantly with time; richness increased gradually toward a stabilization in the last years. Abundance and biomass decreased after the sixth year, and size decreased linearly through time. No temporal trend of fish abundance and biomass was found for piscivorous species; however, these variables increased until the sixth and seventh years for planktivorous and invertivorous fishes. Abundance and biomass of the most frequent fish families changed differentially with time, with a gradual decreasing after the first year for Carcharhinidae and Triakidae, a unimodal response for Carangidae and

Haemulidae, and a steady increment for Sciaenidae and Ariidae. This study reveals the importance of long-term studies of fish assemblages associated with artificial reefs to conduct management actions to effectively improve the captures by local inshore fishermen and minimize the risks of detrimental effects, as fishery failure or overfishing.

INTRODUCTION

Artificial reefs have been deployed worldwide in shallow marine environments to aid in environmental rehabilitation (i.e., providing new suitable habitats for invertebrates and fish) but primarily to enhance attraction and catch of fish resources (Seaman and Sprague, 1991; Seaman, 2000). Few studies on artificial reefs, however, have been properly conducted to include the temporal changes of attracted fishes over large scales, such as a decade or more (but see Spanier, 2000; Relini et al., 2002; Relini et al., 2007; Santos and Monteiro, 2007), while long-term investigations of fish stocks associated with artificial reefs on the South Atlantic coast were virtually unknown (Baine and Side, 2003). Consequently, data on fishery resources variations with artificial reef age are still lacking despite the potential application of artificial reefs to improve inshore artisanal fisheries in the overall low-productivity of coastal Brazilian waters (Amaral and Jablonski, 2005).

This study examined the decadal (1996–2006) changes in the fishery resources associated with artificial reefs deployed on the north coast of Rio de Janeiro. The specific objectives of this work, which has a larger goal to provide long-term data on fish use of artificial reefs in the South Atlantic, are (1) to address the changes of fish richness, abundance, biomass, and size with the age of artificial reefs, and (2) to model the responses of these community descriptors, as well as of the major fish stocks with time. The potential contribution of artificial reefs to local inshore fisheries is also appraised with respect to the utility of artificial reefs to attract and affect the prevalent taxa.

MATERIALS AND METHODS

Study Site

The north coast of Rio de Janeiro (southeastern Brazil; Figure 8.1) is naturally depleted of rock substratum or other hard substrates, and it is covered by extensive sandy beaches with variable amounts of mud and calcareous nodules (i.e., rhodolites; Zalmon et al., 2002). This area is located in a transitional zone between warm and oligotrophic waters of the Brazil Current from the north and cold and nutrient-rich upwelling of the South Atlantic Central Water from the south (Valentin and Monteiro-Ribas, 1993). Primary productivity (chlorophyll a) is low, Secchi depth does not exceed 4 m, and strong bottom currents are common (Krohling and Zalmon, 2008). Although dominated by oligotrophic waters and homogeneous bottom relief, the north coast of Rio de Janeiro is often exploited by local inshore artisanal fishermen (Zalmon et al., 2002).

Together with oceanic circulation, the north coast of Rio de Janeiro is also strongly influenced by weather and freshwater runoff. The outflow of the Paraíba do Sul River (the largest river in Rio de Janeiro state) is especially heavy during the rainy period (December to February), when a turbid (Secchi depth <0.5 m) and polyhaline (18–33 psu) estuarine plume spreads over 15 km north from the river mouth, covering most of the continental shelf up to ca. 10 km distant from the shore (Godoy et al., 2002). This plume does not, however, reach the sea bottom during the rainy period, because the local trade winds lead to the intrusion of clearer and saline bottom waters. During the dry period (April to November, but mostly during winter), the intensity of southwest winds increases, stratification ceases and, consequently, water turbidity significantly increases near the bottom (Godoy et al., 2002).

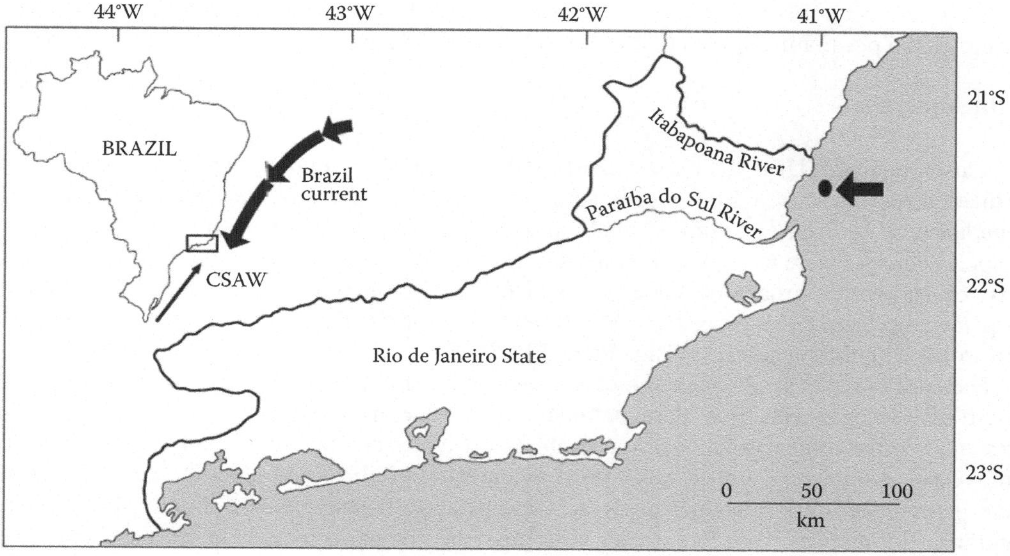

Figure 8.1 Geographic location of the north coast of Rio de Janeiro (southeast Brazil), where the experimental reef complex was deployed.

Since 1996, invertebrate and fish colonization of artificial reefs along the north coast of Rio de Janeiro have been investigated to assess the role of artificial reefs in management and conservation of local fishery resources (Zalmon et al., 2002; Krohling et al., 2006). More than 40 fish species have been recorded associated with artificial reefs of different material and complexity, but mostly in the form of concrete modules, the most effective in attracting and harboring fish (Zalmon et al., 2002). Overall, the deployment of artificial structures on the homogeneous and structureless bottom of the north coast of Rio de Janeiro is regarded as a promising alternative to mitigate local losses of fishery resources and enhance fish populations (Brotto et al., 2006a).

Experimental Design

Artificial reefs were first deployed in March 1996 on a flat and homogeneous bottom, 9 m deep, and 5 km offshore of the Guaxindiba Beach (21°29′S; 41°00′W), northern Rio de Janeiro coast (Figure 8.1). Initially, the reef complex was comprised of experimental modules of concrete pipes (N = 12), tire bundles (N = 12), and brick piles (N = 4), covering approximately 1,500 m^2 of sea bottom (Faria et al., 2001; Zalmon et al., 2002). Subsequently, the reef complex was increased by adding tire bundles (N = 12) and cement prefabricated blocks (N = 7) in February 1997 (Faria et al., 2001), and 36 prefabricated reef balls® (ca. 1.0 m^3) in January 2002 (Brotto et al., 2006a). The precise location of the artificial reefs was marked using global positioning system (GPS).

Artificial reefs were surveyed to assess the associated fish assemblage at 1 month (April 1996), 6 months (September 1996), 1 year (April 1997), 6 years (May 2002), 7 years (April 2003), and 10 years (November 2006) after the initial deployment. During each sampling period, three sets of bottom gillnets (25 × 3 m; 30 mm mesh) were fished over the reef complex at sunset and recovered 24 h later. Gillnets were chosen to assess the potential of artificial reefs for local inshore artisanal fishermen, since this fishing gear has been traditionally used along the northern Rio de Janeiro coast (Zalmon et al., 2002), and had the same technical features as those used by local fishermen. The nets were deployed on top of the reef modules and about 100 meters apart from each other. The position of the three sets of gillnets inside the reef complex was chosen according to the current direction. Again, global positioning system (GPS) was applied for setting gillnets within the reef complex.

Data Analysis

Cluster analysis (UPGMA clustering method) and nonmetric multidimensional scaling (MDS) ordination, performed on Bray–Curtis similarity measures derived from $\log_{10}$ transformed fish abundance, were used to examine trends in temporal changes of fish species composition. The temporal changes were tested using similarity analysis (ANOSIM), while Bray–Curtis similarities between and within groups of replicates in each sampling period were calculated using similarity percentage breakdown procedures (Clarke and Green, 1988; Clarke, 1993). These analyses were performed with the statistical package PRIMER® (V.6).

Fish species richness, abundance, biomass, and total length were used as descriptors of the changes in fish assemblages with time. Permutational multivariate analyses of variance (PERMANOVA) were applied for comparisons of fish assemblage attributes among the six sampling periods. The Bray–Curtis similarity distance was chosen as the basis of all PERMANOVA analysis, and data were permutated 4999 times per analysis, for tests at an α-level of 0.01 (Manly, 1997). Where significant differences were found, pair-wise post hoc comparisons were performed under 4999 permutations (as in Anderson, 2005).

Generalized additive models (GAMs) were also fitted to appraise the influence of time on fish species richness, abundance, biomass, size, and also on abundance and biomass of trophic guilds (e.g., piscivorous, invertivorous, and planktivorous) and dominant fish families. GAMs are an extension of generalized linear models that, unlike more conventional regression methods, do not assume a particular functional relationship between the response variable and the predictor (time in our case) (Lepš and Šmilauer, 2003). The model complexity of GAMs was chosen by the stepwise selection procedure using the Akaike information criterion (AIC) with the software CANOCO® (V.4.5). AIC considers not only the goodness of fit but also parsimony, penalizing very complex models (Burnham and Anderson, 1998).

RESULTS

A total of 961 fishes in 41 species and 17 families were caught over the six surveys at artificial reefs in the north coast of Rio de Janeiro (Table 8.1). Most species were invertivorous ($N = 22$) or piscivorous ($N = 15$ species), whereas a few species ($N = 4$) were planktivorous. Sciaenidae was the richest family ($N = 9$ species), followed by Haemulidae ($N = 5$ species) and Ariidae ($N = 4$ species). No single species was captured in all the survey periods (Table 8.2), even *Chloroscombrus chrysurus* (Carangidae) and *Conodon nobilis* (Haemulidae), the dominant species in number ($N =$ 135 and 117, respectively). Other 11 species, such as the ariid *Genidens genidens;* the sciaenids *Stellifer brasiliensis*, *Larimus breviceps*, *Paralonchurus brasiliensis,* and *Cynoscion virescens;* and the sharks *Rhizoprionodon lalandii* and *Mustelus higmani*, showed variable relative abundances (1%–78%) per survey period, but most species ($N = 28$) contributed individually with less than 1% of total abundance, overall occurring in only one or two sampling periods.

The dendrogram revealed three clusters (at the 20% level; Figure 8.2) based on the similarity of fish assemblage composition among sampling periods. The samples of the two first surveying periods (1 and 6 months after artificial reef deployment) formed a cluster, with a similarity of 37.9% (Group 1). The fish samples of 1 to 6 years after artificial reef deployment clustered with a similarity of 28.9% (Group 2), whereas those of 7 to 10 years after reef deployment clustered with a similarity of 22.6% (Group 3).

Table 8.1 Fish Species Recorded Over 10-Years of Surveys at Artificial Reefs in the North Coast of Rio de Janeiro, Southeast Brazil

Fish Species	Trophic Guild
Triakidae	
Mustelus higmani (Springer and Lowe, 1963)	Piscivorous
Carcharhinidae	
Carcharhinus porosus (Ranzani, 1839)	Piscivorous
Rhizoprionodon lalandii (Muller and Henle, 1839)	Piscivorous
Rhizoprionodon porosus (Poey, 1861)	Piscivorous
Clupeidae	
Harengula clupeola (Cuvier, 1829)	Planktivorous
Odontognathus mucronatus (Lacépède, 1800)	Planktivorous
Opisthonema oglinum (Lesueur, 1818)	Planktivorous
Pristigasteridae	
Pellona harroweri (Fowler, 1917)	Planktivorous
Ariidae	
Genidens genidens (Valenciennes, 1839)	Invertivorous
Sciadeichtys luniscutis (Valenciennes, 1840)	Invertivorous
Bagre bagre (Linnaeus, 1766)	Invertivorous
Bagre marinus (Mitchill, 1815)	Invertivorous
Batrachoididae	
Porichthys porosissimus (Cuvier, 1829)	Piscivorous
	Pomatomidae
Pomatomus saltatrix (Linnaeus, 1766)	Piscivorous
Carangidae	
Caranx crysos (Mitchill, 1815)	Piscivorous
Caranx latus (Agassiz, 1831)	Piscivorous
Chloroscombrus chrysurus (Linnaeus, 1766)	Invertivorous
Oligoplites saliens (Bloch, 1793)	Invertivorous
Selene setapinnis (Mitchill, 1815)	Piscivorous
Gerreidae	
Eucinostomus argenteus (Baird and Girard, 1855)	Invertivorous
Haemulidae	
Anisotremus virginicus (Linnaeus, 1758)	Invertivorous
Conodon nobilis (Linnaeus, 1758)	Invertivorous
Haemulon aurolineatum (Cuvier, 1829)	Invertivorous
Haemulon steindachneri (Jordan and Gilbert, 1882)	Invertivorous
Orthopristis ruber (Cuvier, 1830)	Invertivorous
Sciaenidae	
Ctenosciaena gracilicirrhus (Metzelaar, 1919)	Invertivorous
Cynoscion jamaicensis (Vaillant and Bocourt, 1883)	Piscivorous
Cynoscion virescens (Cuvier, 1830)	Piscivorous
Isopisthus parvipinnis (Cuvier, 1830)	Piscivorous
Larimus breviceps (Cuvier, 1830)	Piscivorous
Micropogonias furnieri (Desmarest, 1823)	Invertivorous
Paralonchurus brasiliensis (Steindanchner, 1875)	Invertivorous
Stellifer brasiliensis (Schultz, 1945)	Invertivorous
Stellifer rastrifer (Jordan,1889)	Invertivorous

(continued on next page)

Table 8.1 (continued) Fish Species Recorded Over 10-Years of Surveys at Artificial Reefs in the North Coast of Rio de Janeiro, Southeast Brazil

Fish Species	Trophic Guild
Ephippidae	
Chaetodipterus faber (Broussonet, 1782)	Invertivorous
Polynemidae	
Polydactylus oligodon (Günther, 1860)	Invertivorous
Polydactylus virginicus (Linnaeus, 1758)	Invertivorous
Labridae	
Halichoeres poeyi (Steindachner, 1867)	Invertivorous
Trichiuridae	
Trichiurus lepturus (Linnaeus, 1758)	Piscivorous
Scombridae	
Scomberomorus brasiliensis Collette, (Russo and Zavala-Camin, 1978)	Piscivorous
Stromateidae	
Peprilus paru (Linnaeus, 1758)	Invertivorous

Note: The trophic guild of each species is also shown, according to stomach analyses of some individuals or data compiled from literature.

The MDS ordination also revealed three groups in fish species composition among <1 year, 1–6 year, and 7–10 year time intervals (Figure 8.3), which were confirmed by global ($R = 0.88$; $P < 0.01$) and pair-wise (dissimilarity among groups ≥84.1%) ANOSIM results. The average similarity within the Group 1 (<1 year samples) was 49.9%, with *R. lalandii* showing the highest individual contribution (70.2%). The average similarity was of 41.8% for the Group 2 (1–6 year samples) and 33.7% for the Group 3 (7–10 year samples), with greatest individual contributions of *C. chrysurus* (70.4%) and *C. nobilis* (62.1%) for each group, respectively.

Fish richness, abundance, biomass, and size changed significantly with time (PERMANOVA; $P = 0.03$ for richness; $P < 0.01$ for the other attributes) but not in the same way among the community descriptors (Figure 8.4). Richness was higher within 6–10 sampling years than before ($F_{5,18} = 2.9$; PERMANOVA's post hoc tests; $P < 0.01$). Although higher during the 1–6 years after deployment than after 10 years, fish abundance peaked at the seventh year and reached a minimum value within the first 6 months after artificial reef deployment ($F_{5,18} = 18.2$; PERMANOVA's post hoc tests; $P < 0.01$). Biomass was higher 6–7 years after artificial reef deployment than in the other time intervals ($F_{5,18} = 10.8$; PERMANOVA's post hoc tests; $P < 0.01$), while fish size were highest within the first 6 months after artificial reef deployment, although a secondary peak was recorded in the sixth year ($F_{5,18} = 13.6$; PERMANOVA's post hoc tests; $P < 0.01$).

The response curves (GAMs) of these four community descriptors also varied with time (Figure 8.5). AIC revealed an increase of richness with time and an apparent stabilization after the sixth year (nonlinear $F_{1,17} = 5.2$; $P < 0.05$), whereas fish size decreased monotonously through time (nonlinear $F_{1,460} = 5.8$; $P < 0.05$). However, the response curves indicated unimodal trends for abundance (AIC; nonlinear $F_{1,17} = 14.5$; $P < 0.01$) and biomass (AIC; nonlinear $F_{1,17} = 40.6$; $P < 0.01$) with time.

The response curves (GAMs) showed a sharp decrease on the abundance and biomass of sharks with time (AIC; linear $F_{1,17} = 5.5$ and 7.1; $P < 0.05$ for both), but largely after 6 months since the artificial reef deployment (Figure 8.6). AIC revealed that abundance of both ariids and sciaenids increased exponentially with time (nonlinear $F_{1,17} = 4.4$ and 7.9; $P < 0.05$ and $P < 0.01$, respectively), but biomass increased linearly for the former (linear $F_{1,17} = 7.2$; $P < 0.05$) and exponentially for the latter family (nonlinear $F_{1,17} = 3.5$; $P = 0.08$). A more unimodal response was found for the

Table 8.2 Absolute (*N*) and Relative (%) Number of Fishes Recorded Over 10-Years of Surveys at Artificial Reefs in the North Coast of Rio de Janeiro, Southeast Brazil

	1 Month		6 Months		1 Year		6 Years		7 Years		10 Years		Total	
Fish Species	***N***	**%**	***N***	**%**	***N***	**%**	***N***	**%**	***N***	**%**	***N***	**%**	***N***	**%**
C. chrysurus	11	44.0	1	2.7	87	80.6	35	35.7	1	0.8	—	0	135	29.3
C. nobilis	1	4.0	1	2.7	—	0	1	1.0	102	83.6	12	16.9	117	25.4
R. lalandii	7	28.0	29	78.4	—	0	—	0	—	0	—	0	36	7.8
G. genidens	—	0	—	0	—	0	9	9.2	1	0.8	20	28.2	30	6.5
O. oglinum	1	4.0	—	0	1	0.9	20	20.4	2	1.6	—	0	24	5.2
S. brasiliensis	—	0	—	0	—	0	3	3.1	—	0	9	12.7	12	2.6
O. ruber	—	0	—	0	11	10.2	—	0	—	0	—	0	11	2.4
L. breviceps	—	0	—	0	—	0	3	3.1	2	1.6	5	7.0	10	2.2
P. brasiliensis	—	0	—	0	—	0	—	0	1	0.8	5	7.0	6	1.3
P. saltatrix	—	0	1	2.7	—	0	1	1.0	4	3.3	—	0	6	1.3
T. lepturus	—	0	—	0	—	0	6	6.1	—	0	—	0	6	1.3
C. virescens	—	0	—	0	—	0	—	0	—	0	5	7.0	5	1.1
M. higmani	1	4.0	3	8.1	1	0.9	—	0	—	0	—	0	5	1.1
B. marinus	—	0	—	0	—	0	—	0	—	0	4	5.6	4	0.9
C. crysos	4	16.0	—	0	—	0	—	0	—	0	—	0	4	0.9
C. porosus	—	0	—	0	—	0	4	4.1	—	0	—	0	4	0.9
H. steindachneri	—	0	—	0	—	0	2	2.0	1	0.8	1	1.4	4	0.9
I. parvipinnis	—	0	—	0	—	0	—	0	—	0	4	5.6	4	0.9
M. furnieri	—	0	—	0	—	0	3	3.1	—	0	1	1.4	4	0.9
P. paru	—	0	—	0	—	0	4	4.1	—	0	—	0	4	0.9
S. brasiliensis	—	0	—	0	—	0	3	3.1	—	0	—	0	3	0.7
S. rastrifer	—	0	—	0	—	0	—	0	2	1.6	1	1.4	3	0.7
C. faber	—	0	—	0	2	1.9	—	0	—	0	—	0	2	0.4
H. aurolineatum	—	0	—	0	2	1.9	—	0	—	0	—	0	2	0.4
O. saliens	—	0	—	0	—	0	2	2.0	—	0	—	0	2	0.4
P. oligodon	—	0	—	0	2	1.9	—	0	—	0	—	0	2	0.4
S. setapinnis	—	0	—	0	—	0	—	0	2	1.6	—	0	2	0.4
A. virginicus	—	0	—	0	1	0.9	—	0	—	0	—	0	1	0.2
B. bagre	—	0	1	2.7	—	0	—	0	—	0	—	0	1	0.2
C. latus	—	0	—	0	—	0	—	0	1	0.8	—	0	1	0.2
C. gracilicirrhus	—	0	—	0	—	0	—	0	—	0	1	1.4	1	0.2
C. jamaicensis	—	0	—	0	—	0	1	1.0	—	0	—	0	1	0.2
E. argenteus	—	0	—	0	1	0.9	—	0	—	0	—	0	1	0.2
H. poeyi	—	0	—	0	—	0	—	0	1	0.8	—	0	1	0.2
H. clupeola	—	0	—	0	—	0	—	0	1	0.8	—	0	1	0.2
O. mucronatus	—	0	—	0	—	0	—	0	—	0	1	1.4	1	0.2
P. harroweri	—	0	—	0	—	0	1	1.0	—	0	—	0	1	0.2
P. virginicus	—	0	—	0	—	0	—	0	—	0	1	1.4	1	0.2
P. porosissimus	—	0	—	0	—	0	—	0	1	0.8	—	0	1	0.2
R. porosus	—	0	—	0	—	0	—	0	—	0	1	1.4	1	0.2
S. luniscutis	—	0	1	2.7	—	0	—	0	—	0	—	0	1	0.2
Total	25	100.0	37	100	108	100	98	100	122	100	71	100	461	100

Note: Cumulative abundances and catches by sampling period after artificial reef deployment are shown.

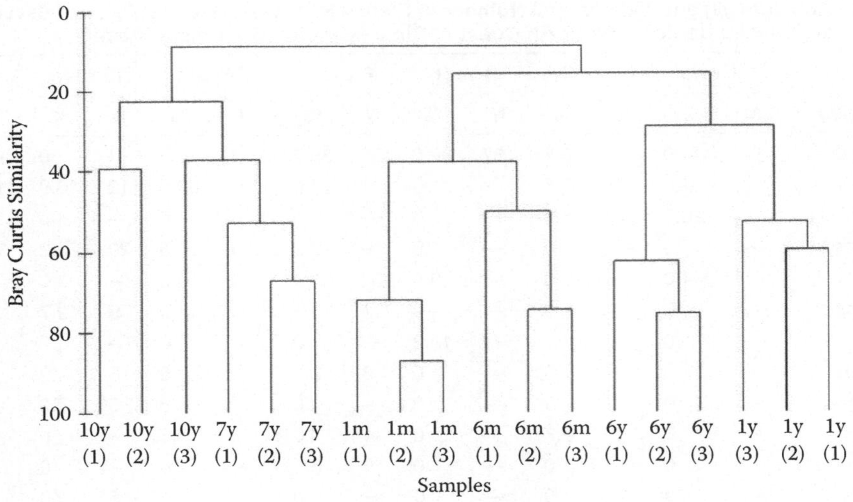

Figure 8.2 Similarity (Bray–Curtis distance) among time intervals based on cluster analysis (UPGMA) of the fish assemblages ($\log_{10}$-transformed abundances) associated with artificial reefs (*x*-axis legend: number before the letter = time after artificial reef deployment; letters = time unity in months (m) or years (y); number between brackets = sample replicate).

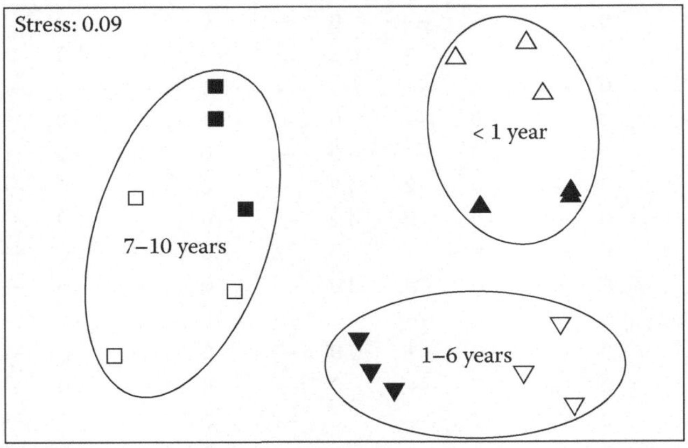

Figure 8.3 Nonmetric multidimensional scaling (MDS) ordinations of fish composition ($\log_{10}$-transformed abundances) with time intervals after artificial reef deployment (▲ = 1 month; △ = 6 months; ▽ = 1 year; ▼ = 6 years; ■ = 7 years; □ = 10 years).

abundance and biomass of Carangidae (AIC; nonlinear $F_{1,17} = 3.8$ and 3.7, respectively; $P = 0.07$ for both) and Haemulidae (AIC; nonlinear $F_{1,17} = 5.8$ and 7.0, respectively; $P < 0.05$ for both), with peaks occurring in the first year for the former family and in the seventh year for the latter.

No GAM was selected by AIC for responses of the abundance and biomass of piscivorous with time (Figure 8.7). A nonlinear response was found for the abundance (AIC; nonlinear $F_{1,17} = 6.1$; $P < 0.05$) and biomass (AIC; nonlinear $F_{1,17} = 7.9$; $P = 0.01$) of invertivorous fishes, with values initially increasing with time and then becoming variable after the seventh year. A similar but more unimodal relationship was found for the abundance (AIC; nonlinear $F_{1,17} = 8.0$; $P = 0.01$) and biomass (AIC; nonlinear $F_{1,17} = 7.4$; $P < 0.01$) of planktivorous fishes.

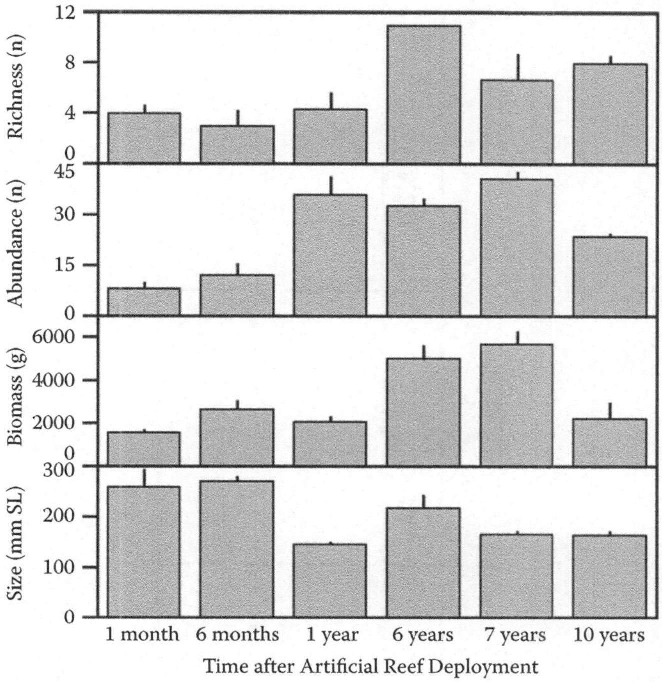

Figure 8.4 Mean fish richness, abundance, biomass, and size recorded in the artificial reefs at different time intervals.

DISCUSSION

Fish Assemblage Structure and Reef Age

The long-term assessment of the artificial reefs on the northern coast of Rio de Janeiro revealed that fish assemblages are still not stabilized after a decade since the first deployment of the artificial structures in 1996. Our results apparently contrast to the general premises of Bortone and Kimmel (1991), who stated that fish assemblages associated with artificial reefs would achieve a relative stable structure within 1 to 5 years after reef deployment. Fabi et al. (1999), monitoring an artificial reef closed to fishery in the Adriatic Sea during a decade, found a gradual increase on the richness of reef-dwelling fishes within the first 4–5 years after reef deployment, following by a stabilization in the subsequent years. Although fish richness have apparently leveled off in the late 6–10 years since the deployment of the first artificial reefs on the northern coast of Rio de Janeiro, the high species turnover rates across the entire 10-year period (as reflected in the lack of a single species that occurred in all sampling intervals) indicated that fish assemblage composition and structure continue to change dynamically with artificial reef age. Relini et al. (2002) attributed the nonstabilizing fish assemblages within 10 years after the deployment of a Mediterranean artificial reef to a slow and gradual process of reef colonization, which is modulated in some degree by changes in the benthic community. Together with the influence of macroinvertebrate communities, as previously noted by Brotto and Zalmon (2007), the harsh environmental conditions (e.g., strong bottom currents, turbid waters, and the presence of a polyhaline plume) are probably the key-factors to affect the

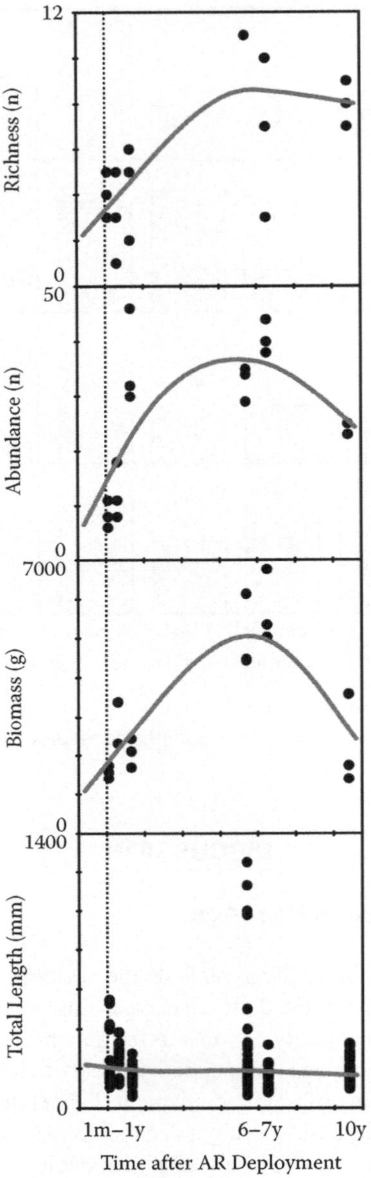

Figure 8.5 Response of fish richness, abundance, biomass, and size with artificial reef age. Lines are the generalized additive models selected by the Akaike information criterion (*x*-axis legend: m = month; y = year).

colonization patterns of fish assemblages in artificial reefs deployed on the northern coast of Rio de Janeiro (Krohling and Zalmon, 2008). However, we cannot determine whether the new structures added subsequently, or natural and anthropogenic variations on fish stocks might have interfered with fish colonization of the artificial reefs.

Three major groups were distinguished from cluster analysis and MDS ordination, and stressed the differences on fish species composition among particular time intervals. Such groups, as supported by ANOSIM and response curves (GAMs) results, also reflected increases or new

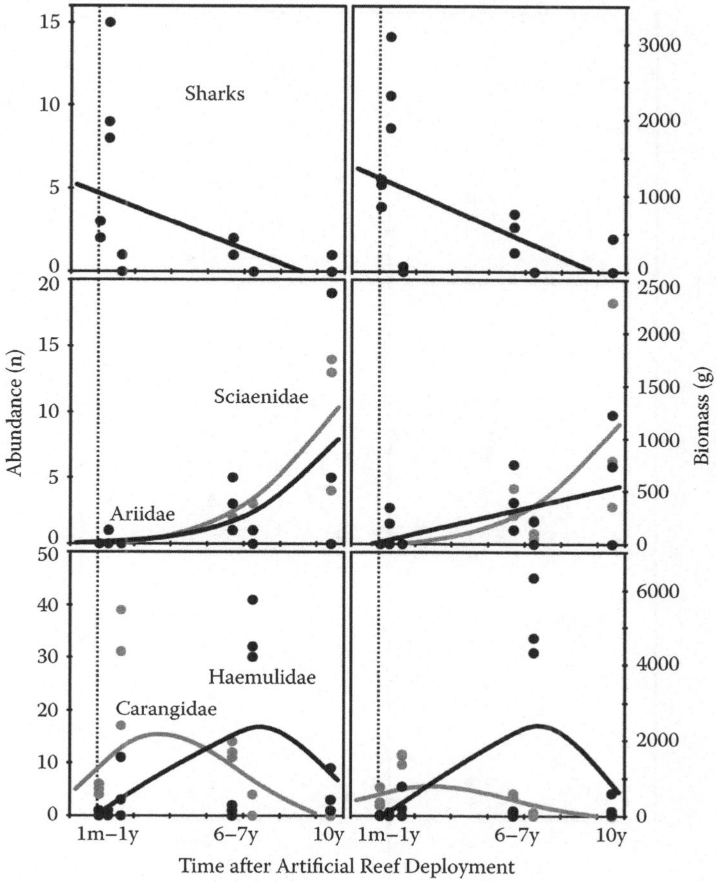

Figure 8.6 Relationship of the abundance and biomass of fish families with artificial reef age. Lines are the generalized additive models selected by the Akaike information criterion (*x*-axis legend: m = month; y = year).

appearances of some species and declines or disappearances of others over time. The clear chronological sequence, as indicated in the <1 year, 1–6 year, and 7–10 year clusters, suggests that temporal changes in fish assemblages were largely influenced by the age of the reefs. The monthly gillnet data on valid control areas to the present study, relative to the first 24 months after the deployment of this artificial reef (Zalmon et al., 2002), also indicated that the new structures deployed in 1997 had no or minor importance on fish use of artificial reefs. However, the fish assemblage composition and structure within the reefs were influenced by large-scale (e.g., regional) variations of fishery resources. Unfortunately, the lack of concurrent surveys on control-structureless sites after 1997 precluded any definite inference on whether the changes on fish assemblage composition and structure during the subsequent periods were actually due to ecological processes within the reef or to regional variations of fish stocks.

Implications on Local Fisheries

Assuming that artificial reef usage on the northern coast of Rio de Janeiro were largely influenced by reef age and, at some degree, by regional changes on fish stocks, there are important issues arising from our findings with direct applied interest on fisheries management. Perhaps the most important

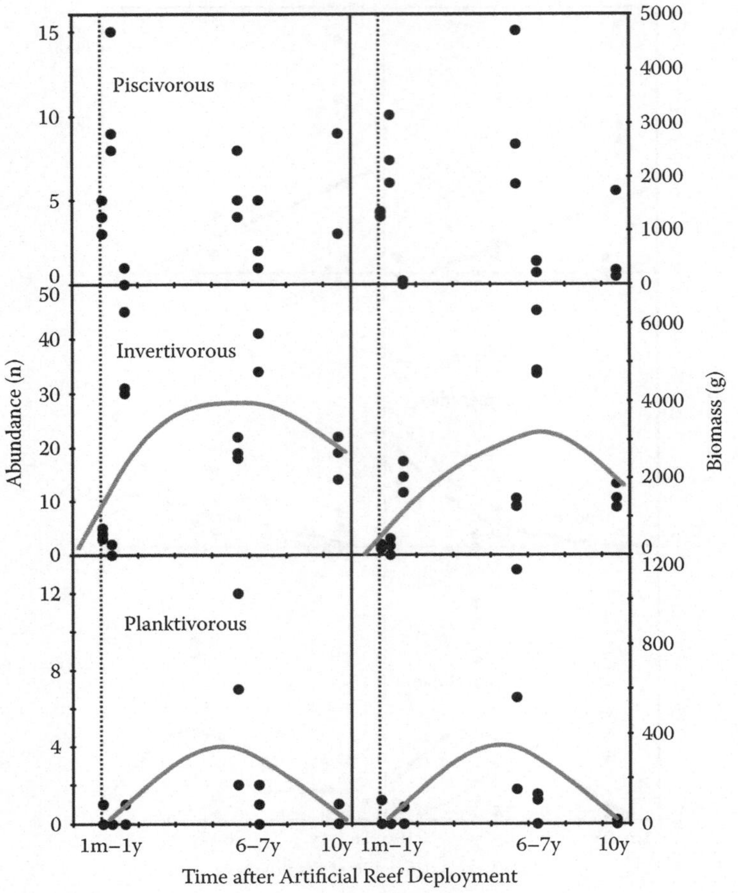

Figure 8.7 Relationship of the abundance and biomass of trophic guilds with artificial reef age. Lines are the generalized additive models selected by the Akaike information criterion (x-axis legend: m = month; y = year).

outcome, considering the general trends recorded for abundance and biomass, is that artificial reefs become more profitable for exploitation by local fishermen within the 6–7 years after the deployment of the structures. If richness and fish size were added to the previous analysis, the most productive peak for fisheries could be even narrowed to the 6th year after artificial reef deployment. Again, the gradual process of reef colonization by benthic organisms and fishes (Brotto et al., 2006b; Krohling et al., 2006), which seems to be critically influenced by the locally harsh environmental conditions (Krohling and Zalmon, 2008), must be taking into account by managers concerned to improve fishery resources along the northern coast of Rio de Janeiro through the use of artificial reefs. It is important to notice that the reef balls added in 2002, by providing additional substrate and shelter for colonization of new invertebrates and reef-dwelling fishes, probably contributed to the increased fish abundance and biomass within the 6–7 years after the deployment of the first structures. It is largely improbable, however, that the temporal variations on fish composition and structure were related to changes on fishing effort, since the artificial reefs were never regulated to commercial fisheries but, still, they have been seldom exploited by local fishermen (I.R. Zalmon, personal observation).

The potential applied use of artificial reefs for local fishermen is, however, relatively more complex when temporal changes on fish families and trophic guilds are analyzed. For example, the sharks, represented by triakids and carcharhinids, had a strong contribution for the captures within the first year after artificial reef deployment, but they were replaced by carangids, namely *C. chrysurus*, during the first to sixth year, and the later was replaced by sciaenids in the last seventh to tenth years. Such variations probably explain why no temporal response (GAMs) was selected for piscivorous, since most sharks, carangids, and sciaenids prey predominantly on fishes. However, the major implication for fishermen is that a higher-prized resource (e.g., sharks) was replaced by others of relative lower value (carangids and sciaenids). Thus, if fishermen wish to maximize their profits by capturing the most available resource at each time interval, they should use less-selective gears (e.g., trammel nets) that are fairly effective in catch all the groups or, alternatively, they could change fishing gears over time to improve the catches over the dominant group (e.g., changing from longlines during shark phases to gillnets or trammel nets for carangids–sciaenids phases).

Together with sciaenids, the haemulids and ariids contributed more to the total fish captures in the artificial reefs after the sixth year, with some species, particularly haemulids, having a high value in the local market. Haemulids and ariids also displayed the greatest temporal response among invertivorous, since both families were largely composed by species that forage preferentially on benthic invertebrates. In addition to the augmented food-resource availability, benthic invertebrates may have played another important role as modeling agents of the fish assemblages associated with artificial reefs. There are evidences that marine macroinvertebrates can increase the complexity of submersed structures through biofouling (Krohling et al., 2006; Krohling and Zalmon, 2008), expanding the functional role of the reefs to yield fish. Other studies have documented a numerical dominance of haemulids at artificial reefs of varied materials and complexity (Godoy and Coutinho, 2002; Cunningham and Saul, 2004; Freitas et al., 2006), in which the intensive use of complex structures was largely related to shelter or nursery purposes. The haemulids, which are likely to be more reef associated and have the narrower home range than the other prevalent families in our study (Ferreira et al., 2001; Godoy and Coutinho, 2002), can be used as an indicator group to evaluate whether the changes on the complexity of artificial are dictated by biofouling.

Despite the greatest captures of haemulids after the sixth year, this family experienced an apparent decline in the last survey period (e.g., tenth year), coinciding with the peaks of ariids and sciaenids. We cannot entirely discard that this decrease could be a result of competitive or predatory relationships with ariids and sciaenids, respectively, but a recent study, performed on the same period after artificial reef deployment, suggested that rather than a true decreasing of haemulids, there is a seasonal alternating dominance between haemulids and ariids–sciaenids (Santos et al., 2010). Such changes on dominant families were also attributed to a cyclical switching of the more clear and oceanic waters near the bottom during the rainy periods, when reef-associated species largely dependent on vision to obtain food are prevalent (i.e., haemulids) in relation to the more turbid and brackish waters during the dry periods, when species not relying on complex habitats or clear waters to find food predominate (i.e., ariids and sciaenids). Therefore, if local inshore fishermen wish to take the best advantage of the fish stocks associated with artificial reefs, they should adjust their fishing gears and capture methods to exploit the haemulids during the periods of clear-saline waters and the ariids–sciaenids during the seasons of more turbid-brackish waters. Again, the reef balls added in 2002, which are quite different in shape and complexity from the other modules formerly deployed, could have contributed to the dominance of sciaenids, haemulids, and ariids after the 6th year within the artificial reef.

Suggestions for Assessing Artificial Reefs in Fishery Management

Our study demonstrated that long-term assessments of fish assemblages associated with artificial reefs are essential to conduct management actions that effectively improve capture by local fishermen and minimize the risks of detrimental effects (i.e., fishery failure or overfishing). For instance, our one-decade assessment was decisive to detect significant temporal changes on the composition and structure of the reef-associated fish stocks, which still continue to occur dynamically over a decade since the first artificial reefs were deployed. Despite their undoubtedly applied importance for local inshore fishermen, if large-scale programs based on the deployment and management of artificial reefs will be carried out on the northern coast of Rio de Janeiro, our findings also argue for the necessity of a well-planned sampling design to improve the scientific outcomes of long-term assessments.

Perhaps the major restriction of our study is the absence of comparable control sites without artificial reefs after the second year of samplings. Unfortunately, the lack of concurrent surveys at control-structureless sites after 1997 precluded any definite inference on whether the deployment of reef balls in 2002, the regional variations (both naturally induced or human-induced) of fish stocks, and those ecological processes occurring especially within the reef complex would have led individually to our long-term findings. The level of fishing effort should be also taken into account with regard to the frequency and effectiveness of experimental samplings (Seaman, 2000) but also in relation to the amount of fish caught in local fisheries (Bombace et al., 1994). As already commented before, we are not aware of significant variations on the level of fishing effort applied by local fishermen to the artificial reefs, but the apparent few number of fishes and species recorded through gillnets could suggest that the experimental sampling effort was not so effective. However, this appears to be not the case when comparing the total number of species recorded in this study ($N = 41$) to that found ($N = 45$) by Zalmon et al. (2002) through an extensive 2-year gillnet monthly sampling at the same reef complex, or when the relatively high mean values of CPUE (1.4 ± 0.2 S.E. fish or 178.7 ± 22.8 S.E. gr per 100^2 gillnet. h^{-1}) recorded in our study were considered. Nevertheless, it is expected that our results would be largely improved, if the surveys were performed in shorter time intervals, since the temporal changes on fish stocks could be thus assessed at more refined scales.

In summary, our results suggest that adding artificial reefs along the north coast of Rio de Janeiro can be an interesting management tool to improve the catches of fish stocks by local inshore fishermen. Our study further revealed that important changes, relative to the abundance and biomass of the fish assemblages and to the dominance of the major fish stocks, occurred mostly with ecological processes within the reef complex, but also due to regional variations of fishery resources and to the further deployment of new reef modules. Therefore, assessing control sites simultaneously with other manipulated areas with artificial reefs are critical to appraise the individual contribution of these factors on the associated fish stocks over extensive temporal series.

ACKNOWLEDGMENTS

This work was funded by North Rio de Janeiro State University—UENF, Brazil (postdoctoral research grant to L.N.S.; code 36193-1), Brazilian Research Foundations—FAPERJ (E26/152.540/2006), and CNPq (479546/2008-8).

REFERENCES

Amaral, A. C. Z. and S. Jablonski. 2005. Conservation of marine and coastal biodiversity in Brazil. *Cons. Biol.* 19: 625–631.

Anderson, M. J. 2005. Permanova: A Fortran computer program for permutational multivariate analysis of variance. Department of Statistics, University of Auckland, New Zealand.

Baine, M. and J. Side. 2003. Habitat modification and manipulation as a management tool. *Rev. Fish. Biol. Fish.* 13: 187–199.

Bombace, G., G. Fabi, L. Fiorentini, and S. Speranza. 1994. Analysis of the efficacy of artificial reefs located in five different areas of the Adriatic Sea. *Bull. Mar. Sci.* 55: 559–580.

Bortone, S. A. and J. J. Kimmel. 1991. Environmental assessment and monitoring of artificial habitats. Pages 177–236 in Seaman, W. and L. Sprague, Eds. *Artificial Habitats for Marine and Freshwater Fisheries*. Academic Press, San Diego.

Brotto, D. S., W. Krohling, and I. R. Zalmon. 2006a. Usage patterns of an artificial reef by the fish community on the northern coast of Rio de Janeiro. *ICES J. Coast. Res.* 39: 1277–1281.

Brotto, D. S., W. Krohling, and I. R. Zalmon. 2006b. Fish community modeling agents on an artificial reef on the northern coast of Rio de Janeiro–Brazil. *Braz. J. Oceanography* 54: 205–212.

Brotto, D. S. and I. R. Zalmon. 2007. The effect of artificial reef structural complexity and benthic colonization on gill net fish assemblages. *Trop. Oceanography* 35: 1–16.

Burnham, K. P. and D. R. Anderson. 1998. *Model Selection and Inference*. Springer-Verlag, New York, 353 pp.

Clarke, K. R. 1993. Non-parametric multivariate analyses of changes in community structure. *Aust. J. Ecol.* 18: 117–143.

Clarke, K. R. and R. H. Green. 1988. Statistical design and analysis for a "biological effects" study. *Mar. Ecol. Prog. Ser.* 46: 213–226.

Cunningham, P. T. M. and A. C. Saul. 2004. Spatial partition of artificial structures by fish at the surroundings of the conservation unit—Parque Estadual da Ilha Anchieta, SP, Brazil. *Braz. Arch. Biol. Technol.* 47: 113–120.

Fabi G., F. Grati, F. Luccarini, and M. Panfili. 1999. Indicazioni per la gestione razionale di una barriera artificiale: studio dell'evoluzione del popolamento necto-bentonico. *Biol. Mar. Medit.* 6: 81–89.

Ferreira, C. E., J. E. Gonçalves, and R. Coutinho. 2001. Community structure of fishes and habitat complexity on a tropical rocky shore. *Environ. Biol. Fish* 61: 353–369.

Freitas, L. E., C. V. Feitosa, and M. E. Araújo. 2006. Mangrove oyster (*Crassostrea Rhizophorae*) (Guilding, 1928) farming areas as artificial reefs for fish: A case study in the State of Ceará, Brazil. *Braz. J. Oceanography* 54: 31–39.

Godoy, E. A., T. C. Almeida, and I. R. Zalmon. 2002. Fish assemblages and environmental variables on an artificial reef north of Rio de Janeiro, Brazil. *ICES J. Mar. Sci.* 59: 138–143.

Godoy, E. A. S. and R. Coutinho. 2002. Can artificial beds of plastic mimics compensate for seasonal absence of natural beds of *Sargassum furcatum*? *ICES J. Mar. Sci.* 59: 111–115.

Krohling, W., D. S. Brotto, and I. R. Zalmon. 2006. Functional role of fouling community on an artificial reef at the northern coast of Rio de Janeiro state, Brazil. *Braz. J. Oceanography* 54: 183–191.

Krohling, W. and I. R. Zalmon. 2008. Epibenthic colonization on an artificial reef in a stressed environment off the north coast of Rio de Janeiro State, Brazil. *Braz. Arch. Biol. Technol.* 51: 215–223.

Lepš, J. and P. Šmilauer. 2003. *Multivariate Analysis of Ecological Data Using CANOCO*. Cambridge University Press, Cambridge, 269 pp.

Manly, B. F. J. 1997. *Randomization, Bootstrap, and Monte Carlo Methods in Biology*. Chapman & Hall, London, 428 pp.

Relini, G., M. Relini, G. Torchia, and G. Palandri. 2002. Ten years of censuses of fish fauna on the Loano artificial reef. *ICES J. Mar. Sci.* 59: 132–137.

Relini, G., M. Relini, G. Palandri, S. Merello, and E. Beccornia. 2007. History, ecology, and trends for artificial reefs of the Ligurian Sea, Italy. *Hydrobiologia* 580: 193–217.

Santos, L. N., D. S. Brotto, and I. R. Zalmon. 2010. Fish responses to increasing distance from artificial reefs on the Southeastern Brazilian Coast. *J. Exp. Mar. Biol. Ecol.* 386: 54–60.

Santos, M. N. and C. C. Monteiro. 2007. A fourteen-year overview of the fish assemblages and yield of the two oldest Algarve artificial reefs (southern Portugal). *Hydrobiologia* 580: 225–231.

Seaman, W. 2000. *Artificial Reef Evaluation with Application to Natural Marine Habitats*. CRC Press, Boca Raton, 246 pp.

Seaman, W. and L. Sprague. 1991. Artificial habitat practices in aquatic systems. Pages 1–29 in Seaman, W. and L. Sprague, Eds. *Artificial Habitats for Marine and Freshwater Fisheries*. Academic Press, San Diego.

Spanier, E. 2000. Changes in the ichthyofauna of an artificial reef in the southeastern Mediterranean in one decade. *Sci. Mar.* 63: 279–284.

Valentin, J. L. and W. M. Monteiro-Ribas. 1993. Zooplankton community structure on the east-southeast Brazilian continental shelf (18–23°S latitude). *Cont. Shelf Res.* 13: 407–424.

Zalmon, I. R., R. Novelli, M. P. Gomes, and V. V. Faria. 2002. Experimental results of an artificial reef program on the Brazilian coast north of Rio the Janeiro. *ICES J. Mar. Sci.* 59: 83–87.

CHAPTER 9

Artificial Reefs in North Cyprus

An Opportunity to Introduce Fishermen to Sustainable Development

Burak Ali Çiçek

CONTENTS

The reason for low fishing productivity in North Cyprus seas manifests as unwanted pressure on sensitive habitats, nursery grounds, and nontarget species, some of which are protected by international regulations and legislation. It is assumed that artificial reefs contribute positively to the productivity of fisheries in areas with bare substratum in their natural setting. Artificial reefs to be deployed on the North Cyprus bare substratum biotopes (rocky, sandy, and muddy) should result in an increase in productivity and have positive effects on fisheries. Making the fisheries more productive would increase support from fishermen, which is often considered to be a rare event under these conditions. Artificial reefs, therefore, can be used to achieve sustainability of the fisheries sector in North Cyprus. The economic, social, and environmental components of the proposed project "Sustainable Fisheries and Artificial Reefs on the İskele Coasts—North Cyprus" should facilitate the establishment of integrated management of fisheries in the area.

INTRODUCTION

Sustainable Development and Sustainable Fisheries

The concept of sustainable development is the integration of short- and long-term economic, social, and environmental effects and values in all decision making (Fletcher et al., 2005). The concept of sustainable development has become accepted worldwide after "Our Common Future" (Brundtland, 1987) was published by the World Commission on Environment and Development. Over the past 20 years, the gradual development of both theoretical and methodological approaches of sustainable development has led to vast disagreements due to different interpretations. The World Bank, as an example of a multinational NGO (nongovernmental organization), would approach an environmental project differently from an environmental organization that works in the field of nature conservation and preservation (Hamdouch and Zuindeau, 2010).

The principles of sustainable development are highly relevant to fisheries management (Fletcher et al., 2005). Fishing is an important activity throughout the world, contributing to the livelihoods of 200 million people and provides in excess of 100 million tons of fish and fish products annually for which over a billion people are dependent for their protein, cultural, and social needs (FAO, 1999). However, the direct impacts of fishing (e.g., target stocks), combined with other human-induced changes to the environment (e.g., pollution and habitat removal), and significant advances in technology have led to many situations where these activities have clearly not been sustainable (Mace, 1997).

While the high-level objectives of sustainable development are relatively simple in concept, the translation of high-level objectives into operational objectives at the fishery management plan level has proven difficult to achieve (Garcia, 2000).

Sustainable development of fishing has also been evaluated and incorporated into their agenda by the Europeans. Consequently, new policies and strategies have been adopted (Symes, 2005) in Europe. Currently, in the European Union (EU), a sustainable balance between resources and fishing capacity is one of the main objectives of the common fisheries policy (CFP) (Madau, Idda, and Pulina, 2009). Moreover, to improve the status of the exploited living resources and achieve sustainability in European marine regions requires constructive efforts at all levels (Rätz et al., 2010).

Mediterranean fisheries play an important socioeconomic role in Europe's fishing industry. In Europe, fishing represents 42% of employment in the EU-catching sector and contributes 12% to EU catches. Because of the narrow Mediterranean continental shelf, this includes mostly coastal waters (Farrugio et al., 1993). Although Mediterranean coastal fisheries are basically traditional, small-scale, and low-investment activities, they are also ecologically and socially complex (Morales-Nin, Grau, and Palmer, 2010).

Given that there is ecological and social complexity, this new pilot project (i.e., "Sustainable Fisheries and Artificial Reefs on the İskele Coasts—North Cyprus") seeks to deploy artificial reefs to achieve sustainability in North Cyprus. To integrate fisheries management into a broad range of economic, social, and environmental objectives in North Cyprus, this pilot project was initiated in the İskele Region by the Eastern Mediterranean University Underwater Research and Imaging Center (EMU-URIC). For the first phase of the project, a €300.000 grant was received from the European Union (Rural Development Program—Council Regulation (EC) No 389/2006 establishing an instrument of financial support for encouraging the economic development of the Turkish Cypriot community). The project will be implemented in collaboration with the İskele–Boğaz Fishermen Solidarity Association and İskele Municipality.

Ecological Characteristics of North Cyprus Coastal Zone

The Eastern Mediterranean is known for its low productivity and extreme oligotrophy (Azov, 1991). This is especially descriptive of the Levant Basin, which includes the island of Cyprus, as it is defined as ultra-oligotrophic (Krom et al., 2005) or exceptionally oligotrophic (Herut et al., 2000). Consequently, the basin is considered to have “clean and clear waters” and is favored by the tourism sector. However, the clearness of these waters indicates low levels of biological productivity, and fisheries productivity is almost three times lower when compared with Western Mediterranean waters (Turley, 1999).

Even though Mediterranean waters, in comparison to other seas, are low in productivity, some of the species listed in the biological inventory have international importance. Some of these species are whales, dolphins, monk seal (Gücü et al., 2004), sea turtles, various fish species, invertebrates, and aquatic plants (Çiçek, 2007). Additionally, monk seal, sea turtles, and some of the fish species frequent the shores and coastal areas of the basin. From a fisheries point of view, many of these species are considered nontarget species and are not preferred by the fishermen (Çiçek and Yerli, 2007).

However, due to the use of traditional fishing gears (trawl and purse seines have been banned since 1999) in the North Cyprus coastal zone, conflicts among the fishermen and the species have developed with regard to usage of the area. These conflicts are in use of the area as a fishing ground and an area where species nest, feed, and find shelter. Consequently, death to organisms and damage to habitat can be directly linked to the intense fishing activities by the fishing community (Çiçek and Yerli, 2007).

The Mediterranean ecosystem is sensitive and has its own characteristics. There is evidence that there has been dense human usage in the area throughout history. Notably, the density of human populations and the damage inflicted as a consequence on nature are factors still affecting the environment today. Therefore, the biotopes in the region, which are providing a living environment for several important species, must be protected (Bianchi and Morri, 2000).

Interestingly, there is a greater effort in fishing activities in Levant Basin due to the low productivity in the Eastern Mediterranean and the seas of North Cyprus. Moreover, as a result of low productivity in the region, there is a higher possibility of sensitive habitat-fishing ground conflicts and more illegal fishing activities. This situation increases pressure on juveniles and nontarget species.

Rural Areas in North Cyprus

The current unresolved political situation in Cyprus has resulted in the isolation of the North Cyprus from development by Europe and the rest of the world (Ghosh and Aker, 2006; Yaşarata et al. 2010). In relation to this, the economic recession in the northern part of Cyprus resulted in all the sectors, especially the rural sectors, to economically lag behind other neighboring areas. This state of affairs has also had negative effects on the social and cultural status of the people living in the rural areas. Over many years, it has been observed that there has been a serious migration of individuals from rural communities to more urban areas for monetary, social, and cultural fulfillment. Many rural areas have been abandoned and are now considered as “weekend getaways.”

Additionally, authorities and other administrative bodies failed to satisfy the needs of various public sectors to encourage business development in the rural areas due to the lack of funding, inadequate technical infrastructure, and not being able to employ enough personnel. Therefore, the rural sectors could not advance economically or develop themselves. As a result of these negative factors, the sectors in the rural areas could not succeed in catching up with international standards, could not implement proper management methods in terms of administration and marketing. Moreover, the residents of the rural areas failed to develop their technical capacity in various sectors and resorted to using “inherited techniques” in farming, fishing, and other areas. In other words, the rural communities failed to employ modern and up-to-date techniques, and have closed themselves to international developments. Subsequently, they lost their competitive edge in the market and became more dependent on the central government. Eventually, because the rural communities failed to develop themselves, the system implemented was far from ever becoming sustainable.

Instead of solving the problems of the local people, the method of "avoiding the problem" was adopted, and the activities in the rural areas were abandoned. The rural sectors have become secondary (second job or as a hobby) to more financially rewarding areas such as civil service. In time, the rural areas were abandoned and migration toward cities or more densely populated areas was initiated. This unfortunate reality, keeping in mind the fact that the resources are limited, would undoubtedly create major socioeconomic problems in the near future.

The results of a problem analysis conducted with the residents and stakeholders from the rural areas revealed that the first step to take would be to increase the affluence of the people in these areas. The social and cultural development of the rural areas can only be established through the people living in the area and by creating an ownership mentality among themselves. This can only be achieved in societies that have attained a certain level of affluence or wealth.

The Fisheries and the Fishermen in North Cyprus

Fishing can be described as "hunting" through the use of traditional fish nets in the northern part of Cyprus. Methods additional to fish nets such as trawl, purse seine, etc., have been banned since 1999 because of the damage these methods inflict on the environment and to fishery resources. The fishing sector has failed to catch up with the international standards in terms of sanitation, productivity methods, hygiene, marketing, environmental conservation, etc. It is estimated that about 500 families earn their living solely through fishing in North Cyprus. Nevertheless, it should be noted that there is a high number of families who have abandoned the occupation of fishing because fishing does not provide adequate income for the fishermen nor their families. Though there are neither adequate nor reliable statistical data on the amount of seafood consumed (which is one of the healthiest food sources) or the amount of fish caught in the northern part of Cyprus. The very low productivity results in high retail prices in the local market.

THE PROJECT

Starting Point: Why Artificial Reefs?

The idea for the project came about through meetings held with the fishermen (İskele–Boğaz Fishermen Solidarity Association) in the region, for the purpose of protecting the nontarget species from hunting. It was observed through sea turtle conservation and monitoring surveys, conducted by Eastern Mediterranean University Underwater Research and Imaging Center (EMU-URIC), that sea turtles and other sensitive species are negatively impacted when they come in contact with fishermen. Meetings were conducted with the fishermen to come up with ideas to help minimize the mortality or injuries to these species.

It was through these interactions that fishermen voiced their concerns that the authorities do not pay enough attention to the problems faced by them or that the authorities do not have adequate resources to do so. As a result it became known that local residents whose primary income is generated through fishing had been seriously affected because the fishing nets, which are damaged by these species, were not being replaced through government intervention and the fishermen were forced to find their own solutions to the problem.

Ultimately, the government decided to assist the fishermen in governmental procedures and agreed that all new projects would be conducted in compliance with the principles of sustainable fisheries. Artificial reefs project was selected as the first phase of implementation because of the known positive effects they have on fisheries productivity as witnessed in the Mediterranean (Santos and Monteiro, 1997).

Deploying artificial reefs is a common practice around the world. Goals for these projects include the stabilization of the benthic substrate, prevention of sand movement, increase in fish productivity, and creation of visual attractions for underwater tourism. Many different materials can be used in the making of artificial reefs; however, concrete is generally the most preferred material. Concrete structures are suitable for the marine environment. They are durable, environmentally friendly, and can last for a long time. Concrete is easy to obtain and can be shaped into different designs. Moreover, concrete is a suitable product for the development of prefabricated blocks (Lukens and Selberg, 2004).

Japan is the only country with standards for reef construction although some provincial and state jurisdictions have established local standards. The first detailed guide to artificial reef construction was published in 1978 as part of the Coastal Fisheries Development Program (CFDP) as the "Structural Design Guide." These guidelines were further refined in 1984 in which standards were established for artificial reefs composed of steel, concrete, plastic, and similar materials. The guidelines were published as part of the Japan Coastal Fisheries Promotion Association (Grove et al., 1989).

The artificial reefs deployed in Turkey have been made out of concrete. However, in order to decrease the manufacturing cost of concrete blocks, additional materials (such as rubble, plastic, mud, and ash, etc.) could be included in the structure (Düzbastılar and Lök, 2004).

Algal reproduction and development is enhanced on reef blocks. Subsequently, attached algae attract fish populations. Concrete provides feeding grounds and shelter for invertebrates and fish, in addition to serving as a substrate for fouling organisms—allowing them to settle and to grow (Lukens and Selberg, 2004).

As a result of discussions with the fishing community, researchers, and the government, it was decided that an initial step toward resolution of the situation in North Cyprus would include the deployment of artificial reefs in the İskele region. It is presumed that this initial pilot project, and its subsequent application, will eventually spread to North Cyprus through cooperation and collaboration among the parties concerned. The main objective of the project is to minimize the pressure, caused by development on natural habitats in the North Cyprus, and at the same time provide the structure whereby the fishermen and their families would reach a level of wealth both economically and socially. The foundation for realizing the main objective for the project is to establish sustainable fisheries whereby the balance among ecological conservation and usage is maintained.

CURRENT SITUATION IN ISKELE FISHERIES (PILOT PROJECT AREA)

During the past two years, the current project proposal attained a level whereby it can be presented to potential supporting organizations. In the initial phase of this process, contact with the fishermen was established, and the current status of each sector was identified and summarized. After an initial meeting was conducted at the İskele–Boğaz fishermen shelter, a meeting with fishermen from the area was held at one of the meeting rooms of the Eastern Mediterranean University in February 2007. Following that meeting, additional public meetings were organized and questionnaires were distributed, to be able to reach the other fishermen, in May 2007, July 2007, February 2008, and April 2008. During these consultations with the stakeholders, the completed questionnaires were collected. The analysis detailed the current situation and the problems faced by the fishermen. Additionally, "marine surveys and evaluation of the area structure" were conducted as agreed after the first meeting. Since then, the EMU Underwater Research and Imaging Center (EMU URIC) organized activities toward identifying the current problems of the fishermen. In attempting to solve these problems, various investigations were conducted and contacts were made to maintain a liaison among the related organizations.

During the meetings and stakeholder consultations, the fishermen identified and prioritized their problems:

- Insurance firms not insuring the fishermen's boats and their equipment against theft, fire, and/or similar situations (50% of respondents).
- Lack of storage and sales facilities resulting in not being able to sell their catch directly to the consumer and therefore needing a "middleman" (30%).
- The middlemen actually making more than 100% profit on the sales of the catch, which is reflected as a "loss" for the fishermen (90%).
- The catch, with low profit margins, not being bought by the middlemen on the basis of "lack of space in the transporting vehicle" and therefore that catch not being presented to the consumer (90%).
- Administration not compensating the fishermen for damaged or destroyed nets, even though it is the same administration that had taken species like sea turtles, dolphins, and monk seals under protection (100%).
- The existence of illegal fishing activity (net, multihooked fishing line, harpoon) and the related administrative body not taking any action to prevent those activities (20%).
- Related administrative body not supporting the sector (technically, legally, financially) (80%).
- Low productivity in the region (50%).
- The fishermen cannot find any organization to demonstrate new fishing techniques, international standards, and marketing techniques (40%).
- People not being familiar with the species caught. These fish generally did not reach the retail markets and, therefore, people do not consume the catch (20%).
- Lack of public awareness about fish/seafood consumption and the preference of frozen or canned fish/seafood as opposed to fresh in retail stores (70%).

Further detailed analysis of the problems confirms that these were important issues proffered by the fishermen. As an example, after the meeting, there were two fires on boats, and a storm damaged a boat on another occasion. None of the fishermen were compensated because insurance firms did not insure them or their boats. A partial compensation was made through the Association to these fishermen as the fishermen did not have additional income other than from fishing. Based on records, the fishing nets were damaged by nontarget species in 345 different incidents in 2008. During field trips with the fishermen, it was observed that the middlemen (i.e., distributors) did not purchase certain species. These fish were at marketable size and taste (4.6 kg per catch). Additional field surveys were conducted at different points of the bay, in different seasons, and at different depths to independently measure fishery productivity by setting nets. Additionally, the bottom structure was surveyed and habitat was classified using SCUBA. This investigation revealed that the catch levels in the area are at least 30% lower compared to Kyrenia, Morphou, and Karpaz coasts, which feature rocky bottom structure (Çiçek and Yerli, 2007).

Project Components

The concept of deploying "artificial reefs" and practicing sustainable fisheries management was first introduced to the fishermen, with detailed objectives and activities, during the February 2008 meeting. At this time, fishermen were asked to provide input and approval. The fishermen took a vote, and the resulting vote was unanimously positive. The participants stated that they would like to see activities initiated that would allow the project to succeed. The fishermen, who were not present at the meeting, were subsequently asked to state their opinions, and none of them objected to the idea.

For this project, the initial groups were identified as people living in the area. Members of this sector include fishermen, hunters, officials, and assistant personnel. The primary target group within this sector was the fishermen who make use of the İskele–Boğaz fishermen shelter, fishermen from the surrounding areas who frequent the same fishing area as the İskele–Boğaz fishermen, support personnel working in the sector, and their families. Moreover, people who had to abandon fishing activities due to low income were considered as part of the target group as well. This project aims to develop sustainable fishing in İskele, Boğaz, and in all nearby villages (villages and rural areas in the Famagusta district, Kumyali, and surrounding villages) as pilot project areas.

The secondary target group was identified as individuals from the sector who do not live in the area. However, it was agreed that the people in the secondary target group would be continually kept informed as to the progress of the project. Moreover, they would be encouraged to participate in the project activities because the problems faced by the primary target group are similar if not identical to those faced by people working in the same sector in the northern part of Cyprus. Therefore, it can be said that all the stakeholders face similar problems. The third and final target group would include the population residing in the northern part of Cyprus. These people are also identified as "consumers."

The proposed project is consistent with the priorities set out in the rural development strategy of EU in North Cyprus and includes almost all conditions, set and prioritized in the strategy. In this respect, the European Union, through the European Commission, had enacted a "financial aid program for the Turkish Cypriot community" and set out a sustainable development program in areas of priority. These areas include vast sectors such as rural development, information communication technologies (ICT), small- and medium-sized enterprises (SME), competitiveness, road safety, health, etc. There are two critical key points to the financial aid program. The first one is to increase the local capacity (in terms of personnel, technical, and structural) and to make the Turkish Cypriot community more competitive in the international market, and the second key reason for implementing the program is to be able to bring the current level of living up to the level of the Greek Cypriot community as a preparation for the day when a solution is established to conclude the "Cyprus issue." For example:

- Create economic development in the rural areas.
- Provide technical assistance to current economic activities in the rural areas.
- Progress in the effort to improve compatibility of European Standards in the fisheries sector.
- Create cooperation among different groups (municipalities—sector representatives—civil society organizations—public) in the rural areas.
- Have a positive effect on the sociocultural level of the people and promote the quality of life through economic development in the rural areas.
- Increase sustainable productivity. The artificial reef project will result in a long-term increase in fish landings. Additionally, the technical assistance provided is a concrete basis for sustainability.

When the project is analyzed from the view point of providing sustainability, it has more than one component. Some of these components will be introduced at the beginning of the project, and some would be introduced with other components of the project. These components, which will be introduced in the field, which have been listed below, are shown on the map in Figure 9.1.

- Fishing reefs
- Nursery (protecting) reefs
- Provide protection zones on the shoreline up to 10 m depth
- Improving marketing techniques (fish market) and cooperative system
- Improvement of fishermen shelter infrastructure (service area for boats, new fishing nets, closed-circuit cable TV (CCTV) and restoration of the shelter)
- Improvement of research and documentation capability (monitoring systems, sampling instruments, nets)
- Capacity building (increasing the capacity of the fishermen to international standards, in terms of hygiene, storage, and different fishing techniques)
- Raise awareness of the consumers

One of the most fundamental elements of the proposed project is the artificial reefs. Artificial reef implementation will feature different types and designs for both fishing and nursery grounds.

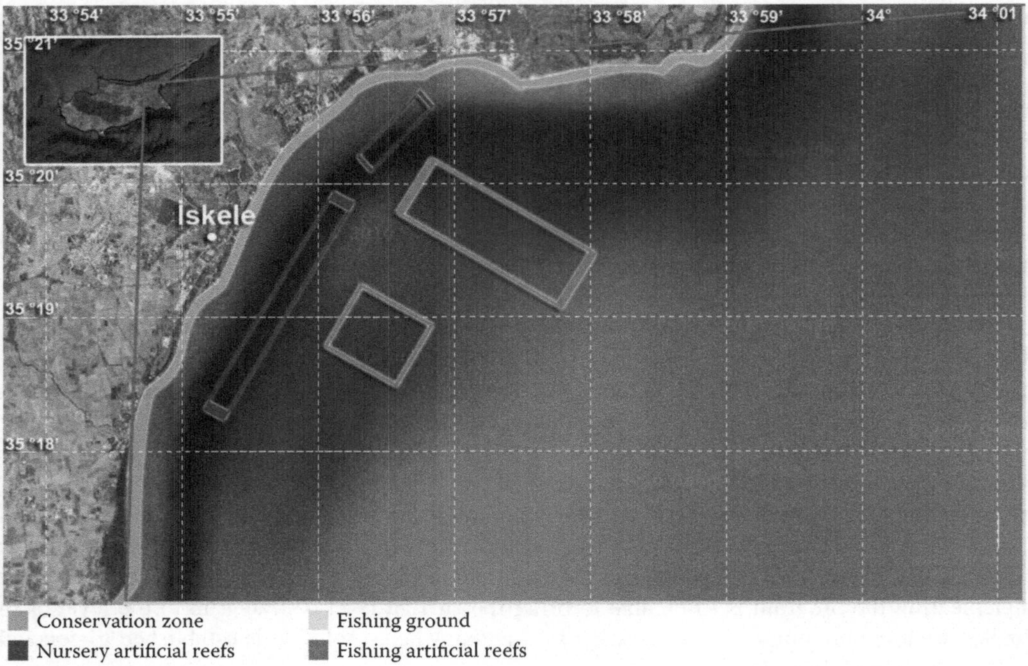

Figure 9.1 ***See color insert.*** Site, fishing ground, artificial reefs, and the conservation zone. Redrawn from images in Google Earth.

Molded (concrete) blocks and briquette models will be used as artificial reef units. Over 2000 block units have been planned to be deposited on the sea bottom to form the artificial reef as part of the first stage. During the next stages, the plan is to increase the reef to include up to 10,000 modules. Eastern Mediterranean University Underwater Research and Imaging Center (EMU-URIC) conducted field research in the area and has identified the most suitable locations for depositing the artificial reef blocks. Additionally, administrative permits for building artificial reefs have been obtained from government authorities who have responded positively to the project.

The blocks will be loaded on "carrier boats" and the artificial reefs will be built under the supervision of the İskele–Boğaz Fishermen Solidarity Association and the İskele Municipality. The manufacturing and technical monitoring of the blocks will be the responsibility of experts working with Eastern Mediterranean University Underwater Research and Imaging Center (EMU-URIC). In this regard, meetings with experts have been conducted, and their contribution and commitment has been confirmed. The installation technique and specific positioning of the artificial reef units will be identified by Eastern Mediterranean University Underwater Research and Imaging Center (EMU-URIC). All work will be visually inspected.

The fishing artificial reefs and nursery artificial reefs, which will be built through the project, should increase the direct income of local fishermen. In addition, these reefs will serve to protect nontarget species and nursery grounds in the area. As part of this project, conservation zones will be established for nontarget species. This includes sea turtles *Chelonia mydas* (Linnaeus, 1758) and *Caretta caretta* (Linnaeus, 1758) and monk seal *Monachus monachus* (Hermann, 1779) as they use the same area for feeding and nesting. The application of "Establishing Conservation Zones" has also proven valuable in other areas for the continuation of the fishing activity because conservation zones lower the fishing pressure put on juveniles through capture. During the meetings with stakeholders, it was agreed that the nursery artificial reefs would be placed in such a way that the reefs would not allow fishing and, therefore, conservation areas would be created.

The proposed project would organize fishermen under the umbrella of a cooperative agreement system. At the moment, the status of the association allows only direct sales of the catch, but after an analysis of the legislation, it was resolved that the association will be converted into a cooperative organization. The firmest support mechanism for the proposed project would be the creation of a "fish market." The fish market concept that would allow fishermen to store (cold storage units and an ice machine) and sell wholesale or retail directly to the consumer. It is anticipated that the utilization of the fish market will increase the fishermen's income level by at least 70%–80% after expenses have been deducted. In addition to the increase in the income levels of the fishermen, a financial pool system would be created, which will serve the fishermen as an insurance system. This pool would be utilized to solve the problems of the fishermen, such as not being able to insure their boats against storms or disasters, or nets being damaged by nontarget species, etc. The pool system would also cover the administrative costs (administration, follow-ups, lobbying, etc.), and will allow fishermen to obtain the latest technology, learn of international standards, and facilitate further investments for and by the fishermen. It is expected that the cooperative system would be highly profitable and, therefore, would be a sustainable approach to business.

Until the cooperative system with the insurance component is operational, fishing nets will be distributed to the fishermen on a "one-time basis" to allow fishermen to immediately realize an increase in fish catches. The distribution of the nets among the fishermen would be conducted and controlled by the association. Additionally, scientific research will be conducted on the relative to the association between distributed nets and the catch obtained using the nets. This research should indicate the relative size of the local fish population along with the frequency of catching the nontarget species.

Building a new service area for fisherman vessels will help decrease maintenance expenses for fishing. Additionally, the new service area would be used during the installation of the artificial reefs. The proposed project also contains a component for harbor security. For the harbor to have a centralized control, close-circuit cable TV (CCTV) will be installed. By doing so, the losses as experienced in the past owing to theft and fire should be minimized. The installed close-circuit cable TV (CCTV) will be monitored by the İskele Municipality and the fishermen association.

The proposed project includes the purchase of diving gear and underwater imaging systems so that the information obtained to determine the effectiveness of the artificial reef system can be shared with the local stakeholders and the public at large. Additionally, with the scuba gear and the imaging systems, the project activities will be documented. It is envisaged that all visual materials (i.e., photographs, video, etc.) obtained through the project will be edited and presented as a documentary to the national and international audiences. Moreover, the research conducted will be used in scientific presentations and analyses to determine if the reef has met its fishery objectives. The data obtained through monitoring work will be publicized through popular communication media. Additionally, the decrease in competition between the nontarget species and the fishermen, which is most often presented as an environmental problem, is an added benefit that should have a larger impact on local fish community structure.

The project also proposes to offer a series of seminars under capacity building component. Presentations of the training program will help increase the level of understanding of the target group and the partners in technical and administrative issues. Experts will be invited to provide technical assistance to stakeholders on the "cooperative system," its applications, administrative responsibilities, and day-to-day operations. The training program will include hygiene sanitation, fishing technology, and nature conservation issues also. The seminars will also help these stakeholders to become aware of international standards and increase the target group's awareness of the importance of conservation in maintaining sustainability. Furthermore, all the activities will be offered to other sector groups in different areas and will be shared with them through booklets, posters, and brochures. This would eventually lead to an increased fishing capacity in the sector in North Cyprus. An effective program such as this, based on very simple principles, would create great excitement and enthusiasm among stakeholders throughout North Cyprus.

The public at large—the ultimate beneficiaries, would have access to healthier, protein-rich seafood source. Eventually, the public would realize this food at cheaper prices through the increase in fishing capacity. Additionally, public awareness of seafood would be promoted through books, booklets, and brochures.

DISCUSSION

Hamdouch and Zuindeau (2010) argue in a detailed evaluation study of sustainable development that any good sustainable development application must successfully combine "theory—methodology and practical" aspects. However, in practice, people, in general, are reluctant to move forward from classical methodologies. A project or an initiative such as this must be multifaceted and must introduce new techniques and innovative methodologies in order to be successful. In this respect, the artificial reef deployment in the İskele area proposes to enrich the productivity of local fisheries and allows the fishing section access to an additional dimension. In the larger view, the project will help fishermen understand and appreciate sustainable development principles. Because, as Garcia (2000) states, sustainable development theories are quite different from applications of sustainable development principles, the situation arises when there is interest to be gained for a certain party. The viewpoints of group(s) that are earning an income directly from the sector become more important and grow to be key factors for providing a basis for sustainability. In other words, it can be simply concluded that successful sustainability depends on these groups.

Nevertheless, it should be noted that a sustainable fisheries system is not going to be created for the fishermen only. As Hamdouch and Zuindeau (2010) argue, one of the biggest difficulties in any sustainable development application is that almost all the stakeholders have different perspectives on the issue. The priorities of these stakeholders, their interests, and their general approach could differ from one another. Due to these factors, any negative approach to the sustainable development application can slow down the entire process, and/or the desired solution may not be achieved. As mentioned previously, the artificial reefs project, at the initial stages, will use the positive effect of the implementation of artificial reefs on the fish biomass as a way of creating a continuum in the sector. Nevertheless, successful sustainability will be achieved through implementing radical changes in the administration and execution of the fisheries in addition to increasing the productivity. Similar improvements and changes had been suggested by universities, environmentalist civil society organizations, and other stakeholders such as the Animal Husbandry Department before; however, due to lack of support from the fishermen in the area, the suggestions remained as suggestions and never applied. The main reason for not applying the suggestions into everyday practice was a decrease in the fishermen's income at the initial implementation phase. Anyhow, it has not been possible for the fishery sector to focus on long-term applications because of the low productivity levels and limited resources in fishing. It is also because of this reason that the authorities could not implement minimum mesh size measures for short-term periods or could not prevent illegal hunting. It should also be noted that even though there are relevant rules and regulations about these issues, neither the relevant authority (Animal Husbandry Department) nor the stakeholders, which are responsible for the implementation of the law upon notification from the competent authority such as the police, municipalities, and the coast guard, have adequate technical and human resources. Moreover, even though there are rules and regulations in place in addition to conservation measures, it has not been possible to regulate marketing principles or collect data on sales.

However, in order to be able to create a sustainable system in fisheries, it is imperative to establish a balanced development in economic, ecological, and social aspects of the sector. Sustainable Fisheries and Artificial Reefs Project, therefore, has incorporated different components into the project methodology in order to both increase the fish population in the coastal area and to meet with the economic and social demands of the local residents. From the economical aspect, it is aimed that, through the artificial reefs, there will be an increase in the productivity in hunting in the short term, while the nursery artificial reefs will safeguard the juveniles and establish protection zones and therefore ensuring the long-term sustainability. Moreover, implementing marketing strategies and modernizing the infrastructure in the area would also contribute to economic development. The ecological benefits of the project will be realized through the nursery artificial reefs, which would create conservation zones and would safeguard the juveniles and nontarget species in addition to protecting the habitats of these species.

On the other hand, the cooperative insurance system would allow the fishermen to offset any by-catch damage and therefore preventing any competition between the fishermen and the nontarget species and, as a result, killing of those species would be avoided. Social development would be ensured through the implementation of the cooperative system, which would enable a mechanism to ensure the sustainability and further development of the sector. Through this system, the fishermen will be able to operate in unity, to finance themselves (without being dependent on any other external sources), to increase their technical capacity, and eventually to create a wider public awareness about the sector.

Finally, the gap in the monitoring mechanism would be overcome through the cooperative system, where the fishermen would be gathered under. By doing so, the expectations of the other stakeholders would be met, and the fishing practice would be in compliance with existing rules and regulations. Moreover, the system would enable further research in the sector and would open up new opportunities for future collaborations or at least would support researches through partnerships. And, eventually, the fisheries sector would be managed by the local administrations rather than the central authorities. As a result of the sustainable development concept, stakeholder participation and democratization of the process would increase the project's success and sustainability of the sector as it is mentioned by Hamdouch and Zuindeau (2010).

Artificial reefs, in this respect, have the primary responsibility to establish cooperation and allow the primary stakeholder to undertake positive duties and tasks for sustainability. As a result, the group of fishermen, who were hesitant to discuss the matter in the initial stage, has embraced the conservation rules and started to implement them. Even though the project is still at its initial stages, public awareness meetings and stakeholder consultations have proved to be productive, whereas no illegal hunting activities or intentional killing incidents (apart from accidents) of nontarget species have been reported. Another development, which has been observed, in the secondary target group (fishermen from other areas) and the members of this group have contacted the fishermen association in İskele and requested that similar applications to be introduced at their fishing grounds as well.

As a conclusion, as Fletcher et al. (2005) identifies sustainable fisheries can be established through protection and conservation of biodiversity of the area and by maintaining critical ecoprocesses and by enhancing the welfare of the current and future residents through a sustainable and continuous economic development system, which would be provided through effective legal and economic frameworks for ecologically sustainable development. In other words, sustainable development can only be maintained by creating a balance between economic welfare of the fishermen, where they will not have to abandon their professions due to low incomes and nature conservation, where nontarget species will have space for sheltering, feeding, and nesting. Once the balance is achieved and components of sustainable development are fulfilled, İskele region initially and then the entire North Cyprus can adopt sustainable development principles.

ACKNOWLEDGMENTS

I would like to acknowledge S. Bortone and C. H. Çicek for their valuable suggestions in writing and editing this article, to İ. Beyoğlu for providing valuable and continuous data used in this article, and I would like to thank K. D. Tözer and G. Kozak for assisting me in producing the image used in this article and for their valuable help.

REFERENCES

Azov, Y. 1991. Eastern Mediterranean: A marine desert?. *Mar. Pollut. Bull.* 23: 225–232.

Bianchi, C. N. and Morri, C. 2000. Marine biodiversity of the Mediterranean Sea: Situation, problems, and prospects for future research. *Mar. Pollut. Bull.* 40: 367–376.

Brundtland, G. H. 1987. *Our Common Future: Report of the World Commission on Environment and Development.* Oxford University Press. United Kingdom. 373 pp.

Çiçek, B. A. 2007. Evaluation of vertebrate species of North Cyprus coastal zone according to national and international protection regulations. Pages 21–28 in B. A. Çiçek and H. Öniz, Eds. *Proceedings of the International Symposium on Underwater Research.* Eastern Mediterranean University Press. Famagusta. North Cyprus. 113 pp.

Çiçek, B. A. and Yerli, S. V. 2007. Ecology of North Cyprus Coastal Zone. Pages 19–20 in B. A. Çiçek and H. Öniz, Eds. *Proceedings of the International Symposium on Underwater Research.* Eastern Mediterranean University Press. Famagusta. North Cyprus. 113 pp.

Düzbastılar, O. F. and Lök, A. 2004. Primary components used in building artificial reefs. *Ege University Journal of Aquaculture. Turkey.* 21: 181–185.

FAO Fishery Resources Division. 1999. FAO Technical Guidelines for Responsible Fisheries No. 8—Indicators for Sustainable Development of Marine Capture Fisheries. Food and Agriculture Organization of the United Nations. Rome. 68 pp.

Farrugio, H., Oliver, P., and Biagi, F. 1993. An overview of the history, knowledge, recent and future research trends in the Mediterranean fisheries. *Sci. Mar.* 57: 105–119.

Fletcher, W. J., Chesson, J., Sainsbury, K. J., Hundloe, T. J., and Fisher, M. 2005. A flexible and practical framework for reporting on ecologically sustainable development for wild capture fisheries. *Fish. Res.* 71: 175–183.

Garcia, S. M. 2000. The FAO definition of sustainable development and the code of conduct for responsible fisheries: An analysis of the related principles, criteria, and indicators. *Mar. Freshwater. Res.* 51: 535–540.

Ghosh, B. N. and Aker, S. L. 2006. Future of North Cyprus: An economic-strategic appraisal. *Futures.* 38: 1089–1102.

Grove, R. S., Sonu, C. J., and Nakamura, M. 1989. Recent Japanese trends in fishing reef design and planning. *Bull. Mar. Sci.* 44: 984–996.

Gücü, A. C., Gücü, G., and Örek, H. 2004. Habitat use and preliminary demographic evaluation of the critically endangered Mediterranean Monk Seal (*Monachus monachus*) in the Cilician Basin (Eastern Mediterranean). *Biol. Conserv.* 116: 417–431.

Hamdouch, A. and Zuindeau, B. 2010. Sustainable development, 20 years on: Methodological innovations, practices, and open issues. *J. Environ. Plan. Manage.* 53: 427–438.

Herut, B., Almogi-Labin, A., Jannink, N., and Gertman, I. 2000. The seasonal dynamics of nutrient and chlorophyll a concentrations on the SE Mediterranean shelf-slope. *Oceanol. Acta.* 23: 771–82.

Krom, M. D., Woodward, E. M. S., Herut, B., Kress, N., Carbo, P., Mantoura, R. F. C., Spyres, G., Thingstad, T. F., Wassmann, P., Wexels-Riser, C. et al. 2005. Nutrient cycling in the south east Levantine basin of the eastern Mediterranean: Results from a phosphorus starved system. *Deep-Sea Res. (II Top. Stud. Oceanogr.).* 52: 2879–2896.

Lukens, R. R. and Selberg, C. 2004. Guidelines for Marine Artificial Reef Materials. A joint Publication of Gulf and Atlantic States Marine Fisheries Commissions. No 121. USA. 198 pp.

Mace, P. M. 1997. Developing and sustaining world fisheries resources: The state of the science and management. Pages 1–20 in D. A. Hancock, Ed. *Developing and Sustaining World Fisheries Resources Second World Fisheries Congress.* CSIRO Publishing. Collingwood. Victoria.

Madau, F. A., Idda, L., and Pulina, P. 2009. Capacity and economic efficiency in small-scale fisheries: Evidence from the Mediterranean Sea. *Mar. Policy*. 33: 860–867.

Morales-Nin, B., Grau, A. M., and Palmer, M. 2010. Managing coastal zone fisheries: A Mediterranean case study. *Ocean Coast. Manage.* 53: 99–106.

Rätz, H. J., Dörner, H., Scott, R., and Barbas, T. 2010. Complementary roles of European and national institutions under the Common Fisheries Policy and the Marine Strategy Framework Directive. *Mar. Policy*. 34: 1028–1035.

Santos, M. N. and Monteiro, C. C. 1997. The Olhao artificial reef system (South Portugal): Fish assemblages and fishing yield. *Fish. Res.* 30: 33–41.

Symes, D. 2005. Altering course: Future directions for Europe's fisheries policy. *Fish. Res.* 71: 259–265.

Turley, C. M. 1999. The changing Mediterranean Sea: A sensitive ecosystem? *Prog. Oceanogr.* 44: 387–400.

Yaşarata, M., Altınay, L., Burns, P., and Okumuş, F. 2010. Politics and sustainable tourism development—Can they co-exist? Voices from North Cyprus. *Tourism Manage.* 31: 345–356.

CHAPTER 10

The Role of Artificial Reefs in Fisheries Management in Turkey

Altan Lök, F. Ozan Düzbastılar, Benal Gül, Aytaç Özgül, and Ali Ulaş

CONTENTS

INTRODUCTION

Small-scale artificial reef applications in the Gulf of İzmir and İstanbul Bosporus in the 1970s are the first records on the subject. However, there is no documentation of these efforts. Subsequent artificial reef applications became widespread after announcement of the results of scientific studies which began in the 1990s.

Until recently, there were no legal regulations in relation to artificial reefs; it was sufficient merely to inform the Ministry of Agriculture and Rural Affairs about planned projects until October 1999 (Lök and Tokaç, 2000). After arranging a meeting with reef stakeholders in September 1999, a "Project Guide for Artificial Reef Applications" was prepared, and recommendations were put forward to establish rules for the planning and implementation of artificial reef projects. The artificial reef development guidelines are still in use today and consist of six parts: (1) description, (2) procedures, (3) objectives, (4) site-selection criteria, (5) construction materials, and (6) design aspects. Although still far from fully covering all biological and engineering aspects, the guidelines represent an important starting point (Lök et al., 2002).

Although there have been 22 artificial reef projects planned in Turkey, to date, only 14 of these projects have been completed. The main objectives of those projects were (1) to support small-scale and traditional fisheries, (2) to create new sites for recreational fishing and diving, (3) to protect biodiversity, especially in the littoral zone, (4) to protect fish-spawning and nursery areas (e.g., *Posidonia* meadows) from illegal trawling, and (5) to conduct scientific experiments (Lök et al., 2002).

GENERAL INFORMATION ON TURKISH SEAS AND FISHERIES

The total length of the Turkish coastline, including islands, is 8,333 km of which 1,067 km are island shores (Figure 10.1). The distribution of this total coastline, relative to Turkey's four seas is the Black Sea: 1,701 km (20.4%), the Sea of Marmara: 1,441 km (17.3%), the Aegean Sea: 3,484 km (41.8%), and the Mediterranean: 1,707 km (20.5%) (Gunay, 1987). There are 28 coastal provinces that have a coastline along at least one of the four seas. Of these, only seven provinces have small-scale artificial reef sites.

According to the fisheries statistics published by the Food and Agricultural Organization (FAO, 2009), Turkey is one of the leading countries in terms of fisheries production in the region surrounding the Mediterranean and the Black Seas. Total production, which includes capture and aquaculture production (Figure 10.2), is approximately 650,000 t annually (TÜİK, 2008). Turkey accounts for 0.7% of world fisheries production (FAO, 2009).

There are 17,824 registered commercial fishing vessels in Turkish waters. The majority of those fishing vessels belong to the small-scale fisheries (longline, trammel net, gillnet; Table 10.1). Even though small-scale fisheries contribute to only a limited proportion of the total catch, this fraction has great importance in terms of traditional fisheries and socioeconomic assessment.

There are many problems in Turkish fisheries. But the most important issues are

- Ineffective governmental organizational structure
- The lack of fish stock assessment
- Overfishing
- Sharing of fishing grounds between different fishing activities
- Illegal trawling

Figure 10.1 Turkish seas and artificial reef sites.

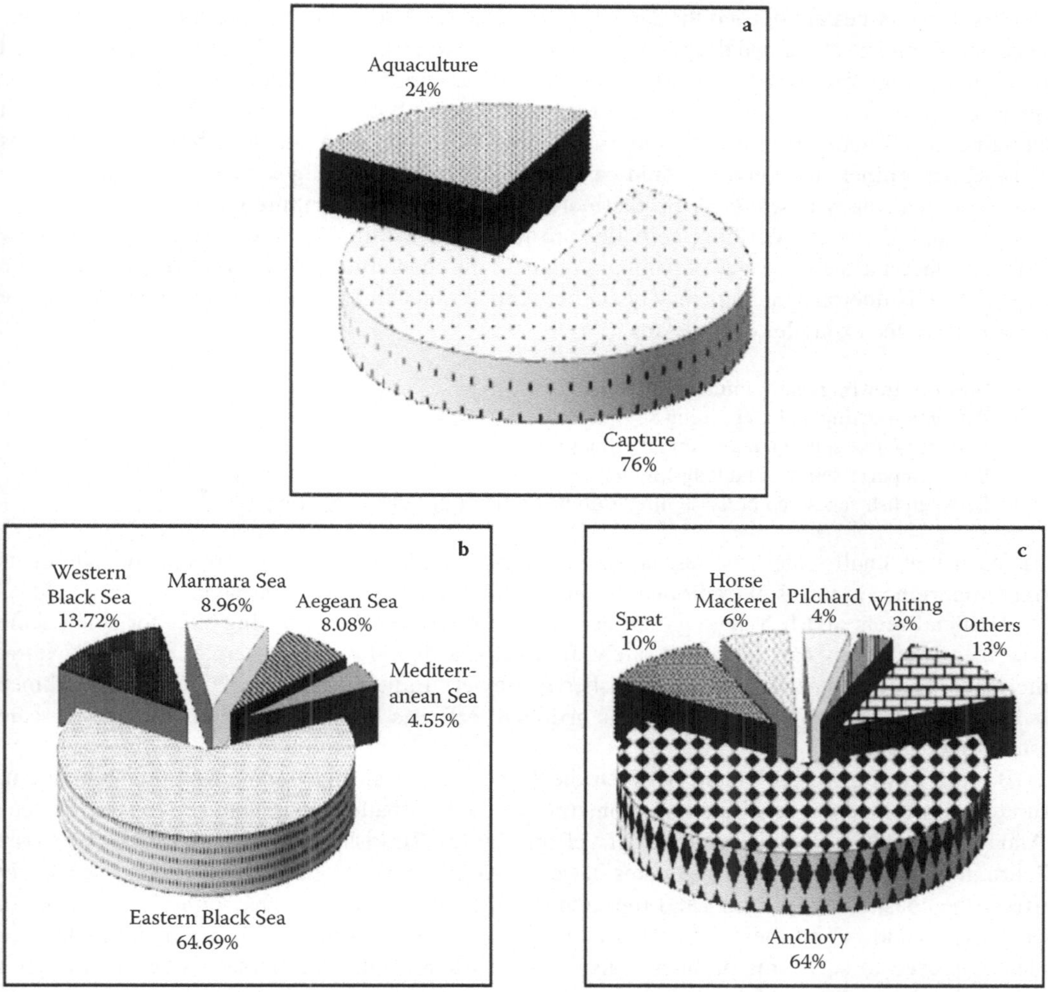

Figure 10.2 Fish production statistics in Turkey: (a) distribution of fisheries production, (b) marine capture productions according to seas, (c) the most captured marine fish species.

Table 10.1 Number of Fishing Vessels in Turkish Seas

Region	Trawler	Purse Seiner	Trawler/ Purse Seiner	Carrier	Small-Scale Fishing Vessels
Black Sea	231	199	229	103	5,865
Sea of Marmara	134	187	98	51	2,580
Aegean Sea	122	112	59	35	5,614
Mediterranean Sea	238	45	8	13	1,900
Total	725	544	394	202	15,959

Source: TÜİK, 2006; Sea Product Statistics. Turkish Statistical Institute, Ankara.

The first three issues are beyond the subject of this chapter. The last two issues have a great importance and should be explained deeply because of relation to artificial reefs. Legal regulations related to sharing fishing grounds between different fishing activities are not clear in Turkey. Generally, purse seining is allowed at depths deeper than 18 m, while bottom trawling is forbidden between shoreline and 3 nautical miles off shore (Anonymous, 2008). Small-scale fishing gears such as trammel net, gillnet, and longline could be used in all places excepting some closed areas (marine protected areas, navy bases, etc.). Furthermore, fish cages for aquaculture cover large areas close to shore, and all types of fishing activities are forbidden within 500 m of these cages. Another restrictive factor affecting usage of fishing grounds is the economic exclusive zone, especially in the Aegean Sea. Numerous conflicts may occur between various kinds of fishing activities under these conditions. Some examples of these are

- Between trawling and trammel/gillnetting
- Between trawling and long-lining
- Between purse seining and trammel/gillnetting
- Between purse seining and long-lining
- Between fish cages and purse seining, trammel/gillnetting, and long-lining

It is clear that small-scale fisheries have a big problem with other fishing activities in spite of its great importance in terms of traditional fisheries and socioeconomic assessment.

Artificial reefs and fish aggregation devices can offer alternative fishing sites for small-scale fishermen where they can catch fish safely. Ironically, artificial reef areas are aimed at providing alternative fishing areas for small-scale fisheries are very limited in Turkey. For example, Ürkmez and Gümüldür (Figure 10.1; 6 and 7 sites) artificial reef sites are only 2 km² area for small-scale fishing operations.

Bottom trawling is prohibited between the 1 and 3 nautical mile zones from the shoreline in Turkey with the actual legal distance sometimes based on bathymetric features of specific seas (Anonymous, 2008). Despite enforcement efforts by the Turkish fisheries authorities and Coast Guard, illegal bottom trawling still takes place in coastal areas. Illegal bottom trawling negatively affects the coastal environment, and repeated trawling often results in heavy physical damage to sea grass meadows and their associated biota (Ramos-Espala et al., 2000). Artificial reefs have been employed to solve this problem. They can protect juvenile fish, nursery areas, and coastal and marine biocoenoses from the mechanical impact of illegal trawling. Examples of this type of construction can be found in the Mediterranean coastal areas of France (Barnabe et al., 2000) Spain (Ramos-Espala et al., 2000), and Italy (Bombace et al., 2000). The usage of artificial reefs as a deterrent to illegal trawling is common in Turkey, mainly in the coastal areas of the Aegean Sea. Artificial reef arranged along 5 nautical miles in Selçuk (Figure 10.1; 11 site), and 3 nautical miles in Ürkmez-Gümüldür (Figure 10.1; 6 and 7 sites) provide protection against illegal trawling Nevertheless, there has been no study assessing of the effect of antitrawl artificial reefs.

DESCRIPTION OF THE ARTIFICIAL REEFS IN TURKEY

The first scientific artificial reef project was conducted by Ege University, Faculty of Fisheries in Hekim Island, İzmir Bay, in 1991. Thirty cubic concrete blocks were assembled in six pyramids (four blocks forming the base, with one on top). The apparent success of this project resulted in increased interest in reef technology, and new cooperative projects were initiated among local governments, fishery cooperations, and universities. Local governments provided financial support universities provided scientific and technical knowledge, and fisheries provided practical information on local conditions (Lök and Tokaç, 2000). These collaborative projects were conducted on the Turkish coast bordering the Aegean Sea, whereas few projects have been completed on the Mediterranean and Black Sea coasts (Figure 10.1).

Reinforced concrete material has been used to construct artificial reef units because it meets most engineering and biological requirements. After 2005, old steel ships and an airplane were also used to construct an artificial reef to create new diving sites. Steel and HDPE (high-density polyethylene) materials have been used to construct Fish aggregation devices (FADs). Figure 10.3h and 10.3i shows artificial reef and FAD designs used in Turkey. The main characteristics of artificial reef projects are summarized in Table 10.2.

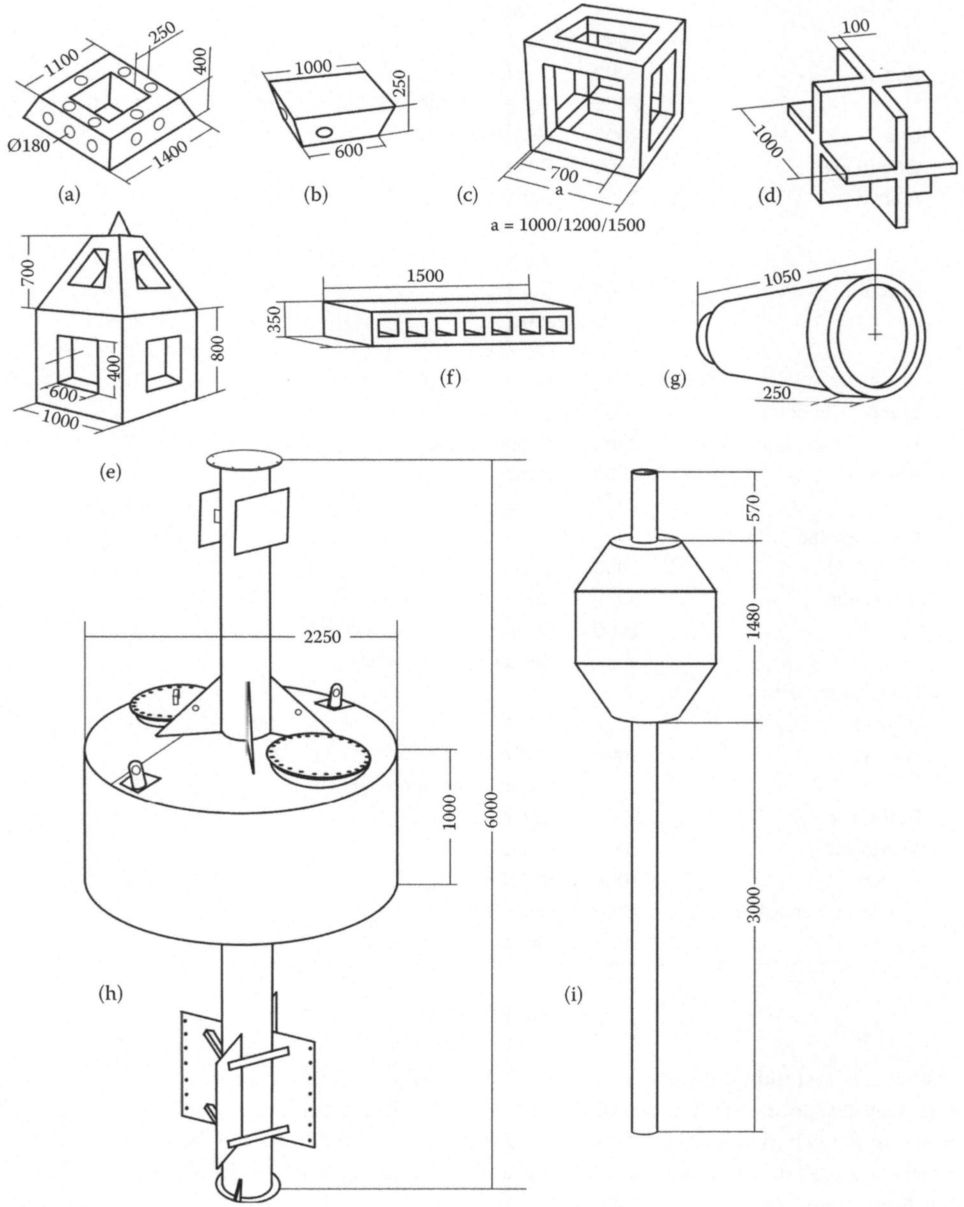

Figure 10.3 Artificial reef and FAD designs used in Turkey.

Table 10.2 Basic Features of Artificial Reef Projects Conducted in Turkish Seas

Locality/Projects	Date	Design	Number	Depth (m)
Izmir Inner Bay	1989	Old trolleys	10	16–20
Hekim Island	1992	Cubic	30	9 and 18
Foça	1994	Plus shape	20	17
Dalyanköy	1995	Cubic and plus shape	100	21
İzmir Inner and Outer Bay	1997	Multi holes	100	15–25
Ürkmez	1998	Pentagon dome	160	14–21
Gümüldür	1998	Cubic	180	16–21
Urla	1999	Octo-reef (with 16 holes)	30	10–15
Selçuk	2002	Cubic	480	14–16
Urla	2005	Octo-reef (with 4 holes)	80	14–20
Urla	2005	Plus shape	10	14
Gümüldür	2006	Culvert	60	18–40
Alanya	2006	Cubic	400	20–25
		Old ship	1	26
		Jugs	200	20–25
Bodrum	2007	Culvert	100	15–25
		Jugs	250	20–25
		Old ship	2	18 and 30
		Airplane (C49)	1	28
Kemer (Üçadalar)	2007	Old ship	1	—
Düzce (Akçakoca)	2009	Airplane (C47)	1	27
Adana	2009	Cubic	120	20–25
	2009	Cubic	100	20–25
Uncompleted projects				
Zonguldak	2000	Cubic	50 (226)	15
Marmaris	2001	Cubic	75 (2070)	24
Kuşadası	2006	Cubic	30 (375)	20
		Old airplane and ship	1	—
Planned projects				
Gökova	2001	Cubic	100	—
Çeşme	2004	Cubic	985	—
		Old airplane and ship	1	—
Burhaniye	2005	Cubic	660	—
Mordoğan	2006	Cubic	225	—
Ayvalık	2006	Old ship	1	—
Antalya-Sıçan Adası	2006	Old ship	1	—
Kaş	2009	Airplane	1	—

RESEARCH

A better understanding of artificial reef functions and the relationship with environmental factors will provide more effective use of these structures (Bohnsack et al., 1991) and facilitate their proper use in fishery management. Scientific research on artificial reefs in Turkey initiated in 1991 and chiefly focused on biological, ecological, and engineering aspects. In this part of chapter, we evaluate briefly how we can use results of researches in artificial reef and fisheries management.

Species richness, abundance, biomass, and biodiversity at most of the artificial reef sites and around fish aggregation devices in Turkey have been well documented (Lök, 1995; Lök et al., 2008,

2009). Seasonal changes of all these variables are also documented (Gül, 2008; Lök and Gül, 2005). These data provide information that serve as a basis to:

- Open and close a season to fisheries of artificial reef sites
- Determine appropriate fishing techniques and gears in artificial reef sites
- Examine the circumstances associated with fish populations at artificial reef sites
- Monitor exotic versus natural species for fisheries management
- Protect endangered and overfished species for sustainable fisheries

The natural shelters for habitat-dependent species are often destroyed by human activities in coastal areas such as through overfishing and illegal fishing. Therefore, species-specific artificial reef applications are designed for habitat-dependent species of major commercial importance such as octopus and lobster. There are many studies relating species and reef characteristics such as size and shape, material, number of holes, design, deployment depth, etc. (Brock and Norris, 1989). However, the number of artificial reefs designed to attract target species are limited (Nakata et al., 2001). Species-specific artificial reef studies have been conducted for *Octopus vulgaris* in Turkey. To design artificial reefs, 168 nests were visited to record important characteristics (water depth, sea bottom structure, nest width and volume, and angle of entrance) of octopus nests characters and habitat associations in nature. Subsequently, then two different types of reef units were designed (Figure 10.3a,b). Our studies show that artificial reefs for octopus succeed to mitigate habitat loss and sustain fishing.

Although FADs have been used in other countries for many years (Gooding and Magnuson, 1967; Safran and Omori, 1990; Durace and Kingsford, 1995), their use in the context of fishing on the Turkish coast has not been reported. No reports have documented the use of FADs in pelagic fisheries on the coasts of the Aegean Sea or eastern Mediterranean. To assess the potential usage of FADs in Turkish fisheries, experimental research has been started since 2008. For this purpose, two FAD units (Figure 10.3h,i) were moored off the east coast of Turkey in the Aegean Sea. As a result, 27 species belonging to 20 families were collected beneath FADs. Of those species, *Seriola dumerili*, *Trachurus mediterranus*, *Coryphaena hippurus*, and *Xiphias gladius* are commercially important for small-scale fisheries (Özgül et al., 2009). Another important result was the acceptance of employing FADs by local fishermen (especially those using line fishing and pelagic longline). In fisheries management, FADs can be employed to separate user groups in the same region to help reduce user conflicts and to increase the overall productivity of the area by aggregating fish.

Hydrographic factors, such as depth, wave height, currents, and tides, among other factors, must be considered in selecting sites for artificial reef deployments to improve the performance and increase the service life of artificial reefs (Grove et al., 1991; Grace, 2001; Sheng, 2000; Tseng et al., 2001). The reef design should be based on numerical modeling using empirical data from monitored reef sites (Ingsrisawang et al., 1995; Düzbastılar and Şentürk, 2009), and physical tests in wave flumes should also be conducted (Kimura et al., 1994; Düzbastılar et al., 2006). We performed laboratory experiments and numerical calculations to determine the effect that design weight, deployment depth, slope of the sea bottom, size and shape of the reef, and orientation has on artificial reef stability. The laboratory study suggests a minimum deployment depth of 20 m for the investigated area in the Black Sea (Figure 10.4).

In Turkey, fishing activities limit the deployment of artificial reefs to deeper waters due to depth regulations or distance restrictions. So, as not to limit fisheries and avoid conflicts among artificial reef stakeholders, depths shallower than about 20 m should be selected for deployment in Turkish waters. Deployments in shallow water may cause a potential loss of reef stability and integrity. Ironically, artificial reefs deployed in extremely deep waters increase problems during deployment and installation. These factors may render it difficult to assess the efficiency of artificial reef after deployment.

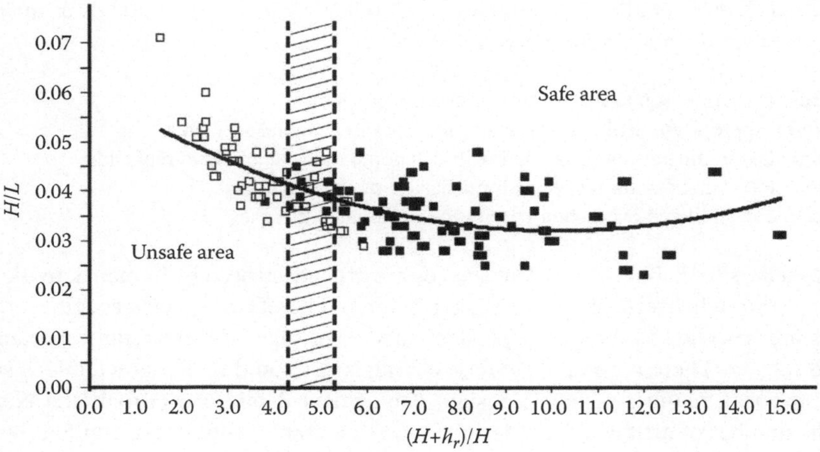

Figure 10.4 Wave steepness (*H/L*) versus the ratio of total deployment depth and reef height to wave height ($[h + h_r]/H$), used to determine the deployment area necessary for reef stability in an area with wave steepness between 0.000–0.075. Lined area indicates critical deployment depths. (From Düzbastılar et al., 2006. *Bulletin of Marine Science* 78 (1): 195–202. With permission.)

NATIONAL ARTIFICIAL REEF PROGRAM

The Ministry of Agriculture and Rural Affairs decided to prepare a National Artificial Reef Program in 2008. This meeting was attended by decision-makers and scientists. The first discussion described the purposes of the program. Potential artificial reef grounds and projects to be conducted at these sites were discussed according to the stated objectives to:

- Create new fishing grounds in minimally productive muddy areas to promote small-scale fisheries
- Conserve biodiversity in the littoral zone

Three provinces along the Black Sea coasts and all provinces on the Aegean Sea and the Mediterranean Sea coasts were included in the program during its first stage due to the importance of small-scale fisheries.

Scientific studies were planned in two stages before reef deployment. In the first stage, information on small-scale fishery structure and hydrographic characteristics (bottom type, slope, etc.) were determined for ten provinces (three on the Black Sea coast, four on the Aegean Sea coast, and three on the Mediterranean coast) between September 2008 and March 2009. Two provinces on the Black Sea coasts were excluded after data analysis because of very soft sea bottom and high sedimentation rate due to large rivers. Eight provinces remained after the first selection. The second stage of the evaluation is currently ongoing and includes the collection of data related to biology, oceanography and fisheries techniques and socioeconomic for one year prior to deployment. The data collected will be evaluated in relation to site-selection criteria.

During the site selection for artificial reefs along the Turkish coasts, the major criteria considered were (1) stability and sinking phenomena, (2) bottom slope, (3) water depth regarding wave and current effects, (4) water depth related to legal and illegal fishing, (5) transport and logistic opportunities, (6) maritime traffic, (7) habitat type, (8) tourism and other marine facilities, and (9) socioeconomic evaluation.

The reefs designed for the master plan can be categorized into three main groups: (1) protection units (a), (2) production units (b, c), and (3) mixed units (d), as shown in Figure 10.5.

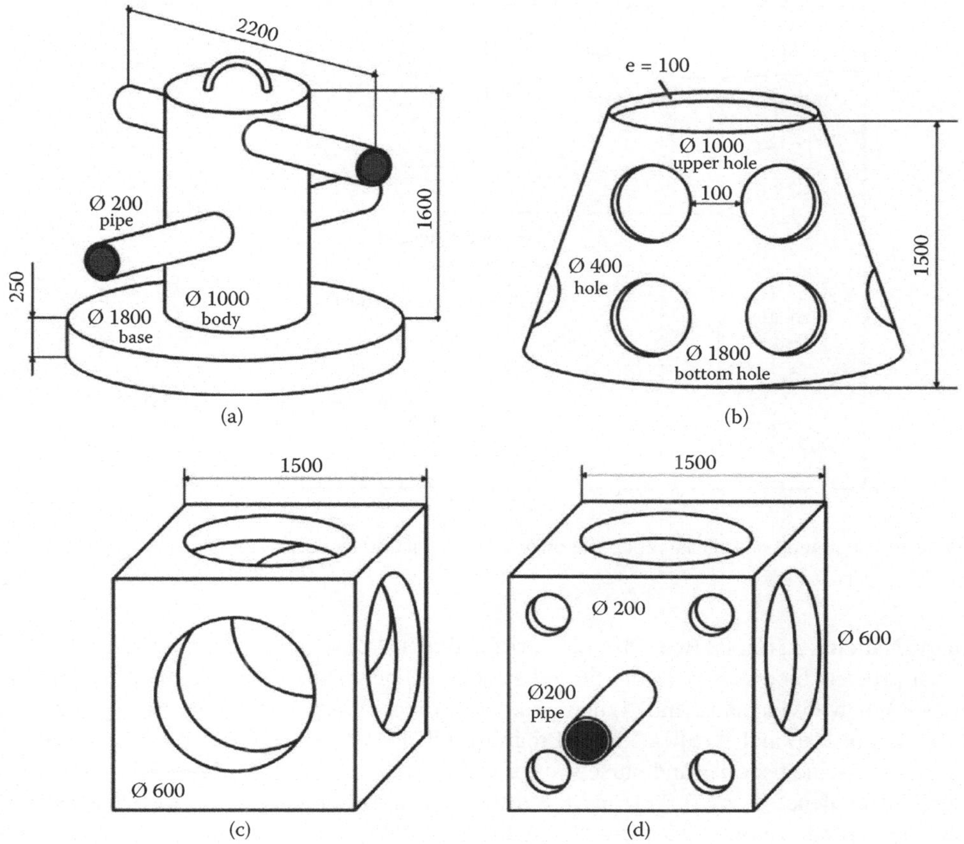

Figure 10.5 Artificial reef designs in the national program.

Thousands of reef blocks will be deployed over the course of the next 10 years in the Black Sea, the Aegean Sea, and the Mediterranean Sea. Therefore, four artificial reef designs and various deployment patterns were planned to protect biodiversity and maintain ecosystems and natural habitats for sustainable fisheries. While planning the artificial reefs, we utilized our experience with fishing and artificial reef applications, as well as other successful Mediterranean applications. For instance, protection artificial reefs (antitrawl) were designed to obtain maximum effects on trawling gear with regard to door distance, trawler route, engine power, and trawl net characteristics.

Monitoring studies of at least 2 years that include assessments of biological, physicochemical, fishery, and socioeconomic parameters are planned for the period after the deployment of each artificial reef site. Those studies will be conducted in collaboration with the ministry and universities, in particular, Ege University.

Preliminary studies were completed between September 2008 and March 2009. The city of Balıkesir located at the northern end of Turkey's Aegean Sea coast was selected as the pilot project area. More extensive studies will be performed in the area. Artificial reef construction and deployment are planned for the second half of 2010. The national program will be finished in 2020. This date depends on the future status of the Turkish national economy because the program's funding is provided entirely by the national budget.

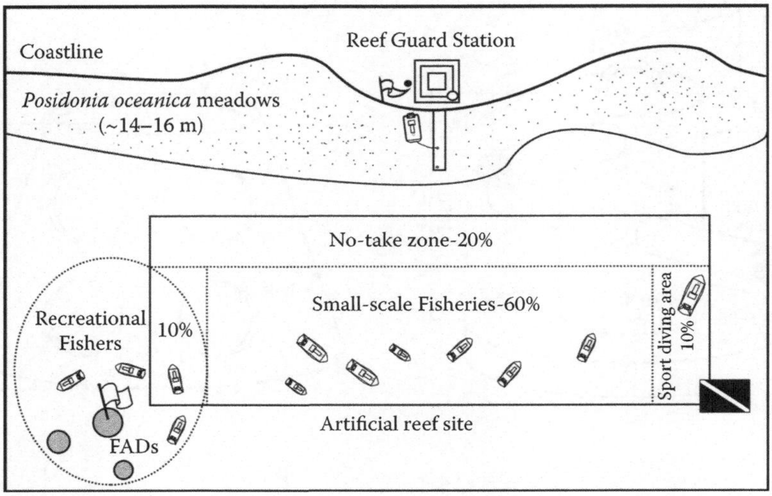

Figure 10.6 Illustration of artificial reef partitioning among pilot project user groups.

In the National Artificial Reef Program, artificial reef sites will be used to provide: (1) a no-take zone that protects biodiversity in the littoral system, (2) new fishing ground areas for small-scale fisheries, (3) new diving sites, and (4) new grounds for recreational fishermen (Figure 10.6).

In addition, National Artificial Reef Program will help to solve the conflicts of fishing area between small-scale fisheries and purse seiners-trawlers.

Previous evidence from Turkey and the rest of the world indicates that artificial reefs should provide the desired outcomes with regard to both biological resources and users. The Artificial Reef Management Plan (ARMP) was prepared by the ministry together with scientists and will be open to users for discussion. Management options such as gear restrictions, no-take zones, and spatial segregation and "reef guard" personnel were accepted in the ARMP. Harvest rotation and closed-season regulations are under discussion. The main idea accepted by all bodies of the ARMP is that management options should be simple and easily applicable.

OUTLOOK

There are small-scale and limited artificial reef applications for fisheries management purposes in Turkey. Together with starting of National Artificial Reef Program, artificial reefs will play important roles in larger scale in management. Scientific monitoring studies should be an essential part of the program to achieve main objectives consisting of creating new and alternative fishing areas for small-scale fisheries and preserve biodiversity in littoral zone.

REFERENCES

Anonymous, 2008. 2/1 numaralı ticari amaçlı su ürünleri avcılığını düzenleyen tebliğ. [The announcement of commercial fishery regulations, no. 2/1.] Sirküler No. 2008/48.

Barnabe, G., Charbonnel, E., Marinaro, J.Y., Ody, D., and Francour, P. 2000. Artificial reefs in France: Analysis, assessment and prospects. Jensen, A.C., Collins, K.J., and Lockwood, A.P.M. (Eds.) *Artificial Reefs in European Seas*. Kluwer Academic Publications, Dordrecht, The Netherlands, 167–184.

Bohnsack, J.A., Johnson, D.L., and Ambrose, R.F. 1991. Ecology of artificial reef habitats and fishes. In Seaman Jr., W. and Spraque, L.M. (Eds.). *Artificial Habitats for Marine and Freshwater Fisheries*. Academic Press, San Diego, California, 61–107.

Bombace, G., Fabi, G., and Fiorentini, L. 2000. Artificial reefs in the Adriatic Sea. In Jensen, A.C., Collins, K.J. and Lockwood, A.P.M. (Eds.) *Artificial Reefs in European Seas*. Kluwer Academic Publications, Dordrecht, The Netherlands, 31–63.

Brock, R.E. and Norris, J.E., 1989. An analysis of the efficacy of four artificial reef designs in tropical waters. *Bulletin of Marine Sciences* 44 (2): 934–941.

Durace, B.E. and Kingsford, M.J. 1995. An experimental investigation of the fishes associated with drifting objects in coastal waters of temperate Australia. *Bulletin of Marine Sciences* 57 (2): 378–392.

Düzbastılar, F.O., Lök, A., Ulaş, A., and Metin, C. 2006. Recent developments on artificial reef applications in Turkey: Hydraulic experiments. *Bulletin of Marine Science* 78 (1): 195–202.

Düzbastılar, F.O. and Şentürk, U. 2009. Determining the weights of two types of artificial reefs required to resist wave action in different water depths and bottom slopes. *Ocean Engineering* Vol. 36, No. 12–13, 900–913.

FAO 2009. The State of World Fisheries and Aquaculture. 2008. Rome.

Gooding, R.M. and Magnuson, J.J. 1967. Ecological significance of a drifting object to pelagic fishes. *Pac. Sci.* (21): 486–497.

Grace, R.A. 2001. The factors and processes that influence artificial reef longevity. *MTS Journal* 35 (2): 3–13.

Grove, R.S., Nakamura, M., and Sonu, C.J. 1991. Design and engineering of manufactured habitats for fisheries enhancement. In: Seaman Jr., W., Spraque, L.M. (Eds.). *Artificial Habitats for Marine and Freshwater Fisheries*. Academic Press, San Diego, California, 109–152.

Gül, B. 2008. A study on fish community structure of Ürkmez and Gümüldür artificial reefs. Ph.D. thesis (In Turkish). Ege University Graduate School of Natural and Applied Science Bornova–İzmir, 100.

Gunay, T. 1987. The management of coastal areas: Development perspectives of Turkey. United Nations Economic Commission for Europe, 14th Session, Portugal, November 1987.

Ingsrisawang, V., Ban, M., and Kimura, H., 1995. Comparative study on the sinking of artificial reefs by local scour between laboratory and field experiments. *Fisheries Engineering* 32 (2): 95–103.

Kimura, H., Ingsrisawang, V., and Ban, M. 1994. A study on local of cylindrical artificial fish reefs. *Fisheries Engineering* 31 (1): 33–40.

Lök, A. 1995. A Study on Artificial Reef Applications in Turkey. Ph.D. thesis (In Turkish). Ege University Graduate School of Natural and Applied Science Bornova–İzmir, 55.

Lök, A. and Tokaç, A. 2000. Turkey: A new region for artificial habitats. Jensen, A.C., Collins, K.J., and Lockwood, A.P.M. (Eds.) *Artificial Reefs in European Seas*. Kluwer Academic Publications, Dordrecht, The Netherlands, 21–30.

Lök, A., Metin, C., Ulaş, A., Düzbastılar, F.O., and Tokaç, A. 2002. Artificial reefs in Turkey. *ICES Journal of Marine Science*, 59 S: 192–195.

Lök, A. and Gül, B. 2005. Evaluation of fish fauna associated with experimental artificial reefs in Hekim Island in Izmir Bay (Aegean Sea, Turkey). *Ege Üniversity Journal of Fisheries and Aquatic Sciences*, Volume: 22 Suppl. (1–2): 109–114.

Lök, A., Gül, B., Ulaş, A., Düzbastılar, F.O., and Metin, C. 2008. Diel variations on the fish assemblages at artificial reefs in two different environments of the Aegean sea (Western Coast of Turkey) Turkish. *Journal of Fisheries and Aquatic Sciences* 8: 79–85.

Lök, A., Gül, B., and Özgül, A. 2009. Comparison on fish community structure at artificial reefs deployed different depths in the Aegean Sea coast of Turkey. *9th International Conference on Artificial Reefs and Related Aquatic Habitats*, November 8–13, Brazil.

Nakata, K., Hamano, T., Hayashi, K-I., and Kawai, T., 2001. Artificial burrow preference by the Japanese crayfish *Cambarodies japonicus*. *Fisheries Science*, 67: 449–455.

Özgül, A., Lök, A., and Gül, B. 2009. Fish assemblages associated with fish aggregation devices in Aegean Sea coast of Turkey. *9th International Conference on Artificial Reefs and Related Aquatic Habitats*, November 8–13, Brazil.

Ramos-Espala, A.A., Guillen, J.E., Bayle, J.T., and Sanchez-Jerez, P. 2000. Artificial anti-trawling reefs off Alicante, south-eastern Iberian Peninsula: Evolution of reef block and set designs. Jensen, A.C., Collins, K.J., and Lockwood, A.P.M. (Eds.) *Artificial Reefs in European Seas*. Kluwer Academic Publications, Dordrecht, The Netherlands, 195–218.

Safran, P. and Omori, M. 1990. Some ecological observations of fishes associated with drifting seaweed off Tohoku coast, Japan. *Marine Biology* 105: 395–402.

Sheng, Y.P. 2000. Physical characteristics and engineering at reef sites. In *Artificial Reef Evaluation with Application to Natural Marine Habitats,* ed. Seaman Jr., W., CRC Press LLC, Boca Raton, FL, 51–94.

Tseng, C.-T., Chen, S.-C., Huang, C.-S., and Liu, C.-C. 2001. GIS-assisted site selection for artificial reefs. *Fisheries Science* 67: 1015–1022.

TÜİK. 2006. Sea Product Statistics. Turkish Statistical Institute, Ankara.

TÜİK. 2008. Sea Product Statistics. Turkish Statistical Institute, Ankara.

CHAPTER 11

Artificial Reefs in the Management of Mediterranean Sea Fisheries

Gianna Fabi and Alessandra Spagnolo

CONTENTS

ABSTRACT

The Mediterranean Sea can be considered as typical in the use of artificial reefs for fish stock and fishery management. Artificial reefs have been used over 40 years in the Mediterranean, mostly to deter illegal trawling in coastal areas and other sensitive habitats, which generally include important spawning and nursery areas for many commercial species. Additionally, artificial reefs are used to reduce conflicts between different fishing activities, mainly illegal trawling and small-scale fisheries operating with set gears. The objectives of most artificial reef deployments in the Mediterranean are to enhance overexploited fish stocks and improve small-scale fisheries, one of the most important activities for the coastal communities. At present, artificial reefs have been constructed in ten Mediterranean bordering countries: Cyprus, France, Greece, Israel, Italy, Malta, Principality of Monaco, Spain, Tunisia, and Turkey, excluding Portugal, which is located in the western part of the Gibraltar Strait. Research on scientific, engineering, legal, and socioeconomic aspects has strongly contributed to the success of artificial reefs around the Mediterranean and has led to drawing up specific guidelines to assist countries in artificial reef construction and avoid dumping.

However, despite the large use of artificial reefs for fishery management in the Mediterranean Sea, there is still a need for additional research on the effective role of artificial reefs in stock enhancement and on the relationship between the artificial substrates and the fish species living inside and/or around the reefs. Moreover, in some Mediterranean countries, there is a lack of management measures directed toward rationally exploiting the artificial reef resources and regulating access by the different users to the reefs.

The Mediterranean Sea is an "inland sea" surrounded by Europe, Asia, and Africa, and covering only 0.7% of the world's oceans. Due to its morphological, chemical, and biotic features, it represents one of the richest seas for biodiversity in the world hosting 7.5% of the world's animal taxa and 18% of the world's marine flora. The Mediterranean marine flora and fauna constitute a unique mixture of temperate and subtropical elements, with about 28% of endemic species.

Around 150 million people, corresponding to a third of the human population that live in the bordering Mediterranean countries, live in the coastal regions and islands. Economic activities in coastal zones continue to expand and fisheries represent an important and vital sector for the human communities in these areas (UNEP-MAP-RAC/SPA, 2003).

The Mediterranean continental shelf is generally narrow and fishing grounds are usually located close to the coasts. This fact, together with various political considerations, might explain why exclusive economic zones (EEZs) have not yet been established in the Mediterranean, and only fisheries protection zones have been declared in some cases. Most of Mediterranean countries have a territorial sea of 12 nautical miles and, consequently, the extent of waters under national jurisdiction in comparison with international waters is more limited than elsewhere in the European Community (European Commission, 2002).

Mediterranean fisheries are a typical case of small-scale fisheries. About 80% of the EU Mediterranean vessels are smaller than 12 m and practice multispecific fishing (European Commission, 2002). Despite their important economic and social role in local communities, there are few suitable strategies and policies for promoting and managing small-scale fisheries in the Mediterranean regions. Besides political reasons, the formulation of appropriate policy measures by national and international institutions is also complicated because of the great diversity of small-scale métiers within and across countries and regions.

Most of small-scale fisheries occur in coastal areas. To protect nursery areas and sensitive habitats and enhance the social sustainability of small-scale fisheries using selective gears, the use of trawl nets is prohibited in the Mediterranean Sea within 3 nautical miles of the coast or within the 50 m isobath where that depth is reached at a shorter distance from the coast. The use of trawl nets is, however, forbidden within 1.5 nautical miles of the coast (EC Regulation 1967/2006). In spite of this, trawlers often do not observe this prohibition, damaging sensitive habitats and set gears while creating conflicts with small-scale fishermen for space and resources.

ARTIFICIAL REEFS IN FISHERIES MANAGEMENT

The deploying artificial structures to improve local fisheries by attracting fish in the Mediterranean Sea goes back to 3,000 years ago. At that time, the rocks used as anchors for the tuna-fishery nets were left on the seabed at the end of each fishing season. These accumulated over time and made new rocky habitats, populated by benthic fauna and fish that were exploited by local fishermen between the tuna fishing seasons (Riggio et al., 2000).

Despite this ancient tradition, the modern concept of artificial reefs was adopted only after the second half of 1900s. A great number of artificial reefs have been deployed in many Mediterranean countries to date and several programs continue to this day.

In this chapter, only the countries bordering the Mediterranean basin are considered, excluding Portugal which is located in the western part of the Gibraltar Strait. At present, artificial reefs

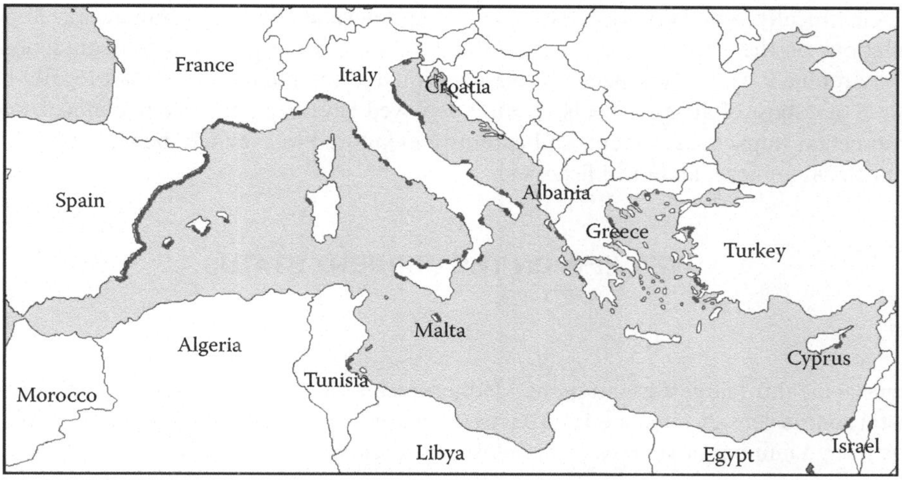

Figure 11.1 ***See color insert.*** Artificial reefs in the Mediterranean Sea.

have been constructed in ten Mediterranean countries: Cyprus, France, Greece, Israel, Italy, Malta, Principality of Monaco, Spain, Tunisia, and Turkey (Figure 11.1).

Artificial reefs have been deployed for purposes of fisheries management in the Mediterranean to enhance small-scale fisheries using set gears (e.g., gillnet, trammel net, traps,), reduce conflicts between different fishing activities, and to protect either coastal spawning and nursery areas or other sensitive habitats such as sea grass meadows against illegal trawling. These general uses are applicable for most countries with the exception of the Principality of Monaco and Malta. The Principality of Monaco deployed reefs to protect the *Posidonia* meadows and the only red coral (*Corallium rubrum*) slope existing within its territorial waters. Additionally, Monaco deployed reefs for experimental cultivation of red coral, while research and potential creation of recreational diving sites were the main purposes in Malta.

Artificial reef deployment in the Mediterranean Sea falls under some general regulations concerning the protection of the sea against pollution due to dumping of unsuitable materials, such as the Convention on the Prevention of Marine Pollution by Dumping of Wastes and Other Matter (London Convention, 1972; replaced by the Protocol in 1996) and the Mediterranean Action Plan (UNEP-MAP, 1975). MAP is a regional cooperative effort involving the European Community and 21 countries bordering the Mediterranean Sea. Objective of the MAP is to protect the marine and coastal environment through regional and national plans to achieve sustainable development. This plan led to the adoption of the convention for the Protection of the Mediterranean Sea against Pollution (Barcelona Convention, 1977).

Due to the increasing interest toward artificial reefs, specific guidelines referring to the placement of these structures in the marine environment were recently adopted as part of MAP (UNEP-MAP, 2005). These guidelines define an artificial reef as "a submerged structure deliberately placed on the seabed to mimic some functions of a natural reef such as protecting, regenerating, concentrating, and/or enhancing population of living marine resources." The objectives of an artificial reef may also include fisheries protection and production, habitat protection and restoration, and research and recreation. Oppositely, the definition does not include submerged structures such as artificial islands, cables, pipelines, platforms, mooring, and breakwaters established for coastal defense purposes. Materials, design, placement, administrative action, monitoring, scientific experiments, management, and liabilities are also addressed. The overall purpose of these guidelines is to assist contracting countries in assessing proposals for the placement of artificial reefs on the basis of scientifically sound criteria, developing an appropriate regulatory framework, implementing regulations on the artificial reef construction, and preventing pollution or degradation of the marine environment as a consequence of waste. Although the guidelines are not legally binding the result is that most of artificial reefs recently deployed have been accurately planned, subjected to environmental impact assessment, and carefully monitored to evaluate their effects on benthic fauna, finfish assemblage, and local fisheries.

OVERVIEW ON THE CURRENT STATUS

Cyprus

Cyprus is the third largest island in the Mediterranean, situated in the eastern part of the basin. The coastal zone is densely populated and environmentally vulnerable, being subjected to increasing pressures from a number of sources (e.g., industrial development, urban expansion, exploitation of marine resources, tourism).

Fish production is chiefly from small-scale fisheries, trawling in territorial and international waters, and purse seine fishery, as well as from aquaculture. The production of fish from Cyprus was around 5,000 t in 2007 (www.fao.org). Trawling has been banned in the northern part of the island (Turkish area) since 1998. The Cypriot fishing fleet mainly consists of small vessels operating with set gears (mainly trammel nets and gillnets) in the coastal areas, and small-scale fisheries represent a necessary activity for the coastal communities.

Historically, the national and local authorities in Cyprus failed to answer the needs of the fishing community due to the lack of funds and inadequate technical infrastructures. Consequently, the fishing industry could not progress and develop its capacity and, hence, it was not a sustainable business enterprise for the participants. Individuals often pursued fishing as a secondary (second job or a hobby) income generator. Moreover, the lack of sustainable management plan has lead to the overexploitation of fishing resources and the destruction of sensitive habitats (seagrass meadows) and nursery areas due to illegal trawling in coastal areas (<50 m depth) with consequent loss of marine biodiversity. Cyprus' national legislation was amended in 2005 to manage all fisheries (both commercial and recreational) and control fishing effort to better manage the marine resources and ecosystems, and to ensure compliance with the EU Common Fisheries Policy. The UNEP-MAP program "Integrated Coastal Area Management in Cyprus: Biodiversity Concerns" was also developed in 2007, and a specific strategy for artificial reef deployment has been recently established by the Department of Fisheries and Marine Research of the Ministry of Agriculture, Natural Resources and the Environment of Cyprus (DFMR) with the participation of the Cyprus Tourism Organization and the Cyprus Diver Center Associations.

Artificial reef deployments are considered a mechanism to increase marine biodiversity, impede illegal trawling, enhance fish stocks, and increase marine productivity to improve small-scale coastal fisheries and developing recreational diving.

Thus far, one artificial reef was deployed in the Amanthus area (southern Cyprus), and there is a plan for the deployment of other three artificial reefs in the near future. The Iskele Fishermen Solidarity Association for Development and Iskele Municipality has promoted deployment of an artificial reef off Iskele, in northern Cyprus (Figure 11.2). These artificial reef projects will be cofinanced by the European Community.

The reef sites off Cyprus are selected based on the results of specific studies that evaluate the ecological features of the area, substrate stability, biocoenoses, habitat suitability, and local biodiversity. An assessment of the possible impacts of the reef presence and its composition on the marine environment is also conducted in the projected areas of deployment.

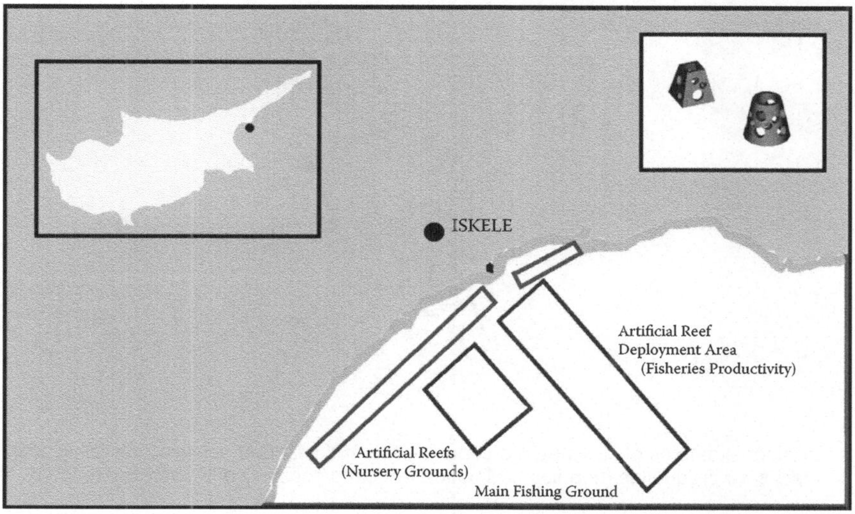

Figure 11.2 ***See color insert.*** Cyprus: Plan of Iskele artificial reef with a particular of the used concrete modules. (Modified from and courtesy of B.A. Cicek.)

Off Cyprus, modules are commonly constructed of concrete with holes and are positioned either individually in a haphazard manner or piled in pyramids A mandatory monitoring program includes ecological, socioeconomic, and fishery aspects after deployment to evaluate the reef effects on the environment and its effectiveness in meeting the management objectives. Importantly, a ministerial order regarding the management of the existing artificial reef off Amanthus is in preparation. Under the provision of this order, no fishing activity is allowed and no vessel transit and anchorage permitted near the reef except at two mooring sites.

France

The eastern side of the French Mediterranean coast is characterized by an irregular shoreline that falls off abruptly into deep water. The sublittoral band is narrow due to the almost total absence of a continental shelf. *Posidonia oceanica* beds are abundant in shallow water between 30 and 38 m deep. Oppositely, a wide continental shelf, with isolated rocky outcrops, occurs on the western side of France's Mediterranean coast. Due to the western Ekman deflection, the Rhone River's waters are pushed along the shore to the west, making this area more productive than along the eastern coast (Barnabé et al., 2000).

In the eastern part of France's Mediterranean coast, the main function of artificial reef deployments has been to facilitate the restoration of marine fauna and *Posidonia* meadows damaged by coastal developments. By contrast, on the western side, artificial reefs have been mainly used to protect the set gears of the small-scale fisheries and shellfish cultures from damage caused by illegal trawling in the coastal areas. These reefs also serve to enhance fish stocks exploited by fisheries, as well.

Starting earlier in 1968, experiments by French researchers on artificial reefs made of waste materials (car bodies) represented the first evaluations of the utility of artificial reefs in Europe. It was only in the mid of 1980s that a concerted program was developed, with 30,000 m^3 of concrete modules; specially designed to be assembled in situ into predetermined configurations. About 90,000 m^3 of artificial reefs have been deployed to date at 20 sites along the French Mediterranean coast.

The artificial reef placement off France involves different agencies and organizations having diverse interests and includes administrators, professional fishermen, scientists, and environmental associations.

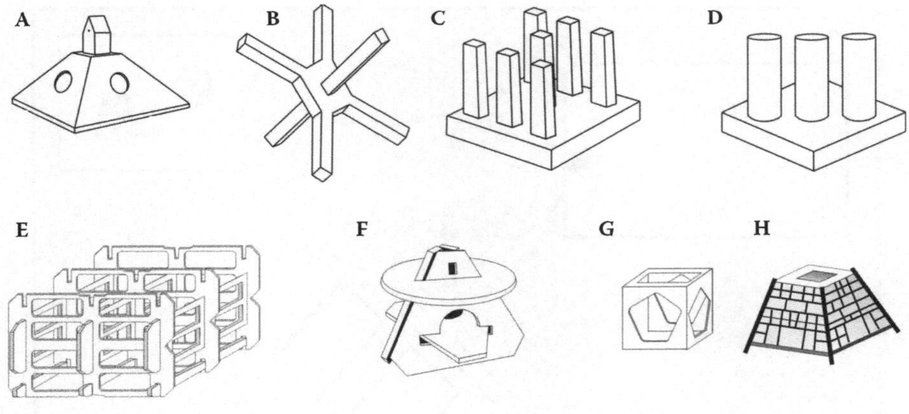

Figure 11.3 France: examples of concrete modules used for artificial reef construction (a–d: protection modules; e–h: production modules). (Modified from and courtesy of E. Charbonnel.)

Most of artificial reefs in France are specially designed and made of concrete. Several kinds of modules have been tested for both production and protection effects (Figure 11.3). Production reefs are usually composed of small cubic modules of 1 to 2 m^3, deployed in chaotic piles of 50–150 m^3. Larger modules (158 m^3) have also been deployed, and these are based on Japanese technology.

Several kinds of modules have been used to serve as antitrawling reefs. It has been estimated that, to be efficient, these protection units must have a minimum weight of 8 t. They are deployed individually, separated from each other by 50 to 200 m, and can be deployed either regularly within the overall area or deployed perpendicular to the coast. These reef deployment projects most often include monitoring surveys aimed at evaluating the reef effects on the marine environment and the resources.

Management of these artificial reefs varies from one area to another. In Languedoc, access is allowed, and all reefs are open to fishermen, while in Provence-Alpes-Côte d'Azur most of the reefs are in protected areas where fishing is prohibited. In the Cote Bleue Marine Park, the situation is intermediate, with some reefs open to fishing (both professional and recreational), while others are subjected to protective management where fishing is prohibited or limited. However, where open access to these artificial reefs is allowed, fishermen, financiers, and contractors generally prefer to have some level of management regulation.

Greece

The Greek coast line is a long and fragmented, with more than 3,000 islands and a large number of natural harbors. Since ancient times, these features allowed the spread of the fishing fleet over an extensive area. Except for some shallower areas in the north Aegean Sea and inside the gulfs, the seas around the Greek peninsula are generally deep, with a narrow continental shelf and steep bottom profile. The high diversity of oceanographic and geologic conditions creates a large amount of diverse habitats where many different stocks and species live.

Fishing is one of the most important economic and social activities in coastal zones. Together with tourism, fishing also represents the main activity in distant islands or isolated areas where usually there is not the alternative of agricultural employment.

Greece has typical multigear and multispecies fisheries that depend to a great extent upon coastal resources. The main fishing gears for the offshore fisheries are bottom trawls and purse seines, while gillnets, trammel nets, longlines, and traps are used primarily for near coastal fisheries (www.fao.org). These fishing grounds are exploited by the majority of the fishing fleet but are not uniformly located along the coastal zone. The narrowness of the continental shelf favors the concentration of the fleet to this area, creating conflicts between fishers using different gears, and increasing pressure on the coastal resources.

The construction of artificial reefs represents a recent activity in Greece, initiated in 2000. The deployment of artificial reefs is authorized within the framework of the Multiyear Program for Fishery Development, which includes a specific action relative to the use of artificial reefs to protect and manage fishery resources and small-scale fisheries. The Ministry of Agriculture is the official agency responsible for the construction and monitoring of artificial reefs and provides the necessary funds. Local fishermen's associations, usually through the local prefectures, take the initiative to promote new reef locations.

A multiyear monitoring study is required for reef deployments and normally includes oceanographic investigations, sampling of the fish assemblage (both inside the reef and in the surrounding areas), and studies on the colonization of the artificial structures by benthic organisms. Surveys of the fishery landings at the nearest fishing ports to collect data on the commercial catch are also included in the monitoring study. Each kind of fisheries (both commercial and recreational) and diving are forbidden during the monitoring period. At the end of the monitoring study, a fishery management plan is drafted, establishing rules and limitations for the small-scale fisheries that operate near the reef. For example, these regulations may include temporal closures that coincide with the presence of heavy juveniles' recruitment.

Four artificial reefs, each covering an area of 8–10 km^2, have been constructed from 2000 to 2006, and six additional sites have been already proposed (Figure 11.4). Greek authorities have established that 10 km^2 shall be the minimum area for each of the new reefs.

Off Greece, different concrete modules have been employed. Mixed-function modules (protection and production), consisting of concrete cubic blocks provided with holes and deployed alone on the seabed or assembled in pyramids, are the most commonly deployed. Production modules, such as bulky cement bricks on a concrete base and concrete pipes assembled in pyramids, have been also used (Figure 11.5).

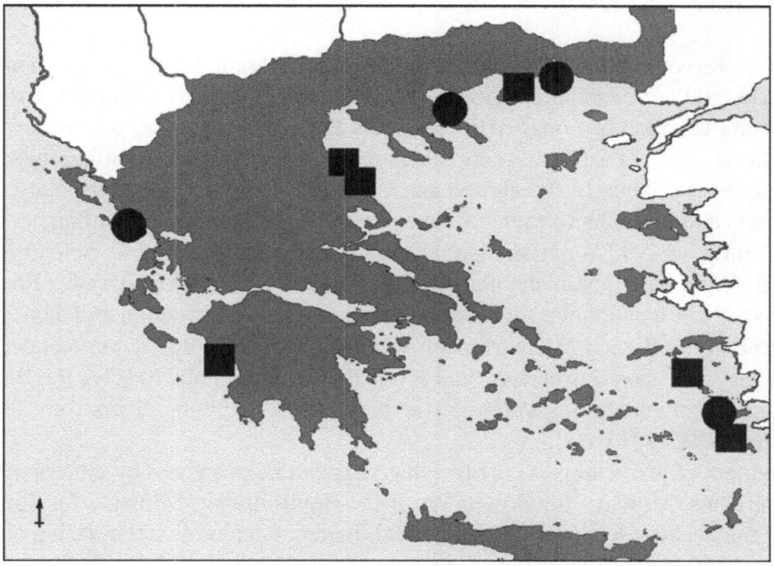

Figure 11.4 Greece: Locations of deployed (black circles) and planned artificial reefs (black squares). (Modified from and courtesy of A. Kallianiotis.)

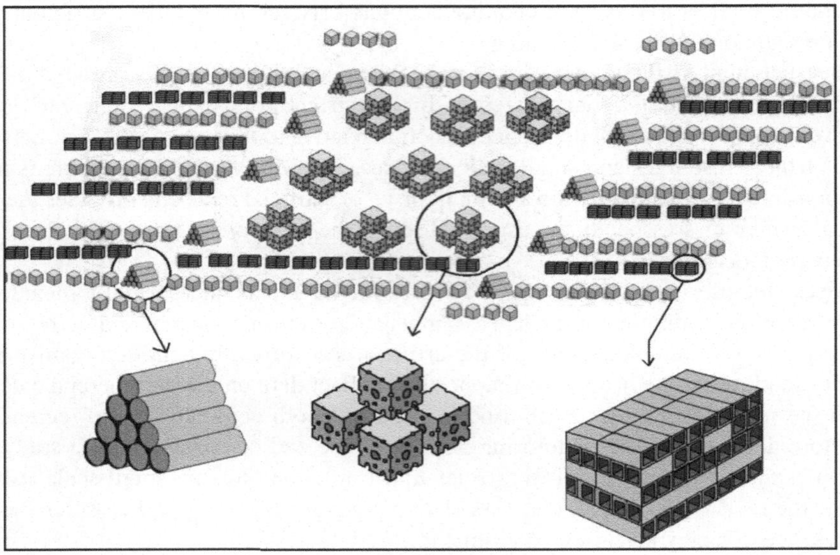

Figure 11.5 Greece: Plan of an artificial reef made of four different types of modules in order to increase the complexity of the structure. (Modified from and courtesy of A. Kallianiotis).

Israel

Similar to the overall southeastern Mediterranean region, the Mediterranean Sea off the Israel continental shelf is characterized by oligotrophic waters and low productivity. Nevertheless, the few rocky substrates provide preferred habitat of many commercially important fish species. Consequently, these habitats are heavily exploited either by professional and sport fishermen or scuba divers, which, consequently, results in a decrease in the macrofauna inhabiting these limited habitats.

Israel may represent a special case where artificial reefs are used for management of both professional fisheries and recreational activities. The first artificial reefs off Israel were deployed in the early 1980s when the Israeli National Artificial Reefs Project for the Mediterranean was initiated to evaluate the effectiveness of artificial reefs as a potential solution to overcoming the limited habitat and food resources available. In the subsequent years, the lack of a national policy for the placement of new reefs and of a management of the existing reefs led to rising conflicts among different users. Consequently, in 1993, a national committee (Israeli National Public Scientific Committee) was nominated by the Israeli Ministry for the Environment (current Ministry for Protection of the Environment) to define a national policy and establish general construction and deployment criteria for artificial reefs in the Israeli Mediterranean waters. This committee recommended establishing artificial reefs for professional fishermen and scuba divers, separately, to solve these user conflicts. Following this recommendation, several new artificial reefs for scuba divers were deployed along the Mediterranean coast off Israel.

The deployment of artificial reefs off Israeli coasts can be promoted by either national authorities (e.g., universities, Ministry for Protection of the Environment, Ministry for Tourism, etc.) or diving clubs. Projects have to be approved by local, regional, and national building committees, by the navy and port authorities, the Fishery Department, and the Authority for Preservation of Nature. After these, the consent of the Committee for Approval of Dumping into the Sea is also necessary.

A monitoring program is required for reef deployments in Israel. It usually includes scientific studies aimed at evaluating the reef's effects on the biodiversity, biomass, and succession of local

Color Figure 3.3 The "pallet" reef type with age-0 red snapper September 9, 2009.

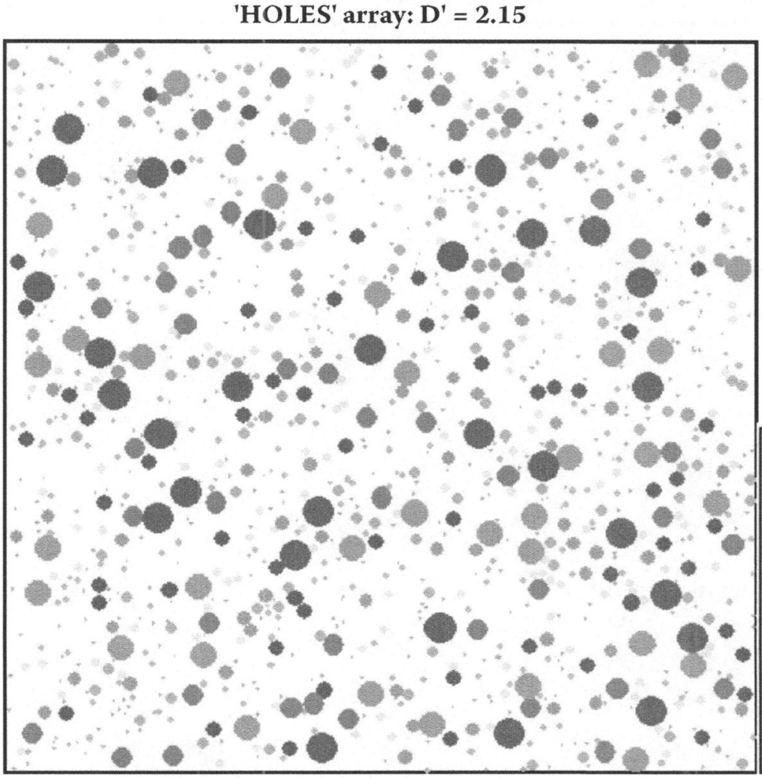

Entry diam. (cm)	# crevices
5	500
10	225
15	141
20	102
25	79
30	64
40	46
50	35
60	29

Color Figure 4.1 Visualizing the relative frequency of small and large holes in two arrays corresponding to (a) $D' = 2.15$, and (b) $D' = 3$. The predicted size distributions of holes are shown in the accompanying tables. Arrays created by "Holes for Windows" (Stamatopoulos and Caddy, 1991).

'HOLES' array: D' -> 3.0

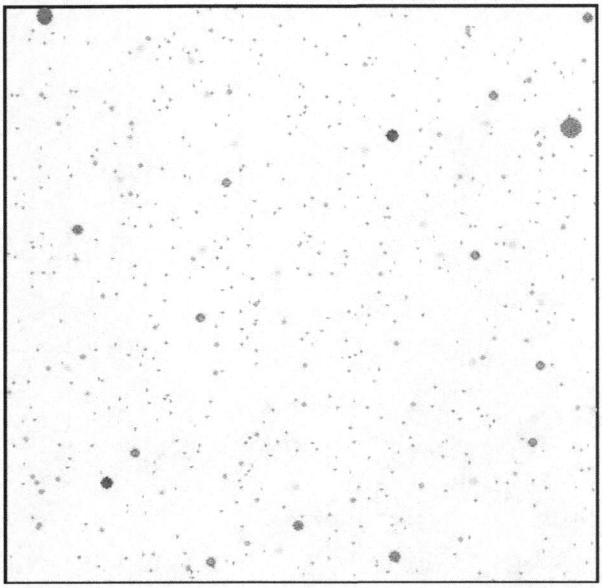

Entry diam. (cm)	# crevices
5	500
10	63
15	19
20	8
25	4
30	2
40	1
50	1
60	0

Color Figure 4.1 (continued) Visualizing the relative frequency of small and large holes in two arrays corresponding to (a) $D' = 2.15$, and (b) $D' = 3$. The predicted size distributions of holes are shown in the accompanying tables. Arrays created by "Holes for Windows" (Stamatopoulos and Caddy, 1991).

D' = 3.0, With Artificially Enhanced Numbers for 3 Largest Crevice Sizes.

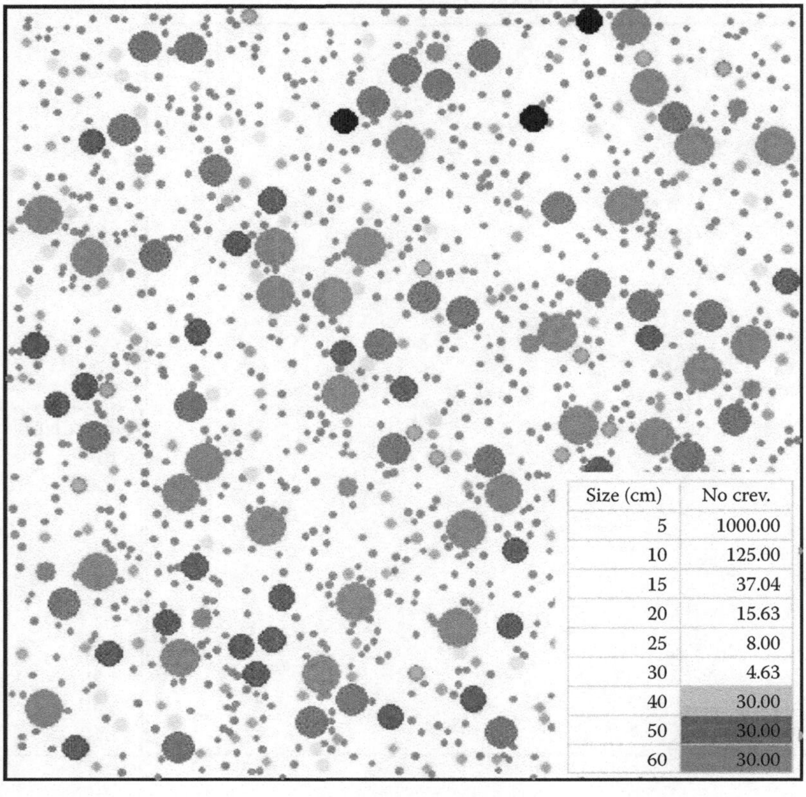

Size (cm)	No crev.
5	1000.00
10	125.00
15	37.04
20	15.63
25	8.00
30	4.63
40	30.00
50	30.00
60	30.00

Color Figure 4.11 Illustrating a reef surface with enhanced numbers of crevices in the size range 40–60 cm superimposed on a natural reef with fractal coefficient $D' = 3.0$.

Color Figure 7.1 Sandfish, *Arctoscopus japonicus*.

Color Figure 7.11 Egg lumps of sandfish attached to each egg collectors, *S. fulvellum* (a), polypropylene rope (b), and polyethylene fishing net (c).

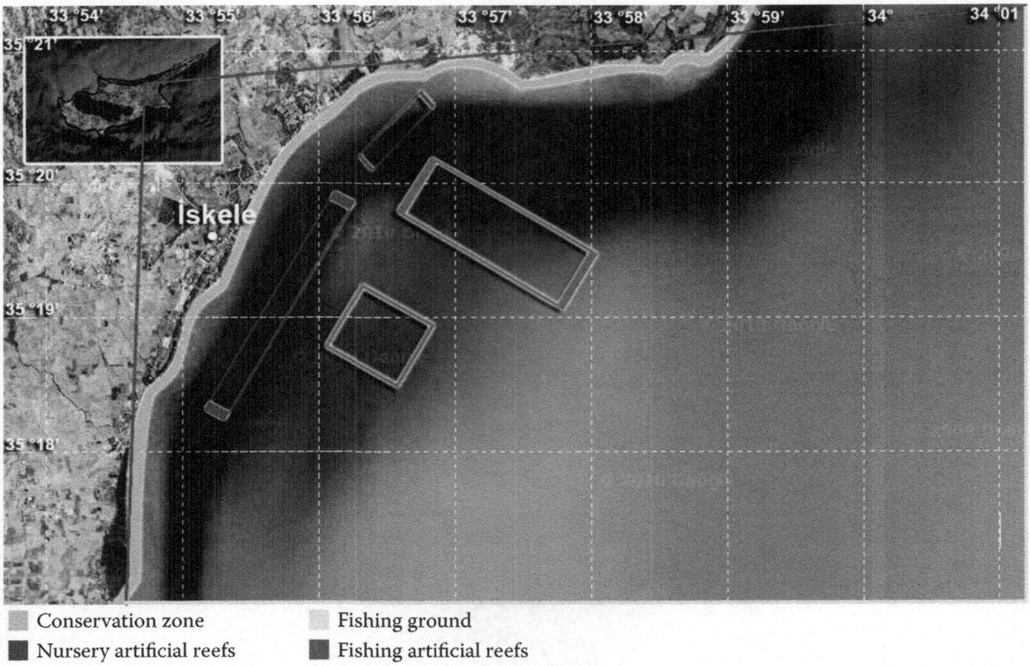

Color Figure 9.1 Site, fishing ground, artificial reefs, and the conservation zone. Redrawn from images in Google Earth.

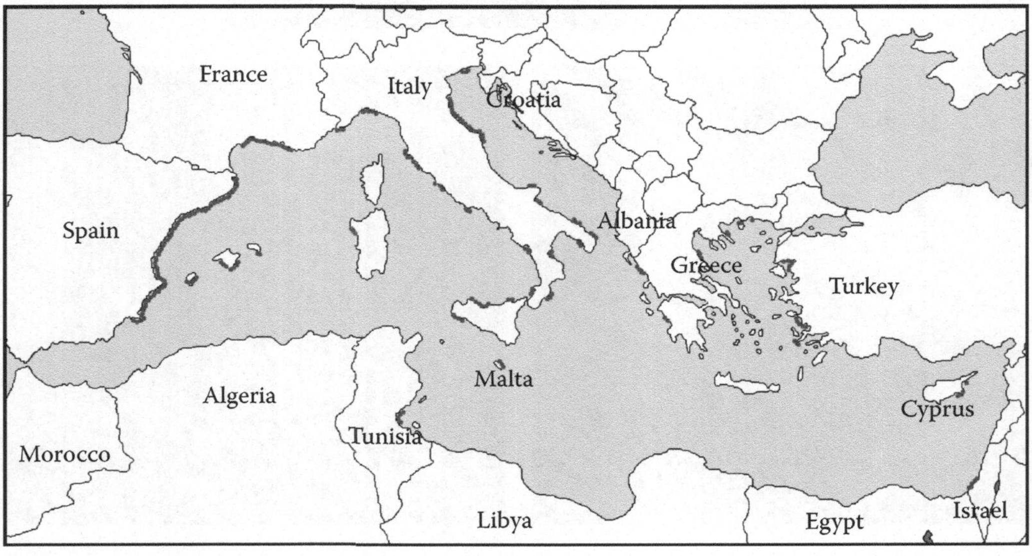

Color Figure 11.1 Artificial reefs in the Mediterranean Sea.

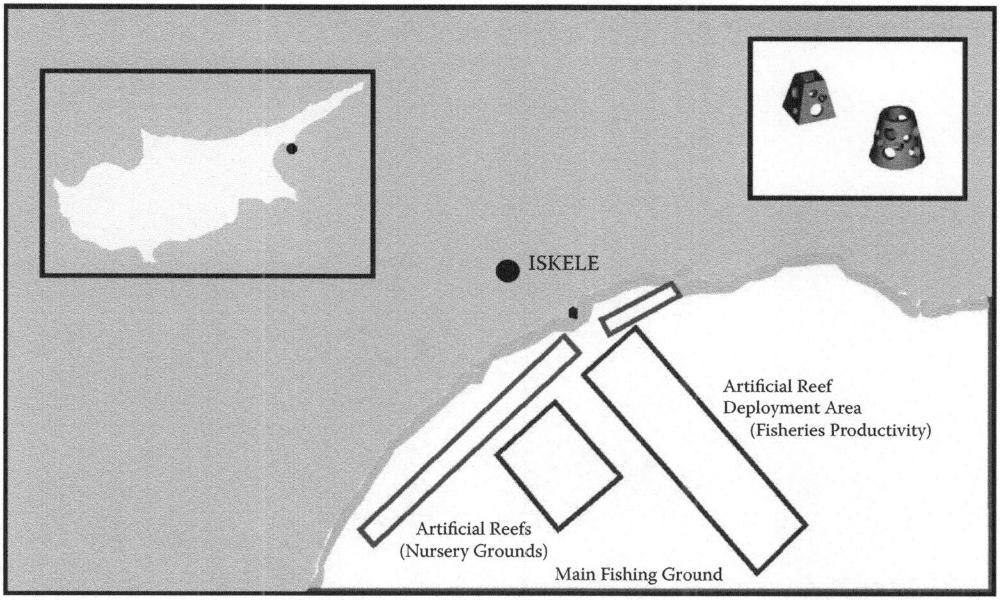

Color Figure 11.2 Cyprus: Plan of Iskele artificial reef with a particular of the used concrete modules. (Modified from and courtesy of B.A. Cicek.)

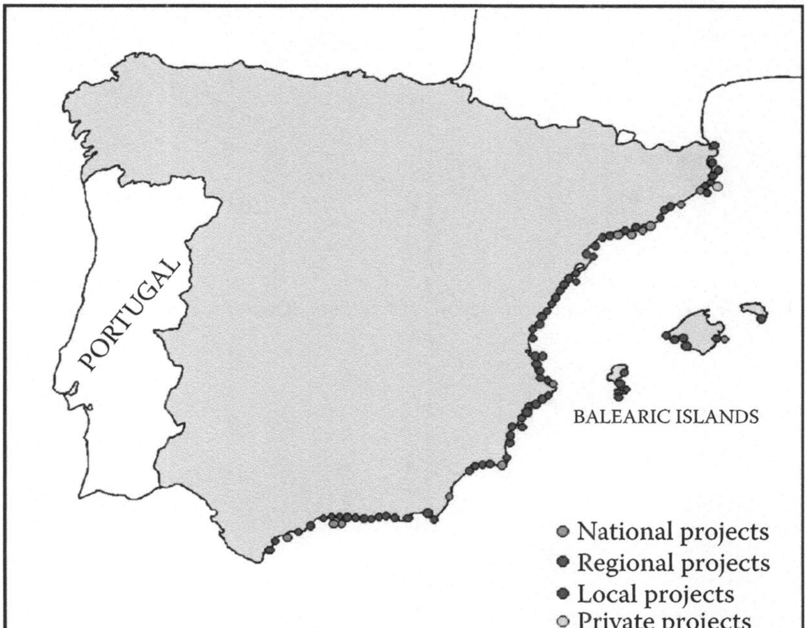

Color Figure 11.10 Spain: Artificial reefs deployed along the Mediterranean coast. (Courtesy of J.J. Goutayer Garcia.)

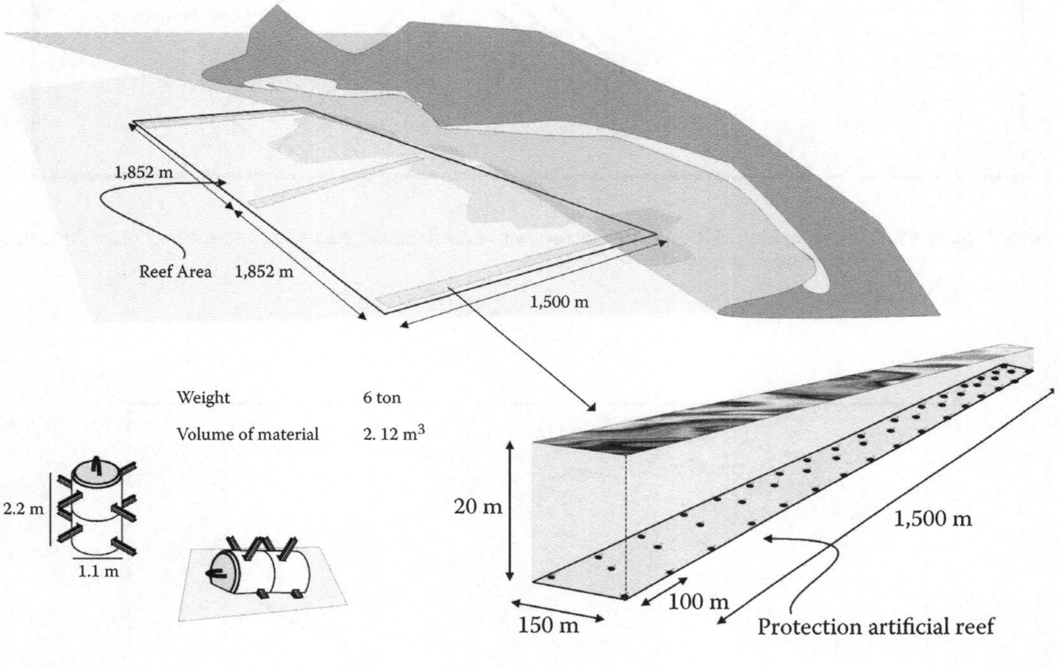

Color Figure 11.12 Spain: Scheme of a protection artificial reef. (Courtesy of J.J. Goutayer Garcia.)

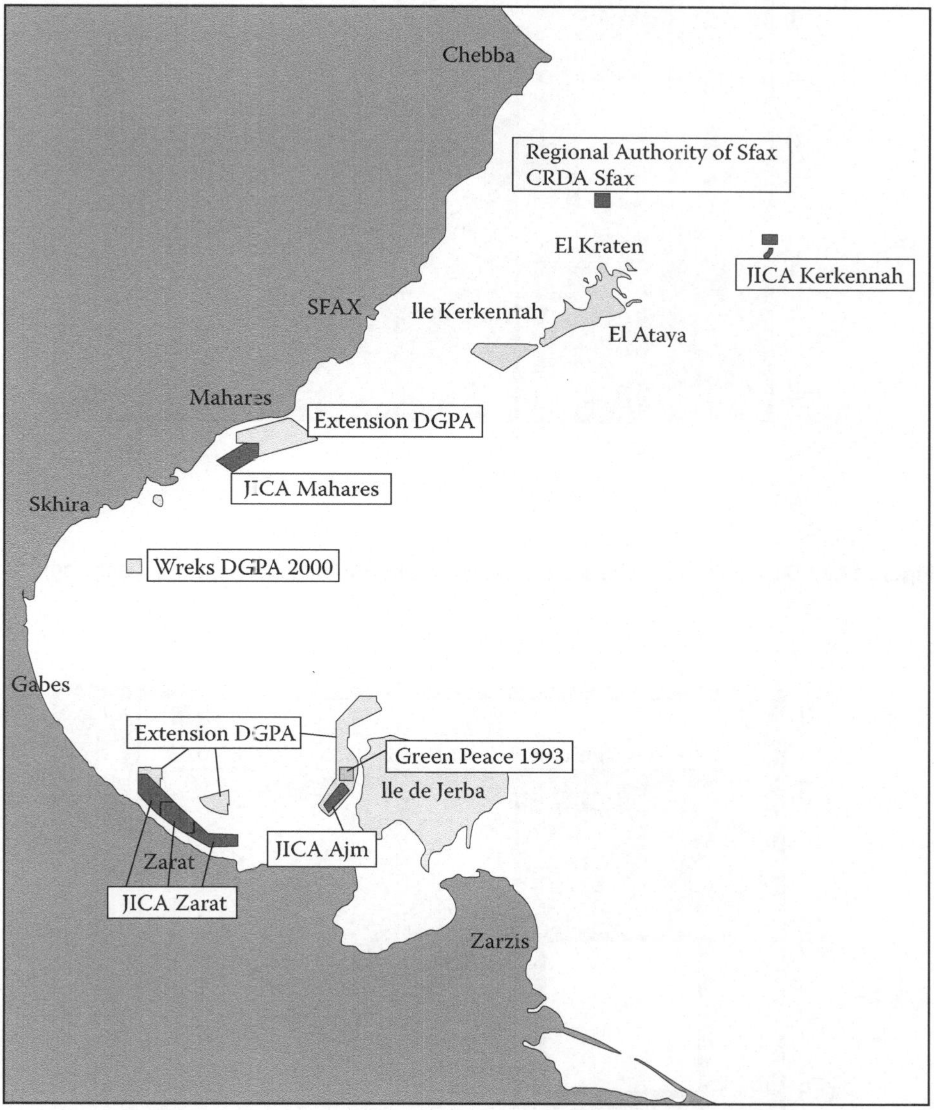

Color Figure 11.13 Tunisia: Artificial reefs deployed in the Gulf of Gabes DGPA: General Direction for Fisheries and Aquaculture; CRDA: Regional Commissariat for Agriculture Development. (Modified from JICA, 2009. Project de Coopération Technique avec le Japon sur la Gestion Durable des Ressources de la Pêche Côtière en République Tunisienne. Rapport des Activités du Projet. 216 pp).

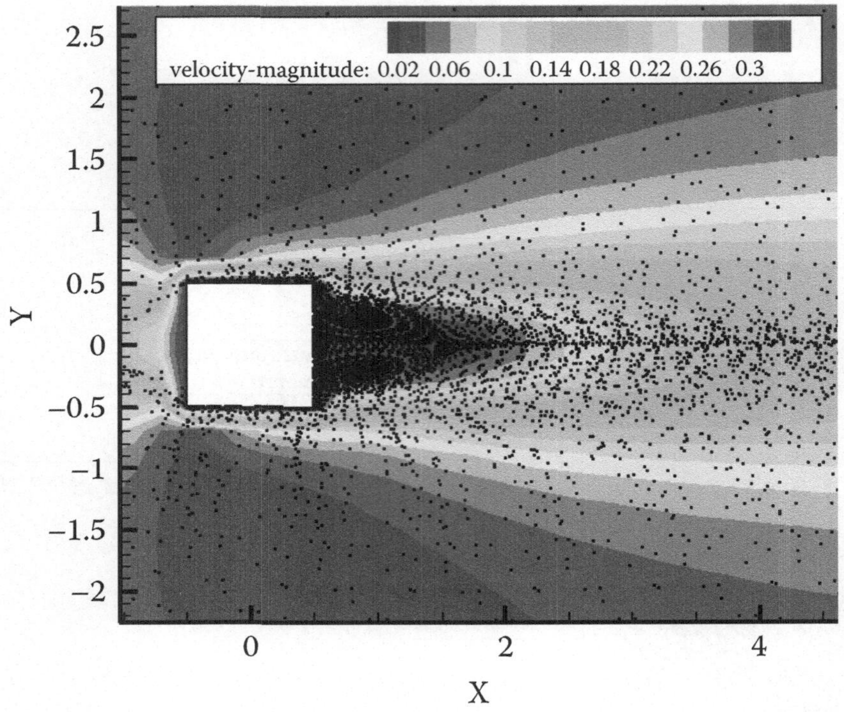

Color Figure 17.8 Distribution of streamline by counter flow behind a square body (Re = 300,000).

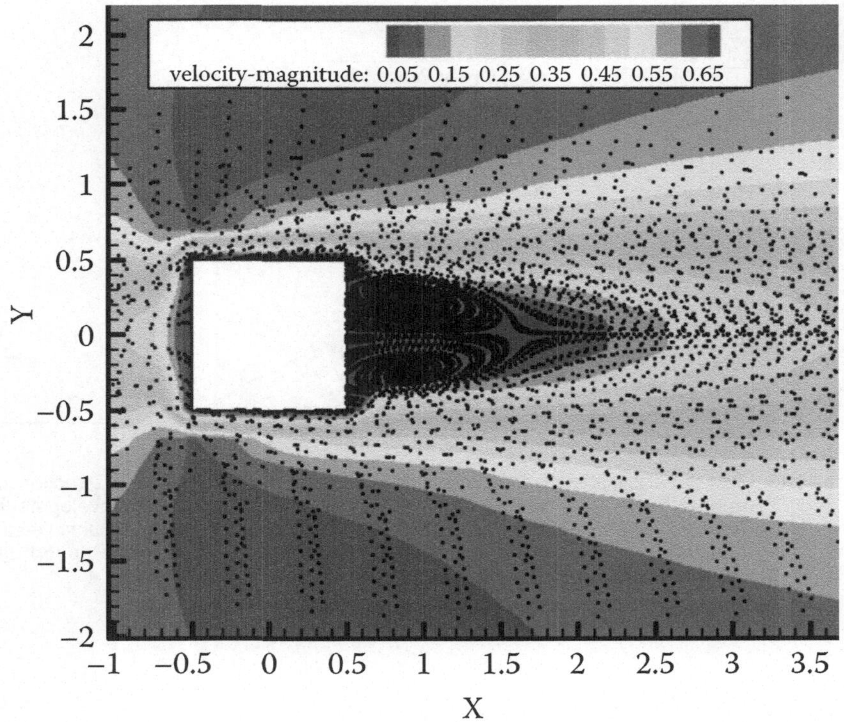

Color Figure 17.9 Distribution of streamline by counter flow behind a square body (Re = 600,000).

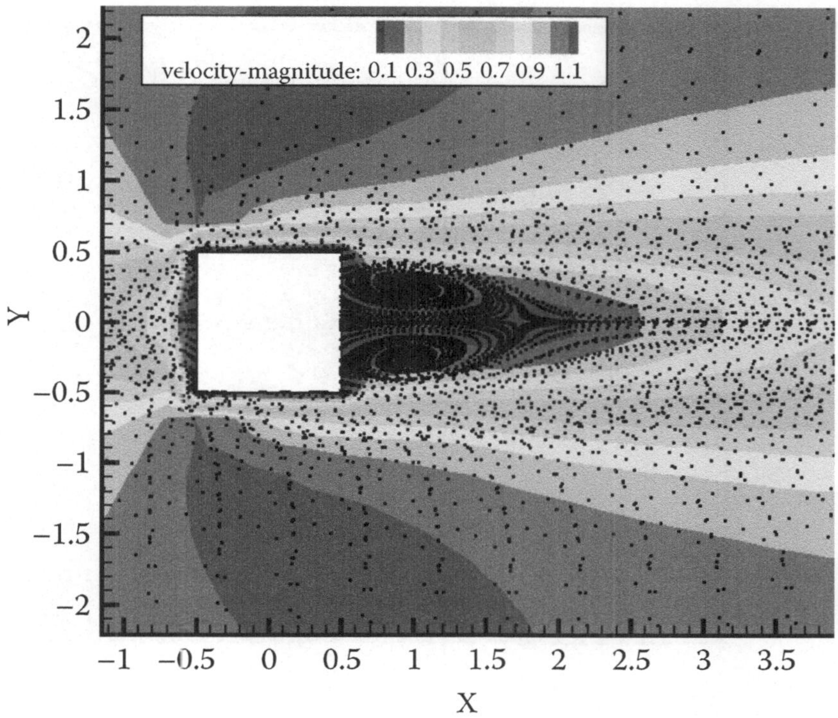

Color Figure 17.10 Distribution of streamline by counter flow behind a square body (Re = 1,000,000).

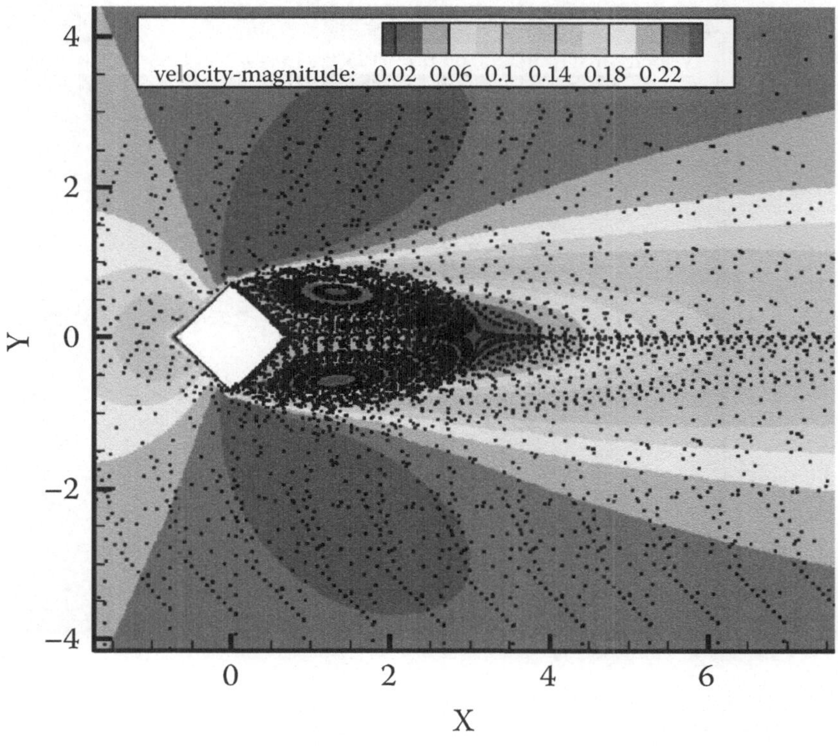

Color Figure 17.11 Distribution of streamline by counter flow behind an inclined square body (Re = 300,000).

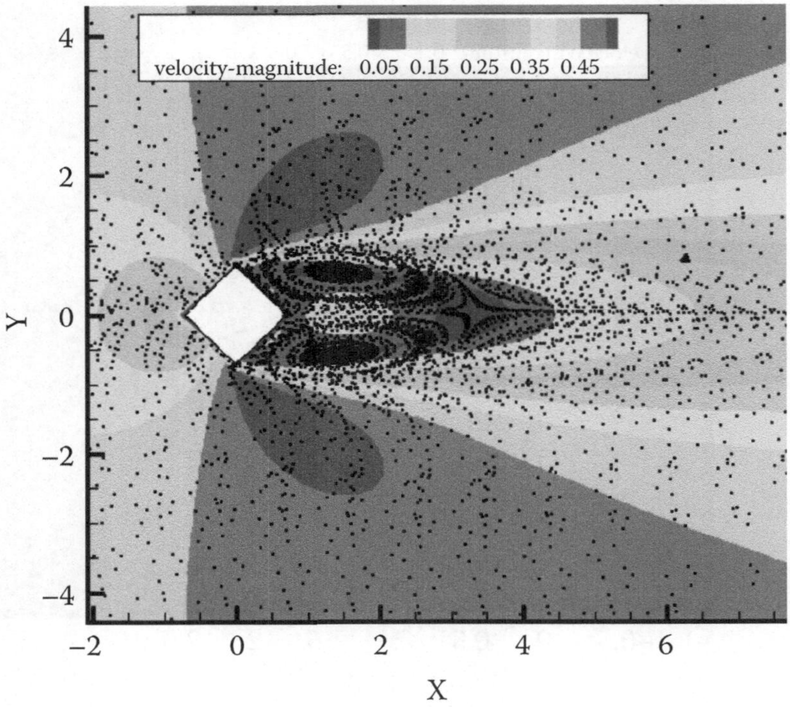

Color Figure 17.12 Distribution of streamline by counter flow behind an inclined square body (Re = 600,000).

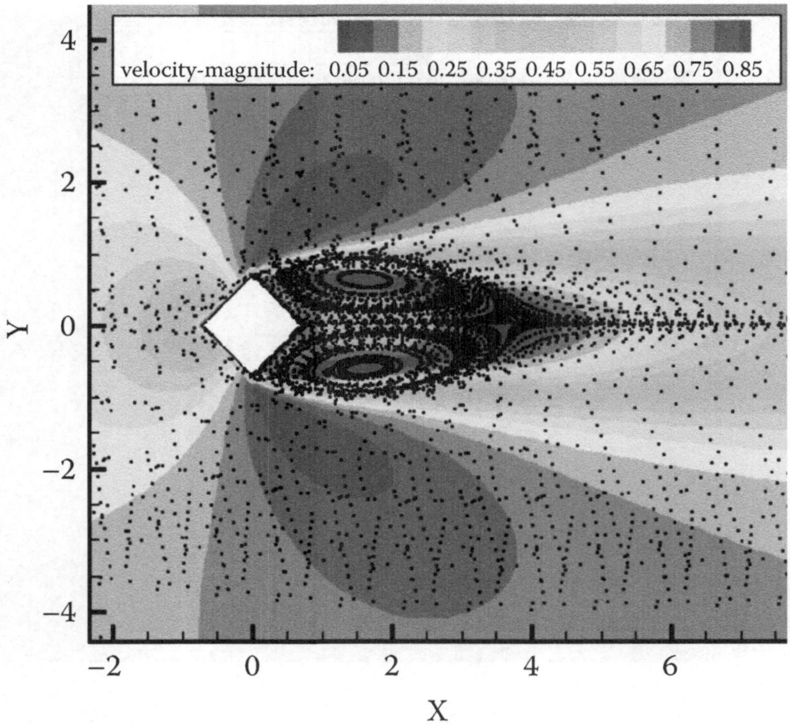

Color Figure 17.13 Distribution of streamline by counter flow behind an inclined square body (Re = 1,000,000).

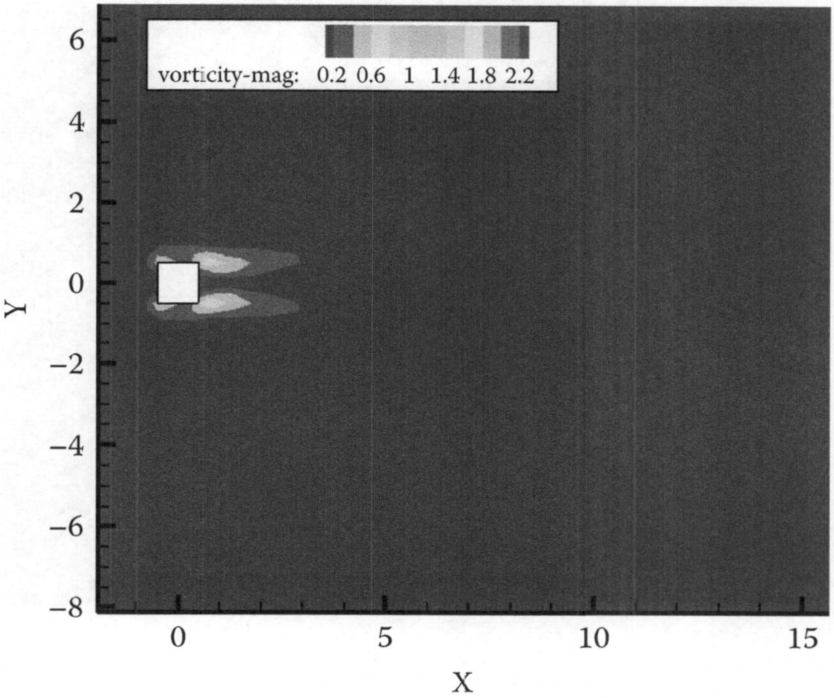

Color Figure 17.14 Distribution of vorticity magnitude behind a square body (Re = 300,000).

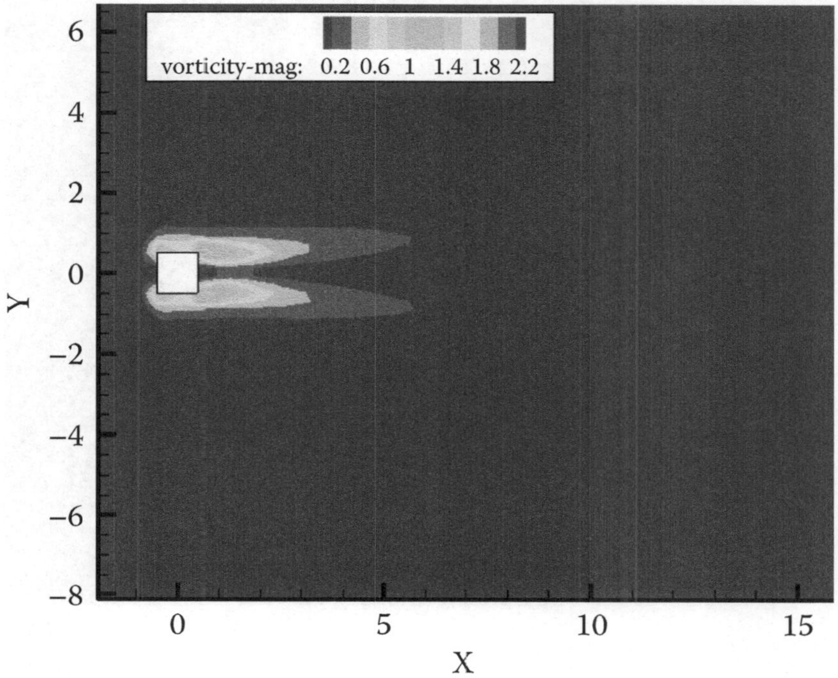

Color Figure 17.15 Distribution of vorticity magnitude behind a square body (Re = 600,000).

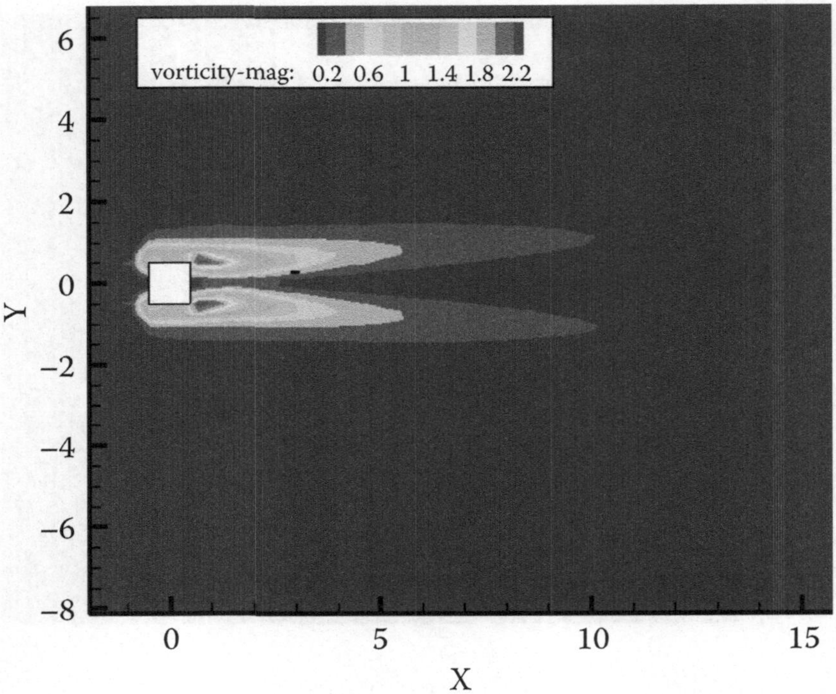

Color Figure 17.16 Distribution of vorticity magnitude behind a square body (Re = 1,000,000).

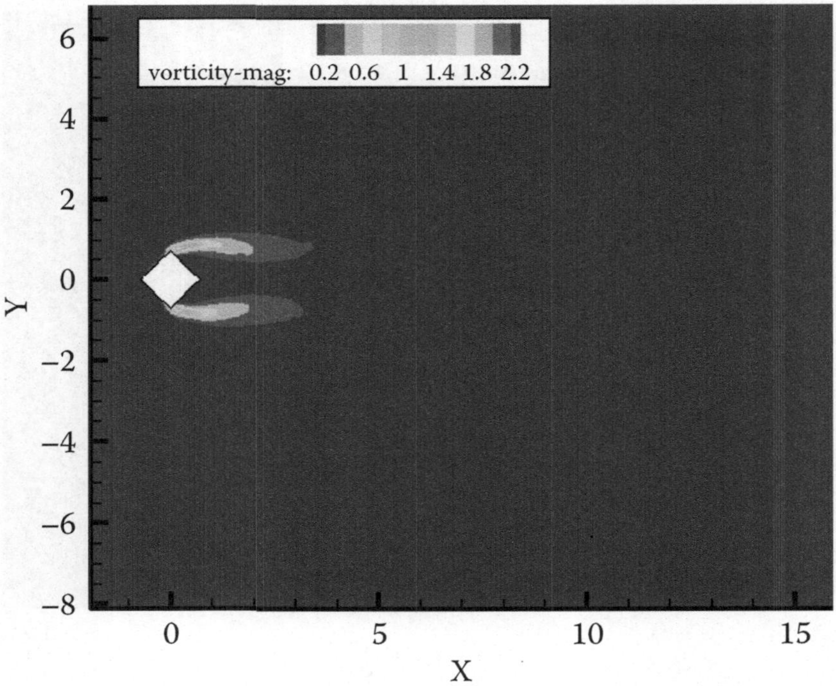

Color Figure 17.17 Distribution of vorticity magnitude behind an inclined square body (Re = 300,000).

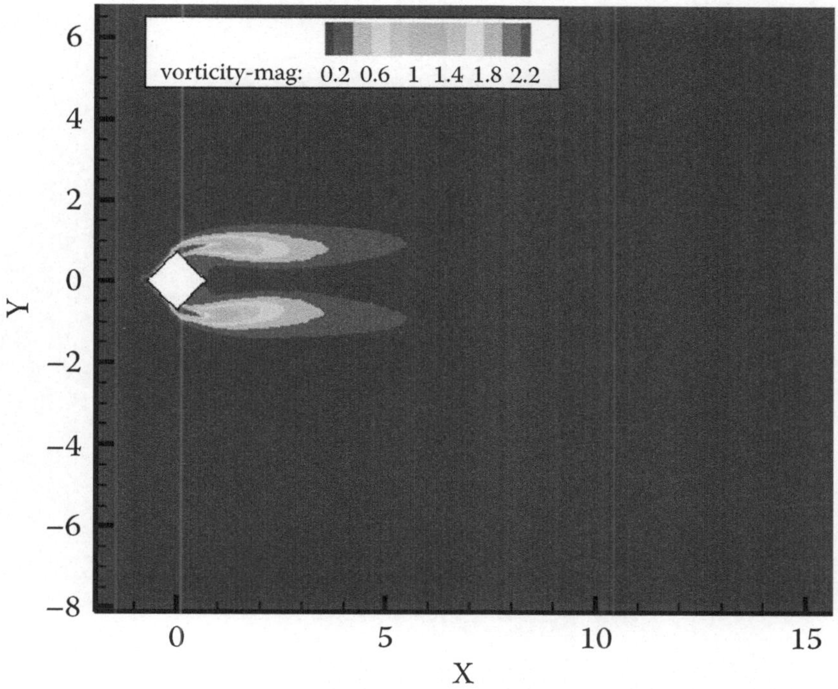

Color Figure 17.18 Distribution of vorticity magnitude behind an inclined square body (Re = 600,000).

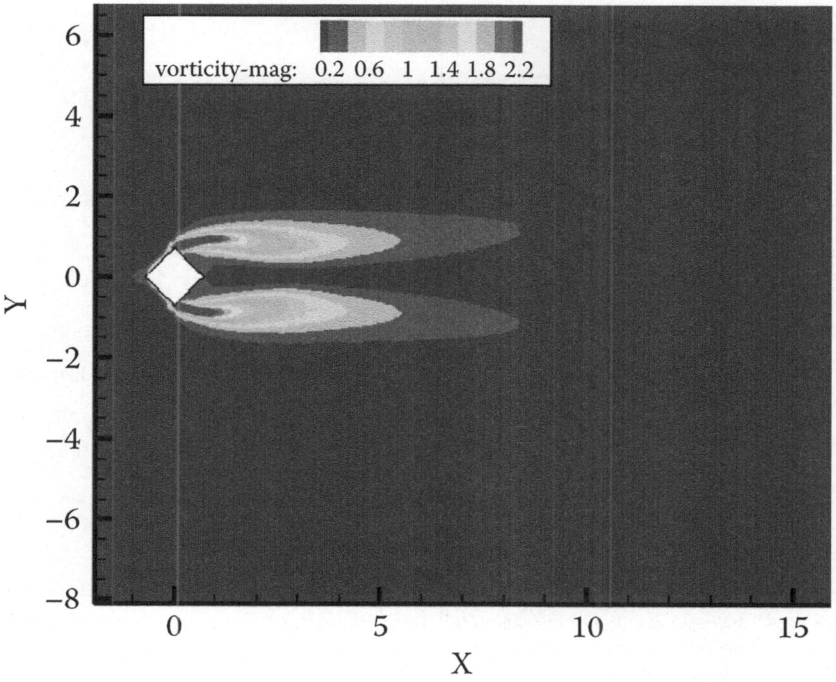

Color Figure 17.19 Distribution of vorticity magnitude behind an inclined square body (Re = 1,000,000).

Color Figure 18.3 Depiction of artificial seamount constructed on the seabed at a depth of 82 m off Ikitsuki Island (height, 12 m; length, 120 m; conic bottom radius, 30 m; number of "Ashcrete" blocks, 4,860 (29,160 t)).

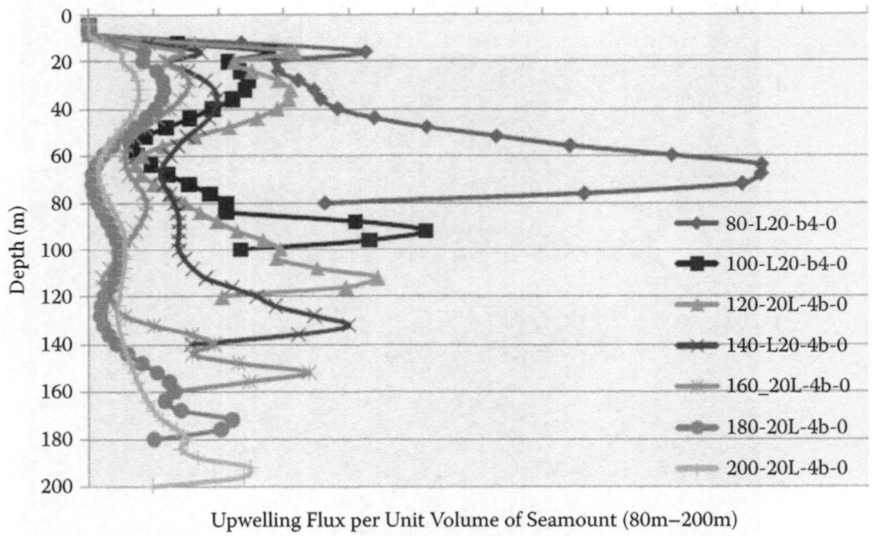

Color Figure 18.5 Vertical profiles of upwelling flux of nutrients per volume versus construction depth of the artificial seamount. The lowest point of each line shows the depth where the seamount was constructed.

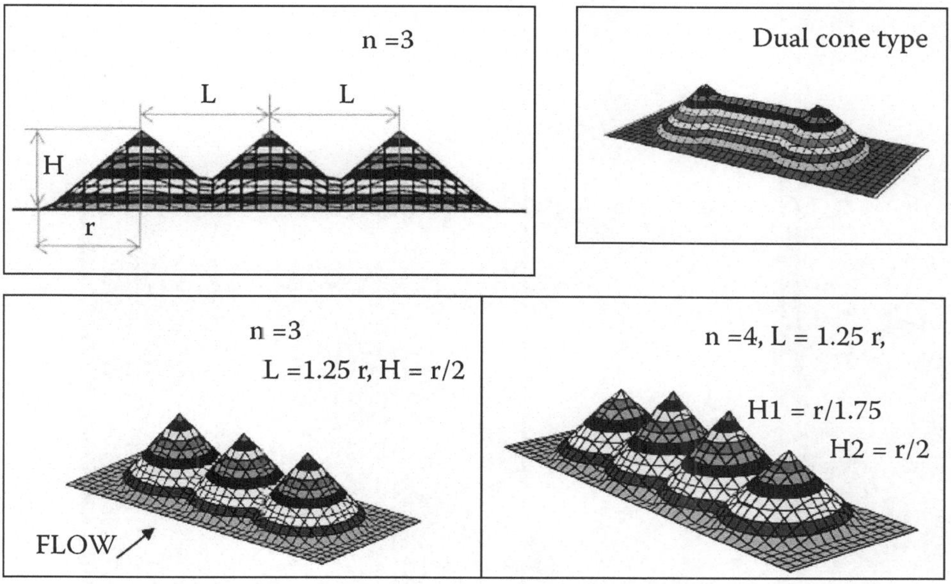

Figure 18.6 Upper left are the parameters of an artificial seamount: *r*, conic bottom radius; *H* = *r*/2, conic height; *L*, conic top interval; and *n*, conic number. Lower left is a bird's-eye view of an artificial seamount consisting of three cones. Lower right is an artificial seamount consisting of four cones with deferent heights. Upper right is the previous dual-cone seamount.

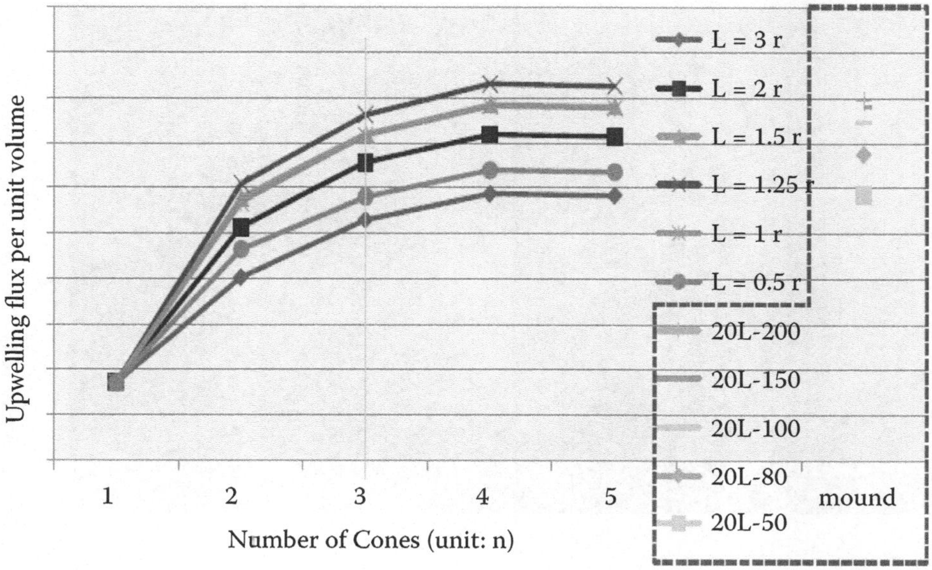

Color Figure 18.7 Unit upwelling flux of nutrients versus number of cones composing artificial seamount and versus various types of conventional levee-type seamount varying peak distance from 50 m to 200 m. (Dots on the right side surrounded by the dashed line.)

Color Figure 18.11 Depiction of cone-type artificial seamount made of Ashcrete blocks.

Figure 11.6 Israel: Example of artificial reef modules made of old car tires. (From Spanier, E. et al., 1985. *Bull. Mar. Sci.* 37(1): 356–363.)

benthic communities and fish populations. To lesser extent, the frequencies of diving on the artificial reef are also assessed.

At present, in Israel, there is not a specific plan to manage the deployed artificial reefs in order to avoid overexploitation of the reef resources.

Old car tires, assembled in different configurations, were employed for the construction of the first artificial reefs in 1980s (Figure 11.6), while the more recent ones mainly consisted of concrete structures or sunk vessels (Spanier, 2000; Edelist and Spanier, 2009). These reefs are under the responsibility of the Israeli Diving Federation (Spanier, 2000).

Italy

Italy was among the first countries to make extensive use of artificial reefs in the Mediterranean Sea. The first scientifically planned reef designed specifically to impede illegal trawling and improve local small-scale fisheries was established off Italy in 1970s. More than 70 reefs have been deployed along Italian coasts. Most of these are medium- or large-scale reefs, while others are small-scale experimental reefs used for research.

The main purpose for the use of artificial reefs off Italy is fisheries management. Here, fisheries management includes (1) protection of coastal nursery areas or other sensitive habitats from illegal trawling, (2) reduction in conflicts between different fishing sectors (e.g., small-scale fishers and trawlers), and (3) enhancing small-scale fisheries by increasing local finfish populations and, in eutrophic waters such as those in the Adriatic Sea, by establishing new populations of edible bivalves (i.e., mussels).

Despite the large use of artificial reefs, national programs have been never developed in Italy. More recently, artificial reef deployment has been managed by the regional authorities. Most of the projects have been and continue to be supported with public funds: 50% from the European Community, and with support from both local authorities and fishermen associations. Projects can be also proposed by private subjects (diving clubs, recreational fishers, professional fishermen, etc.), but these require nongovernmental funding and permission from the regional authority to lease the seabed.

Artificial reefs off Italy are generally made of different types of concrete modules/structures: antitrawling, production, and mixed (antitrawling and repopulation) (Bombace et al., 2000; Fabi and Spagnolo, 2001a, b; Fabi, 2006; Figure 11.7).

Artificial reefs off Italy can be classified as to three general types based on their function: protection, production, and mixed. Protection and mixed reefs are usually large-scale reefs placed at about 3 nm off the coast and cover a seabed surface ranging between 0.5 and 365 km^2. The former are simply deployed to protect the coastal areas and other sensitive habitats against illegal trawling and are made of simple antitrawling bodies (e.g., tetrapods, concrete cubic blocks). The latter are

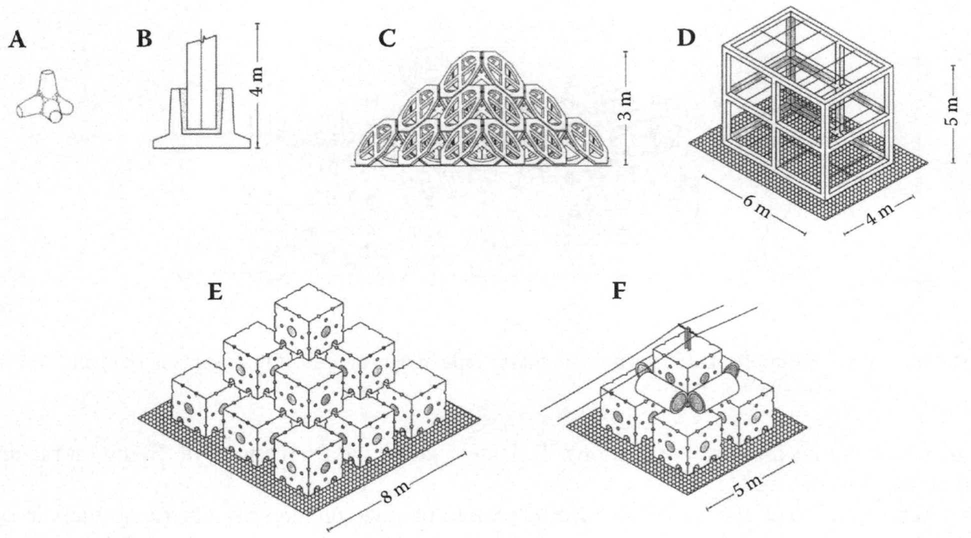

Figure 11.7 Italy: Concrete modules used for artificial reef construction (a,b: protection modules; c,d: production modules; e,f: mixed modules).

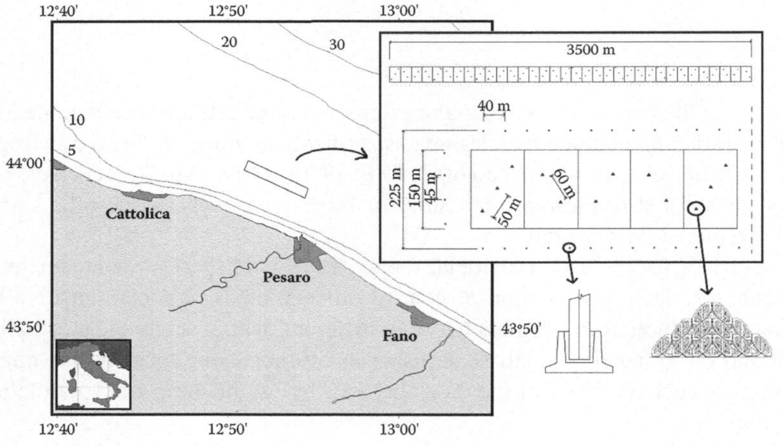

Figure 11.8 Italy: Location and schematic view of a mixed large-scale artificial reef in the northern Adriatic Sea with a perspective view of the concrete pyramids (production modules) and concrete piles (protection modules) that constitute the reef.

used to both impede illegal trawling and enhance fish populations; hence, they can consist of either antitrawling, mixed, and production modules (Figure 11.8). Production reefs are medium-sized reefs placed within the 3 nm coastal area and occupy a seabed surface between 0.04 and 0.08 km^2. They are usually made of mixed or production modules with the aim of increasing fish stocks and exploitable biomass by providing shelter and food, improving local small-scale fisheries and, in eutrophic waters, developing extensive mussel culture (Figure 11.9).

Multiyear monitoring is required for artificial reef deployments off Italy. Monitoring usually begins 1 year before the deployment of the man-made substrates and ends 3 to 5 years after

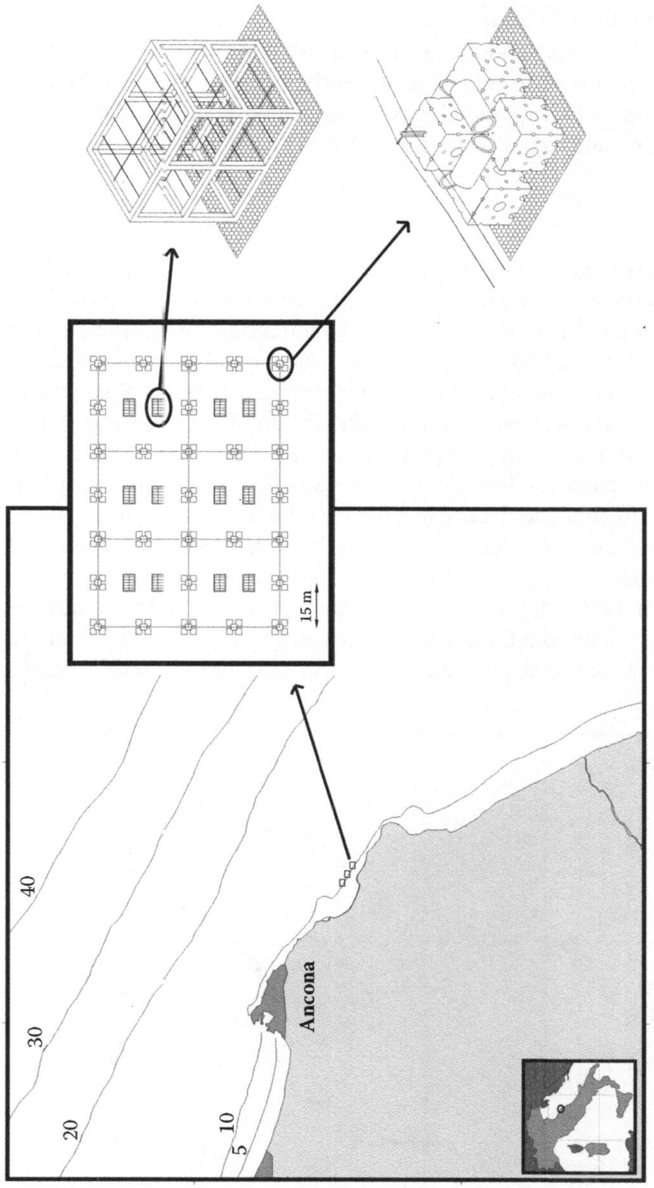

Figure 11.9 Italy: Map of a small-scale artificial reef in the northern Adriatic Sea with a perspective view of the concrete pyramids (mixed modules) and cages for shellfish culture (production modules) that constitute the reef.

deployment. Monitoring includes: (1) acoustic surveys to assess the stability of the structures (scouring, burial, subsidence); (2) studies on the natural soft-bottom benthic community inside and outside the reef; (3) research on the benthic community settled on the hard substrates; (4) sampling the finfish assemblage inside and outside the reef area to assess the deployment effects, both on the fish community and on the yields of small-scale fisheries. During the monitoring period, each kind of fishery is forbidden within the reef zone.

Being constructed with public funds, most of reefs are part of the public domain and cannot be subjected to exclusive exploitation rights. Consequently, once the monitoring investigations have been completed, the reef may be exploited by everyone, and this often creates conflicts among different users and/or overexploitation of the reef resources.

Spain

Spain is a leader in artificial reef use in Europe with more than 103 reefs having been deployed since 1982, with most of these located off Spain's Mediterranean coast (Figure 11.10). Spain is also the second highest consumer of fish in the European Union (after Portugal) with an average of 40 kg of fish consumed per capita in 2005 (Failler, 2007). Concomitantly, the Spanish fishing industry has had to address some problems relative to the decreasing activity of Spain's distant water fleet caused by the increase in international fishing regulations and the increasing competition for space and resources between the fishing fleets operating in Spanish waters. The need to resolve these problems and the high demand for fish has lead the Spanish government to promote a policy of deploying artificial reefs to enhance fishery resources and manage fisheries.

The first legal and policy regulations for artificial reef deployment dates to 1982, but no public funding was committed to support reef construction at that time. After Spain's entry into the European Community in 1986, the Spanish government promoted a policy of deploying artificial reefs with the public funds from the European Community and from national and regional Spanish governmental agencies. A national program and some regional programs (Andalusia, Valencia

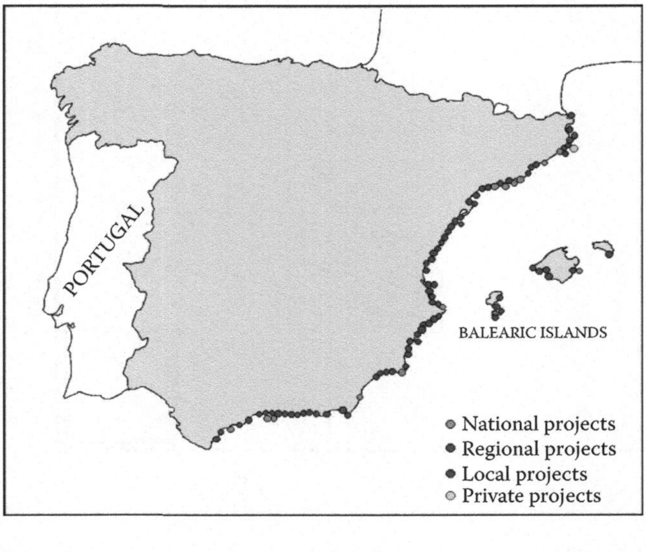

Figure 11.10 ***See color insert.*** Spain: Artificial reefs deployed along the Mediterranean coast. (Courtesy of J.J. Goutayer Garcia.)

Region, Catalonia, Balearic Islands) are currently active, and artificial reefs are specifically mentioned in Spanish fishery laws as an instrument for fisheries management that may be directed toward: (1) protecting and enhancing fish stocks, (2) impeding illegal trawling, (3) protecting sensitive ecosystems (e.g., *Posidonia* beds) and natural habitats important for fisheries, and (4) reducing conflicts between different commercial fisheries.

The placement of artificial reefs in Spain is regulated by the Royal Decree 798/1995, and the "Methodological Guidelines for Artificial Reefs Placement" published by the Spanish Ministry of the Environment and Rural and Maritime Affairs in 2008. These documents analyze the different categories of artificial reefs employed in Spain and their environmental effects, establish the methodology for the installation of the structures and the administrative procedure to be followed in order to get the authorization for reef placement, and give advice on the postplacement monitoring programs. The projects are promoted by public administrations together with fishermen associations, and sites are selected on the base of fisheries management purposes.

A 5-year monitoring program has to be conducted after the reef deployment. It includes: (1) side-scan sonar surveys for general structural quality and monitoring of *Posidonia* meadows, (2) biological studies on the benthic biocoenoses and fish communities associated to the reef, (3) fisheries control studies, and (4) surveys among fishermen and other users.

Being constructed with public funds, artificial reefs are automatically part of the public domain and, hence, they cannot be subjected to exclusive exploitation. However, special use regulations can be applied in some areas. The physical presence of the reef is considered sufficient to regulate the type of fishing activities that can be conducted in its vicinity; hence, no additional rules are usually associated with the structures (Revenga et al., 2000).

Three types of modules are commonly employed: protection, production, and mixed units. The protection modules are simple concrete units of different shapes but heavy enough to obstacle illegal trawling. They are often provided with iron beams aimed to rip trawl nets. Production modules are larger structures, having holes and shelters for vagile organisms. Finally, mixed modules are protection units having a certain structural complexity (Figure 11.11). These modules are employed for the construction of two typologies of artificial reefs: protection reefs and production reefs. The former usually consists of protection and mixed modules and is aimed to: (1) protect fisheries resources; (2) reduce conflicts between users, above all professional fishermen; and (3) protect ecosystems and natural habitats important for fisheries (Figure 11.12). The production reefs are usually made of production and/or mixed units, and their purposes are (1) to increase the fish stocks and exploitable biomass by providing shelter and food, (2) to attract marine organisms, and (3) to spatially redistribute the fish populations.

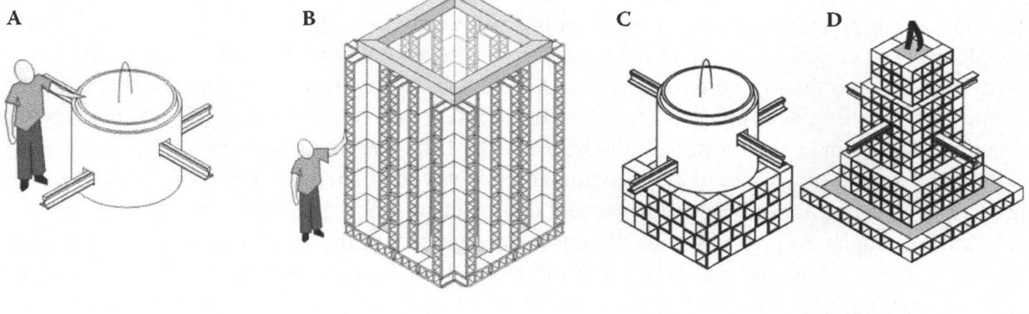

Figure 11.11 Spain: Examples of concrete modules used for artificial reef construction (A: production module; B: protection module; C and D: mixed modules.) (Courtesy of J.J. Goutayer Garcia.)

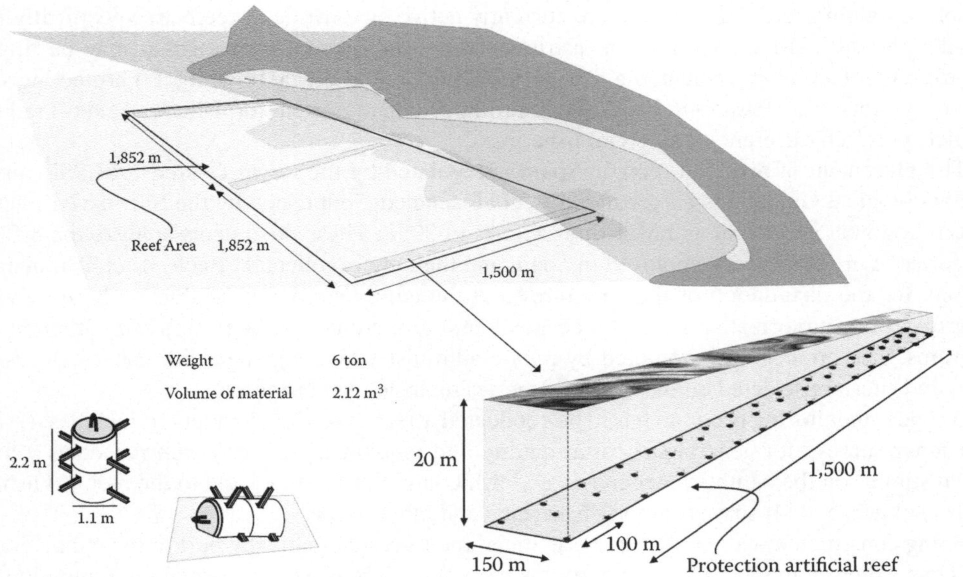

Figure 11.12 ***See color insert.*** Spain: Scheme of a protection artificial reef. (Courtesy of J.J. Goutayer Garcia.)

Tunisia

In Tunisia, there has been an increase in overfishing due to the introduction of large trawlers and the progressive destruction of the seagrass beds (mainly *Posidonia oceanica* meadows) resulting from the illegal activity of small-sized trawlers in coastal areas. Over the past decade, these activities have lead to a gradual decrease of the demersal resources and, consequently, the reduction in the income to small-scale fishers. In the southern region of Tunisia, the production by small-scale fisheries was reduced by about 50% in 10 years, decreasing from around 46,000 t in 1989 to 26,000 t in 2000. Interestingly, it has been estimated that around 90% of the seagrass beds has disappeared in the Gulf of Gabes, and this represents an important spawning and nursery area for several coastal fishing resources (JICA, 2009).

To solve these problems, starting from 2002, the Tunisian government adopted a management policy directed toward regulating fishing practices and fishing effort to maintain the equilibrium between fishing pressure and exploitable fishing resources. In addition, the Tunisian government strongly recommended adopting active measures for the protection and restoration of the marine environment on the fishing grounds, the enhancement of fishing resources, and the diversification of small-scale fishing activities. Artificial reefs represent one of the most important management measures adopted in coastal waters in the framework of the Tunisian program.

The first Tunisian project involving artificial reefs was conducted by Greenpeace© in 1993. This initial project was followed by a governmental project in 1996. Subsequently, in 2006, a multiyear Tunisian–Japanese project for the "Durable management of the coastal fishing resources in the Tunisian Republic" was initiated in the Gulf of Gabes. This project was based on a global approach to manage fishing resources through a number of actions (i.e., impeding illegal trawling, fish stock enhancement, protection of the environment, and involvement of fishermen communities). It included, among other activities, the placement of antitrawling and/or repopulation artificial reefs.

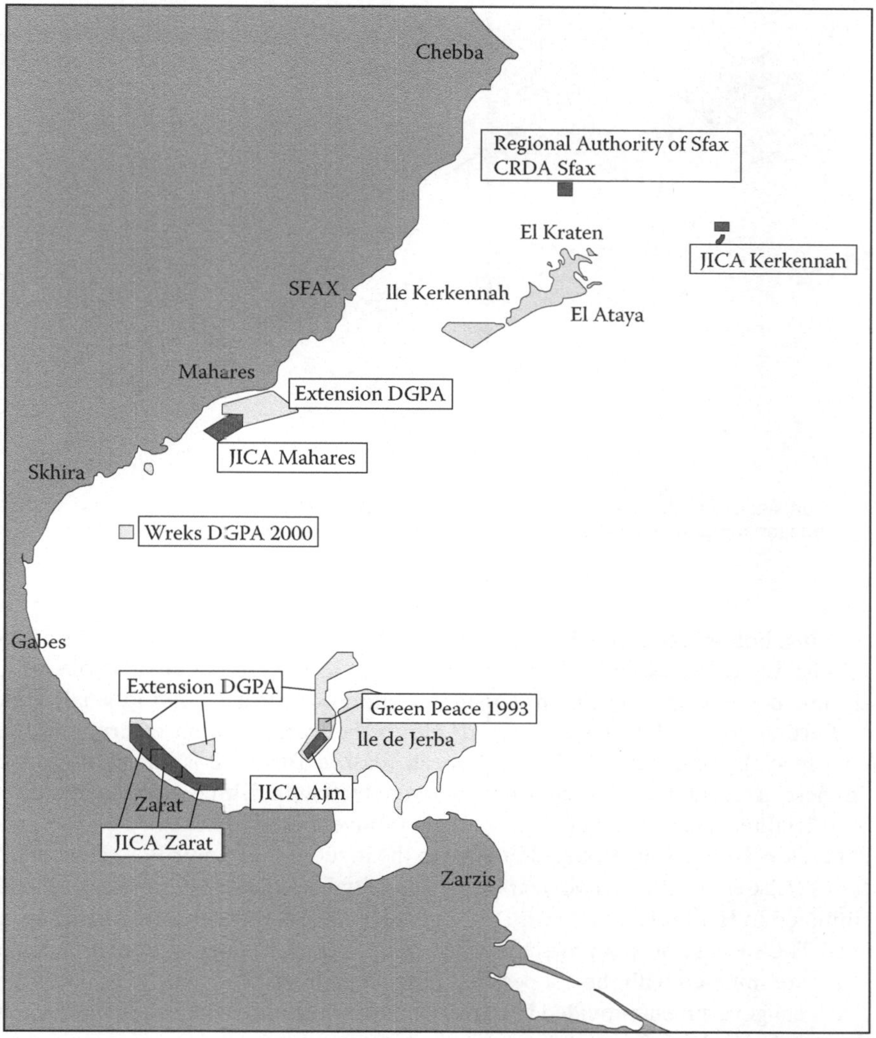

Figure 11.13 ***See color insert.*** Tunisia: Artificial reefs deployed in the Gulf of Gabes DGPA: General Direction for Fisheries and Aquaculture; CRDA: Regional Commissariat for Agriculture Development. (Modified from JICA, 2009. Project de Coopération Technique avec le Japon sur la Gestion Durable des Ressources de la Pêche Côtière en République Tunisienne. Rapport des Activités du Projet. 216 pp)

Fourteen reefs were deployed in six sites of the Gulf of Gabes from 2006 to 2009 with funds provided by the Japanese International Cooperation Agency (JICA), the General Directorate of Fisheries and Aquaculture (DGPA), and fishermen associations (Figure 11.13). The goals of this project were to impede illegal trawling in coastal areas, to improve small-scale fisheries, and to protect the seagrass beds. Simple antitrawling units with iron bars attached and occasional incorporation of pots for *Octopus vulgaris*, and production chambered concrete modules were employed. These modules were placed on the seabed in a geometrical arrangement or in haphazard piles (Figure 11.14).

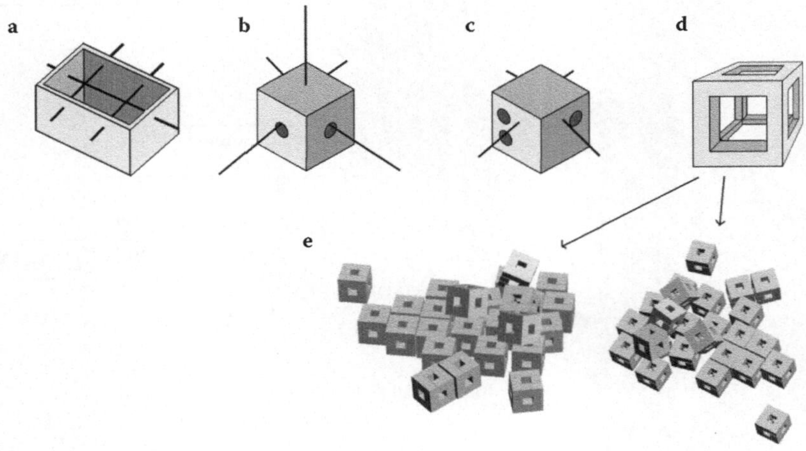

Figure 11.14 Tunisia: Concrete modules used for artificial reef construction (a–c: protection modules; d: production module; e: examples of unit arrangement on the seabed).

Turkey

Purse seining, bottom otter trawling, and small-scale fisheries using trammel nets and longlines are the main fishing activities in Turkey. Illegal trawling in the coastal areas is one of the major problems in this country as it results in damage to littoral ecosystems and to small-scale fishing equipment. Furthermore, a lot of structures to aid in the transportation of crude oil have been constructed, especially in the eastern Mediterranean coast of Turkey. It is illegal to conduct fishing activities in these areas, and this has contributed to a reduction in fishing grounds upon small-scale fisheries. Artificial reefs have been deployed to help resolve these problems.

The first reef in Turkey was deployed in 1989 in the inner part of Izmir Bay. This reef consisted of 10 old buses (Lök et al., 2002). Two years after this initial artificial reef project, a second deployment was initiated in Hekim Island, Izmir Bay. This reef consisted of 30 cubic blocks assembled in six pyramids. The success of this project resulted in an increased interest in reef technology, and new projects were initiated with the cooperation of local authorities, universities, and fishery associations. The local government provided financial support, while universities provided scientific and technical knowledge and fishermen offered practical information on the local conditions. The chief objectives for these artificial reef projects were (1) to protect coastal spawning and nursery areas, sensitive ecosystems (i.e., *Posidonia* beds), and the set gears of the small-scale fisheries from illegal trawling, (2) to create new sites for recreational fishing and diving, and (3) to conduct research on artificial reefs.

In 2001, a "Project Guide for Artificial Reef Application," including guidelines for reef construction, was developed, and a national plan for reef construction was instituted in 2009. In the framework of this plan, artificial reefs are indicated as a way to support small-scale fisheries and to protect marine biodiversity.

Before 2010, 14 reefs had been deployed (Figure 11.15). Moreover, eight new sites have been identified in 2009 as possible reef locations as part of the framework in the new national plan.

Before 2009, 2-year monitoring programs (postdeployment) were obligatory on the reefs constructed or planned. With regard to the artificial reefs planned for deployment as part of the national plan, a 1-year monitoring survey is projected at each new site in 2010. This monitoring will include biological, engineering, socioeconomic, and fishing aspects. It is anticipated that the construction and deployment of these reefs will be conducted using public funds in 2011 and, minimally, 2 years of scientific investigation will be required after deployment.

Figure 11.15 Turkey: Location of the deployed artificial reefs in Turkey. (Modified from and courtesy of A. Lök.)

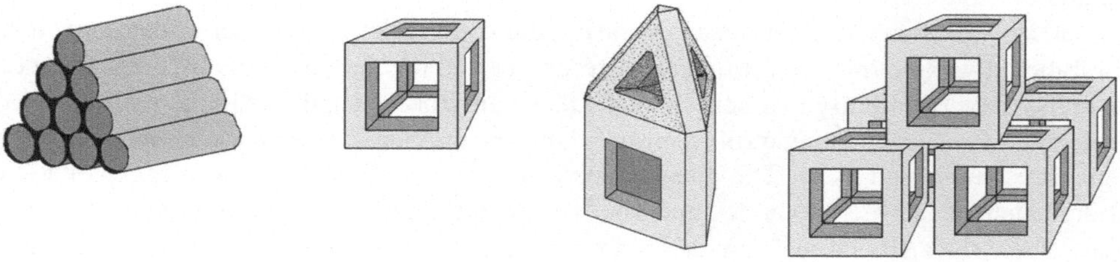

Figure 11.16 Turkey: Concrete modules used for artificial reef construction.

No fishery management plans currently incorporate existing artificial reefs, but this issue is under discussion in the framework of the national plan.

Different materials and configurations have been employed thus far. These construction materials and configurations include chambered concrete modules and pipes, old buses, steel and wooden vessels, and aircrafts. However, specific concrete modules, simply amassed on the seabed or geometrically assembled, are most commonly used (Figure 11.16).

CONCLUSIONS

The Mediterranean Sea can be considered as typical in the use of artificial reefs for fish stock and fishery management. Artificial reefs have been used for over 40 years in the Mediterranean and mostly to deter illegal trawling in coastal areas and other sensitive habitats. These sensitive habitats generally include important spawning and nursery areas for many commercial species. Additionally, artificial reefs are used to reduce conflicts between different fishing activities, mainly, illegal trawling and small-scale fisheries operating with set gears. The objectives of most artificial reef deployments in the Mediterranean are to enhance overexploited fish stocks and improve small-scale fisheries, one of the most important activities for the coastal communities.

Italy and France were the first countries to utilize artificial reefs for fishery management along their coasts between the end of the 1960s and the beginning of the 1970s. These countries were followed by Spain and Israel in the early 1980s in the use of artificial reefs. However, Spain makes more use of artificial reefs than any other Mediterranean country, and they are officially considered as a management option in the management of fisheries both nationally and regionally. A similar policy has been also adopted in Turkey in the recent years.

Research on scientific, engineering, legal, and socioeconomic aspects has strongly contributed to the success of artificial reefs around the Mediterranean, providing a basis of information and experience that have been very useful for a better understanding of the many challenges offered by artificial reefs for the conservation and sustainable use of the marine environment and exploitable resources, as well as for fishery management. In this context, the European Artificial Reef Research Network (1995–1998) has played an important role (Jensen, 1998).

Based on previous research experience, specific guidelines to assist countries in artificial reef construction and avoid dumping in the Mediterranean waters have been developed during the last decade (UNEP MAP, 2005). Although not binding, they have been acknowledged by several countries. These guidelines give an unambiguous definition of artificial reef, thus helping to avoid past misunderstandings. Moreover, these guidelines provide for common protocols for reef deployments and their assessment of effectiveness and environmental impacts.

In spite of the large use of artificial reefs for fishery management in the Mediterranean Sea, there is still a need for additional research on the effective role of artificial reefs in stock enhancement and on the relationship between the artificial substrates and the different fish species living inside and/or around the reefs.

Importantly, in some Mediterranean countries, there is a lack of management measures directed toward rationally exploiting the artificial reef resources and regulating access by the different users to the reefs. In fact, similarly to other types of aquatic environments, the artificial reefs may require some degree of management control to ensure that they provide the desired outcomes for both the biological resources and users. The open access of artificial reefs can lead to overexploitation and rapid depletion of the reef resources and conflicts within and between user groups. This usually happens where the artificial reefs are created with public funds in public waters, such as in most of Mediterranean countries, without effective restrictions on access by the different user groups (Milon, 1991), or where there is a lack of control to ensure that the restrictions are respected. Indeed, the grouper decline observed at an artificial reef deployed off Haifa (Mediterranean Israeli cost) was likely due to the uncontrolled activity of scuba divers and spear fishermen (Edelist and Spanier, 2009).

User conflicts on artificial reefs can be generated by either stock effects and congestion effects (Samples, 1989). The former may occur from overexploitation of all species or particular species at a reef site. The latter occurs when the activities of different users interfere with each other and may result from either incompatible uses (e.g., recreational and commercial fishing), incompatible fishing gears, or too many users in a limited site. Stock and congestion effects are not mutually exclusive.

Four basic options for habitat management can be identified (Samples, 1989): (1) selective access controls, (2) gear and catch restrictions, (3) temporal segregation of users, and (4) spatial segregation of users. The first three options are applicable where only one habitat exists, while all four strategies are feasible in multiple site reefs.

Stock effects can be reduced by decreasing harvesting through setting catch limits (size and number), limiting fishing gears, setting closed seasons, and/or taxing harvests. Gear restriction and closed seasons appear as the most practical alternatives in the Mediterranean, as these types of control are easy to apply.

Congestion effects can be reduced by selective access controls, gear restrictions, and temporal or spatial segregation of users. Selective access control usually consists of user fees and limited licensing, but they are often not feasible due to political and institutional constrains that explicitly forbid the application of fees and licenses to discriminate between different users. Gear limitation can be an effective tool for reducing crowding levels especially when a particular type of gear or fishing method is the primary source of conflict. Temporal segregation allows separate user groups allocating specific periods of time to each of them. Times may be chosen on the basis of various factors such as stock availability, weather conditions, market prices, etc. In this way, different user groups can continue to utilize the artificial reef without interaction between each other, but this management measure is easily enforceable only when the different user groups (e.g., recreational and commercial fishermen) are easy to identify. In addition, when fishing seasons are restricted, there may be an increase in congestion within user groups because access opportunities for each of them are compressed into shorter time periods.

The best management approach to reduce stock and congestion effects on artificial reefs would be the spatial segregation of different user groups by creating separate reefs for each. Nevertheless, creating and maintaining multiple artificial reefs in the same area are much more expensive than the other control options. However, no single management control can be optimal for all situations, and the choice of one or more options must be based on the nature of the conflicts and the effectiveness of the management options adopted. Also, in this case, the cooperation among researchers, administrators, stakeholders, and official institutions concerned with policy management issues are essential in order to develop adequate measures that combine the relevant research findings with the users' needs and the sustainable exploitation of the reef resources.

ACKNOWLEDGMENTS

The authors wish to thank D. Bellan-Santini (France), B.A. Cicek (Cyprus), E. Charbonnel (France), J.J. Goutayer Garcia (Spain), A. Kallianiotis (Greece), A. Lök (Turkey), and E. Spanier (Israel) who provided useful information and material on artificial reefs in their own country.

REFERENCES

Barnabé, G., E. Charbonnel, J.-Y. Marinaro, D. Ody, and P. Francour. 2000. Artificial reefs in France: Analysis, assessments, and prospects. Pages 167–184 in A.C. Jensen, K. Collins, and A. Lockwood, Eds. *Artificial Reefs in European Seas*. Kluwer Academic Publ., Dordrecht, The Netherlands.

Bombace, G., G. Fabi, and L. Fiorentini. 2000. Artificial reefs in the Adriatic Sea. Pages 31–63 in A.C. Jensen, K. Collins, and A. Lockwood, Eds. *Artificial Reefs in European Seas*. Kluwer Academic Publ., Dordrecht, The Netherlands.

Edelist, D. and E. Spanier. 2009. Influence of Levantine Artificial Reefs on the fish assemblage of the surrounding seabed. *Medit. Mar. Sci.* 10(1): 35–54.

European Commission. 2002. A Community Action Plan for the Conservation and Sustainable Exploitation of Fisheries Resources in the Mediterranean Sea under the Common Fisheries Policy. Communication from the Commission to Council and the European Parliament, COM (2002) 535 final, Brussels. 37 pp.

Fabi, G. 2006. Le Barriere artificiali in Italia. Pages 20–34 in *Campo Sperimentale in mare: prime esperienze nel Veneto relative a elevazioni del fondale con materiale inerte*. ARPAV, Padova, Italy.

Fabi, G. and A. Spagnolo. 2001a. Le barriere artificiali. Pages 454–466 in S. Cataudella and P. Bronzi, Eds. *Acquacoltura Responsabile*. Unimar-Uniprom, Roma, Italy.

Fabi, G. and A. Spagnolo. 2001b. Artificial reefs and mariculture. Pages 91–98 in J. Coimbra, Ed. *Modern Aquaculture in the Coastal Zone: Lessons and Opportunities*. NATO Science Series: Series A: Life Sciences. Vol. 314. IOS Press, Amsterdam, The Netherlands.

Failler, P. 2007. Future prospects for fish and fishery products. 4. Fish consumption in the European Union in 2015 and 2030. Part 1. European overview. *FAO Fisheries Circular*. No. 972/4, Part 1. FAO, Rome, Italy. 243 pp.

Jensen, A.C. 1998. European Artificial Research Network (EARRN)—Final report and Recommendations—SUDO/TEC/98/11. Final report to the European Community. 235 pp.

JICA. 2009. Project de Coopération Technique avec le Japon sur la Gestion Durable des Ressources de la Pêche Côtière en République Tunisienne. Rapport des Activités du Projet. 216 pp.

Lök, A., C. Metin, A. Ulaş, F.O. Düzbastilar, and A. Tokaç. 2002. Artificial reefs in Turkey. *ICES J. Mar. Sci.* 59: S192–S195.

Milon, J.W. 1991. Social and economic evaluation of artificial aquatics habitats. Pages 237–270 in W. Seaman Jr. and L.M. Sprague, Eds. *Artificial Habitats for Marine and Freshwater Fisheries*. Academic Press, San Diego, California.

Revenga, S., F. Fernández, J.L. Gonzáles, and E. Santaella. 2000. Artificial reefs in Spain: The regulatory framework. Pages 185–194 in A.C. Jensen, K. Collins, and A. Lockwood, Eds. *Artificial Reefs in European Seas*. Kluwer Academic Publ., Dordrecht, The Netherlands.

Riggio, S., F. Badalamenti, and G. D'Anna. 2000. Artificial reefs in Sicily: An overview. Pages 65–73 in A.C. Jensen, K. Collins, and A. Lockwood, Eds. *Artificial Reefs in European Seas*. Kluwer Academic Publ., Dordrecht, The Netherlands.

Samples, K.C. 1989. Assessing recreational and commercial conflicts over artificial fishery habitat use: Theory and use. *Bull. Mar. Sci.* 44: 844–852.

Spanier, E. 2000. Artificial reefs off the Mediterranean coast of Israel. Pages 1–19 in A.C. Jensen, K. Collins and A. Lockwood, Eds. *Artificial Reefs in European Seas*. Kluwer Academic Publ., Dordrecht, The Netherlands.

Spanier, E., M. Tom, and S. Pisanty. 1985. Enhancement of fish recruitment by artificial enrichment of man-made reefs in the southeastern Mediterranean. *Bull. Mar. Sci.* 37(1): 356–363.

UNEP-MAP-RAC/SPA. 2003. Strategic Action Programme for the Conservation of Biological Diversity (SAP BIO) in the Mediterranean Region. UNEP-MAP-RAC/SPA. 101 pp.

UNEP-MAP. 1975. Convention for the Protection of the Marine Environment against Pollution in the Mediterranean. UNEP/WG.3/5. 14 pp.

UNEP-MAP. 2005. Guidelines for the Placement at Sea of Matter for Purpose other than mere Disposal (Construction of Artificial Reefs). UNEP(DEC)/MED WG.270/10. 9 pp.

CHAPTER 12

Coastal Fish Farms as Fish Aggregation Devices (FADs)

Pablo Sanchez-Jerez, Damian Fernandez-Jover, Ingebrigt Uglem, Pablo Arechavala-Lopez, Tim Dempster, Just T. Bayle-Sempere, Carlos Valle Pérez, David Izquierdo, Pål-Arne Bjørn, and Rune Nilsen

CONTENTS

ABSTRACT

Coastal fish farms concentrate large numbers of fish species of both ecological and economic importance. In this respect, the aggregative effect of sea cages on wild fish populations is analogous to the effect of fish aggregation devices (FADs). High food availability, chiefly through loss of aquacultural feed, greatly increases feeding opportunities in the vicinity of farms. This increased food availability leads to increases in condition of farm-associated wild fish. Negative effects of aggregation at fish farms are also possible. These include the modification of fat levels and fatty acid distribution in the tissues of wild fish, which may have consequences for spawning success and subsequent larval survival, and the increase of specific species of parasites. From an ecological perspective, holistic management of coastal waters requires integration of the effects of aquaculture with those of other industries, such as fisheries.

FISH FARMS AS FADs WITH FEEDING ADVANTAGES

The association of wild fish of all developmental stages, from larvae to adults, with artificial floating objects has been widely documented (Freon and Dagorn, 2000). Marine fish are attracted to a high variety of natural and artificial structures in the sea, commonly called *fish aggregation devices* (FADs; Dempster and Taquet, 2004). This behavior leads to large aggregations of wild fish around FADs compared to natural habitats. Fishermen have known of this behavior for centuries and have used FADs extensively to attract commercially and recreationally important pelagic fishes in different parts of the world (Kingsford, 1999).

Fish aggregate at a wide variety of natural objects and structures including natural floating debris gathered in estuarine oceanographic fronts (Kingsford, 1990) and artificial structures such as oil platforms (Helvey, 2002), fish attraction devices designed for fisheries (Massuti et al., 1999), or artificial reefs (Stone et al., 1991). Several factors may influence the choice of floating habitats by fish and, correspondingly, their assemblage structure. Availability of shelter, the abundance of food, and the presence of conspecifics or other species influence assemblage compositions of fish around FADs (Kingsford, 1992). The vast majority of pelagic and demersal teleost fishes have a pelagic larval stage, which can respond to FADs during the different fish development stages. Two widely offered explanations for this behavior are the "meeting-point hypothesis," where fish aggregate around a reference site to form schools and therefore gain protection and other advantages schooling provides; and the "indicator-log hypothesis," where fish aggregate around drifting objects that accumulate in frontal zones where food may be more abundant (Freon and Dagorn, 2000). Factors that induce an individual fish to choose an artificial structure instead of a natural environmental feature remain unclear. This behavior may have important consequences for population dynamics due to potentially differential predation rates or food availability between natural and artificial sites.

During the last 2–3 decades, coastal aquaculture has expanded due to the development of fish farms using sea cages. These cages are confinement systems with floating and submerged components. There are several major environmental concerns regarding coastal aquaculture. Discharges of uneaten food and fish waste may lead to eutrophication and subsequent alteration of benthic communities (Karakassis and Hatziyanni, 2000). Escapees from cultured stocks and their subsequent mixing with wild populations (Thorstad et al., 2008) are another potential concern. Also, parasites and disease may become amplified within farms, and these are later transfered to wild stocks (Krkosek et al., 2007; Ford and Myers, 2008). Lastly, a concern is the increased reliance developed by wild stock on the components of aquafeeds (Naylor et al., 2000). In addition to these effects, fish farms may precipitate ecological changes in wild fish populations with subsequent cascading effects on fisheries. Effectively, sea-cage fish farms act as large FADs, providing structure in the pelagic environment but with higher food availability compared to traditional FADs (Dempster et al., 2002). The unexploited portion of food pellets lost through the cages probably enhances the attractive effect (Bjordal and Skar, 1992). Therefore, fish farms may affect the presence, abundance, residence times, and diets of fishes in a given area and can, consequently, have important effects on local fisheries.

This chapter presents results from several research projects undertaken by the authors in both Spain and Norway at nine farm and control locations during the last decade (Arechavala-Lopez et al. 2010a,b; Bjørn et al., 2007; Dempster et al., 2002, 2004, 2005, 2006, 2009, 2010; Fernandez-Jover et al., 2007a,b, 2008, 2009; Sanchez-Jerez et al., 2008; Valle et al., 2006; Uglem et al., 2008, 2009). Additionally, the projects we review provide an assessment of previously published studies on the subject. The chapter describes and synthesizes the effects of European coastal fish farms on wild fish populations, with a focus on ecological and physiological effects that, in turn, may influence local fisheries.

FISH FARMS PROVIDE EXTENSIVE ARTIFICIAL STRUCTURES IN COASTAL WATERS

Fish farms are now common artificial elements in coastal ecosystems, from cold to tropical regions, producing 2,233,982 tons of fish annually (FAO, 2007) across thousands of sites. The chief countries involved with marine and brackish-water cage aquaculture are Norway (27.5%), Chile (24.8%), China (12.1%), Japan (11.3%), United Kingdom (5.5%), Canada (4.2%), Greece (3.2%), and Turkey (2.9%). Commercial cage culture has been mainly restricted to finfish species of high economic value such as salmonids (Atlantic salmon, coho salmon, and Chinook salmon), Japanese amberjack, red sea bream, yellow croaker, European sea bass, gilthead sea bream, and cobia (Halwart et al., 2007).

In Norway, marine aquaculture is continuously expanding. In 2008, around 1,200 salmon and trout farms existed throughout the country and produced 825,000 tons (Kjønhaug, 2009), while almost 100 farms were licensed to culture cod (Norwegian Directorate of Fisheries, 2008). To sustain this production, 1.2 million tons of fish feed were used by farmers. In 2009, 65 new concessions for salmon production were granted, and an additional 5% increase was scheduled during 2010. All farms in Norway use gravity-type sea cages that retain their shape based on gravity and a series of weights suspended from plastic rings or steel platforms (Lader et al., 2008). Net cage volumes are typically 20,000 to 50,000 m^3, and each individual cage may contain from 50,000 to 400,000 fish. In Norway, fish farms are mostly located in fjords or in sheltered areas amongst islands. True offshore fish farms in open, fully exposed oceanic waters do not exist, due to a high risk of structural breakdowns as a consequence of bad weather. In most farms, feeding systems are comprised of a floating centralized container, supplying feed to the individual cages through plastic pipes via compressed air (Grøttum and Beveridge, 2007).

Spain produced 43,966 tons of marine fish in 2008, mainly consisting of sea bream, sea bass, turbot, meager, and eel (APROMAR, 2009). The production of sea bream increased by 7.2% from 2007, with a total of 23,930 tons. However, sea bass production declined by 6.1% to 9,840 tons. The type of cages used for farming sea bass and sea bream are mainly circular floating cages made with HDPE pipes that are moored as a group, typically within submerged-grid mooring systems. These cages are often composed of two to three floating rings of pipe held together by several stanchions positioned throughout the entire circumference. Fish are fed either manually or by using a manually operated air compressor feeders. During 2008, marine aquaculture consumed a total of 99,000 tons of food pellets (APROMAR, 2009). Commercial cage sizes vary from 6 to 25 m diameter, and net depths of 6 to 20 m. The number of cages is highly variable but usually ranges from 9 to 24, occupying a surface area of up to 15 hectares (Figure 12.1). A single farm can stock up to 800,000 fingerlings with an annual production of 100 to 1,000 tons of fish (Cardia and Lovatelli, 2007).

AGGREGATION OF WILD FISH AROUND FISH FARMS

Several studies have determined the abundance, biomass, and species diversity of wild fish around fish farms, as well as aggregation persistence. These studies are important to understand how farms act as FADs. Globally, around 160 fish species, belonging to 60 families, have been observed in the close proximity of fish farms, although few studies have been conducted in areas other than the Mediterranean Sea and the Norwegian coastline (Table 12.1). Strong evidence of association of wild fish with FADs exists for just 20 species. Aggregations of fish species that are typically targets of fisheries (e.g., carangids, mugilids, and sparids) in a concentrated area may

Figure 12.1 Left: Typical structure of a coastal fish farm used for farming sea bass and sea bream with circular floating cages made with HDPE pipes, which are moored as a group, typically within submerged grid mooring systems (P. Sanchez-Jerez). Right: Under a cage, a dense shoal of mugilids are aggregated and feeding upon the lost aquafeed (P. Arechavala-Lopez).

affect local fisheries in several ways (Figure 12.1). In addition to sea-cage farms, fish aggregations have also been described to occur around bivalve aquaculture rafts and longline installations (Brehmer et al., 2003; Laffargue et al., 2006; Morrisey et al., 2006).

Abundance and assemblage composition of wild fish around farms vary significantly across geographical areas. For example, using an underwater video camera system, Dempster et al. (2009) described 15 fish species around salmon farms in Norway. The most common families observed at both farm and control locations were Gadidae (6 species) and Lotidae (2 species), with a higher species richness close to the bottom. The saithe (*Pollachius virens*) and the Atlantic cod (*Gadus morhua*) were the only species observed in all depth strata. Eight species were restricted to the bottom or immediately above the bottom, while three predominantly pelagic species (i.e., *Scomber scombrus*, *Trachurus trachurus*, and *Clupea harengus*) were only observed from midwater to the surface. Combined farm-aggregated biomass of the four dominant species averaged 10.2 tons per farm. Dominance of species varied depending on geographical region and depth strata (Figure 12.2). Saithe populations varied highly across regions and depths depending on the farm. In general Atlantic cod and haddock (*Melanogrammus aeglefinus*) were significantly more abundant close to the bottom. Abundance of Atlantic mackerel (*S. scombrus*) around farms in southern Norway attained densities of 10 ind. 500 m^{-3}, while only a few individuals were recorded at farms in mid- and northern-Norway; likely related to the biogeography of this species.

In the Mediterranean, rapid visual counts were employed around sea-cage installations and recorded large aggregations of up to 40 tons of fish composed of up to 33 fish taxa belonging to 17 families (Dempster et al., 2002, 2004, 2005; Fernandez-Jover et al., 2008). The most common families observed were the Clupeidae, Sparidae, Mugilidae, and Carangidae. Several pelagic planktivorous fish species (*Boops boops*, *Oblada melanura*, *Trachurus mediterraneus*, *Trachinotus ovatus*, *Sardinella aurita*) or several species belonging to the family Mugilidae were numerically dominant in assemblages, depending on both the farm and season (Fernandez-Jover et al., 2008 Figure 12.2).

SETTLEMENT PROCESS AROUND FISH FARMS

The influence of FADs on settlement processes is a relevant feature of these systems since natural variation in recruitment success may affect the magnitude of postrecruitment processes (Jones

Table 12.1 Global Synthesis of Wild Fish Species That Have Been Observed Closely Aggregated Around Coastal Sea-Cage Fish Farms

Family	Species	Source
Acanthuridae	*Acanthurus grammoptilus*	m
Ammodytidae	*Ammodytes tobianus*	c
Anarchichadidae	*Anarchichas lupus*	g
Apogonidae	*Sphaeramia orbicularis*	m
Atherinidae	*Atherina hepsetus*	d,i
	Atherinidae	j
	Atherinomorus lacunosus	m
Atherinopsidae	Atherinopsidae	l
Balistidae	*Balistes capriscus*	i
	Balistoides sp.	m
Belonidae	*Belone belone*	d
	Tylosorus sp.	m
Blenniidae	*Aspidontus taeniatus*	m
	Petroscirtes breviceps	m
Bothidae	*Bothus podas*	n,p
Caesionidae	*Pterocaesio tile*	m
Carangidae	*Carangoides fardau*	m
	Caranx papuensis	m
	Gnathanodon speciosus	m
	Lichia amia	d,i
	Naucrates ductor	f,i
	Pseudocaranx dentex	e
	Seriola dumerilii[a]	d,f,i,o
	Seriola lalandi	l
	Seriola sp.	e,j
	Trachinotus ovatus[a]	d,f,i,j
	Trachurus mediterraneus[a]	f,i,j,o
	Trachurus sp.	d
	Trachurus symmetricus[a]	l
	Trachurus trachurus[a]	g
Centracanthidae	*Spicara maena*	d,i,o
	Spicara smaris	i
Chaetodontidae	*Heniochus acuminatus*	m
Clinidae	*Heterostichus rostratus*	l
Clupeidae	*Sardina maderensis*	e
	Sardina pilchardus	e
	Sardinella aurita[a]	d,f,i
	Sardinops sagax	l
Coryphaenidae	*Coryphaena hippurus*	d,f,i
Cottidae	*Taurulus bubalis*	c
Dasyatidae	*Dasyatis centroura*	e
	Dasyatis pastinaca	e,i
	Dasyatis spp.	n

(continued on next page)

Table 12.1 (continued) Global Synthesis of Wild Fish Species That Have Been Observed Closely Aggregated Around Coastal Sea-Cage Fish Farms

Family	Species	Source
	Gymnura altivela	e,n
	Taeniura grabata	n
Embiotocidae	*Brachystis frenatus*	l
	Cymatogaster aggregata	l
	Embiotoca jacksoni	l
	Hypsurus caryi	l
	Phanerodon furcatus	l
	Rhacochilus toxotes	l
	Rhacochilus vacca	l
Engraulidae	*Engraulis encrasicolus*	f,i
	Engraulis mordax	l
Ephippidae	*Platax teira*	m
Gadidae	*Gadus morhua*	c,g,h[a]
	Melanogrammus aeglefinus	g,h[a]
	Merlangius merlangus	g
	Pollachius pollachius	g
	Pollachius virens	a,c,g,h[a]
	Trisopterus minutus	g,h[a]
Gasterosteidae	*Gasterosteus aculeatus*	c
	Spinachia spinachia	c
Gobiesocidae	*Lepadogaster candollei*	c
Gobiidae	Gobiidae	l
	Gobius bucchichi	f
	Gobius niger	c
	Gobiusculus flavescens	c
	Lythrypnus dalli	l
	Pomatoschistus minutus	c
	Pomatoschistus pictus	c
	Rhinogobiops nicholsii	l
Haemulidae	*Anisotremus davidsonii*	l
Heterodontidae	*Heterodontus francisci*	l
Hexagrammidae	*Oxylebius pictus*	l
Kyphosidae	*Girella nigricans*	l
	Kyphosus sp.	e
	Medialuna californiensis	l
Labridae	*Cheilinus undulatus*	m
	Coris julis	f
	Crenilabrus melops	c
	Ctenolabrus rupestris	c
	Halichoeres semicinctus	l
	Labrus viridis	i
	Oxyjulis californica	l
	Semicossyphus pulcher	l
	Symphodus cinereus	f
	Symphodus roisalli	j
	Xyrichthys novacula	e,n

Table 12.1 (continued) Global Synthesis of Wild Fish Species That Have Been Observed Closely Aggregated Around Coastal Sea-Cage Fish Farms

Family	Species	Source
Liparidae	*Liparis liparis*	c
Lophiidae	*Lophius piscatorius*	g
Lotidae	*Brosme brosme*	g
	Molva molva	g
Malacanthidae	*Caulolatilus princeps*	l
Monacanthidae	*Aluterus scriptus*	e,m
	Monacanthus chinensis	m
Monodactylidae	*Monodactylus argenteus*	m
Moronidae	*Dicentrarchus labrax*	d,f,i,j,o,p
Mugilidae	*Chelon labrosus*[a]	j
	Liza aurata[a]	j
	Liza ramada[a]	j
	Mugil cephalus[a]	k
	Mugilidae[a]	d,f,i,j,n,p
Mullidae	*Mullus barbatus*	p
	Mullus surmuletus	e
Myliobatidae	*Myliobatis aquila*	e,i,n
	Myliobatis californica	l
	Taeniura grabata	e
Nemipteridae	*Nemipterus hexodon*	m
Pholidae	*Pholis gunnellus*	c
Pleuronectidae	*Platichthys flesus*	c
	Pleuronectes platessa	c, g
Pomacentridae	*Abudefduf vaigiensis*	m
	Chromis chromis	d
	Chromis punctipinnis	l
	Hypsypops rubicundus	l
	Neopomacentrus violascens	m
Pomatomidae	*Pomatomus saltatrix*[a]	d,f,i,j,o
Salmonidae	*Oncorhynchus mykiss*	c
Sciaenidae	*Cheilotrema saturnum*	l
	Seriphus politus	l
Scombridae	*Auxis rochei*	f,i
	Sarda sarda	e,i
	Scomber japonicus	l
	Scomber scombrus[a]	g
	Scomberomorus sierra	l
	Thunnus thynnus	d,i
Scorpaenidae	*Pterois* sp.	m
	Scorpaena scrofa	f
	Sebastes atrovirens	l
	Sebastes serriceps	l
Sebastidae	*Sebastes viviparus*	g

(continued on next page)

Table 12.1 (continued) Global Synthesis of Wild Fish Species That Have Been Observed Closely Aggregated Around Coastal Sea-Cage Fish Farms

Family	Species	Source
Serranidae	*Epinephelus* spp.	m
	Paralabrax clathratus	l
	Paralabrax nebulifer	l
	Serranus atricauda	e
	Serranus cabrilla	f,i
	Serranus hepatus	f
	Serranus scriba	j
Siganidae	*Siganus* spp.	m
Sparidae	*Boops boops*[a]	b,d,e,f,i,j,o
	Diplodus annularis	b,d,p
	Diplodus cervinus	d,i,j
	Diplodus puntazzo	d,i,j
	Diplodus sargus	d,f,i,j,o
	Diplodus vulgaris[a]	d,f,i,j,o,p
	Lithognathus mormyrus	d,p
	Unidentified sparid	j
	Oblada melanura[a]	d,f,i,j,o,p
	Pagellus acarne[a]	b,d,e,f
	Pagellus erythrinus	e,i
	Pagellus sp.	n
	Pagrus pagrus	e
	Pagrus sp.	n
	Sarpa salpa[a]	d,i,j,n,p
	Sparus aurata[a]	d,i,j,n,o
	Spondyliosoma cantharus	d,e,i,n,o
Sphyraenidae	*Sphyraena argentea*	l
	Sphyraena barracuda	m
	Sphyraena sphyraena	d,f,i,o
	Sphyraena viridensis[a]	b,e
Squatinidae	*Squatina californica*	l
	Squatina squatina	n
Syngnathidae	*Nerophis lumbriciformis*	c
	Syngnathus acus	c
Synodontidae	*Synodus* sp.	n
Tetraodontidae	*Arothron hispidus*	m
	Canthigaster capistrata	n
	Sphoeroides marmoratus	n
Trachinidae	*Trachinus draco*	n
Triglidae	*Chelidonichthys lucernus*	e
Zoarcidae	*Zoarces viviparus*	c

Note:

a	Bjordal and Skar, 1992 (Norway)	Salmon
b	Boyra et al., 2004 (Canary Islands)	Sea bass, sea bream
c	Carss, 1990 (Scotland)	Salmon
d	Dempster et al., 2002 (SW Mediterranean)	Sea bass, sea bream
e	Dempster et al., 2005 (Canary Islands)	Sea bass, sea bream

Table 12.1 (continued) Global Synthesis of Wild Fish Species That Have Been Observed Closely Aggregated Around Coastal Sea-Cage Fish Farms

Family	Species	Source
f	Dempster et al., 2005 (SW Mediterranean)	Sea bass, sea bream
g	Dempster et al., 2009 (Norway)	Salmon
h	Dempster et al., 2010 (Norway)	Salmon
i	Fernandez-Jover et al., 2008 (SW Mediterranean)	Sea bass, sea bream
j	Fernandez-Jover et al., 2009 (SW Mediterranean)	Sea bass, sea bream
k	Katz et al., 2002 (Red Sea)	Sea bream
l	Oakes and Pondella, 2009 (South California, USA)	White sea bass
m	Sudirman et al., 2009 (South Sulawesi, Indonesia)	Tiger grouper, humpback grouper, rabbitfish
n	Tuya et al., 2006 (Canary Islands)	Sea bass, sea bream
o	Valle et al., 2006 (SW Mediterranean)	Sea bass, sea bream
p	Vita et al., 2004 (SW Mediterranean)	Sea bass, sea bream

[a] Indicates species where strong evidence exists for significantly higher concentrations at fish farms than at adjacent control locations. Species are ordered by family. Strictly benthic species are not included. Count methodology of studies was visual observations or visual censuses (as described in Dempster et al., 2002), except for Bjordal and Skar (1992) and Carss (1990) in which fishing nets were used and Dempster et al. (2009, 2010) in which an underwater video system was used.

1990). Fish recruit to a wide variety of natural environments, but they also recruit to artificial structures such as docks, jetties (Rilov and Benayahu, 2000), oil platforms (Love et al., 1994), fish attraction devices (FADs; Sinopoli et al., 2007), and artificial reefs (Beets, 1989). The majority of small juvenile fish that associate with artificial habitats only do so for a specific period of their life history and, as such, spawning periods are thought to regulate the appearance of these species around FADs (Dempster and Taquet, 2004). Information on the role of fish farms as settlement habitat is scarce. For Mediterranean fish farms, Fernandez-Jover et al. (2009) found that 20 juvenile fish species settle at farms throughout the year, mainly belonging to the families Sparidae, Mugilidae, and Atherinidae. The abundance of postlarvae and juveniles around a single cage of 12 m diameter may include tens of individuals of *Diplodus* spp. to thousands of individuals of *Atherina* spp. and *Mugil* spp. The influence of fish cages on the pelagic postlarval stage could affect the connectivity between recruits and fishing stocks, through a spatial modification of the available settlement habitat, alteration of mortality, and modification of trophic resources (e.g., increase of particulate organic matter or zooplankton abundance). Bivalve aquaculture structures also affect fish settlement. Algal and epibiotic growth on the bottom mesh used in bivalve aquaculture (as practiced in North Carolina and elsewhere) can enhance the nursery habitat for the many species that preferentially associate with sea grass habitat, at least as juveniles (Powers et al., 2007).

FISH FARMS ATTRACT PREDATORS

The attraction of predators to FADs, such as tuna, is well known by fishers. Tuna aggregate around drifting FADs and are captured by purse seining in tropical waters (Moreno et al., 2007). Fish farms, due to the high concentrations of wild and reared fish, also attract numerous predator species. Fin fish, sharks, and marine mammals are attracted to the farm structures mainly as a consequence of the increase in foraging opportunities. In the United States and Canada, the Californian sea lion (*Zalophus californianus*), the harbor seal (*Phoca vitulina*), and Steller sea lion (*Eumatopias jubatus*) all interact with coastal fish farms by predating upon salmonids inside the cages and damaging netting in the process. On the Atlantic coast, harbor seals and gray seals (*Halichoerus grypus*) cause similar problems (Nash et al., 2000). In Chile, attraction of sea lions (*Otaria flavescens*) to

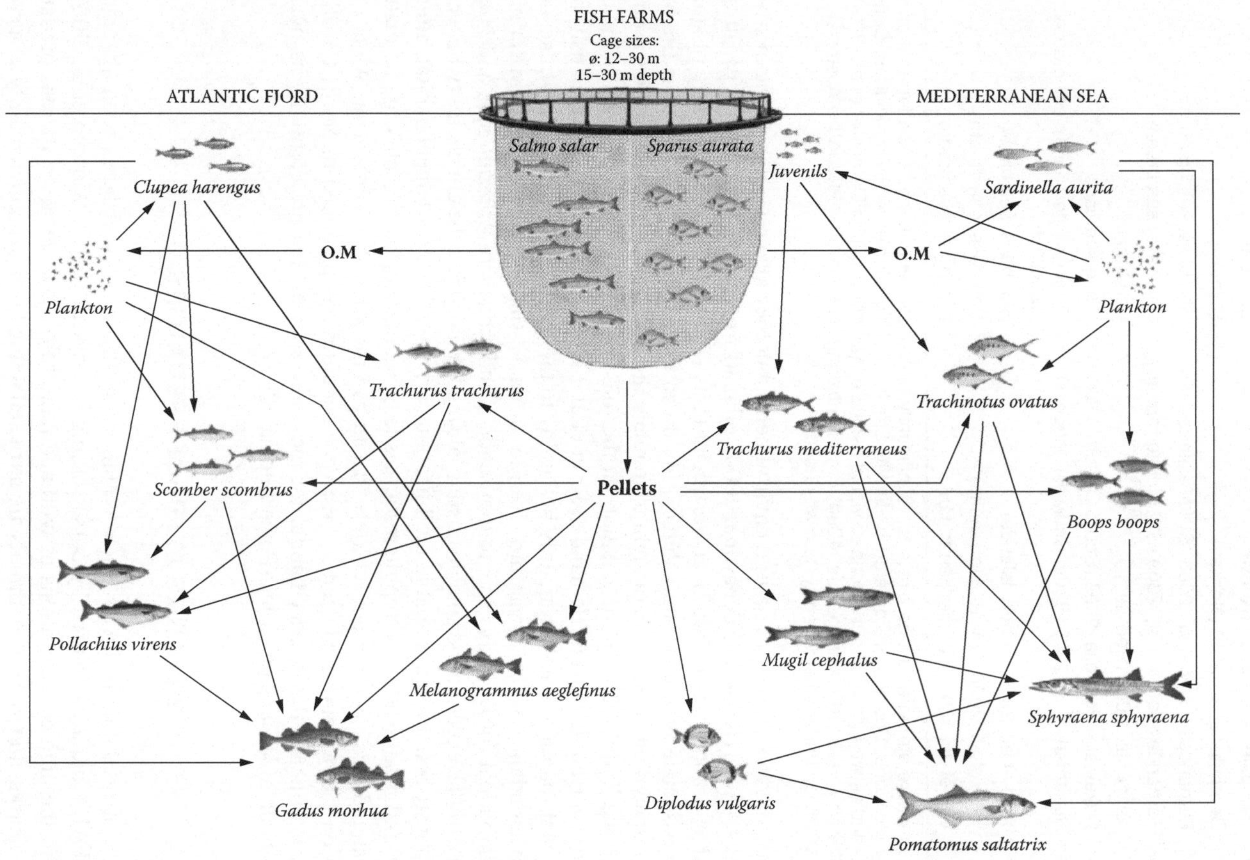

Figure 12.2 Schematic representation of the typical wild fish aggregations around Norwegian and Mediterranean fish farms and trophic relationships among main fish species. Organic matter (O.M.) can be consumed for plankton, as well as for several plankton-feeders fish. Many different species make use of the lost pellets and are strongly aggregated to cages. Predators, such as bluefish, use the fish farms as feeding areas because the high concentration of potential prey.

salmon farms has been described (Sepúlveda and Oliva, 2005). Otters have also caused conflicts with production in specific regions (e.g., Freitas et al., 2007), and sharks are a common cause of cage damage and loss of fish in tropical and subtropical areas. In particular, great white sharks have been detected around tuna farms in the Mediterranean Sea. In Norway, seals (*Phoca vitulina* and *Halichoerus grypus*), otters (*Lutra lutra*), and dogfish (*Squalus acanthias*) are attracted to salmon farms, especially if dead fish occur in the bottom of cages.

Along the Mediterranean coast, marine mammals and birds occur in relatively low numbers around fish farms since natural and human-induced disturbances have greatly reduced Mediterranean populations. However, in spite of this reduced abundance compared with other geographical areas, predatory interactions with aquaculture occur. For example, bottlenose dolphins (*Tursiops truncates*) aggregate around farms along the coast of Italy (Díaz-López and Bernal-Shirai, 2007), with some deaths among the dolphins occurring as a consequence of accidental net entanglements. Monk seals (*Monachus monachus*) have attacked fish at several marine fish farms in the Turkish Aegean Sea resulting in damage to both the net cages and, on most occasions, fish also escaped as a result of the attacks (Güçlüsoy and Savas, 2003).

The assemblages of small wild fish concentrated in large numbers around fish farms attract larger predatory fish species, such as *Coryphaena hippurus*, *Seriola dumerili*, *Pomatomus saltatrix*, *Dentex dentex*, and *Thunnus thynnus* (Dempster et al., 2002). The attraction of *P. saltatrix* (bluefish) to Mediterranean fish farms is of particular interest (Sanchez-Jerez et al., 2008) because it is an aggressive predator of economic importance. In some farms, bluefish intrude into cages, where they may kill or harm large numbers of farmed fish. This is a serious problem for farmers in terms of economic loss and technical difficulties in the production process. Bluefish appear to use farms as a new and productive feeding habitat, which may be related to a reduction in trophic resources for these predators due to overfishing of their normal pelagic fish prey stocks. As bluefish are widely distributed, increased development of marine net pen farms in coastal and offshore areas will most likely also involve an increasing level of interaction between fish farms and bluefish populations.

Despite the attention given to the interaction of predators with aquaculture, there is little evidence of positive or negative interactions of these aggregations of predators with local fishermen. A higher concentration of predators, such as dolphins or bluefish, in coastal waters where fisheries operate could result in economic distress for fishers (Bearzi, 2002). However, few studies have addressed conflicts between fishers and predators in areas where coastal aquaculture has developed. Díaz-López (2005) suggests that interactions between gillnet fisheries and bottlenose dolphins in Sardinia (Italy) may be increasing in the vicinity of fish farms because dolphins and fishers target the same prey.

AVAILABILITY OF FOOD FOR AGGREGATED FISH AND POTENTIAL EFFECTS ON FISHERIES

Food availability in specific natural habitats may indicate habitat quality. Traditionally, increased fish abundance at artificial reefs has been attributed to the presence of additional food, increased feeding efficiency, and the presence of shelter to reduce predation or enhance recruitment (Bohnsack, 1989). To determine the influence of artificial habitats on trophic pathways, feeding behavior and the diet of fish living in such habitats have been widely investigated (Nelson and Bortone, 1996; Sanchez-Jerez et al., 2002).

Fish farms may represent habitats of even higher quality than traditional artificial reefs. Great amounts of artificial food with high energy levels are introduced into the environment that directly feeds the aggregated wild fish (Fernandez-Jover et al., 2008). The increase of particulate organic matter around fish farms is also substantial (Sara et al., 2004) and can be directly used as food by planktivores, such as *Sardinella aurita*, or can stimulate the growth of fouling communities.

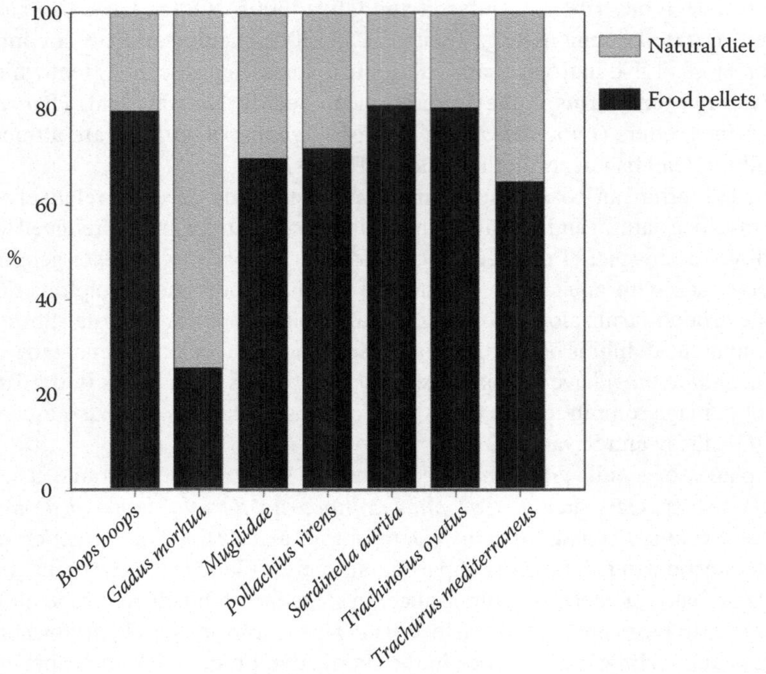

Figure 12.3 Percentage of diet by weight composed of food pellets or natural dietary items eaten by several species that aggregate around fish farms.

Potential trophic resources for aggregated fish, such as amphipods, mollusks, or polychaetes occurred in high numbers and/or biomass within the fish farm fouling community (Greene and Grizzle, 2007). In addition, soluble nitrogenous wastes from fish farms should be enhanced by phytoplankton growth (Islam, 2005). However, several studies have failed to establish a relationship between enhanced nutrient concentrations and phytoplankton growth around fish farm facilities (Beveridge, 1996).

As most fish species show great flexibility in their trophic ecology (Dill, 1983), these new resources are perhaps the most important factor driving the huge aggregation of wild fish around fish farms (Tuya et al., 2006). Utilization of the food resources derived from fish farms has been described by relating the amount of artificial and natural feed items consumed by the fish assemblages (Fernandez-Jover et al., 2007a, 2008; Dempster et al., 2009). Fish aggregated around farms take advantage of lost food; however, the dependence of fish predators on pellets rather than natural prey items depends on the fish species (Figure 12.3). This dependency can range from 20% of all food consumed for Atlantic cod and up to 80% for bogue (*Boops boops*).

The high food availability at farms translates to enhanced foraging efficiency compared to control areas. For example, stomachs of saithe caught around fish farms had significantly greater content by weight than saithe caught away from farms, and the incidence of empty stomachs (16%) was less than in control areas (31%). Farm-associated saithe diets were dominated by food pellets (71% by weight), whereas control fish diets spanned a broad trophic niche eating mainly fish, zooplankton and ophiuroids (Dempster et al., 2009). Similarly, the horse mackerel (*T. mediterraneus*) consumed a variety of food sources, principally juvenile fish, crustaceans, and cephalopods. Nevertheless food pellets were the main item found in the stomach of horse mackerel aggregated around Spanish fish farms presumably because they were readily available (Fernandez-Jover et al., 2007a).

The consumption of food pellets by aggregated fish changes their biological conditions due to the different availability of food and its composition compared to natural resources. Aquafeeds are composed of fish meal and fish oil, as well as vegetable-based ingredients such as soy beans. They contain a high-protein content (40%–70%), are highly digestible, and have low amounts of ash, salts, total volatile nitrogen, and dimethylnitrosamine (Autin, 1997). As a consequence, farm-associated fish usually have a significantly higher Fulton's condition index and hepatosomatic index than control individuals, as has been described for saithe, Atlantic cod, and bogue (Dempster et al., 2009; Arechavala-Lopez et al., 2010a). This enhanced biological condition may be a typical marker of higher spawning success.

However, the fat content and fatty acid composition of commercial aquafeeds may differ so greatly from typical natural fish diets that negative effects may occur. The fat concentration in food pellets used to feed sea bass and sea bream are from 17% to 24% (Fernandez-Jover et al., 2007a). In addition, due to difficulties in obtaining fish oil and fish feed and their elevated prices, vegetables oils of terrestrial origin are used in the formulation. These vegetable oils include high concentrations of other ingredients such as oleic acid (18:1ω9), linoleic acid (18:2ω6), and α-linolenic acid (18:3ω3). The introduction of this source of food to the marine environment could modify the fatty acid (FA) composition of wild fish that feed on lost pellets, as demonstrated for *Trachurus mediterraneus* (Fernandez-Jover et al., 2007a) and has been also detected in *Pollachius virens* (Skog et al., 2003) and *Gadus morhua* (Fernandez-Jover et al., unpublished data). Whether these modified fatty acid compositions alter egg composition and larval survival and thus, also reproductive success rates, remains to be investigated.

MOVEMENTS OF FARM-ASSOCIATED FISH

Wild fish attracted to fish farms might move among farms and also to other areas of ecological and commercial interest. Such movements may affect the local fish populations and, implicitly, the fisheries in several ways. For instance, diseases and parasites are persistent problems in marine fish farming (e.g., Bergh, 2007), and wild fish moving among farms and to other areas might carry pathogens. Movement patterns of several farm-associated fish species have been studied by using acoustic telemetry methodology, which involves tagging fish with acoustic transmitters that emit unique sound signals that are recorded by automatic listening stations positioned throughout a study area (Uglem et al., 2009; Arechevala-Lopez et al., 2010b). These studies have shown that saithe in Norway and mullet (*Liza aurata* and *Chelon labrosus*) in Spain that were captured at farms and subsequently equipped with transmitters move rapidly and repeatedly among fish farms in typical farming areas (Figure 12.4). Tagged fish were also detected on local traditional fishing areas close to the fish farms. Similar studies on farm-associated Atlantic cod have shown that cod repeatedly move from fish farms to local spawning areas for cod and also to commercial fishing areas (Uglem et al., 2008). Therefore, these species exhibit movement patterns that make them potential vectors for transmission of diseases and parasites both to farms and from farms into wild fish populations. The possibility that wild fish might spread diseases or parasites occurring on cultured fish does, however, assume that wild fish share pathogens with the farmed fish and that these pathogens can be transferred among wild and farmed species under natural conditions. In this context, Fernandez-Jover et al. (2010) found that reared sea bass and sea bream did not share macroparasites with farm-associated wild fish (bogue and Mediterranean horse mackerel). Similarly, no effect of farms on the total parasite community was detected when it was compared farm-associated and not farm-associated wild bogue and mackerel neither a host-range enlargement that has been detected in some other works. The percentages shown in Figure 12.4 represent the proportions of the tagged fish that moved from one site to another during the study period. In Guardamar Bay, two mugilids

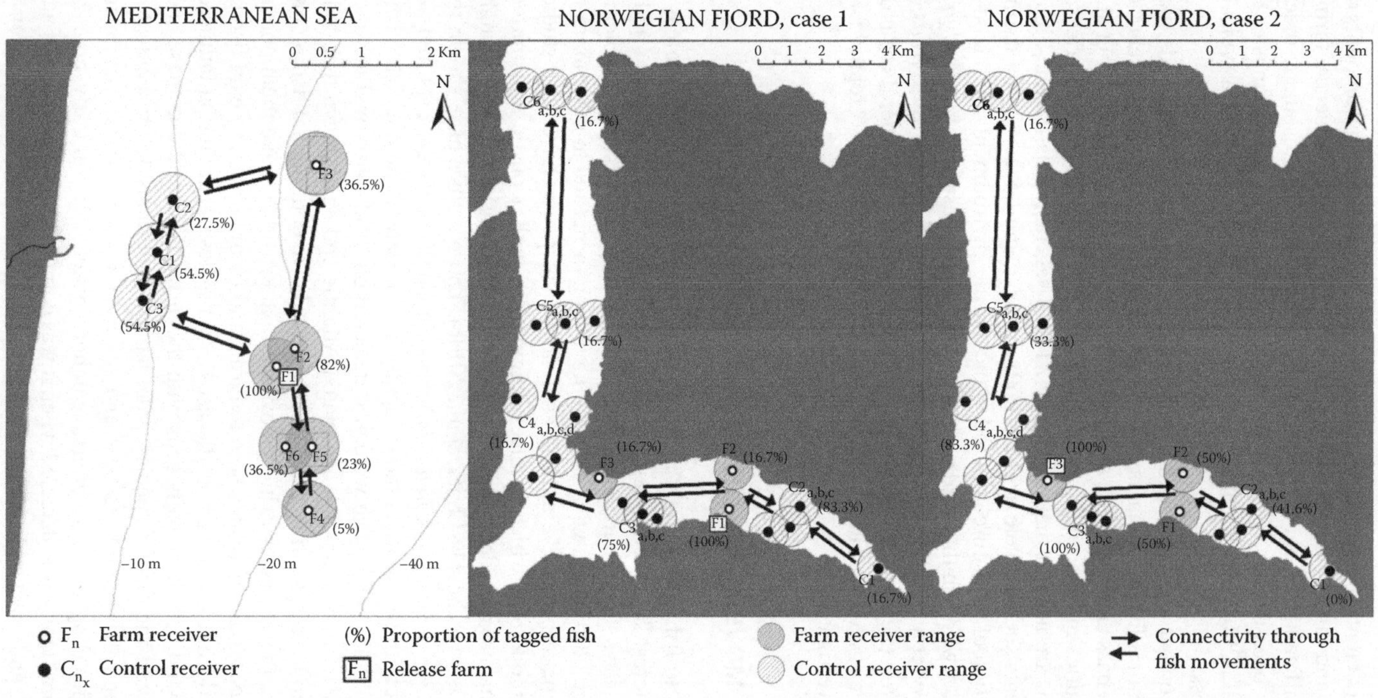

Figure 12.4 Connection among fish farms and other marine areas in the Guardamar Bay in southeast Spain and in Øksfjord in northern Norway. Receivers located at farms and other control sites are indicated with white and black circles, respectively, while the approximate detection ranges around the farm and control receivers are indicated with dark and light striped circles.

species (14 *Liza aurata* and 8 *Chelon labrosus*) were tagged and tracked. Mugilids released from a farm move to all other farms more than 2 km away and also to fishing areas (C_n; 27.5% to 54.5% of detected fish). In Norway, 24 *Pollachius virens* were tagged and tracked at two salmon farms in a typical farming fjord. In one release, all of the tagged fish were observed at other farms, while 25% of the released fish at the second farm were detected at other farms. Farm location might therefore affect the movement patterns of wild farm-attracted fish.

In contrast to this potentially negative effect, as the high abundance of artificial waste feed at fish farms implicitly involves an increased biomass of wild fish, movements of fish from farms to other areas in the sea may correspond to an export of "added biomass" to the fisheries. Little is known about the extent of such biomass export, but tag/recapture studies of Atlantic cod caught at fish farms have shown that a high proportion (32%) of externally tagged fish was recaptured at local traditional cod fishing areas (Bjørn et al., 2007). Farm-associated fish might also leave the fish farms during their reproductive period to spawn. This possibility has hitherto received little attention. If and how this might affect the reproductive ability of wild fish is unclear. However, it has been shown that acoustically tagged, farm-associated cod may move rapidly and frequently between a fish farm and local spawning grounds during the natural spawning season (Uglem et al., 2008).

ARTIFICIAL INPUT OF FISH TO COASTAL ENVIRONMENTS: ESCAPEES FROM SEA CAGES

The escape of cultured fish is a constant possibility if operational or technical failures occur at fish farms. A single fish farm may hold hundreds of thousands to millions of cultured fish, many of which are nonnative species. In the Mediterranean Sea, approximately 500 million sea bass and 450 million sea bream are held in sea cages, with wild stock numbers believed to be considerably lower (ICES, 2006). Today, over 94% of all adult Atlantic salmon (*Salmo salar*) are held in aquaculture. Wild salmonid populations are in decline (Ford and Myers, 2008), while numbers of salmon held in aquaculture settings increase. Due to the large numerical imbalances of caged compared to wild populations, escapement raises important concerns about ecological and genetic impacts.

Almost all species presently cultured across Europe, including Atlantic salmon, Atlantic cod, rainbow trout, Arctic charr, halibut, sea bream, sea bass, and meagre, have escaped from sea cages. In comparison to Atlantic salmon, knowledge of the extent and effect of escapees of other species such as Atlantic cod, sea bream, sea bass, and meagre is limited (Dempster et al., 2007), although Moe et al. (2007) estimate that between 1% to 6% of cultured cod escape in any given year.

Because fish farms areas are typically located close to wild fish habitats, escapees mix with their wild conspecifics. Consequently, the potential exists for escapees to interact negatively with wild populations through competition, transfer of diseases and pathogens, and interbreeding (Thorstad et al., 2008). Interbreeding may be particularly problematic for wild salmonid populations, as farmed fish are genetically divergent from wild stocks (Triantafyllidis, 2007). For Atlantic salmon, escaped salmon have lower fitness in the wild, as measured by survival and reproductive success, than wild salmon (Weir and Grant, 2005).

If escapees produce negative effects, local fisheries could suffer reduced catches as local wild fish stocks decline (Svåsand et al., 2007). However, in spite of this long-term risk, local fisheries can take advantage of accidental escapes, or deliberate release of fish by restocking programs, in the short term (Soto et al. 2001; Sánchez-Lamadrid, 2004). The use of natural resources, good growth rates, and condition indices of escapees indicate that the released fish adapt to life in the wild and suggests that populations of wild fish could also be increased by released fish. Dempster et al. (2002) found few sea bream or sea bass near sea cages in which they were being reared, suggesting either low levels of escape or that escapees move rapidly away from farms to other more favorable habitats. Alternatively, these escaped fish may become more readily captured by sport and professional fisheries. For example, correlative evidence exists of a substantial increase in wild populations of sea bream after fish farming began in the Messolonghi Lagoon, Greece (Dimitriou et al., 2007). In the Canary Islands, sea bass is a nonindigenous species, yet it has established populations that are increasing in coastal areas, although its abundance depends strongly on distance from fish farms (Toledo-Guedes et al., 2009). This distribution could mean that escaped individuals show site fidelity or mortality rates are high because of sport and professional fisheries and, therefore, population size might be highly dependent on a continual supply of escapees. Capture rates of Atlantic cod escapees in the local commercial and recreational fisheries is high (approximately 40%), indicating that local fisheries capture may receive an unintentional boost through escapees.

ECOSYSTEM-BASED MANAGEMENT OF FISH FARMING AND LOCAL FISHERIES

Incorporating aquaculture into integrated coastal zone management, such that significant ecological changes to surrounding wild stocks do not precipitate, requires an understanding of the interaction of the ecological impacts of aquaculture with those of other coastal users (Soto et al. 2008). In our case, particularly impacts that alter habitats and populations of species that are important to fisheries should be taken into account.

Fish Farms Can Help Local Fisheries

In Spain, increased commercial and recreational fishing pressure around fish farms has been noticed by the farm managers. This is especially apparent because of the deployment of gillnets and purse seines that capture large quantities of wild fish when they move away from the farm, seasonally migrate, or escape events occur. Previously, farm-associated fish have been identified in samples from local fish markets through their distinct farm-modified fatty acid profiles (Fernandez-Jover et al., 2007a; Arechavala-Lopez et al., 2010a). In addition, local fishermen along the Norwegian coast report that relatively high amounts of saithe with salmon pellets in their stomach are being caught in fjords with intensive fish farming. In general, farm-associated saithe are significantly fatter and have much larger livers than nonassociated fish (Skog et al., 2003). Previous studies have also shown that saithe caught, tagged, and released at a salmon farm later occurred in the catches of commercial fishermen (Bjordal and Skar, 1992). In this regard, most of the aggregated species can be considered as type B, following the model proposed by Bortone (2007; species attracted to artificial reeds but also taking some production benefit from the reef), because the use of food resources provided by the artificial habitat (i.e., lost food pellets). This feature indicates that marine farms can provide one of the functions of marine-protected areas (Forcada et al., 2009) by increasing the export of fish biomass (Machias et al., 2006). If restrictions on fishing are applied within farm leasehold areas, it has been suggested that coastal sea-cage fish farms may act as small (up to 160 000 m^2) pelagic marine-protected areas (Dempster et al., 2002; Figure 12.5).

Fish Farms Can Function as Ecological Traps to Reduce Capture Quality

In the marine environment, artificial structures that aggregate fish have been suggested to act as ecological traps by serving as a super stimulus and misleading fish to make inappropriate habitat selections (Hallier and Gaertner, 2008). Additionally, the structures may make fish more susceptible to capture, thus causing higher rates of mortality. Aggregation of fish in a restricted area can make a stock more vulnerable and increase the possibility of overfishing. In addition, the attractive nature of fish farms may lead to modifications to migration patterns and distributions of fish in coastal waters with no increase in biomass. In addition, evidence exists that wild fish may have the potential to horizontally transfer pathogens and parasites among farmed and wild fish (e.g., Sepúlveda

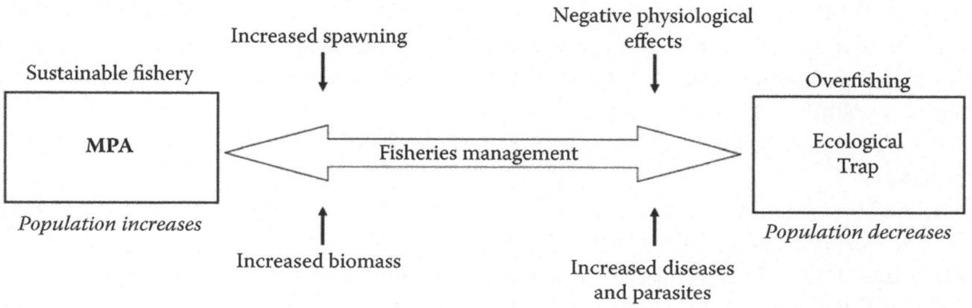

Figure 12.5 Model representing the extremes in the interaction of aquaculture and fisheries. Depending on the management, sea-cage fish farms have the potential to act as marine reserves or ecological traps.

et al., 2004; Bergh, 2007; Uglem et al., 2009), although direct evidence of this effect is at present limited. If management strategies are not enacted to avoid these possibilities, fish farms could function as "ecological traps," continuously attracting fish from surrounding waters with their populations persistently diminished through fishing (for a review of ecological traps, see Battin, 2004; Figure 12.5).

Saithe in some Norwegian fjords constitute an important local fishery. The interaction of saithe with fish farms has created conflicts between farmers and fisheries as saithe, which have resided around farms and consumed food intended for salmon, are believed to have inferior flesh quality. This has led to some local fishermen in Norway to avoid fishing in farm areas as they claim that the flesh quality of farm-associated fish is inferior related to nonassociated fish (Bjørn et al., 2007). The assumed negative relationship between association with fish farms and inferior flesh quality is, however, only partially supported by scientific studies (Skog et al., 2003; Bjørn et al., 2007; Otterå et al., 2009).

In conclusion, coastal aquaculture directly affects coastal wild fish populations, effectively acting as highly-attractive FADs due to their additional significant inputs of food. Through their modifications of fish abundances, diets, condition, and physiology, fish farms have the potential to influence local fisheries at the scale of kilometers. Where fish farms are concentrated in coastal waters, these effects are likely to be amplified and may interact with fisheries at a regional scale. Therefore, sea-cage aquaculture should be taken into account in fisheries management as they may have an effect on spatial distribution and demographic processes of a range of important fisheries species.

ACKNOWLEDGMENTS

Funding was provided by the Spanish Ministry of Science to the ACUFISH project (MYCT-REN2003-00794) and FATFISH project (CTM2009-14362-C02-01), by the Norwegian Research Council's "Havet og kysten" program to the CoastACE project (173384) and to the project "The Environmental role of Marine Aquaculture (ECoMA)," and by the European Union's 7th Research Framework to the Prevent Escape project (226885).

REFERENCES

APROMAR. 2009. La Acuicultura Marina de Peces en España 2009 (Marine Aquaculture in Spain 2009). 68 pp. (www.apromar.es)

Arechavala-Lopez, P., P. Sanchez-Jerez, J. Bayle-Sempere, D. Fernandez-Jover, L. Martinez-Rubio, J.A. Lopez-Jimenez, and F.J., Martinez-Lopez. 2010a. Direct interaction between wild fish aggregations at fish farms and fisheries activity at fishing grounds: a case study with *Boops boops*. *Aquaculture Research* (online): 1–15.

Arechavala-Lopez, P., P. Sanchez-Jerez, J. Bayle-Sempere, D. Fernandez-Jover, L. Martinez-Rubio, J.A. Lopez-Jimenez, and F.J., Martinez-Lopez. 2010b. Movements of grey mullet *Liza aurata* and *Chelon labrosus* associated with coastal fish farms in the western Mediterranean Sea. *Aquaculture Environment Interactions* (online) 1: 127–136.

Autin, M. 1997. Commercial aquafeed manufacturer and production. *CIHEAM—Options Mediterraneennes* 22: 79–104.

Battin, J. 2004. When good animals love bad habits: Ecological traps and the conservation of animal populations. *Conservation Biology*, 10(6): 1482–1491.

Bearzi, G. 2002. Interactions between cetacean and fisheries in the Mediterranean Sea. In G. Notarbartolo di Sciara (Ed.), Cetaceans of the Mediterranean and Black Seas: State of Knowledge and Conservation Strategies. A report to the ACCOBAMS Secretariat, Monaco, February 2002. Section 9, 20 pp.

Beets, J. 1989. Experimental evaluation of fish recruitment to combinations of fish aggregating devices and benthic artificial reefs. *Bulletin of Marine Science*, 44(2): 973–983.

Bergh, Ø. 2007. The dual myths of the healthy wild fish and the unhealthy farmed fish. *Diseases of Aquatic Organisms*, 75: 159–164.

Beveridge, M.C.M. 1996. *Cage Aquaculture*. Fishing News Books, Blackwell, Oxford.

Bjordal, A. and A.B. Skar. 1992. Tagging of Saithe (*Pollachius virens* L.) at a Norwegian Fish Farm: Preliminary Results on Migration. ICES Council Meeting Papers, 1992/G:35.

Bjørn, P.A., I. Uglem, B.S., Sæther, T. Dale, S. Kerwath, F. Økland, R. Nilsen, K. Aas, and T. Tobiassen. 2007. Videreføring av prosjektet "Behavioural Responses in Wild Coastal Cod Exposed to Salmon Farms: Possible Effects of Salmon Holding Water—A Field and Experimental Study." Fiskeriforskning/NOFIMA Report series 6/2007 http://www.nofima.no/publikasjon/FCD42684E7D6FE0DC12572AC00407D8B.

Brehmer, P., F. Gerlotto, J. Guillard, F. Sanguinede, Y. Guennegan, and D. Buestel. 2003. New applications of hydroacoustic methods for monitoring shallow water aquatic ecosystems: The case of mussel culture grounds. *Aquatic Living Resources* 16: 333–338.

Bohnsack, J.A. 1989. Are high densities of fishes at artificial reefs the result of habitat limitation or behavioural preference? *Bulletin of Marine Science*, 44 (2): 631–645.

Bortone, S.A. 2007. Coupling fisheries with ecology through marine artificial reef deployments. *American Fisheries Society Symposium* 49: 587–594.

Boyra A, P. Sanchez-Jerez, F. Tuya, F. Espino, and R. Haroun. 2004. Attraction of wild coastal fishes to Atlantic subtropical cage fish farms, Gran Canaria, Canary Islands. *Environmental Biology of Fishes* 70: 393–401.

Cardia, F., and A. Lovatelli. 2007. A review of cage aquaculture: Mediterranean Sea. In M. Halwart, D. Soto, and J.R. Arthur (Eds). Cage Aquaculture—Regional Reviews and Global Overview, pp. 156–187. FAO Fisheries Technical Paper. No. 498. Rome, FAO. 2007. 241 pp.

Carss, D.N. 1990. Concentrations of wild and escaped fishes immediately adjacent to fish farm cages. *Aquaculture* 90: 29–40.

Dempster T., P. Sanchez-Jerez, J.-T. Bayle-Sempere, F. Giménez-Casalduero, and C. Valle. 2002. Attraction of wild fish to seacage fish farms in the south-western Mediterranean Sea: Spatial and short-term temporal variability. *Marine Ecology Progress Series* 242: 237–252.

Dempster, T. and M. Taquet. 2004. Fish aggregation device (FAD) research: Gaps in current knowledge and future directions for ecological studies. *Reviews in Fish Biology and Fisheries* 14(1): 21–42.

Dempster, T., P. Sanchez-Jerez, J. Bayle-Sempere, and M.J., Kingsford. 2004. Extensive aggregations of wild fish at coastal sea-cage fish farms. *Hydrobiologia* 525(1–3): 245–248.

Dempster, T., D. Fernandez-Jover, P. Sanchez-Jerez, F. Tuya, J. Bayle-Sempere, A. Boyra, and R.J. Haroun. 2005. Vertical variability of wild fish assemblages around sea-cage fish farms: Implications for management. *Marine Ecology Progress Series* 304: 15–29.

Dempster,T., P. Sanchez-Jerez, F. Tuya, D. Fernandez-Jover, J. Bayle-Sempere, A. Boyra, and R. Haroun. 2006. Coastal aquaculture and conservation can work together. *Marine Ecology Progress Series* 314: 309–310.

Dempster, T., H. Moe, A. Fredheim, and P. Sanchez-Jerez. 2007. Escapes of marine fish from sea-cage aquaculture in the Mediterranean Sea: Status and prevention. *CIESM Workshop Monograph* 32: 55–60 (www.ciesm.org/online/monographs/Lisboa.html).

Dempster, T., I. Uglem, P. Sanchez-Jerez, D. Fernandez-Jover, J. Bayle-Sempere, R. Nilsen, and P.A. Bjørn. 2009. Coastal salmon farms attract large and persistent aggregations of wild fish: An ecosystem effect. *Marine Ecology Progress Series* 385: 1–14.

Dempster, T., P. Sanchez-Jerez, I. Uglem, and P. Bjørn. 2010. Species-specific aggregations of wild fish around fish farms. *Estuarine, Coastal and Shelf Science* 86: 271–275.

Díaz-López, B. 2005. Interactions between Mediterranean bottlenose dolphins (*Tursiops truncatus*) and gill-nets off Sardinia, Italy. *ICES Journal of Marine Science* 63: 946–951.

Díaz-López, B. 2005. Interactions between bottlenose dolphins with trammel nets in the Sardinia Island. ICES Document CM 2005/X 1.

Díaz-López, B. and J.A. Bernal-Shirai. 2007. Bottlenose dolphin (*Tursiops truncatus*) presence and incidental capture in a marine fish farm on the north-eastern coast of Sardinia (Italy). *Journal of the Marine Biological Association of U.K.* 87: 113–117.

Dill, L.M. 1983. Adaptive flexibility in the foraging behaviour of fishes. *Canadian Journal of Fisheries and Aquatic Sciences* 40: 398–408.

Dimitriou, E., G. Katselis, D.K. Moutopoulos, C. Akovitiotis and C. Koutsikopolous. 2007. Possible influence of reared gilthead sea bream (*Sparus aurata*, L.) on wild stocks in the area of the Messolonghi lagoon (Ionian Sea, Greece). *Aquaculture Research* 38: 398–408.

FAO. 2007. The state of World Fisheries and Aquaculture; 2006. Rome. http://www.fao.org/docrep/009/a0699e/a0699e00.htm.

Fernandez-Jover, D., J.A. Lopez Jimenez, P. Sanchez-Jerez, J. Bayle-Sempere, F. Gimenez Casalduero, F.J. Martinez Lopez, and T. Dempster. 2007a. Changes in body condition and fatty acid composition of wild Mediterranean horse mackerel (*Trachurus mediterraneus*, Steindachner, 1868) associated to sea cage fish farms. *Marine Environmental Research* 63: 1–18.

Fernandez-Jover, D., P. Sanchez-Jerez, J. Bayle-Sempere, A. Carratala, and V.M. Leon. 2007b. Addition of dissolved nitrogen and dissolved organic carbon from wild fish faeces and food around Mediterranean fish farms: Implications for waste-dispersal models. *Journal of Experimental Marine Biology and Ecology* 340: 160–168.

Fernandez Jover, D., P. Sanchez-Jerez, J. Bayle-Sempere, C. Valle, and T. Dempster. 2008. Seasonal patterns and diets of wild fish assemblages associated with Mediterranean coastal fish farms. *ICES Journal of Marine Science* 65: 1153–1160.

Fernandez-Jover, D., P. Sanchez-Jerez, J.T. Bayle-Sempere, P. Arechavala-Lopez, L. Martinez-Rubio, J.A. Lopez Jimenez, and F.J. Martinez Lopez. 2009. Coastal fish farms are settlement sites for juvenile fish. *Marine Environmental Research* 68: 89–96.

Fernandez-Jover, D., E. Faliex, P. Sanchez-Jerez, P. Sasal, and J. Bayle-Sempere. 2010. Coastal fish farming does not affect the total parasite communities of wild fish in SW Mediterranean. *Aquaculture* 300: 10–16.

Forcada, A., C. Valle, P. Bonhomme, G. Criquet, G. Cadiou, P. Lenfant, and J.L. Sanchez-Lizaso. 2009. Effects of habitat on spillover from marine protected areas to artisanal fisheries. *Marine Ecology Progress Series* 379: 197–211.

Ford, J.S. and R.A. Myers. 2008. A global assessment of salmon aquaculture impacts on wild salmonids. *PLoS Biology* 6: 411–417.

Freitas, D., J. Gomes, T. Sales-Luis, L. Madruga, C. Marques, G. Baptista, L.M. Rosalino, P. Antunes, R. Santos, and M. Santos-Reis. 2007. Otters and fish farms in the Sado estuary: Ecological and socio-economic basis of a conflict. *Hydrobiologia* 587: 51–62.

Freon, P. and L. Dagorn. 2000. Review of fish associative behaviour: Toward a generalisation of the meeting point hypothesis. *Reviews in Fish Biology and Fisheries* 10: 183–207.

Greene, J.K. and R.E. Grizzle. 2007. Successional development of fouling communities on open ocean aquaculture fish cages in the western Gulf of Maine, USA. *Aquaculture* 262 (2–4): 289–301.

Grøttum, J.A. and M. Beveridge, 2007. A review of cage aquaculture: Northern Europe. In Halwart, M., Soto, D., Arthur, J.R. (Eds.). Cage aquaculture—Regional reviews and global overview. FAO Fisheries Technical Paper. No. 498. Rome, FAO. 129–154 pp.

Güçlüsoy, H. and Y. Savas. 2003. Interaction between monk seals *Monachus monachus* (Hermann, 1779) and marine fish farms in the Turkish Aegean and management of the problem. *Aquaculture Research* 34: 777–783.

Hallier, J. and D. Gaertner. 2008. Drifting fish aggregation devices could act as an ecological trap for tropical tuna species. *Marine Ecology Progress Series* 353: 255–264.

Halwart, M., D. Soto, and J.R. Arthur. (Eds.). 2007. Cage aquaculture—Regional reviews and global overview. FAO Fisheries Technical Paper. No. 498. Rome, FAO. 241 pp.

Helvey, M. 2002. Are southern California oil and gas platforms essential fish habitat? *ICES Journal of Marine Science*, 59: S266–S271.

ICES. 2006. Report of the Working Group on Environmental Interactions of Mariculture (WGEIM), April 24–28, 2006, Narragansett, Rhode Island, USA. ICES CM 2006/MCC:03. 195 pp.

Islam, M.S. 2005. Nitrogen and phosphorus budget in coastal and marine cage aquaculture and impacts of effluent loading on ecosystem: Review and analysis towards model development. *Marine Pollution Bulletin* 50: 48–61.

Jones, G.P. 1990. The importance of recruitment to the dynamics of a coral reef fish population. *Ecology* 71 1691–1698.

Karakassis, I. and E. Hatziyanni. 2000. Benthic disturbance due to fish farming analyzed under different levels of taxonomic resolution. *Marine Ecology Progress Series* 203: 247–253.

Katz, T., B. Herut, A. Genin, and D.L. Angel. 2002. Gray mullets ameliorate organically enriched sediments below a fish farm in the oligotrophic Gulf of Aqaba (Red sea). *Marine Ecology Progress Series* 234 205–214.

Kingsford, M.J. 1990. Linear oceanographic features: A focus for research on recruitment processes. *Australian Journal of Ecology* 15: 391–401.

Kingsford, M.J. 1992. Drift algae and small fish in coastal waters of northeastern New Zealand. *Marine Ecology Progress Series* 80: 41–55.

Kingsford, M.J. 1999. Fish attraction device (FADs) and experimental designs. *Scientia Marina* 63 (3–4) 181–190.

Kjønhaug, A.F. 2009. *Kyst og havbruk*. Chapter 3. 128–130.

Krkosek, M., J.S. Ford, and A. Morton. 2007. Declining wild salmon populations in relation to parasites from farm salmon. *Science* 318: 1772–1775.

Lader, P., T. Dempster, A. Fredheim, and Ø. Jensen. 2008. Current induced net deformations in full-scale sea-cages for Atlantic salmon (*Salmo salar*). *Aquacultural Engineering* 38(1): 52–65.

Laffargue, P., M.L. Bégout, and F. Lagardère. 2006. Testing the potential effects of shellfish farming on swimming activity and spatial distribution of sole (*Solea solea*) in a mesocosm. *ICES Journal of Marine Science*, 63: 1014–1028.

Love, M., J. Hyland, A. Ebeling, T. Herrlinger, A. Brooks, and E. Imamura. 1994. A pilot study of the distribution and abundances of rockfishes in relation to natural environmental-factors and an offshore oil and gas-production platform off the coast of southern California. *Bulletin Marine Science* 55, 1062–1085.

Machias, A., M. Giannoulaki, S. Somarakis, C.D. Maravelias, C. Neofitou, D. Koutsoubas, K.N. Papadopoulou and I. Karakassis. 2006. Fish farming effects on local fisheries landings in oligotrophic seas. *Aquaculture* 261(2): 809–816.

Massuti, E., B. Morales-Nin, and S. Deudero. 1999. Fish fauna associated with floating objects sampled by experimental and commercial purse nets. *Scientia Marina* 63(3–4): 219–227.

Moe, H., T. Dempster, L.M. Sunde, U. Winther, and A. Fredheim. 2007. Technological solutions and operational measures to prevent escapes of Atlantic Cod (*Gadus morhua*) from sea-cages. *Aquaculture Research* 38 90–99.

Moreno, G., L. Dagorn, G. Sancho, and D. Itano. 2007. Fish behaviour from fishers' knowledge: The case study of tropical tuna around drifting fish aggregating devices (DFADs). *Canadian Journal of Fisheries and Aquatic Sciences* 64: 1517–1528.

Morrisey D.J., R.G. Cole, N.K. Davey, S.J. Handley, A. Bradley, S.N. Brown, and A.L. Madarasz. 2006 Abundance and diversity of fish on mussel farms in New Zealand. *Aquaculture* 252: 277–288.

Nash, C.E., R.N. Iwamoto, and C.V.W. Mahnken. 2000. Aquaculture risk management and marine mammal interactions in the Pacific Northwest. *Aquaculture* 183: 307–323.

Naylor, R.L., R.J. Goldburg, J.H. Primavera, N. Kautsky, M.C.M. Beveridge, J. Clay, C. Folke, J. Lubchenco H. Mooney, and M. Troell. 2000. Effect of aquaculture on world fish supplies. *Nature* 405: 1017–1024.

Nelson, B.D. and S.A. Bortone. 1996. Feeding guilds among artificial-reef fishes in the northern Gulf of Mexico. *Gulf of Mexico Science* 2: 4–5.

Norwegian Directorate of Fisheries. 2008. www.fiskeridir.no/statistikk/akvakultur.

Oakes, C.T. and D.J. Pondella. 2009. The value of a net-cage as a fish aggregating device in southern California *Journal of the World Aquaculture Society* 40(1): 1–21.

Otterå, H., O. Karlsen, E. Slinde, and R.E. Olsen. 2009. Quality of wild-captured saithe (*Pollachius virens* L.) fed formulated diets for 8 months. *Aquaculture Research* 40: 1310–1319.

Powers, M.J., C.H. Peterson, H.C. Summerson, and S.P. Powers. 2007. Macroalgal growth on bivalve aquaculture netting enhances nursery habitat for mobile invertebrates and juvenile fishes. *Marine Ecology Progress Series* 339: 109–122.

Rilov, G. and Y. Benayahu. 2000. Fish assemblage on natural versus vertical artificial reefs: The rehabilitation perspective. *Marine Biology* 136 (5): 931–942.

Sanchez-Jerez, P., Gillanders, B.M., Rodrıguez-Ruiz, S., and Ramos-Espla, A.A. 2002. Effect of an artificial reef in Posidonia meadows on fish assemblage and diet of *Diplodus annularis. ICES Journal of Marine Science*, 59: S59–S68.

Sanchez-Jerez, P., D. Fernandez-Jover, J. Bayle-Sempere, C. Valle, T. Dempster, F. Tuya, and F. Juanes. 2008. Interactions between bluefish *Pomatomus saltatrix* (L.) and coastal sea-cage farms in the Mediterranean Sea. *Aquaculture* 282: 61–67.

Sánchez-Lamadrid, A. 2004. Effectiveness of releasing gilthead sea bream (*Sparus aurata,* L.) for stock enhancement in the bay of Cádiz. *Aquaculture* 231: 135–148.

Sara, G., D. Scilipoti, A. Mazzola, and A. Modica. 2004. Effects of fish farming waste to sedimentary and particulate organic matter in a southern Mediterranean area (Gulf of Castellammare, Sicily): A multiple stable isotope study (d13C and d15N). *Aquaculture* 234: 199–213.

Sepúlveda, F., S.L. Marín, and J. Carvajal. 2004. Metazoan parasites in wild fish and farmed salmon from aquaculture sites in southern Chile. *Aquaculture* 235(1–4): 89–100.

Sepúlveda, M. and D. Oliva. 2005. Interactions between South American sea lions *Otaria flavescens* (Shaw) and salmon farms in southern Chile. *Aquaculture Research* 36: 1062–1068.

Sinopoli, M., G. DÁnna, F. Badalamenti, and F. Andaloro. 2007. FADs influence on settlement and dispersal of the young-of-the-year greater amberjack (*Seriola dumerili*). *Marine Biology* 150: 985–991.

Skog T.E., K. Hylland, B.E. Torstensen, and M.H.G. Berntssen. 2003. Salmon farming affects the fatty acid composition and taste of wild saithe *Pollachius virens* L. *Aquaculture Research* 34(12): 999–1007.

Soto, D., F. Jara, and C. Moreno. 2001. Escaped salmon in the inner seas, southern Chile: Facing ecological and social conflicts. *Ecological Applications* 11: 1750–1762.

Soto, D., J. Aguilar-Manjarrez, and N. Hishamunda (Eds.). 2008. Building an ecosystem approach to aquaculture. FAO/Universitat de les Illes Balears Expert Worshop. May 7–11, 2007, Palma de Mallorca, Spain. *FAO Fisheries and Aquaculture Proceedings* 14. Rome, FAO. 221 pp.

Stone, R.B., J.M. McGurrin, L.M. Sprague, and W. Seaman Jr. 1991. Artificial habitats of the world: Synopsis and major trends. Pages 31–60 in W. Seaman, Jr., and L.M. Sprague, Eds. *Artificial Habitats for Marine and Freshwater Fisheries*. Academic Press, San Diego. 285 pp.

Sudirman, H.H., J. Jompa, I. Zulfikar, and A.D. McKinnon. 2009. Wild fish associated with tropical sea cage aquaculture in south Sulawesi, Indonesia. *Aquaculture* 286: 233–239.

Våsand, T.D., E. Crosetti, E. García-Vázquez, and E. Verspoor (Ed.). 2007. Genimpact—Evaluation of Genetic Impact of Aquaculture Activities on Native Populations. A European network. (http://genimpact.imr.no/__data/page/7649/genetic_impact_of_aquaculture.pdf).

Thorstad, E.B., I.A. Fleming, P. McGinnity, D. Soto., V. Wennevik, and F. Whoriskey. 2008. Incidence and Impacts of Escaped Farmed Atlantic Salmon *Salmo salar* in Nature. Report from the Technical Working Group on Escapes of the Salmon Aquaculture Dialogue, January 2008. 108 pp.

Toledo Guedes, K., P. Sanchez-Jerez, G. Gonzalez-Lorenzo, and A. Brito Hernandez. 2009. Detecting the degree of establishment of a non-indigenous species in coastal ecosystems: Sea bass Dicentrarchus labraxescapes from sea cages in Canary Islands (Northeastern Central Atlantic). *Hydrobiología* 623: 203–212.

Triantafyllidis, A. 2007. Aquaculture Escapes: New DNA Based Monitoring Analysis and Application on Sea Bass and Sea Bream. CIESM Workshop Monograph 32: 67–71 (www.ciesm.org/online/monographs/Lisboa.html).

Tuya, F., P. Sanchez-Jerez, T. Dempster, A. Boyra, and R. Haroun. 2006. Changes in demersal wild fish aggregations beneath a sea-cage fish farm after the cessation of farming. *Journal of Fish Biology* 69: 682–697.

Uglem, I., P.A. Bjørn, T. Dale, S. Kerwath, F. Økland, R. Nilsen, K. Aas, I. Fleming, and R.S. McKinley. 2008. Movements and spatiotemporal distribution of escaped famed and local wild Atlantic cod (*Gadus morhua* L.) in a Norwegian fjord. *Aquaculture Research* 39: 158–170.

Uglem, I., T. Dempster, P.A. Bjørn, and P. Sanchez-Jerez. 2009. High connectivity of salmon farms revealed by aggregation, residence, and repeated migrations of saithe (*Pollachius virens*) among farms. *Marine Ecology Progress Series* 384: 251–260.

Valle, C., J. Bayle, T. Dempster, P. Sanchez-Jerez, and F. Gimenez-Casalduero. 2006. Temporal variability of wild fish assemblages associated with a sea-cage fish farm in the south-western Mediterranean Sea. *Estuarine, Coastal, and Shelf Science* 72: 299–307.

Vita, R., A. Marin, J.A. Madrid, B. Jimenez-Brinquis, A. Cesar, and A.L. Marin-Guirao. 2004. Effects of wild fishes on waste exportation from a Mediterranean fish farm. *Marine Ecology Progress Series* 277: 253–261.

Weir, L.K. and J.W.A. Grant. 2005. Effects of aquaculture on wild fish populations: A synthesis of data. *Environmental Reviews* 13: 145–168.

CHAPTER 13

Artificial Reefs in Artisanal Fisheries

Erick R. Baqueiro Cárdenas

CONTENTS

ABSTRACT

In this chapter, consideration is given to the deployment of artificial reefs expressly for artisanal fisheries, with examples from the state of Campeche on the Yucatan Peninsula, Mexico with its state artificial reef program, the main objective being to support and enhance artisanal fisheries. The first such reef in the Yucatan was created out of a need to dispose of 70 out-of-use shrimp trawlers. At the time, there was no regulation or official rules for the construction of artificial reefs in Mexican waters; therefore, a code of norms was proposed that latter became part of national regulations on the subject. This first reef generated an outstanding increase in fish yields at nearby localities. Subsequently, there was a near-chaotic installation of small reefs by fishermen, which led to a local reef program that established regulations for the disposal of materials in Mexican waters. Three additional localities were identified for the construction of artificial reefs close to the main fishing towns in the state of Campeche. Production (i.e., yields by local fishermen) was enhanced during the first 5 years after deployment of the reefs. However, owing to misuse and abuse from fishermen, yields have declined to levels recorded prior to the deployments.

INTRODUCTION

Artificial reefs have potential to increase fish landings or yields among artisanal fisheries, especially those in coastal areas. With this contribution, the experience on the creation of an artificial reef program for coastal fisheries in Campeche is presented to further contribute to our understanding of the potential utility that artificial reefs may have.

A reef as described by *Webster*'s is a submerged ridge of rock or coral near the surface of the water. In nautical parlance, a reef is a rock, sandbar, or other feature beneath the surface of the water shallow enough to be a hazard to ships (http://www.websters-online-dictionary.org/definition/reef). Given their origin, these reefs have been classified as natural or artificial. Natural reefs are further classified as organic and inorganic, among which are sand bars, rocky ridges, shell banks, and corals. In contrast, artificial reefs, or man-made reefs, can be of three types. *Casual reefs* include shipwrecks, lost anchors, buoys, light houses, buildings, and any coastal structure destroyed by earthquakes or severe storms. *Consequential reefs* are any structure built in the sea for a utilitarian purpose, such as piers, brake waters, bridges, etc. Thirdly, reefs can be *intentional*, which are all those structures built with a specific purpose. Artificial reefs have been built to protect the seashore, navigation channels, and, specifically, to enhance fish catches, either as fish aggregation devices, habitat enhancement, nursery areas, etc. (Nakamura, 1991; Stone et al., 1991). Fishers have long had the impression that their best catches occurred along rocky shores and especially in soft bottom areas closest to solid outcrops; and if these are not available within accessible distance, they are created with trees from coastal vegetation or from materials of opportunity.

Artisanal fishers in Mexico are an economically depressed sector of the public, dependent on government assistance or private investment to maintain their activity. Artisanal fisheries are an alternative to subsistence to unemployment. Fishermen use 7 to 8 m outboard motor boats with 40 to 75 hp motors. A typical fishing trip is of 6 to 8 hours duration. Campeche fishers may go up to 50 km from their home ports. The boats generally do not have radio communication. Annually there has been a loss of several boats, with a crew of three to five fishers, before the establishment of the fishing grounds through the artificial reef program.

The current artificial reef program in the state of Campeche, Mexico, began as an alternative way to make use of aging ships. In Campeche, there were over 75 trawlers and other boats made of wood, iron, and concrete, which were obstructing port operations and interfering with the operating fleet. Although every boat that makes use of the port facilities is charged a fee, these derelict vessels paid no fees as the vessels had been declared "on hold" by insurance companies or just abandoned. Generally, the identity of the owner was known.

Mexican law, in 1988, did not allow the outright sinking of these ships. Old derelict vessels had to be dismantled before the ship materials could be reused. The cost for dismantling a standard wooden shrimp trawler was one million Mexican pesos equivalent at the time to $10,000.

Boat owners, insurance companies, government agencies, and other entities involved with the ownership or legal property rights of the boats were brought together by the port authority to offer an acceptable solution to the problem. Those participating in the meeting were fishers, boat administrators, insurance representatives, port administrators, and representatives of the Federal and State fisheries sectors (the later were invited as observers). All propositions and negotiations focused on identifying the responsible party for vessel removal and the entity that would assume financial responsibility for vessel removal. The cost of dismantling vessels seemed high to fishing cooperatives, and boat owners. Regardless, insurance companies were unwilling to bear these costs. After all pros and cons were outlined by the fishery representatives, the alternative of creating an artificial reef for costal artisanal fishers was considered an acceptable alternative. The costs of refloating the vessels and subsequently transporting and sinking them were estimated, on an average, to be $1,000 per vessel. It represented less than 10% of the cost to dismantle the vessel. Nevertheless, this was not considered as acceptable to boat owners until port authorities threatened to charge vessel owners the legal fees for every boat that remained in the fishing port for more than a week.

The option of creating artificial reefs with the discarded vessels was acceptable to all parties. Interestingly, local fishermen had been making small artificial reefs from materials of opportunity, such as boats, since colonial times. Representatives from state and federal fishery authorities were charged with identifying the legal and environmental requirements that would allow for the establishment of the first artificial reef for artisanal fishery enhancement in Mexico. At that time, there were no specific regulations applicable to allow the creation of the Campeche reef program. Those adopted were later used to set the foundation for the latter legal procedures at national level.

The Campeche bank is characterized by its low profile and soft sandy bottom. The banks are also known for their coral reef formations, which have historically attracted artisanal fishers. However, the expansive distance to the coast makes these coral reefs potentially dangerous, especially if fishers are caught by a sudden tropical storm. Consequently, an additional function for artificial reefs is to offer safe fishing grounds by providing accessible fishing locations to coastal artisanal fishers.

PILOT PROJECT IN CAMPECHE

The artificial reef made from discarded shrimp boats was deployed 15 km offshore, at a 10 m depth. Fishermen from Seybaplaya began catching fish at the reef as soon as it was built in 1985. An analysis of fishing yield from five years previous to the deployment of the reef and five years after showed a tenfold increase in landings (Baqueiro and Mendez, 1994).

After the reef was deployed, the number of species registered at the fisheries office increased from 23 to 49 (sharks and skates excluded; Table 13.1), with a notable increase in landings of preferred target species such as *Epinephelus adscensionis* (cabrilla), *E. itajara* (cherna), *Lutjanus campechanus* (huachinango), *L. griseus* (pargo), *Trachiconus glaucus* (palometa), *T. corionus* (pampano), and *Centropomus undecimalis* (robalo). The impact on fish landings at Seybaplaya associated with the artificial reef deployment generated a chaotic situation. In an effort to further "improve" the fishing situation, fishermen along the coast added to the artificial reef by using trees cut from coastal vegetation and other readily available materials such as discarded cars, car tires, and toilets. These materials were not properly ballasted and were easily moved by other fishermen. Much of these materials ended up on nearby beaches after the first winter storms. This circumstance prompted authorities to create an artificial reef program specifically for artisanal fisheries and regulate the oceanic disposal of materials. The location of these reefs was thought to be accessible by fishermen using outboard motor boats (i.e., no more than an hour). Locations A to D were the first stage of the artificial reef programs' four locations (Figure 13.1). This state reef program had the following objectives:

1. To increase coastal fish production and fishermen's productivity.
2. To reduce the risk of fishermen venturing into deep water using small outboard boats in search of better fishing grounds.
3. To promote areas for sport fishing, angling, and diving.
4. To control the chaotic creation of small artificial fishing grounds, which were actually littering the coastal ocean floor.
5. To prevent the deforestation of mangroves and other coastal vegetation.
6. To make use of waste material that represented a disposal problem on land and presented no hazard to the marine environment.

Tab1e 13.1 Common and Scientific Names of Finfish Catches Recorded at Seybaplaya, Campeche, Before and After the First Artificial Reef Was Deployed

Common Name	Scientific Name	1980–1985	1986–1990
Abadejo	*Epinephelus guttatus*	285	1,425
Armado[a]	*Haemulon striatum*		2,062
Atun	*Euthynnus pelamis*	895	4,475
Bagre o Bosh	*Arius felis*	693	3,465
Bandera	*Bagre marinus*	681	3,405
Besugo	*Rhomboplites aurorubens*	457	2,285
Bobo	*Ictalurus* sp.	102	510
Bonito[a]	*Euthynnus alletteratus*		1,185
Boquinete[a]	*Lachnolaimus maximus*		296
Cabrilla[a]	*Epinephelus adscensionis*		266
Canane[a]	*Ocyurus chrysurus*		9,521
Carito o Peto[a]	*Scomberomorus cavalla*		1,128
Chacalcay	*Lutjanus analis*	693	3,465
Chacchi	*Haemulon plumieri*	182	910
Cherna	*Epinephelus itajara*	441	2,205
Chopa[a]	*Lobotes surinamensis*		789
Cojinuda	*Caranx chrysops*	429	2,145
Corcovado[a]	*Vomer setapinnis*		158
Coronado	*Seriola* sp.	182	910
Coruco[a]	*Menticirrhus saxatilis*		659
Corvina	*Cynoscion nebulosus*	1,115	5,575
Esmedregal	*Rachycentron canadum*	859	4,295
Guavina[a]	*Diplectrum formosum*		158
Gurrubata	*Micropogon undulatus*	195	975
Huachinango	*Lutjanus campechanus*	558	2,790
Ixpompol	*Sphoeroides laevigatus*	698	3,490
Jurel	*Biella chrysurus*	945	4,725
Lebrancha	*Mugil curema*	932	4,660
Lisa	*Mugil cephalus*	386	1,930
Manjua[a]	*Anchoa leydois*		820
Mero	*Epinephelus morio*	256	1,280
Mojarra	*Eugerres plumieri*	3,147	15,735
Molpich[a]	*Eucinostomus gula*		18
Negrillo	*Mycteroperca bocans*	785	3,925
Palometa[a]	*Trachinotus glaucus*		325
Pampano[a]	*T. carolinus*		300
Pargo[a]	*Lutjanus griseus*		21,557
Picuda	*Sphyraena barracuda*	132	660
Posta[a]	*Anisotremus virginicus*		298
Raton[a]	*Menticirrhus americana*		105
Robalo	*Centropomus undecimalis*	1,322	6,610
Roncador[a]	*Cheilotrema saturnum*		248
Ronco	*Conodon nobilis*	268	1,340
Rubia[a]	*Gonioplectrus hispanus*		351

Tab1e 13.1 (continued) Common and Scientific Names of Finfish Catches Recorded at Seybaplaya, Campeche, Before and After the First Artificial Reef Was Deployed

Common Name	Scientific Name	1980–1985	1986–1990
Sargo[a]	*Archosargus probatocephalus*		2,890
Sierra	*Scomberomorus maculates*	2,156	10,780
Tambor	*Pogonias cromis*	284	1,420

Note: Numbers indicate mean annual production in kilograms for year groups 1980–1985 and 1986–1990.

[a] *Species listed under "other species" before the installation of the artificial reef.*

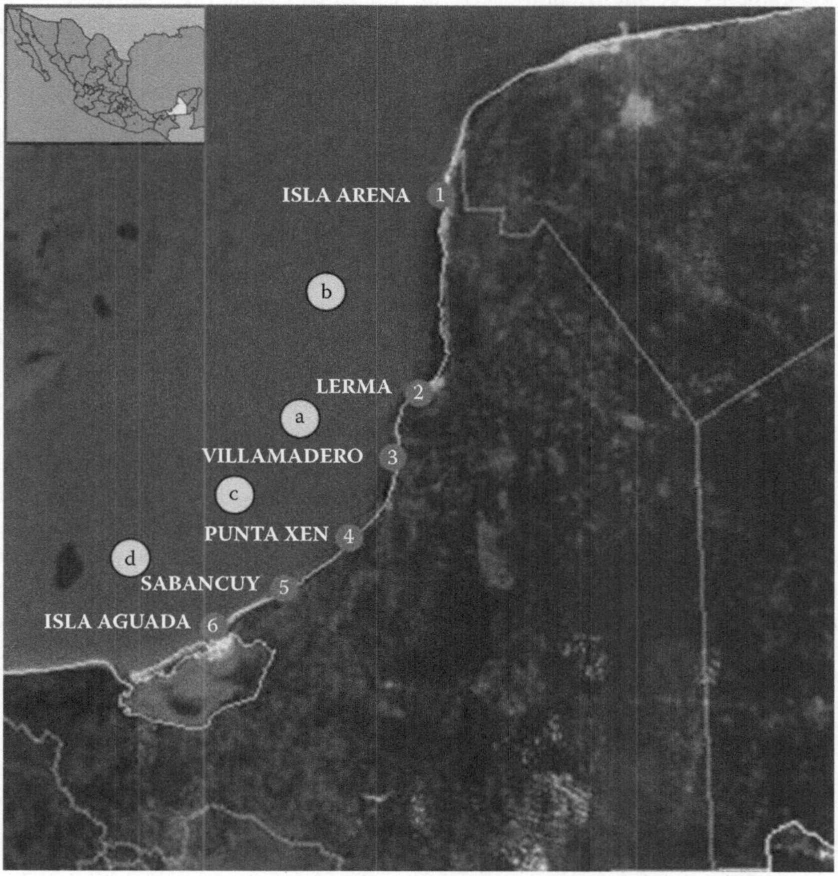

Figure 13.1 Reef sites at the state of Campeche, Mexico: a, Seybaplaya (shrimp boat reef) 19° 42' 32.36″ N. 91° 9' 0.84″ W, depth: 20 m: b, Jaina (tire reef): 19° 42' 32.36″ N, 91° 9' 0.48″ W, depth: 10 m, c, Champoton (La Barcaza) 19° 27' 6.33″ N, 91° 19' 35.18″ W, depth: 25 m and d, Sabancuy (rigs reef) 19° 8' 22.06″ N, 91° 38' 8.11″ W, depth: 25 m. One through 6 coastal reef ball reefs (for position see table N).

LEGAL REQUIREMENTS TO DEPLOY AN ARTIFICIAL REEF IN MEXICO

In theory, environmental requirements for artificial reefs should be the same anywhere in the world, but legal aspects vary from country to country and, quite often, vary between states or provinces within a country. Here, we outline Mexican legal requirements, which are based on international treaties and the national constitution.

In Mexico, environmental activities are normally under the jurisdiction of the Secretaria del Medio Ambiente y Recursos Naturales or SEMARNAT (Secretary of Environment and Natural Resources). However, when the activity takes place at sea, the Secretaria de Marina is involved (Secretary of the Navy) as well as the Secretaria de Agricultura, Ganadería, Desarrollo Rural Pesca y Alimentación SAGARPA (Secretary of Agriculture, Animal Husbandry, Rural Development, Fisheries, and Food), and the Secretaría de Salud (Secretary of Health).

In establishing an artificial reef, the environmental, ecological, social, and economical issues must be considered. When the artificial reef program was proposed for the Campeche Bank, the first objection came as a result of conflicts of interest. Shrimp fishermen objected to the installation of the reef, claiming it would obstruct trawling areas. It was demonstrated that the depth and distance from the coast where the reefs were to be placed were within an area that was forbidden to trawlers. Hence, the reefs would serve one additional function: preventing illegal trawling within coastal waters.

Another claim of conflict of interest came from the Mexican oil company PEMEX. The company argued that the reefs would become navigation hazards. This argument was overcome by placing the reefs out of shipping lanes and by proper public notification of the reef location and the addition of navigation signals. Moreover, environmentalists claimed the artificial reefs would only act as fish aggregation devices and increase fishing mortality. An extensive report was compiled from the existing literature showing the benefits of artificial reefs to juvenile fish survival and increase in productivity, particularly in soft bottom biotopes, as is the case for the coastal Campeche. Environmental authorities complained about the possible pollution caused by discarded materials. This led to the removal of fuel tanks from every vessel sunk as an artificial reef. The reef was finally authorized, and the vessels were deployed at the location in July 1985.

Currently, the agency responsible for the environment, SEMARNAT, has a specific guidebook for impact assessment, including artificial reefs: Impact Assessment for Fishery and Aquatic Activities. It specifies every environmental and social topic to be considered. The impact assessment has to consider several environmental dispositions stated on the "ley general del equilibrio ecologico y protección al ambiente" (general law for ecological equilibrium and environmental protection) ruled by the following dispositions:

- *Reglamento de la ley general del equilibrio ecológico y la protección al ambiente en materia de ordenamiento ecológico* (code of laws for ecological equilibrium and environmental protection in the matters of ecological order). *Capitulo V de ordenamiento ecológico marino* (Chapter V on the ecological marine order).
- *Reglamento de la ley general del equilibrio ecológico y la protección al ambiente en materia de impacto ambiental* (code of laws for ecological balance and environmental protection in ecological order in the matters of impact assessment).
- *Reglamento para prevenir y controlar la contaminación del mar por vertimiento de desechos y otras materias* (code of laws to prevent and control the pollution of the sea by drainage and other materials).

The impact assessment for fishery and aquatic activities required by SEMARNAT has to be complemented with authorization from the Secretaria de Marina (Secretariat of the Navy) through Form H that requires similar information for the impact assessment study.

The Secretariat of the Navy, responsible for the preservation of Mexican seas, has the following requirements:

1. Justification and approval from the municipal authorities, adjacent to the reef location.
2. Detailed analysis, characterizing the materials used, from a certified laboratory.
3. Technical report from the Secretariat of Health certifying the integrity of all materials relating to the environment and animal and human health.
4. Technical report from SEMARNAT on the possible impacts in the short and long terms of the structures and materials used.
5. Technical report from the Secretariat of Communications and transport on the impact on marine traffic.
6. In the case of ships and other sea vessels, a certificate of no contamination should be authored by a committee composed of designated officials from the secretariats of health, communications and transports, SEMARNAT, and the navy.

GUIDELINES STATED FOR THE CAMPECHE STATE ARTIFICIAL REEF PROGRAM

The location for the artificial reefs for artisanal fisheries was chosen based on the distance to fishing communities and bottom type. The technical considerations that must be taken into account when establishing an artificial reef have been well documented in the literature (Bellan-Santini, 1983; Mathews, 1983, 1985; Carter et al., 1985; D'Itri, 1985; Gutierrez, 1988; Dewitt et al., 1989; Grove et al., 1991):

- Distance from the coast versus depth.
- Depth and wavelength and frequency.
- Wave length versus density and shape of structure materials.
- Currents.
- Bottom type.
- Structure architecture and materials.

Tire Reef

There is an ample literature on the use of automobile tires to build artificial reefs (Bellan-Santini, 1983; Mathews, 1984; Grove and Sonu, 1989), suggesting module shape and the amount of ballast in proportion to tire mass. Given the availability of discarded tires, they were the first materials considered for construction in the artificial reef program of the state of Campeche. But the report by the Rubber Manufacturers Association (Radian Corp., 1989) resulted in a reconsideration of the use of tires until it could be ascertained that there is no environmental risk from pollutants leaching from the tires. With regard to module construction, the works of several authors were consulted resulting in the use of two truck tires, a triangle of three truck tires and a ring of 10 car tires (Figure 13.2). These configurations had the advantage of portability of individual tires and construction of the modules on board the boat used for transportation. Because of the shallower depths, tires were ballasted with 12 kg of concrete, attaining a density of 300 kg/m^2, slightly greater than that recommended by Dewitt et al. (1989).

The reef was finally completed in 1998; it was deployed at 10 m depth. After a month it was colonized by juvenile grouper, *Epinephelus morio,* about 3–5 fish per tire. Unfortunately, funding was not available to evaluate pollutant leaching from the tires. Laboratory tests have shown that under extreme pH conditions some materials leach heavy metals at pH 3 to 5 and polycyclic aromatic hydrocarbons (PAHs) leach at pH 8 to 9 (Radian Corp., 1989; Twin City Testing Corp., 1990). These acidic conditions may be present in reduction sediments (pH 3 to 5) (Chester, 1968; Turekian, 1968)

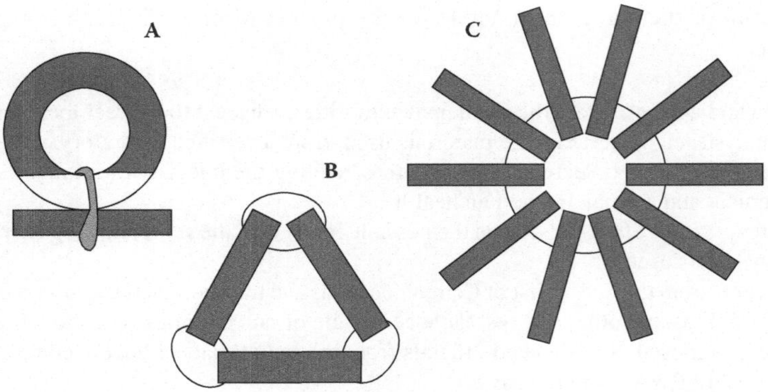

Figure 13.2 Tire modules design for the tire pilot reef: A, two truck tires; B, three truck tires; C, ten car tires.

and in shell-forming organisms (Muzii and Skiner, 1966; Digby, 1968; Nakahara and Bevelander, 1971; Livingstone and Bayne, 1977; Lutz and Rhoads, 1977; Nicol, 1978; Fournie, 1979a, 1979b; Carriker et al., 1980; Wheeler et al., 1981). Consequently, no additional tire reefs were deployed, although, fishermen continued to make tire modules, deploying them in unauthorized locations.

Fishermen who used this artificial reef were interviewed at the landing beaches. These interviews confirmed three types of fishing methods were used: gillnets on the vicinity of the reef, free-diving with harpoon, and hook-and-line over the structures. Although no catch records were available, fishermen classified catches as good, very good, and extraordinary. They emphasized that the closeness to the coast was an outstanding feature that allow them to make it to the coast safely, with such large catches.

La Barcaza Reef

The third reef deployed as part of the Campeche artificial reef program was created by sinking a barge that was damaged during a hurricane in 1996. This site was proposed as a sport diving attraction. Unfortunately, the lack of surveillance by authorities does not prevent frequent fishing, by fishermen who set their gillnets over La Barcaza. These nets frequently entangle with the reef and are subsequently abandoned, resulting in massive kills and ghost fishing of resident and migratory fish. Detonation of explosives had also been reported as ascertained from evidence of detached incrusting organisms from most vertical surfaces.

Although there has not been an evaluation of the impact of the artificial reef program on the catch of coastal fish, there has been a 3-year survey of the increase in use of the reef by fishermen on La Barcaza Reef. Fishermen point out that they prefer fishing on the artificial reefs rather than on traditional fishing grounds as the fishermen believe they are certain to catch more and bigger fish.

An inquiry of fishermen who work regularly on the artificial reefs indicates that their chief fishing activity is from March to June with a mean catch per trip of 924 kg, an average of four trips to each reef during the season, and two boats working during one or two days. It was estimated that 20 fishing groups make regular use of La Barcaza Reef.

At this reef, gillnets are directed toward pelagic fishes like barracuda, rudderfish, and jack. Spear fishing is directed toward schools of fish, which are surrounded with a purselike net and captured with spears by free divers. Free divers or divers using hooks catch Goliath grouper, gray snapper, and groupers with spears. Barracuda comprised the largest portion catch with 134 kg per trip. Averages of 14 species were caught during trips where the three methods were used simultaneously. Fishermen from the ports of Punta Arenas, Campeche, Lerma, Seybaplaya, Champoton, Isla

Aguada, and Carmen regularly fish the artificial reefs and have increased their catch by an average of 30.55% from an annual mean of 4.8 thousand tons a year to 6.3 thousand tons a year.

Rigs to Reefs

The fourth group of reefs that were deployed as part of the Campeche reef program was composed of dismantled material from oil rigs, which was set on the bottom. There is no information on fish yield from these reefs.

Santa Ana oil camp started operation on July 3, 1959. It is located 2.5 to 3 km from the coast of Tabasco state. The last oil rig operated until 1983. Negotiations for permits for dismantling the oil rig were initiated in June 1995, and permission to dismantle was granted in August 1995. In June 1998, negotiations with the government of Campeche and the required permits to use the legs from eight platforms were granted for the construction of artificial reefs at five locations. Only the submerge portion of the rigs was used for the construction of the reefs. Fifteen-meter sections of legs were cut and laid sidewise on location.

Reef Ball Project

The latest effort to enhance artisanal fisheries off Campeche was through a reef ball project focused on those fishermen with smaller boats or without an outboard motor. Six coastal locations were chosen for the construction of border reefs with reef balls: Isla Arena, Lerma, Villamadero, Champoton, Punta Xen, Sabancuy, and Isla Aguada (Figure 13.1). The construction was completed in 2006. The position of these reefs along the coast is given in Table 13.2. Since the completed deployment of this coastal reef system, there has been a constant increase in the number of sport beach anglers, but commercial fishermen also make use of them by placing gillnets on the off shore side of the structure or fishing from the beach or from small boats with hook and line.

CONCLUSIONS

Finfish production in the community of Seybaplaya in Campeche before the first artificial reef was created averaged 19 tons per year; after 1996, with the construction of the first artificial reef, the catch increased to over 137 tons. With the construction of the four reefs, state production increased from an average of 25 thousand tons before 1985 to over 70 thousand tons in 1993. Subsequently, catch dropped (17 thousand tons in 2007) to levels similar to those recorded prior to the reef (Figure 13.3). When the reef system was initially installed, fish yields were dramatically enhanced. As evidenced by the evaluation of the Barcaza reef, these newly deployed reefs provided a refuge for spawning and juveniles of various fish species, and served to facilitate aggregation to other species.

There is little doubt that the abuse and misuse of these reefs has prevented the artificial reef program from fulfilling its main objective of enhancing and maintaining coastal artisanal fisheries. The loss of fishermen at sea has stopped since the reef deployments, consequently there has been a reduction in fishing activity on the more distant natural reefs. Unfortunately, there are no records to evaluate fish populations on these distant, natural reefs to determine if one of the objectives of artificial reefs (i.e., recovery of natural fishing grounds) has been fulfilled.

The effectiveness of artificial reefs to enhance artisanal fisheries has been demonstrated again with the dramatic increase of yields with the Campeche program. The program also accomplished two more of its objectives: The safety of fishers and enhancement of sport angling. But it is evident that lack of management and misuse make useless any effort to enhance artisanal fisheries through artificial reefs.

Table 13.2 Geographic Location of Reef Ball Reefs in the State of Campeche

	Vertex 1		Vertex 2		Vertex 3		Vertex 4	
	W	N	W	N	W	N	W	N
Isla Arena	90°43′22.511″	20°41′49.493″	90°43′8.762″	20°41′42.663″	90°42′44.56″	20°42′25.778	90°42′58.481	20°42′20.908
Lerma	90°54′31.659″	19°50′16.932″	90°54′22.717″	19°50′4.998″	90°53′40.671	19°50′33.198″	90°53′4961″	19°50′45.136″
Villa Dero	91°0′24.072″	19°45′58.152″	91°0′18.039″	19°35′1.879″	90°59′48.847″	19°46′36.633″	91°0′11.011″	19°35′52.376″
Punta Xen	91°30′6.80″	18°48′18.986″	91°29′57.331″	18°48′7.454″	91°29′19.008″	18°48′39.835″	91°29′28.475″	18°48′51.361″
Saban Cuy	91°12′44.743″	19°0′14.095″	91°12′39.11″	19°0′32.868″	91°11′50.946″	19°0′49.628″	91°11′56.228″	19°1′3.371″
Isla AAguada	91°30′6.80″	18°48′18.986″	91°29′57.331″	18°48′7.454″	91°29′19.008	18°48′38.831″	91°29′28.475″	18°48′51.361″

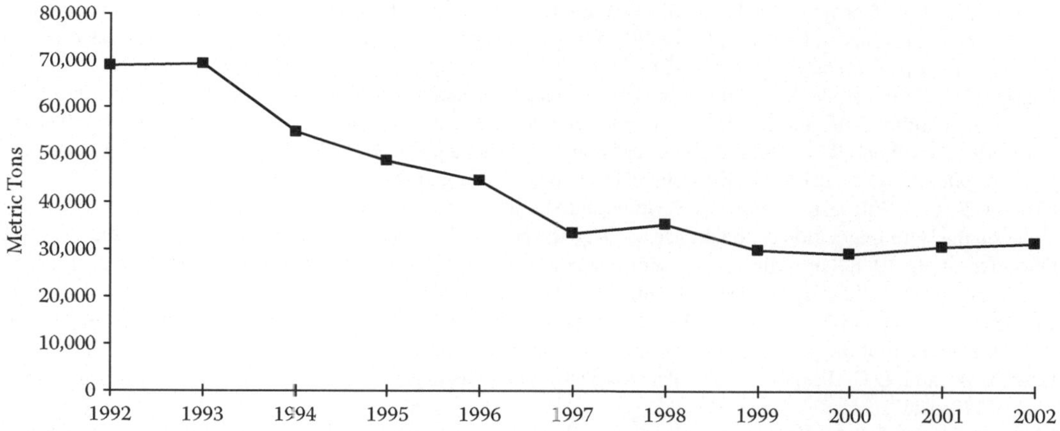

Figure 13.3 Finfish production for the years 1992 to 2007 showing a drastic fall 5 years after the first four modules were installed and stabilization 10 years later on values similar to prereef times.

RECOMMENDATIONS

Based on an examination of the results of the efforts to construct artificial reefs to facilitate the catch by the artisanal fishing community off the Campeche, the following suggestions are offered as recommendations to help guide the program in the future.

1. Establish a committee to support and advise communities on the technical and legal aspects for the planning stage to monitoring of activities on the reef once it has been established.
2. Design a local and regional program with specific uses for each reef based on the participation of every sector involved.
3. Establish reefs with different purposes, for example, commercial fishing, sport line fishing, sport diving, juvenile protection, and reproductive stock protection.
4. Promote the creation of rules, with the participation of every sector, to assure the proper use of the reefs and reinforce the rules with policing by the same fishers and fishing authorities.

REFERENCES

Baqueiro C.E. and R. Mendez, 1994. Artificial reefs: An alternative to enhance Mexican littoral commercial fisheries. *Bull. Mar. Sci.* 55 (2–3): 1014–1020.

Bellan-Santini, D., 1983. (Ed.) Journée d'études sur les aspects scientifiques concernant les récifs artificiale et Mariculture susmpendu. Cannes, CIESM.

Carriker, R.M., R.E. Palmer, L.V. Sick, and C.C. Johnson, 1980. Interaction of mineral elements in sea water and shell of oyster *Crassostrea virginica* (Gmelin) cultured in controlled and natural systems. *J. Exp. Mar. Biol. Ecol.* 46: 279–296.

Carter, J.W., W.N. Jessee, M.S. Foster, and A.L. Carpenter, 1985. Management of artificial reefs design to support natural communities. *Bull. Mar. Sci.* 37 (1): 114–128.

Chester, R., 1968. Elemental geochemistry of marine sediments. En: *Geochemistry of Marine Sediments*, Pettersson, H. (Ed.) Univ. Chicago Press. Cap. 15: 23–80.

D'Itri, F., 1985. (Ed.) Artificial reefs. *Marine and Freshwater Applications*. Lewis Publishers, Chelsea, Michigan.

Dewitt, O.M., E.N. Myatt, and W.K. Figley, 1989. New Jersey tire reef stability study. *Bull. Mar. Sci.* 44 (2): 807-817.

Digby, S.P., 1968. The Mechanism of Calcification of Molluscan Shell. En: *Studies in the Structure, Physiology, and Ecology of Molluscs*, Fretter, V. (Ed.). *The Malacol. Soc. London*, 22: 93–108.

Fournie, J., 1979a. Formation de la coquille des mollusques: Les problemes poses par la presence et le comportement de cellules libres dans la coquille normale et regeneree chez *Agriolimax reticulatus* (Gasteropode Pulmone). *Malacologia.* 18: 543–548.

Fournie, J., 1979b. Estude des cellules libres presentes a la surface interne de la coquille d'*Agriolimax reticulatusts* (Muller): Origine et role dans la mise en place de L'hypostracum. *Ann. Sci. Nat. Zool.* 1: 169–185.

Grove, R.S. and Sonu, C.J., 1989. Fishing reef planning in Japan. En: *Artificial Reefs (Marine and Freshwater Applications).* D'Itri, F.M. (Ed.) Lewis Publishers, Michigan. 187–251.

Grove, R.S., C.A. Wilson, R.M. Buckley, C. Sonu, and M. Nakamura, 1991. Artificial habitats for fisheries. Memories on the Fifth International Conference on "Aquatic Habitat Enhancement," Long Beach, California.

Gutierrez, A.R., 1988, Hidrodinámica y sedimentación en torno a una estructura artificial de llantas, en Bahia Rosrito, B.C. Tesis Prof. Oceanol. Univ. Autón B.C.

Livingstone, D.R. and B.L. Bayne, 1977. Responses of Mytilus edulis L. to low oxygen tension: Anaerobic metabolism of the posterior adductor muscle and mantle tissue. *J. Comp. Physiol.* 114: 1443–155.

Lutz, R.A. and D.C. Rhoads, 1977. Anaerobiosis and a theory of growth line formation. *Science* 198: 1222–1227.

Mathews, H., 1983. Artificial fishing reefs. Material and construction. *Florida Coop. Ext. Mar. Adv. Bull. MAP.* 29: 8 pp.

Mathews, H., 1984. Artificial reefs: Permit application guidelines. *Florida Coop. Ext. Ser. Sea Grant Ext. Prog.* 8 pp.

Mathews, H., 1985. Physical and geological aspects of artificial reef site selection. En: *Artificial Reefs, Marine and Freshwater Applications.* D'Itri, F.M. (Ed.). Lewis Publishers, Chelsea, Michigan. Chap. 6: 187–251.

Muzii, E.O. and H.C. Skiner, 1966. Calcite deposition during shell repair by the aragonitic gastropod *Murex fulvescens. Science.* 151: 201–203.

Nakahara, H. and G. Bevelander, 1971. The formation and growth of the prismatic layer of *Pinctada radiata. Cal. Tiss. Ress.* 7: 31–45.

Nakamura, M., 1991. Artificial habitat technology in Japan, today and tomorrow. En: *Recent Advances in Aquatic Habitat Technology*, Nakamura, M., R.S. Grove, and C.J. Sonu (Ed.) *Proc. Jap.-U.S. Symp. Artif. Habit. for Fishr.* 11–19.

Nicol, D., 1978. Shell-cemented pelecypods. *Florida Scientist.* 41 (1): 39–41.

Radian Corp., 1989. A Report on the RMA TCLP Assessment Project. Technical report for the Rubber Manufacturers Association.

Stone, B.R., R.S. Grove, and C.J. Sonu, 1991. Artificial habitats technology in the United states, today and tomorrow. En: *Recent Advances in Aquatic Habitat Technology*, Nakamura, M., R.S. Grove, and C.J. Sonu (Ed.) *Proc. Jap.-U.S. Symp. Artif. Habit. for Fishr.* 3–9.

Turekian, K.K., 1968. Some aspects of the geochemistry of marine sediments. En: *Geochemistry of Marine Sediments*, Pettersson, H (Ed.) Univ. Chicago Press. Cap. 16: 81–127.

Turner, H.C., 1969. Artificial reefs. En: *The Encyclopedia of Marine Resources.* F.E. Firth (Ed.). Van Nostrand Reinhold Co., New York. 35–39.

Twin City Testing Corp., 1990. Environmental Study of the Use of Shredded Waste Tires for Roadway Sub-Grade Support. Waste Tire Manag. Unit. Technical report for Minnesota Pollution control agency.

Wheeler, A.P., J.W. George, and C.A. Evans, 1981. Control of calcium carbonate nucleation and crystal growth by soluble matrix of oyster shell. *Science* 212: 1397–1398.

CHAPTER 14

The Role of Artificial Reefs in the Sustainability of Artisanal Fisheries

Miguel Neves Santos, Carlos Costa Monteiro, and Francisco Leitão

CONTENTS

ABSTRACT

Artificial reefs have been developed worldwide over the last few decades, generally as a response to different problems concerning coastal activities, marine ecosystems, and fisheries. In the Algarve (southern Portugal), the use of artificial reefs was a response to problems related with fisheries management, as traditional management measures were insufficient in guaranteeing small-scale fisheries sustainability. Here, we describe the pilot project started in 1990 with the deployment of two artificial reefs, which was the baseline for the deployment of large artificial reef complexes in the coastal waters. By 2011, the Algarve artificial reef complexes will contain more than 21,500 modules, occupying a total area of 45 km^2, with an estimated area of influence of 70 km^2. The main results and achievements of a number of multidisciplinary studies are briefly presented, highlighting the usefulness of the artificial reefs on small-scale fisheries management. These summaries include different aspects of reef research that have been carried out during the last 20 years, namely:

(1) artificial reef colonization and fish assemblages, (2) juvenile protection, (3) essential fish habitats (4) predation and stocking, (5) catch and value per unit of effort and reef use, and (6) rebuilding fisheries. The importance of artificial reef management is raised, with particular emphasis on the issues related to open-access areas/fisheries. The need of long-term monitoring and applied research is also discussed, as the role of artificial reefs will be reinforced in the near future, taking into account the growing need to integrate fisheries and environmental policies into fisheries management.

INTRODUCTION

Artisanal Fisheries in the Algarve

The fish fauna in Algarve (southern Portugal) coastal waters reflects the high species richness of this region. Such high species richness indicates the geographic location of the Algarve, between the Mediterranean Sea and the Northeast Atlantic Ocean. The advantage of this geographical location for enhancing species richness is enhanced by the presence of several estuarine-lagoon systems (Ria Formosa, Ria de Alvôr, and the Estuary of Guadiana), which act as nursery grounds for a number of fish populations that cyclically migrate in the nearby coastal waters. Thus, these estuaries and lagoons play a major role supplying the local fisheries (Monteiro et al., 1987, 1990). Juvenile fish from these estuaries and lagoons face a paucity of natural rocky systems (natural reefs) along much of the coast. The rarity of this hard substrate provides only a small portion of the potential food resource available if there were more hard bottom available. These hard bottom substrates also provide shelter for juvenile fish from natural predators (Monteiro and Santos, 2000).

The Portuguese are among the world's top three fish consumers (after Japan and Iceland), with an annual consumption of marine products of around 60 kg per inhabitant. In the Algarve, consumption of marine products is even higher, largely a consequence of gastronomic heritage and because it is Portugal's most important tourism region. This strong demand for fishing products is one of the reasons for the importance of this fishing industry, which has been operating in the Algarve for decades. The southern coast of Portugal can be considered to be privileged in terms of fishing, when compared to other regions along the Portuguese coast. It has calm seas during most of the year, and this favors fishing activity, particularly the use of fishing gear set close to shore. Although there has been a noticeable decrease in the fleet since the mid-1980s (approximately by 50%), the Algarve fishing fleet had around 1,550 vessels, of which around 90% belonged to the artisanal fleet (<12 m) in 2008. The artisanal fleet is chiefly composed of open-deck fiberglass boats, which are licensed to use a wide range of fishing gears as a result of the variety of fish resources available in Algarve coastal waters. Bottom gillnets and trammel nets are the most popular fishing gears (in terms of number of active licenses) in the region. These nets have quiet different characteristics, depending on the target species. The gillnets are made of nylon, monofilament line, while the trammel nets can be either multifilament (used in shallow waters) or monofilament (used in deeper waters). The second most popular fishing gear are traps/pots; having different shapes and made of plastic (e.g., baited *covos* for finfish, cuttlefish, and octopus), steel wire (e.g., baited *murejona* for finfish), or pottery (e.g., *alcatruz* for octopus). Longlines were important historically but currently they are only used in a few villages (e.g., Praia de Faro and Fuzeta). Another important fishing gear is the small purse seine (*rapa*), which is responsible for the highest landings (i.e., *Sardina pilchardus*, *Scomber* spp., *Trachurus* spp., and seabreams). Between 2006 and 2008, these multifisheries fleets landed an annual average of 15.3 MT, with a mean value of €2.58/kg. The main fisheries targeted were octopus, mackerel, sardine, and horse-mackerel (Anonymous, 2009). However, the seabreams species (genus *Diplodus*) were the most valuable, with mean prices on auction varying from €4–15/kg.

The need for a strong fish supply was facilitated by the ease of accessibility to fisheries in inshore waters and the high commercial value of the catch. Inevitably, this lead to an intensive exploitation of most fish populations. This fishery for adult fish, together with some by-catch (and discard) of juvenile fish, combined to cause significant disturbance to local fisheries and ecosystems. Concomitantly, the multispecies characteristic of these artisanal fisheries and the dynamics of these coastal resources, exacerbated problems in the management of these fisheries. It is important to realize that traditional management measures (e.g., minimum sizes, closed seasons, catches limits, closed areas, and effort or gears restrictions, etc.) were insufficient in guaranteeing fisheries sustainability. These conditions generated a need to promote alternative and/or complementary management options directed to facilitating the sustainability of local artisanal fisheries.

Artificial reefs have been developed worldwide over the last few decades, generally as a response to these types of problems concerning coastal resources, ecosystems, and fisheries. Although artificial reefs have been deployed worldwide, only a limited number of examples demonstrate the usefulness of artificial reef technology on the management of artisanal fisheries. These include the cases of: the spiny lobster fishery in the Caribbean's (Arce et al., 1997; Sosa-Cordero et al., 1998; Cruz and Phillips, 2000; Briones-Fourzán and Lozano-Álvarez, 2001; Briones-Fourzán et al., 2007); the restoration program of Hong Kong's marine ecosystems and fisheries, within marine parks and fishery-protected areas (Pitcher et al., 2000; Wilson et al., 2002); and the European programs in the Mediterranean Sea (Bombace et al., 2000; Relini et al., 2008) and off southern Portugal. What follows is a general description of the Portuguese artificial reef program, with particular emphasis on the results and accomplishments regarding the role of the man-made structures on the management of local artisanal fisheries.

The Artificial Reef Pilot Project

The Japanese technology and experience in this field of science inspired IPIMAR (Portuguese Fisheries and Marine Science Laboratory) researchers to propose and develop an artificial reef pilot project for Algarve coastal waters. The project was a response to the previously mentioned fisheries issues off Algarve, and the first two artificial reefs were deployed in 1990, off Faro and Olhão, along the southern Portuguese coast (Figure 14.1). This pilot project was supported by the Integrated Plan

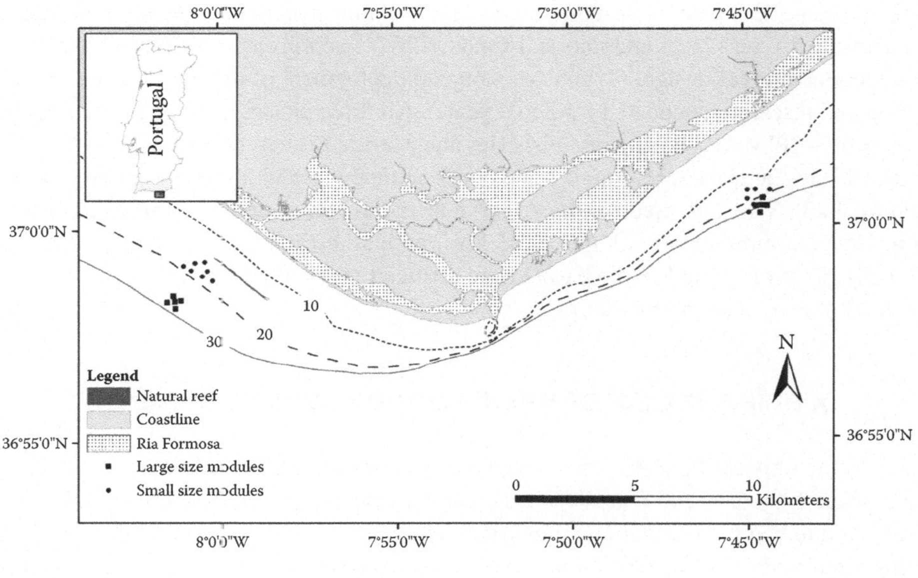

Figure 14.1 Location of the pilot Faro and Olhão artificial reefs systems. Insert indicates Ria Formosa (Algarve, southern Portuguese coast).

for Regional Development. IPIMAR took responsibility for all aspects of the deployments, from site selection to reef design and monitoring. The Faro and Olhão artificial reef systems consisted of sets of small- (*protection reefs*) and large-sized (*exploitation reefs*) concrete modules. The objective for deploying the protection reef was to create a shelter for the juvenile fish populations that seasonally migrate from the lagoon and the continental shelf. These reefs consisted of 735 small concrete cubic units (modules), of 3 MT each with a unitary external volume of 2.7 m^3 (for a total volume of 2017 m^3), distributed in 21 groups, at depths from 15 to 22 m, and occupying a total area of 39 ha (Figure 14.2) for the entire reef complex. The exploitation reef consisted of 20 large, reinforced concrete blocks with total volume of 3036 m^3, comprising two different shapes: rectangular with an unitary external volume of 130 m^3 and weighing 30 MT (Figure 14.2) and octagonal with an unitary external volume of 174 m^3 and weighing 40 MT (Figure 14.2). These large-sized modules were deployed in five groups at depths from 25 to 40 m. Altogether, these large modules occupy a total area of 21 ha for the reef complex. The distance between the protection and the exploitation reef sets varied from 0.9 to 2.7 km in the Olhão and Faro artificial reef systems, respectively.

The objective of these two artificial reefs was to evaluate, within the framework of local geo-ecological conditions, the effects these man-made structures had at the ecological and fishing levels. The program had the following secondary objectives: (1) provide protection for juvenile fish populations, particularly to those of commercial interest; (2) promote biodiversity and fishing resources; (3) increase fishing yields on deteriorated fishing grounds; and (4) test alternative (and innovative) management options for marine environmental and fisheries management. The results achieved during these pilot projects (1990–1996), encouraged the national authorities to prepare a plan for the deployment of large artificial reef complexes in Algarve coastal waters.

The Algarve Artificial Reef Program

Between 1998 and 2003 six larger artificial reef complexes were deployed off Algarve, and another complex is schedule for deployment in 2011–2012. After deployment is completed, the Algarve artificial reef complex will represent an investment of €10M, 20% of which having been allocated to scientific studies and monitoring. The program benefited from the financial support of the European Union (75%) and national funds (25%). Each project has three different phases: (1) site selection and reef design, (2) construction and deployment of the reef modules, and (3) reef monitoring and research (conducted for 5 to 7 years). All three phases are being coordinated by the same authority—IPIMAR (Portuguese Fisheries and Marine Research Institute). Each of the newer artificial reefs have at least 2,940 small-sized concrete modules and 36 octagonal large-sized modules (Figure 14.3). When completed, the Algarve artificial reef complexes will contain more than 21,500 modules, occupying a total area of 45 km^2, with an estimated area of influence of 70 km^2 (Figure 14.4). To our knowledge, the Faro/Ancão artificial reef is the largest complex of this type in Europe (8.2 km long and 1.5 km wide).

A BRIEF REVIEW OF RESULTS AND ACCOMPLISHMENTS

To assess the impact of the Algarve artificial reef program, multidisciplinary studies have been conducted since 1990. The components of this long-tern evaluation and study included water chemistry analyses and benthic invertebrate and fish assemblages, as well as an assessment of the use by the fishing community and an evaluation of its socio-economics impact on local communities.

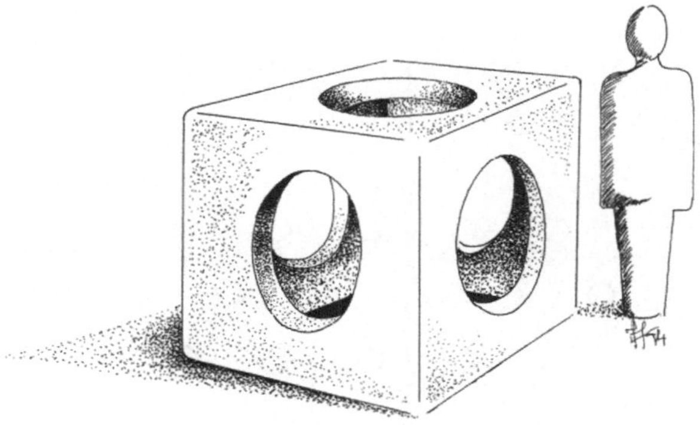

(a)

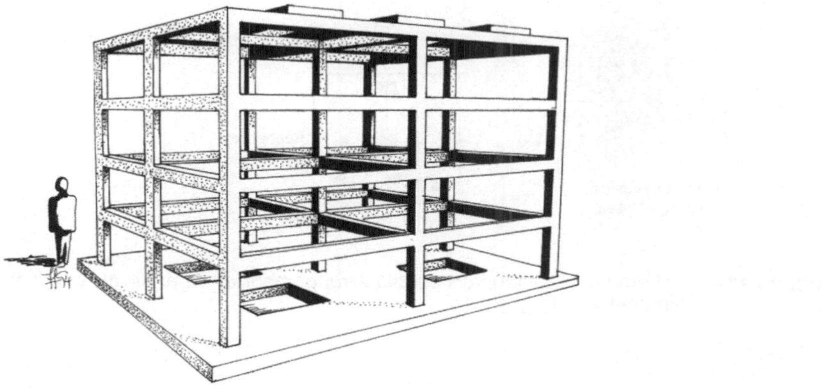

(b)

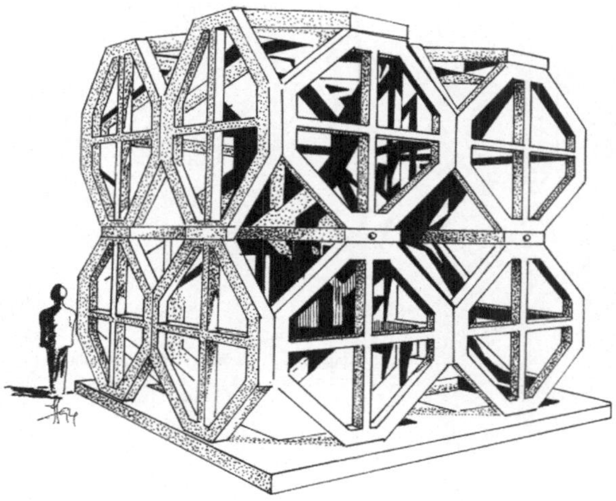

(c)

Figure 14.2 Small cubic concrete reef unit installed as protection reef (a), large-sized rectangular (b), and octagonal (c) concrete blocks as part of the exploitation reef. (Drawings by J.J. Sá e Silva.)

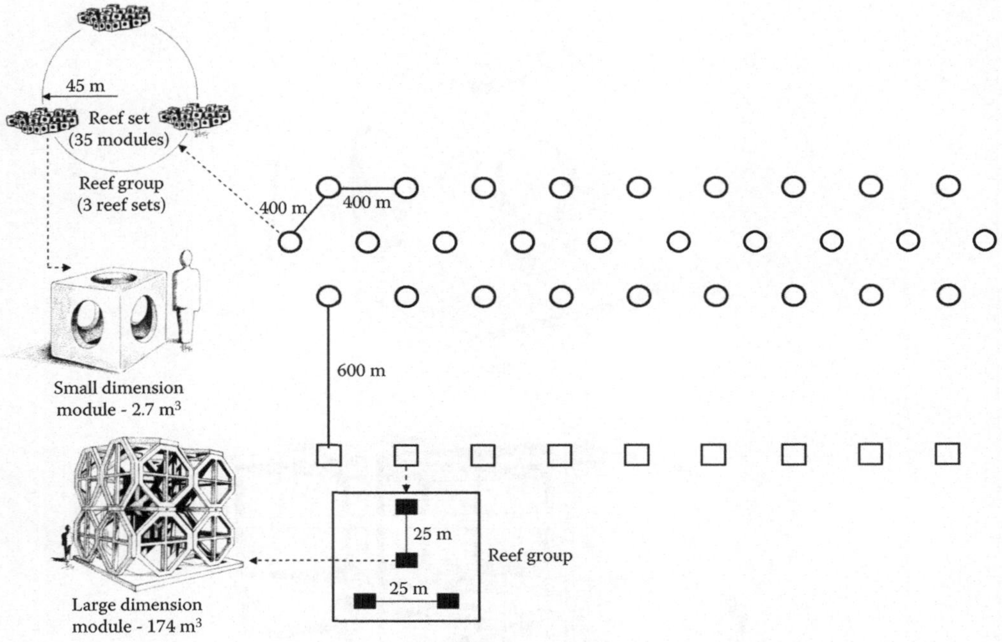

Figure 14.3 Scheme of the standard spatial arrangement of an artificial reef system on the Algarve coast (southern Portugal).

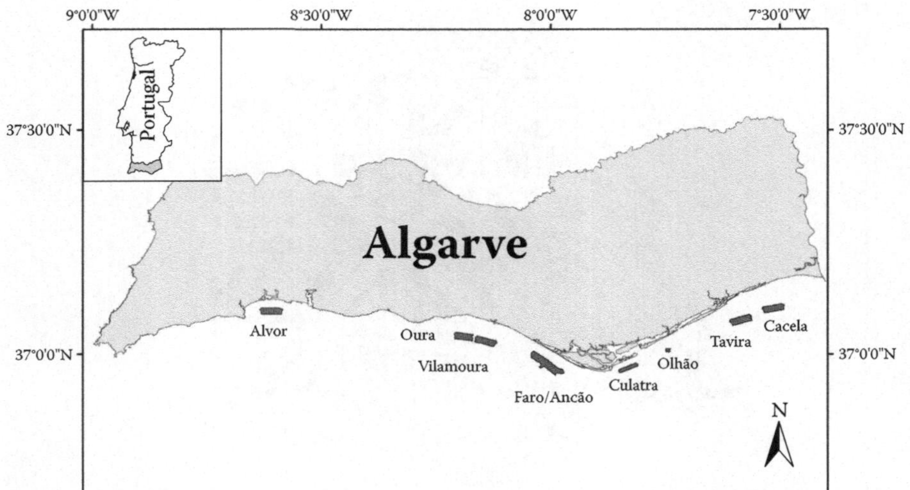

Figure 14.4 Algarve artificial reef complex (covering an area of 45 km^2) deployed in the Algarve coastal waters (southern Portuguese coast). The Culatra artificial reef is scheduled for deployment in 2012.

The Colonization Process and the Fish Assemblages

Artificial reef colonization and fish assemblages were the most important features of the reefs to be studied as it was essential to determine the period of time required for the artificial reef systems to accumulate the fish assemblages that could support commercial harvesting. The rate of fish colonization and measures of the standing stock of fish assemblages were key features for managers, who allocated a significant amount of financial support to enhance local fisheries. Therefore, understanding the progression of artificial reef fish assemblages since reef deployment was a crucial part of the evaluation. These studies were conducted for both the initial pilot project and the secondary enlargement of the Faro artificial reef (Faro/Ancão artificial reef system deployed in 2003).

Advances in SCUBA technology (e.g., Nitrox) since the 1980s permitted divers to more accurately gather information related to artificial reef fish assemblages that included species composition and length-frequency distribution among other parameters. These data provided insight into a broad knowledge of the mechanisms that influence fish assemblages. An analysis of the data obtained from a 2-year study (sampled monthly at each site using visual census) indicated that the deployment of large-sized Faro/Ancão artificial reef system proximate to the pilot reef off Faro was associated with a rapid increase in fish density and biomass (Leitão et al., 2008b), compared to previous reports for fish assemblages in the old, smaller pilot artificial reefs deployed in 1990 (Santos et al., 2005). A rapid increase in the mean number of species per survey was observed during the first four months after the artificial reefs were deployed. After this initial period, the fish assemblage structure remained quit stable (showed high similarity > 73%). In addition, a high level of available (and constant) exploitable biomass was recorded for both the overall fish assemblage and for commercially important seabream species. These accounted for 63% of the resident fish biomass three months after deployment (Leitão et al., 2008a, 2009). Approximately one third of the estimated monthly commercial species biomass available in the artificial reefs corresponded to adult biomass. This means that the artificial reefs had a high capacity to aggregate fish, rapidly becoming viable fishing grounds by providing available exploitable biomass. Such fast colonization and stabilization of the artificial reef fish assemblage are probably linked to habitat connectivity. As mentioned by Fernández et al. (2008 and references therein), the chance of fish locating an artificial reef increases with habitat connectivity. Oppositely, the increase of habitat (i.e., artificial) availability and the immediate instinctive response of fish to move to habitat, affects the rate of fish relocation among habitat patches. Therefore, the degree of connectivity, or isolation, among reefs and other suitable habitats (e.g., Ria Formosa and newly deployed artificial reefs; the old and new artificial reef systems; and coastal areas and artificial reef systems) favors fish movement and post-settlement relocation by individuals. This feature contributes to fast colonization in an area largely characterized by open sandy bottom. In addition, habitat size and degree of benthic colonization also affect fish assemblage settlement. Artificial reef surfaces provide additional critical habitat that increases the local carrying capacity, and therefore the abundance and biomass of marine biota (Polovina, 1991). The rapid colonization of reef fish was also suggested to be related to trophic attraction and seasonal (late summer/autumn) migration of subadults commercial fish from Ria Formosa, a coastal nursery lagoon (Monteiro et al., 1990). Trophic attraction has been found to explain fish assemblages colonization patterns (Leitão et al., 2008b) as the enhancement of fish assemblages matched the time when macrobenthic fauna cover markedly increased (Boaventura et al., 2006). Sale (1969) predicted that fish accumulate in preferred habitats through a process of appetitive exploration, where fishes in an adequate environment will spend little time exploring new environments, compared to fish in a less adequate environment. In addition, enhanced opportunities for food, where natural rocky substratum is scarce, are important since most fish species on artificial reefs are categorized as carnivorous (Leitão et al., 2008b) and are commonly observed foraging on reef-associated macrophytobenthic organisms.

Santos et al. (1995) showed that fish biomass was higher (21.0 kg/250 m^3 water) on artificial reefs than on proximate rocky areas (18.1 kg/250 m^3 water). More recently, Leitão et al. (2008a) reported differences in fish density and biomass when comparing fish assemblages before and after enlargement of the Faro/Ancão artificial reef, with economically important species accounting for 95 kg/reef set. These results denoted two and three times higher density and biomass of fish being recorded from an area with an artificial reef than the same area before reef deployment. This clearly shows that artificial reefs can have an immediate, beneficial effect in enhancing local fisheries. Moreover, if the biomass on the nearby natural reef or substrate still of the same order of magnitude than in the past and so remains the proportion between the different species, to produce the same amount of biomass as a natural area, a much smaller artificial reef area is needed (approximately 3 times less). On the other hand, it suggests that the Faro/Ancão artificial reef system (12.2 km^2) could compensate for the loss (due to fishing) of biomass available on the natural reef. Nevertheless, a time lag between fish standing stock (through immigration, reproduction, and growth) and increases in fishing yield is expected.

Juvenile Protection

Nursery grounds are of major importance in coastal areas for many reasons. One of the most important reasons is that nursery areas supply recruits for local fisheries. This is the situation along the Algarve coast where Ria Formosa, Ria de Alvôr, and the Guadiana Estuary act as nursery grounds for most of the species commercially harvested by the local artisanal fleet. Once these species migrate to shore, juveniles are confronted with a paucity of natural rocky habitats, which would supply them with food and shelter from natural predators (Monteiro and Santos, 2000). Thus, artificial reef systems of the Algarve coast were placed off these nursery grounds, as it was believed they could provide such features, namely increasing survivorship of juveniles and subadults specimens. The decision to deploy artificial reefs in the Algarve area proved correct as up to 88% of the artificial reef fish assemblage is composed of juveniles (Figure 14.5). In situ observations indicate that these juveniles: feed directly on the benthic assemblages of the man-made structure and/or the water column above and near the structures; use the artificial reefs as shelters, hiding from natural predators; and are afforded protection against strong currents and fouling due to sedimentation.

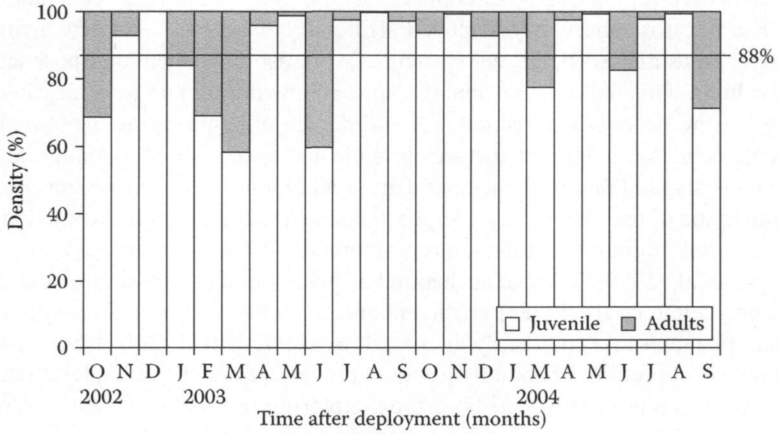

Figure 14.5 Mean monthly adult and juvenile fish density (%) on the Faro/Ancão artificial reef system in Algarve coastal waters. Fish assemblages were assessed over the course of a 2-year period after reef deployment.

Just one month after reef deployment, in an area proximate to a nursery coastal lagoon (Ria Formosa lagoon), juveniles of *Diplodus vulgaris* were found associated with reef structures (Leitão et al., 2008b). Leitão et al. (2009) followed the *D. vulgaris* recruit growth on the artificial reefs. Moreover, the seasonal migration of other seabreams (*Diplodus* spp.) from Ria Formosa to the artificial reefs was previously reported by Santos et al. (2005). Another important example of recruitment to artificial reefs is the year-round arrival of young-of-the-year of several economically important demersal species, such as *Boops boops*, *Trachurus trachurus*, *Pagellus acarne*, and *P. erythrinnus* as, reported by Leitão et al. (2009). Cryptic and sedentary species that occupy a stationary position within the artificial reef sets, make use of the artificial reef microhabitat; using the concavities, holes/space between the reef modules. The macroinvertebrate assemblages also make use of these microhabitat features of the reefs. Demersal, commercially important fish species with small or intermediate home ranges find refuge in the inner part of the reef (e.g., *Trisopterus luscus*, *Scorpaena notata*, *Serranus cabrilla*, and wrasses such as *Coris julis* and species belonging to the Genus *Symphodus*), with many species finding shelter near the blocks.

These findings demonstrate the protection effect provided by artificial reefs to a wide variety of species, most of which are commercially important. Despite the fact that many demersal species are only occasional reef inhabitants and do not spend their entire life cycle on the man-made structures, the role of artificial reefs for the survivorship of these species could still be highly important. Such protection provided by artificial reefs is similar to that provided by nursery and recruitment areas. Thus, one may classify the man-made reefs as recruitment areas, as by definition recruitment means the incorporation of older juvenile fishes into adult populations after their settlement and survival in nurseries (Levin, 1994; MacPherson, 1998). Although, as reported by Harmelin-Vivien et al. (1995) the variation in abundance of adult fish populations is affected by survivorship at the different life stages and success of recruitment may depend on specific requirement of early juveniles for settlement and on the availability of suitable habitats. One such requirement is linked with the connectivity between nursery areas and fishing grounds. Artificial reef deployments can certainly increase such connectivity, promoting the replenishment of adult populations through an increase in juvenile survivorship (e.g., by providing food and shelter) during important phases of a species' life cycle. Such connectivity is likely to depend, not only on the distance between nursery and fishing areas, but also on the presence of corridors or "stepping-stones" to favorable habitats. The lack of such habitats, the failure to protect them and/or the limitation of fish's capacity to move freely among them, may have detrimental effects on adult populations. Therefore, the presence of the man-made structures is highly important for commercial fish species, representing an extension to coastal waters from the protection effect provided by local lagoons, estuaries, and near-shore shallow waters. Additionally, artificial reefs can act as temporary habitat as fish pass through from inshore areas on their way to adult habitat offshore that generally serves as fishing grounds. Moreover, artificial reefs have a role similar to that of local natural reefs, which are scarce along much of the Southern Algarve coast (e.g., The Faro/Ancão artificial reef systems cover 12.21 km^2, in an area of 36 km^2 where natural reefs cover less than 3 km^2).

Essential Fish Habitats

Benaka (1999) defined Essential Fish Habitats (EFH) as "those waters and substrate necessary for fish for spawning, feeding, or growth to maturity." Although, the Algarve artificial reef program did not explicitly mentioned the creation of EFH as one of its main objectives, a number of findings showed that these artificial structures can be classified as EFH for a number of artificial reef assemblage species. In fact, artificial reefs play multiple roles for a number of species; acting to enhance recruitment, growth and nursery areas for juveniles; and spawning/mating areas for adults. For these features, artificial reefs can be considered essential fish habitat.

What follows is a summary of these findings, highlighting the role of man-made structures for fish species that aggregate in large numbers on the Algarve artificial reef systems. Examples can be observed among several demersal species, such as the seabreams belonging to the Genus *Diplodus* and the seabass, *Dicentrarchus labrax* (Linnaeus, 1758). The summary below describes the use of the artificial reefs by the different species as previously reported by Santos et al. (2005) and Leitão et al. (2007, 2008a, 2008b, and 2009).

Diplodus bellottii (Steindachner, 1882), *D. sargus* (Linnaeus, 1758), and *D. vulgaris* (Geoffroy Saint-Hilaire, 1817) are three economically valuable species that are often found associated with artificial reefs. These three species account for 40% and 63% of overall resident reef fish assemblages and 20% and 20% in terms of abundance and biomass, respectively, of the overall artificial reef fish assemblage in Algarve. *D. bellottii* populations on artificial reefs are composed of subadults, which use the man-made structure as nursery areas. Adults that are present year-round use the reefs for reproduction/spawning in late winter and spring. *D. vulgaris* on these reefs are mostly juveniles (namely, young-of-the-year, which account for 90% of the populations present), and they utilize these structures as nursery grounds. Subadults and adults are also present. With recruitment to the artificial reefs occurring in winter at a total length of 3 cm, it was possible (Leitão et al., 2009) to follow the seasonal growth of *D. vulgaris* throughout the year. This highlights the importance of artificial reefs for this species. Regarding *D. sargus,* their populations are composed chiefly of adults (98%), with the presence of subadults restricted to some autumn months (Leitão et al., 2009). High abundance of adults was recorded during their spawning season, suggesting that *D. sargus* use the reefs as mating/spawning areas. Moreover, all three species forage for food on artificial reefs and they feed mostly on sessile benthic invertebrates and algae. *D. labrax* is a highly valuable transient, demersal species. It had a peak abundance on the artificial reefs during winter months which corresponds to its spawning season. During these months, the population's size structure was composed mostly of adults, while juveniles dominated during the remaining months. These juveniles are regularly observed feeding on the artificial reef's macrobenthic community and small pelagic fish species (M. Santos pers. observation). Leitão et al. (2008b) reported that *D. labrax* adults are opportunistic, preying on juveniles (0 and 1 age classes) of several demersal fish species on the artificial reefs. Prey and *D. labrax* abundance were negatively correlated, with the mean numbers of prey per predator increasing with an increase in prey abundance. These results clearly indicate that energy is transferred from the artificial reef fish and macrobenthic assemblages to the seabreams and seabasses on reefs. Thus, the artificial reefs can be classified as Essential Fish Habitats for *Diplodus* spp. and *D. labrax* species. However, it is likely that these structures are acting as EFH for other species that show high site fidelity, such as wrasses, small groupers, and the scorpion fish, to these man-made structures. This, however, requires additional investigation to substantiate this inference.

Other Ecological Aspects

During underwater visual surveys, divers observed the swimming behavior of small pelagic and demersal schools of *Trachurus* spp., *B. boops*, and *P. acarne* (among other demersal) species around the structures. These filter-feeding and bottom-dwelling fishes were often observed feeding in the water column. Feeding behavior studies on *D. labrax* reveal that they do not direct feed on the benthic fauna and flora of artificial reefs (Leitão et al., 2008b). Normally, they predate on juvenile reef demersal species attracted to the man-made structures. Predation, therefore, can be assumed to be a significant feature for this species on artificial reefs, as the structure of the food web is dictated by the biodiversity within the system and/or by high-level predators (top-down control). Whether or not *D. labrax* predation on artificial reefs is similar to that in natural areas is unknown, but the addition of man-made substrate facilitates an increase in prey–predator encounters. Concomitantly, prey consumption and higher prey mortality (recruitment and stock size) increase with the deployment of artificial reefs. Fish prey–predator interactions are important in conservation and management, as well as for the evaluation of the long-term effects of artificial reefs deployment and for the understanding of fish assemblage's successional development.

For some species, artificial reefs can promote biomass increases (somatic growth) due to the production associated with an artificial reef (e.g., *D. labrax* and *D. sargus*). For other species, this may not be true (i.e., fish prey such as *Boops boops*). Results from the studies reviewed above highlight the fact that some demersal species associated with artificial reefs are a potential energy link between secondary production and higher trophic level biomass of commercial and recreationally important species such as the seabass.

Another important aspect raised during our previous research investigations was related to the relationship between fish and macrobenthic assemblages. By simultaneously sampling biota on artificial reefs deployed at different time periods (different days since deployment) in the same area, it was noted that reefs of different age have different fish assemblage density and biomass (Santos et al., 2011). Such fluctuations in fish abundance and biomass were attributed to the different benthic community composition of the artificial reefs.

Some artificial reef designs are deployed to mimic (at least at some degree) natural rocky areas. Rocky biotopes in intertidal areas are described as having reef fish communities with high resilience. This despite the fact that, rocky biotopes have distinctly different faunas in many parts of the world (Almada and Faria, 2004). Excluding seasonally fluctuations in abundance among demersal fishes and interannual peaks in Blennid recruitment described by Santos et al. (2005) and Leitão et al. (2008b), the Algarve artificial reef fish assemblage does not undergo marked fluctuations in species richness and density. This indicates that these man-made habitats serve to aggregate and stabilize fish assemblages as previously suggested by Santos (1997).

A number of stocking experiments using native species already available from local hatcheries have been conducted on the Algarve artificial reefs. These studies used mostly seabreams such as *Sparus aurata* (gilthead seabream), *D. sargus* (common seabream), *D. cervinus* (zebra seabream), and *D. vulgaris* (two-banded seabream). Preliminary results by Santos et al. (2006) showed that both *S. aurata* and *D. sargus* released at the artificial reefs do not stay associated with the artificial structures for a long time. These results were reinforced by study results reported as part of a telemetry study conducted by Lino et al. (2009). Here, Lino et al. (2009) showed that hatchery produced and reared fish have little site fidelity and no consistent daily movement patterns, with fish leaving the release site soon after release. However, dispersal among the hatchery produced and reared fish is mostly limited to neighboring areas (e.g., 6 and 10 nm for *S. aurata* and *D. sargus*, respectively). Wild *D. sargus* move from one habitat to the other, using the artificial reefs on a daily basis, with increased preference for the artificial habitat during day time (Lino et al., 2009). Recent studies (P.G. Lino unpublished) indicate that, after more than a year posttagging, wild *D. sargus* were still using the same area and associated with the same artificial reef, reinforcing the idea of a limited home ranging for this species.

CPUE, VPUE, and Artificial Reefs Use

The fishing yields from the artificial reefs off Portugal have continued to exceed those yields from the control sites in catch per unit of effort (CPUE), even when fished on a daily basis by the commercial fleets. For the first 5 years after the artificial reef systems deployment, the CPUE from a standard gillnet was 1.86–2.28 and 1.11–2.03 times higher than those from the control areas at Olhão and Faro, respectively (Santos and Monteiro, 1997 and 1998). Nine years later, the difference in CPUE over the 14-year period increased to 1.90–2.63 and 1.77–2.06 at Olhão and Faro, respectively (Santos and Monteiro, 2007). Demersal species such as sparids were the fishes that chiefly influenced these results. Interestingly, during the study period CPUE at the control sites decreased in the Olhão area but remained stable at Faro. The results from the monitoring program being conducted at the remaining artificial reefs along the Algarve coast showed similar results (M.N. Santos, unpublished data), although each of these reefs were at different stages of their successional process. Overall, both the protection and exploitation modules of the Algarve artificial reef complex let to increased fishing yield, but did not alter the balance (or the equilibrium) of the fish assemblages. This was because the relative proportion of the different functional groups of fish caught by the gillnets remained stable over the sampling period.

Whitmarsh et al. (2008) showed that catches at the artificial reefs would be expected to earn approximately €13 per unit of effort more than they would at the control sites. After a period of time, however, the value per unit of effort on the reefs (but not at the control sites) increases by €0.177 per month. These results demonstrate the economic benefits of artificial reefs by showing that the revenue from fishing is substantially higher at the reefs than at the control sites. Additionally, there is evidence that the productivity of the artificial reef systems has risen throughout the 15-year period of study (M.N. Santos, unpublished).

Fishers from the commercial fleet are the chief users of the artificial reef systems. Ramos et al. (2006a), showed that at Olhão artificial reef, both components of the commercial fleet fished this area, with the local fleet component (boat length <12 m) showing the highest frequency. Recreational vessels, despite being the second most numerous type of vessel were observed fishing least in the area. About 40% of the commercial fishing gears were set inside the artificial reef complex, with 39% being set on its boarder (<0.5 nm from the reef edge). Bottom set nets (gill and trammel nets) for finfish, traps and pots for octopus, and small purse seines for small pelagic and demersal finfish species are the most popular gears being used on the artificial reef systems. There has been an increase on the use of bottom longlines for seabreams by professional fishers in recent years. Recreational fishers make use mostly use of rod-and-reel, hand-lines, and jigs. Dive operators also make use the artificial reefs for their activities (which includes sites to conduct instructional courses and visits). Ramos et al. (2006b) reported that since the artificial reefs were deployed, recreational diving had increased in the Faro area by 7% (or by 22% if a vessel-reef sunken in 1995 is considered). On the western part of the Algarve coast, the artificial reefs were selected for 9% of total number of dives conducted by PORTISUB between 2005 and 2009 (J. Ramos, unpublished data).

Rebuilding Fisheries

Higher density and biomass of *D. labrax* and *D. sargus* were recorded after the enlargement of the Faro/Ancão artificial reef system. Currently, the exploitable biomass for these two species accounts for 75% (138 g/m^3 or 73 kg/reef set) of the total exploitable fish biomass (Leitão et al., 2008). Such results demonstrate that installment of large-scale artificial reef systems alters the fish assemblages present, namely by increasing the abundance of species which were previously uncommon or rare. From an economic perspective, it does not require a large number of individuals of highly prized (€12–18/kg) species, such as *D. labrax* and *D. sargus*, to be caught and sustain a small-scale fishery. In fact, the abundance of these species could promote the recovery of the local longline fishery, which had declined since the 1980s. Because of the technical developments in the fishing gears, artisanal fishermen have abandoned gears that require more man power (e.g., bottom longlines) in favor of other gears that are easier to operate and provide high yields (e.g., gillnets).

Some fish such as large seabreams and seabass became abundant in the Algarve artificial reefs areas. Interestingly, they were not targeted by the common gill/trammel nets (which make use of small mesh sizes), but the use of bottom longline gear could be profitable for local fishing communities for these fish. In fact, soon after the enlargement of the Faro/Ancão artificial reef (from 0.6 to 12.2 km^2), the local fishing community began using bottom longlines again. As a result, the landings of *D. labrax* and *D. sargus* show an increasing trend since 1995. Thus, one may conclude that the deployment of large-scale artificial reefs allowed the rebuilding of a traditional highly valuable fishery.

MANAGEMENT OF ARTIFICIAL REEF AREAS

The capacity of an artificial reef to attract, maintain, and facilitate production of exploitable biomass for fisheries off the Algarve coast as been demonstrated. It is clear that these structures have the potential to contribute positively to the economy of the region, notably by increasing the incomes of artisanal fishing communities, provided that the direct use benefits can be captured (Whitmarsh et al., 2008). As mentioned by several authors (Pitcher et al., 2002; Milon, 1989 and 1991) the economic benefits of artificial reefs may be potentially large. It is also generally accepted that if a commercial fleet benefits, it may be offset by an expansion in harvesting pressure that can lead to overexploitation. The risks of this happening are greater where the effect of a reef is purely aggregation rather than enhancement. Thus, as mentioned by Whitmarsh and Pickering (1997), in theory it can be shown that in an open-access fishery, stocks will be reduced when an artificial reef raises the efficiency of capture (e.g., by concentrating fish and making them more accessible to harvesting). Such undesirable effects can also affect other components of the artificial reef functioning. This was illustrated by Milon (1989) as the "paradox of artificial habitat development," namely that a technology which is biologically (and ecologically) effective may jeopardize the overall performance if access to the exploitable resources is not controlled. In socioeconomic terms, there is strong evidence that artificial reef deployment without a site-specific management plan for their exploitation cannot be successful (Milon, 1991; Grossman et al., 1997). Thus, management has been recognized as essential in order to guarantee the long-term performance of an artificial reef, both at the fisheries and environmental levels.

As mentioned by Whitmarsh et al. (2008), if the artificial reefs have the capability to increase economic rent, in an open-access scenario this will generate incentives that are likely to alter the level and pattern of fishing effort in a way that imposes additional demands on the regime of management controls. As a consequence, it is important to have a monitoring system in place that accounts for both the harvesting activities on the reef and the ecological succession of the reef-associated assemblages. Fishing activity can be easily monitored, taking advantage of recent developments in tracking systems (e.g., GPS). These new tracking systems allow responsible agencies to record and analyze in real-time data on geographic location and vessel speed. Once these data are recorded, it is possible to create algorithms that allow knowing the fishing gear being used and the fishing effort. Such a scheme, which also involves access to the landing data and sampling on port, are currently being tested by IPIMAR. Self-reporting is another important component of the monitoring system, as the engagement of stakeholders is essential, as public fisheries authorities alone cannot support the costs for monitoring systems. As regards the natural assemblage succession on artificial reefs, it is desirable that a monitoring program include as many as possible reef assemblages, rather than just those that are being harvested. Although most studies only have one or two control sites, it is of utmost importance having more than 10 control sites.

To have higher chances of success, an artificial reef has to be useful to people (Milon et al., 2000). In this sense, well-defined policy objectives are essential. These objectives refer to the potential artificial reefs have in providing different services, and the involvement stakeholders have in the process is of fundamental importance (Whitmarsh et al., 2008). The application of clear objectives to artificial reef projects represents a unique opportunity to progressively promote stakeholders co-responsibility on artificial reef management and, particularly, to engage the small-scale commercial fleet on management issues, which is not a tradition in Portuguese fisheries policy. As mentioned by Whitmarsh et al. (2008), several management options are possible, but the nature of reef-based fisheries would seem to lend themselves to the establishment of property or user rights, in which local fishing communities would be co-responsible with government agencies for regulating access and for monitoring the fisheries. The exploitation strategies must be goal oriented, to allow optimal fishing yields, while avoiding disruptions in the natural succession of the artificial reefs and associated assemblages. It is important to assure that the fishing effort will be spread over all the fish communities, rather than focusing on target species, to avoid irreversible disruptions in the equilibrium and organization of the different functional groups of fish. The exploitation strategies must include the use of different fishing gears, thereby diversifying the catches and, consequently, exploiting all the available resources. Moreover, gear diversification is expected to contribute to the distribution of the fishing effort over different fish sizes (or ages) of the assemblage's, in order to reduce the risk of truncated size (or age) distributions and the decrease of the fish abundance.

Studies on artificial reefs have focused predominantly on fish assemblages and have largely disregarded the development of sessile biota and their structural and functional relationships (Svane and Petersen, 2001). One important argument that highlights the necessity of conservation of high-quality habitats is the recognition of the value of different habitats for determining the structure of exploited fish assemblages. Despite the evidences of important ecological linkages between habitat(s) and fishery production, the management of most commercial resources worldwide has historically concentrated on assessing stock size and controlling fishing mortality. Therefore, to maintain a sustainable fisheries industry, it is important that managers regulate not only the total allowable catch and/or fishing effort, but also to have in mind the role of shallow coastal habitats on the production of fish to the fisheries, that is, the ecosystem services provided by the habitats and thus also regulate society's exploitation of the coastal environment (Costanza et al., 1997; Stål et al., 2007). It is most probable that in the future, fisheries management will focus a great effort directly on the conservation of coastal environments, which are essential fish habitats. This also applies to the near shore artificial reef areas that are known for supporting a high diversity of species compared to offshore areas, offering organisms multiple opportunities (habitats) and food resources. Consequently, management necessarily needs to preserve the entire ecosystems, aiming the long-term sustainability of the fisheries, instead of focus particularly on fish harvesting (ecosystem-based management).

Finally, applied research is a key element in any artificial reef management program, providing assistance in data analysis, helping on monitoring and facilitating the links between the different stakeholders. Therefore, it is important to guaranty the necessary funding in order that reef scientists can be engaged on artificial reef programs, providing the necessary guidance to fisheries and environmental managers.

FINAL COMMENTS

The Algarve artificial reef program has been an opportunity to improve our knowledge in several aspects of marine and fisheries sciences. New findings and interesting results have been achieved, but there are additional steps. Hopefully, new opportunities will arise in terms of research that advances the science and management of marine resources and activities.

The artificial reefs have been fairly considered as important elements (or tools) of integrated coastal fisheries management programs. We believe that the interest in using artificial reef technology will grow worldwide, as a complementary management tool. An example of this is the identification of artificial reefs as essential fish habitats, which can be an important step towards the creation of MPAs based on artificial reef deployment.

The role of artificial reefs will be reinforced in the near future, taking into account the growing need to integrate fisheries and environmental policies into fisheries management. On the other hand the new concepts of governance to be applied on the coming years in fisheries and environmental management will bring artificial reefs to a level they have not attained previously. In Europe, the *Marine Strategy Framework Directive* is imposing restrictions to fishing whenever the established indicators of *Good Environmental Status* do not satisfy the standards. Moreover, concepts such as biological biodiversity, marine food webs, and status of commercially exploited fish and shellfish among others that are highly linked with the fisheries, will assume a much stronger role on spatial planning of coastal areas and, consequently, on new regulations for the fishing activities. Currently, we are facing a lack of effective options to apply integrated management (ecosystem approach) into fisheries. However, there is clear evidence on what can be the role of artificial reefs on the effective implementation of such policies directed toward the sustainability of fisheries and environment. Thus, it is up to managers to deploy effective artificial reef programs and establish clear management strategies, based on the best available artificial reef science. The potential of these man-made structures should also be used as the trigger to definitely involve the fishing sector on participative and co-responsible based management, towards the new governance strategies based on bottom-up inputs. However, as mentioned by S. Bortone (personal commuication) if artificial reefs are to be a management option for altering some aspect of a fishery, they should be designated, regulated, and prescribed in a way that allows an assessment of their effectiveness.

The results present herein on the Portuguese experience on the use of artificial reef technology is, to the authors best knowledge, one of the few suitable examples of the usefulness of such technology for the management of artisanal fisheries, as it is based on a long-term scientific monitoring scheme and experimentally proved. We hope it can help change the skepticism of many fisheries managers to artificial reef technology. But scientists must be ready to give fishery managers the information they need to justify this technology in their management plans. Moreover, as mentioned by S. Bortone (personal comunication), managers and scientists should direct their efforts and studies toward the management of only a few selected species instead of the entire fish assemblage.

ACKNOWLEDGMENTS

This chapter summarizes much of the work that has been conducted by a very active research team at IPIMAR-Olhão. We express our acknowledgments for the dedication and enthusiasm of all team members (J. Ramos, A. Moura, P.G. Lino, J. Cúrdia, S. Carvalho, M. Cerqueira, M. Falcão, M.B. Gaspar, M. Vicente, and D. Serpa). Thanks are also due to our technicians (T. Simões, J.L. Sofia, M.L. Santos, L. Oliveira, and E. Brás); the crews of the research vessels *Donax* and *Diplodus*; as well as colleagues from academia (D. Whitmarsh, K. Erzini, L. Cancela da Fonseca, and D. Boaventura). The Algarve artificial reef deployment program has been partially (75%) supported by a number of EU funding programs (PIDR, PROPESCA, and MARE) and by national funds.

REFERENCES

Almada, V. and C. Faria. 2004. Temporal variation of rocky intertidal resident fish assemblages—patterns and possible mechanisms with a note on sampling protocols. *Rev. Fish Biol. Fish.* 14 (2): 239–250.

Anonymous, 2009. Datapescas: Janeiro a Dezembro 2008. *DGPA* Difusão; 79: 12 pp.

Arce, A.M., Aguilar-Davila, W., Sosa-Cordero, E., and J.F. Caddy. 1997. Artificial shelters (casitas) as habitats for juvenile spiny lobster (*Panulirus argus* Latreille) in the Mexican Caribbean. *Mar. Ecol. Progr. Ser.* 218: 217–224.

Benaka, L.R. 1999. Summary of panel discussions and steps toward an agenda for habitat policy and science. Pages 455–459 in L.R. Benaka, Ed. Fish Habitat: Essential Fish Habitat and Rehabilitation, *Proceedings Sea Grant Symposium of the American Fisheries Society 22*, Bethesda, Maryland.

Boaventura, D., Moura, A., Leitão, F., Carvalho, S., Cúrdia, J., Pereira, P., Cancela da Fonseca, L., Santos, M.N., and C.C. Monteiro. 2006. Macrobenthic colonisation of artificial reefs on the southern coast of Portugal (Ancão, Algarve). *Hydrobiologia* 555: 335–343.

Bombace, G., Fabi, G., and L. Fiorentini. 2000. Artificial reefs in the Adriatic Sea. Pages 31–63 in A. Jensen, K. Collins, and A. Lockwood, Eds. *Artificial Reefs in European Seas*. Kluwer Academic Publishers, London.

Briones-Fourzán, P. and E. Lozano-Álvarez. 2001. Effects of artificial shelters (Casitas) on the abundance and biomass of juvenile spiny lobsters *Panulirus argus* in a habitat-limited tropical reef lagoon. *Mar. Ecol. Prog. Ser.* 221: 221–232.

Briones-Fourzán, P., Lozano-Álvarez, E., Negrete-Soto, F., and C. Barradas-Ortiz. 2007. Enhancement of juvenile Caribbean spiny lobsters: An evaluation of changes in multiple response variables with the addition of large artificial shelters. *Oecologia* 151 (3): 4001–416.

Costanza, R., D'Arge, R., de Groot, R., Farber, S., Grasso, M., Hannon, B., Limburg, K., Naeem, S., O'Neill, R.V., Paruelo, J. et al. 1997. The value of the world's ecosystem services and natural capital. *Nature* 387: 253–260.

Cruz, R. and B.F. Phillips. 2000. The artificial shelters (Pesqueros) used for the spiny lobster (*Panulirus argus*) fisheries in Cuba. Pages 400–419 in B.F. Phillips and J. Kittaka, Eds. *Spiny Lobsters: Fisheries and Culture*, 2nd ed. Fishing News Books, Oxford.

Fernández, T.V., D'Anna, G., Badalamenti, F., and A. Pérez-Ruzafa. 2008. Habitat connectivity as a factor affecting fish assemblages in temperate reefs. *Aquat. Biol.* 1: 239–248.

Grossman, G.D., Jones, G.P., and W.J. Seaman Jr. 1997. Do artificial reefs increased regional fish production? A review of existing data. *Fisheries* 22 (4): 17–27.

Harmelin-Vivien, M.L., Harmelin, J.G., and V. Leboulleux. 1995. Microhabitat requirements for settlement of juvenile sparid fishes on Mediterranean rocky shores. *Hydrobiologia* 301: 309–320.

Leitão, F., Santos, M.N., and C.C. Monteiro. 2007. Contribution of artificial reefs to the diet of the white seabream, *Diplodus sargus* (Linnaeus, 1758). *ICES J. Mar. Sci.* 64 (3): 473–478.

Leitão, F., Santos, M.N., Erzini, K., and C.C. Monteiro. 2008a. Fish assemblage and rapid colonisation after enlargement of an artificial reef off Algarve (Southern Portugal). *Mar. Ecol.* 29 (4): 435–448.

Leitão, F., Santos, M.N., Erzini, K., and C.C. Monteiro. 2008b. Effect of predation on artificial reef juvenile demersal fish species. *Mar. Biol.*153 (6): 1233–1244.

Leitão, F., Santos, M.N., Erzini, K., and C.C. Monteiro. 2009. Artificial reef use by *Diplodus* spp. species: Importance for artisanal near shore fisheries. *Fish. Manag. Ecol.* 16 (2): 88–99.

Levin, P.S. 1994. Fine-scale temporal variation in recruitment of a temperate demersal fish: The importance of settlement versus post-settlement loss. *Oecologia* 97 (1): 124–133.

Lino, P.G., Bentes, L., Abecasis, D., Santos, M.N., and K. Erzini. 2009. Comparative behavior of wild and hatchery reared white seabream (*Diplodus sargus*) released on artificial reefs off the Algarve (Southern Portugal). Pages 23–34 in J.L. Nielsen, H. Arrizabalaga, N. Fragoso, A. Hobday, M. Lutcavage, and J. Sibert, Eds. *Tagging and Tracking of Marine Animals with Electronic Devices*. Rev. Method. Techn. Fish Biol. Fish. 9.

Macpherson, E. 1998. Ontogenetic shifts in habitat use and aggregation in juvenile sparid fishes. *J. Exp. Mar. Biol. Ecol.* 220 (1): 127–150.

Milon, J.W. 1989. Economic evaluation of artificial habitat for fisheries: Progress and challenges. *Bull. Mar. Sci.* 44 (2): 831–843.

Milon, J.W. 1991. Social and economic evaluation of artificial aquatic habitats. Pages 237–270 in W. Seaman and L.M. Sprague, Eds. *Artificial Habitats for Marine and Freshwater Fisheries*. Academic Press, San Diego, California.

Milon, J.W., Holland, S.M., and D. Whitmarsh. 2000. Social and economic evaluation methods. Pages 165–194 in W. Seaman, Eds. *Artificial Reef Evaluation with Application to Natural Marine Habitats*. CRC Press, Boca Raton, Florida.

Monteiro, C., Lam Moi, T., and G. Lassèrre. 1987. Distribution chronologique des poissons dans deux station de la lagune Ria Formosa (Portugal). *Oceanolog. Acta* 10 (3): 359–371.

Monteiro, C., Lassèrre, G., and T. Lam Hoi. 1990. Organisation spatiale des communautés ichtyologiques de la Lagune Ria Formosa (Portugal). *Oceanolog. Acta* 13 (1): 79–96.

Monteiro, C. and M.N. Santos. 2000. Portuguese artificial reefs. Pages 249–262 in A.C. Jensen, K.J. Collins, and A.P.M. Lockwood, Eds. *Artificial Reefs in European Seas*. Kluwer Academic Publishers, London.

Pitcher, T.J., Watson, R., Haggan, N., Guenette, S., Kennish, R., Sumaila, R., Cook, D. Wilson, K., and A.,Leung. 2000. Marine reserves and the restoration of fisheries and marine ecosystems in the South China Sea. *Bull. Mar. Sci.* 66:543–66.

Pitcher, T.J., Buchary, E.A., and T. Hutton. 2002. Forecasting the benefits of no-take human-made reefs using spatial ecosystem simulation. *ICES J. Mar. Sci.* 59: S17–S26.

Polovina, J.J. 1991. Fisheries applications and biological impacts of artificial habitats. Pages 153–176 in W. Seaman and L.M. Sprague, Eds. *Artificial Habitats for Marine and Freshwater Fisheries*. Academic Press, San Diego, California.

Ramos, J., Santos, M.N., Whitmarsh, D., and C.C. Monteiro. 2006a. Patterns of use in an artificial reef system: A case study in Portugal. *Bull. Mar. Sci.* 78(1): 203–211.

Ramos, J., Santos, M.N., Whitmarsh, D., and C.C. Monteiro. 2006b. The usefulness of the analytic hierarchy process to understand reef diving choices: A case study. *Bull. Mar. Sci.* 78(1): 213–219.

Relini, G., Fabi, G., Santos, M.N., Moreno, I., and E. Charbonnel. 2008. Fisheries and their management using artificial reefs in the Northwestern Mediterranean Sea and Southern Portugal. *Amer. Fish. Soc. Symposium* 49: 891–898.

Sale, P.F. 1969. A suggested mechanism for habitat selection by the juvenile manini, *Acanthurus triostegus sanduicensis* streets. *Behaviour* 35: 27–44.

Santos, M.N. 1997. Ichthyofauna of the Artificial Reefs of Algarve Coast. Exploitation Strategies and Management of the Local Fisheries. Ph.D. thesis, Univ. Algarve, Faro, Portugal: 268 pp.

Santos, M.N. and C.C. Monteiro. 1997. Olhão artificial reef system (south Portugal): Fish assemblages and fishing yield. *Fish. Res.* 30: 33–41.

Santos, M.N. and C.C. Monteiro. 1998. Comparison of the catch and fishing yield from an artificial reef system and neighbouring areas off Faro (south Portugal). *Fish. Res.* 39: 55–65.

Santos, M.N. and C.C. Monteiro. 2007. A fourteen-year overview of the fish assemblages and yield of the two oldest Algarve artificial reefs (southern Portugal). *Hydrobiologia* 580: 225–231.

Santos, M.N., Monteiro, C.C., and K. Erzini. 1995. Comparison of natural and artificial reef fish assemblages in Algarve waters (south Portugal). Proc. ECOSET'95, Vol. 1: 210–214.

Santos, M.N., C.C. Monteiro, and G. Lassèrre. 2005. Observations and trends on the intra-annual variation of the fish assemblages on two artificial reefs in Algarve coastal waters (Southern Portugal). *Sci. Mar.* 69 (3): 415–426.

Santos, M.N., Lino, P.G., Pousão-Ferreira, P., and C.C. Monteiro. 2006. Preliminary results of hatched-reared seabream release at artificial reefs off the Algarve coast (southern Portugal): A pilot experiment. *Bull. Mar. Sci.* 78 (1): 77–84.

Santos, M.N., Leitáo, F., Lino, P.G., Moura, A., Cerqueira, M., and C.C. Monteiro. 2011. Diplodus spp assemblages on artificial reefs of different ages: Influence of the associated epibenthic macrofauna. *ICES Journal of Marine Research* 68(1): 87–97.

Sosa-Cordero, E., Arce, A.M., Aguilar-Dávila, W., and Ramírez-González, A. 1998. Artificial shelters for spiny lobster *Panulirus argus* (Latreille): An evaluation of occupancy in different benthic habitats. *J. Exp. Mar. Biol. Ecol.* 229 (1): 1–18.

Stål, J., Pihl, L., and H. Wennhage. 2007. Food utilization by coastal fish assemblages in rocky and soft bottoms on the Swedish west coast: Inference for identification of essential fish habitats. *Estuarine Coastal Shelf Sci.* 71 (3–4): 593–607.

Svane, I.B. and J.K. Petersen. 2001. On the problems of epibiosis, fouling, and artificial reefs: A review. *Mar. Ecol.* 22 (3): 169–188.

Whitmarsh, D.J. and H. Pickering. 1997. Commercial Exploitation of Artificial Reefs: Economic Opportunities and Management Imperatives. University of Portsmouth, CEMARE Research Paper 115: 20 pp.

Whitmarsh, D.J., Santos, M.N., Ramos, J., and C.C. Monteiro. 2008. Marine habitat modification through artificial reefs off the Algarve (Southern Portugal): An economic analysis of the fisheries and the prospects for management. *Ocean Coast. Manag.* 51 (6): 463–468.

Wilson, K.D.P., Leung, A.W.Y., and R. Kennish. 2002. Restoration of Hong Kong fisheries through deployment of artificial reefs in marine protected areas. *ICES J. Mar. Sci.* 59: S157–S163.

CHAPTER 15

Artificial Reef Function in Fishing Grounds off Japan

Yasushi Ito

CONTENTS

ABSTRACT

In 1952, a project to create fishing grounds off Japan was begun using artificial reefs composed of stone. Subsequently, the Coastal Fishing Ground Improvement and Development Law was enacted in 1974, and this artificial-reef fishing area came under control of the social infrastructure of public works. Since 1974 to 2001, development and improvement projects of fishing grounds include about 20,000 sites in Japan. From 2007, under the direct control of the government, a large-scale construction project began with an objective of providing protective-nursery reefs for snow crabs within a larger goal of managing crab resources. The outline of the projects related to the creation of fishing grounds is as follows:

1. The artificial reef construction project targeted mainly fishes. A broad variety of artificial reefs were constructed, from low relief to those over 30 meters tall. Features to enhance the functions as nurseries were added later.
2. The addition of nursery features through a nursery works construction project consisted of two types of nurseries. One type was directed toward littoral marine species, developing seaweed beds, and tidal lands in shallow water to serve as nurseries for abalone, clams, lobsters and kelp, etc. The second type was directed toward the creation and development of spawning and rearing grounds for fry and fishes that migrate to other areas. These included sea bream, flatfish, and squids, etc.
3. Other project objectives included: (1) aquaculture ground construction to create new habitat for aquaculture, (2) fishing ground preservation to recover function of fishing ground, and (3) artificial upwelling mound construction to generate upwelling, etc.

CREATING FISHING GROUNDS IN JAPAN

History of Fishing Grounds Creation

The marine waters around Japan comprise one of the world's largest fishing grounds controlled by a single nation. Because of the productivity of these waters, it has high potential as an expansive fishing ground. In addition, the exclusive economic zone is 12 times the size of Japan, making it the sixth largest fishing ground under management by a single nation in the world. Japan has had a long history with the construction and deployment of artificial reefs. The oldest artificial reef on record was created in 1650. Early in Japan's history, fishermen noted that fish gathered around sunken ships. Consequently, fishermen sank old ships and rocks to actively build artificial reefs. The creation of fishing grounds through the deployment of artificial reefs has been continued since 1952 as part of a national project. Later, beginning in 1976, these projects were conducted as part of a government sponsored activity. In 2001, the creation of fishing grounds through artificial reef deployment was initiated cooperatively between the national government and local fishing ports. In 2007, these projects were under the direct control of the government.

Prototype of an Artificial Reef (Stone Bed)

One of the earliest forms of artificial reef construction off Japan consisted of the deployment of stones or large rocks to forms beds. Figure 15.1 shows the structure of an old artificial fishing reef. Rocks were piled in rivers and fish were caught among the rocks.

Fishing Grounds Creation

There are two major objectives that guide the creation of fishing grounds based on artificial reefs. The first objective is to enable efficient fishing by the fishing industry. The second is to increase fishery resources by creating an environment conducive to the enhancement of aquatic life. The general method followed to achieve these objectives was, first, construct artificial reefs as fishing grounds to facilitate catching fish when fish aggregate around the structures. Second, construct artificial reefs in propagation areas to enhance fish reproduction and growth in early fish life stages.

Figure 15.1 Rock piles serving as artificial reefs.

Fishing Grounds Creation with Artificial Reefs

Types of Artificial Reefs

Artificial reefs come in various forms, and there are over 100 different reef types used off Japan. Some examples of these reefs that have been deployed are indicted below.

1. **Artificial reefs made of concrete.** Figure 15.2 shows a single artificial reef made of concrete. This kind of artificial reef is individually constructed and is installed on the sea bottom using a construction crane. The materials are not merely dropped from above but are purposefully placed in a designated location and arrangement. To assure the return on the investment to deploy these reefs, there are standards for durability as these artificial reefs are configured so that they will have a duration of 30 years or more.
2. **Prefabricated artificial reefs made of concrete (5–15 meters high).** Prefabricated artificial reefs are most often made of concrete (Figure 15.3). Initially these reefs were built on site but subsequently, the construction took place in a factory yard and these were later transported to an offshore site for deployment.

Figure 15.2 Artificial reef modules made of concrete.

Figure 15.3 Prefabricated artificial reef made of concrete.

Figure 15.4 Artificial reef made of steel.

3. **Artificial reefs made of steel.** As with artificial reefs made of concrete, reefs composed of steel (Figure 15.4) are designed and built so that they last for 30 years, even if they become corroded. Corrosion is taken into consideration when constructing these kinds of reefs.
4. **High-rise artificial reefs.** Beginning around 2000, improvements were made in installation techniques. This allowed for the construction and deployment of larger, high-rise artificial reefs. More recently, steel artificial reefs have been built that are more than 20 meters high (Figure 15.5). The photograph on the left in Figure 15.5 shows that the reef is a compound-type reef made from both concrete and steel.
5. **Artificial reefs using wood or shells.** Figure 15.6 shows artificial reefs used for propagation (i.e., reproduction and production) that are made from wood and shells. Artificial reefs made of several different materials appeal to those responsible for reef building as the reefs can make effective use of waste material.
6. **Fish aggregating at artificial reefs.** Different fish species may aggregate or gather around artificial reef structures based on the structure and location of the reef. Figure 15.7 shows examples of fish species that are attracted to and gather around artificial reefs.

Figure 15.5 High-rise artificial reefs.

Figure 15.6 Artificial reefs composed of wood in a concrete base (left) and shells (right) placed in baskets.

Figure 15.7 Fish gathered at an artificial reef off Japan (left: chicken grunt, center: striped beakperch; right: goldeye rockfish).

Example of a Propagation Ground Artificial Reef for a Nursery

Although mentioned previously, the objective of constructing propagation grounds is to enhance reproduction of the associated fishes and other marine life in the area. Off Japan, some of these areas dedicated as propagation grounds are also used as fishing grounds.

1. **Construction of a seaweed-bed nursery ground for young fish: Artificial reefs for abalone and sea urchins (0–20 meters depth).** Seaweed beds are constructed by placing blocks and rocks in places where there is no seaweed in shallow, sandy areas (Figure 15.8). Seaweed beds can serve as nursery grounds for young fish. They also provide habitat for abalone, sea urchins, and young fish. These seaweed beds produce seaweeds that can also be consumed by humans.
2. **Nursery home for Japanese spiny lobster (0–30 meters depth).** Artificial reefs that serve to help propagate Japanese spiny lobsters are depicted in Figure 15.9. Japanese spiny lobsters have an affinity for crevices that take up residence in sandy areas. These areas are propagation grounds for all benthic life stages of Japanese spiny lobsters, from young to adults.
3. **Nursery reef for spawning octopus (0–60 meters depth).** Artificial reefs can be designed to function as a nursery reef for spawning octopus (Figure 15.10). Octopus homes are created by using unglazed, ceramic pots.
4. **Nursery reef for spawning spear squid (10–30 meters depth).** Artificial reefs can provide a propagation area of squid as well. Figure 15.11 shows a nursery reef for spawning spear squid. Since spear squid spawn in areas such as ledges, spawning areas are created artificially to increase the amount of this preferred spawning substrate.
5. **Protective developmental reef for snow crab (200–300 meters depth).** Artificial reefs can serve as protective reefs for snow crabs in deep water (Figure 15.12). These reefs protect snow crabs from

Figure 15.8 Artificial reefs composed of seaweed bed for young fish (left), abalone (center), and sea urchins (right).

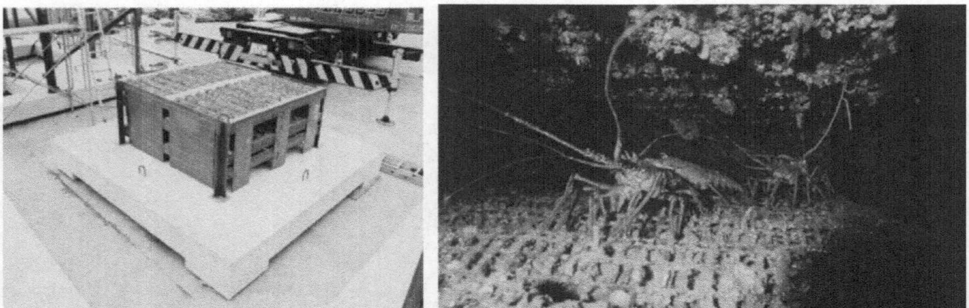

Figure 15.9 Nursery "home" for Japanese spiny lobster.

Figure 15.10 Nursery reef for spawning octopus.

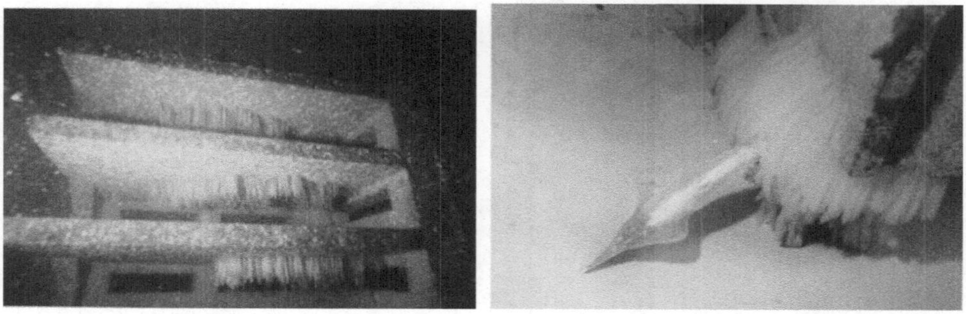

Figure 15.11 Nursery reef for spawning spear squid.

Figure 15.12 Protective developmental reef for snow crab.

trawlers and are used to help achieve the objective of resource management. The construction of protective development reefs has been implemented since 1990. Since 2007, these reefs have been under the direct control of the Japanese national government in its exclusive economic zone.

6. **Artificial upwelling fishing ground (60–150 meters depth).** Figure 15.13 diagrammatically depicts an artificial reef project that includes a concept that differs from that of previous artificial reefs. Artificial upwelling fishing grounds use technology where an artificial mound is deployed on the ocean bottom and the nutrients in the bottom layers of the ocean are forced to upsurge (upwell) based on the driving force of the current. By bringing nutrients into the euphotic zone, the ocean becomes fertilized, and the development of phytoplankton and zooplankton is promoted. The use of artificial reefs to promote upwelling in fishing grounds is a relatively new approach to increase fishery productivity.

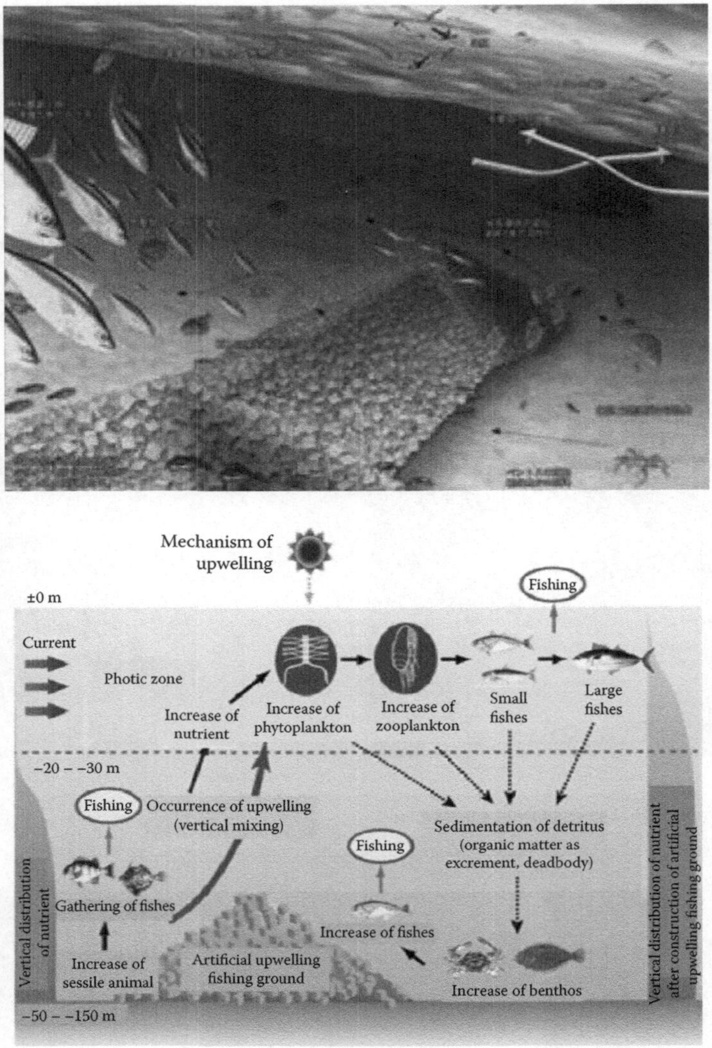

Figure 15.13 Depiction of artificial reefs used to induce upwelling on a fishing ground.

STUDY OF ARTIFICIAL REEF FUNCTIONS

Following are presented research results conducted by the author. A large-scale diagram of the general study area and relationships among the studies is presented in Figure 15.14. The first project involves the use of artificial reefs to enhance the propagation of marbled sole, *Pleuronectes yokohamae* (Gunther, 1877) in extremely shallow waters (i.e., 5 to 10 meters deep). The second project involves the swarming behavior of migratory jack mackerel, *Trachurus japonicus* (Temminck and Schlegel, 1844), around high-rise fishing reefs. The third project examines the protective effects of artificial reefs for snow crabs, *Chionoecetes opilio* (Fabricius, 1788), in deep-sea areas.

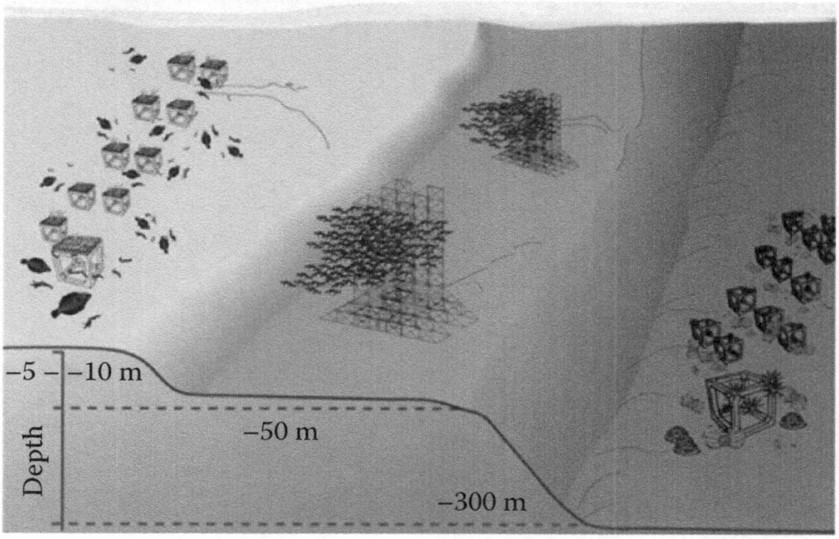

Figure 15.14 Artificial reef serving several functions depending upon location (inshore, marbled sole; midwater, jack mackerel; deepwater, snow crab).

Reef Affinity and "Feeding Ground" Effect for Marbled Sole (5–10-Meter Depth)

Below, the reef affinity and "feeding ground" effect for marbled sole (Figure 15.15) is examined. This species serves as an example of artificial reef use off Japan for the shallow water, inshore fisheries.

Methods

The study location was at a water depth of 5 to 10 meters, in an enclosed inner bay called the Seto Inland Sea (Figure 15.16). The reef module that served as the basis for the artificial reef used in this study is depicted in Figure 15.17 (left panel). The substrate of the reef was filled with shells (Figure 15.17, right panel). These shells served to provide substrate for potential prey for marbled sole. Thirty of these reef modules were deployed at this study location.

Results

1. **Distribution of prey animals around the artificial reef.** Figure 15.18 shows a generalize view of the artificial reef at the conclusion of this project. Around the artificial reef, there are phytal animals, sessile animals, and benthic animals. Phytal animals (phytal animals live on the seaweed or seagrasses, i.e., annelida, arthropoda, etc.) often fall to the substrate and, consequently, the number of benthic animals increases. Marbled sole congregate around the artificial reef to feed. In this study area, 30 of these artificial reef modules were deployed, each 2 meters apart from other modules.
2. **Biomass of seaweed and phytal animals.** The graph presented in Figure 15.19 displays the time associated trends in the standing biomasses of phytal animals and attached epiphytic seaweed (*Sargassum*) on the artificial reef modules. Generally, as found by Ito et al. (2008a), the amount of phytal animals is proportionate to the standing biomass of *Sargassum*.
3. **Composition of phytal animals (annual average).** Among phytal animals, Malacostraca (lobsters) account for more than 80% (Figure 15.20). Ito et al. (2008a) indicated that the guts of phytal animals are full of prey.

Figure 15.15 Marbled sole associated with an artificial reef off Japan.

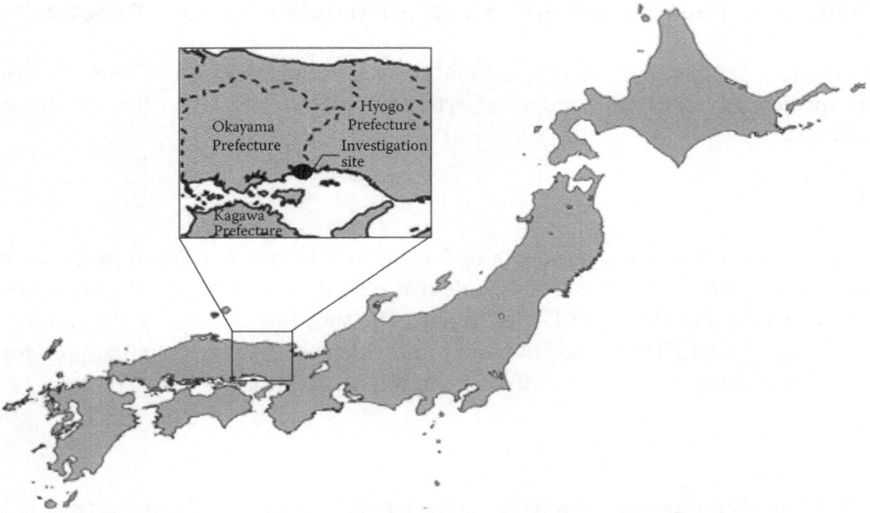

Figure 15.16 Investigation site for feeding study on marbled sole off Japan.

4. **Composition of sessile animals (annual average).** Figure 15.21 shows the gut content composition of sessile animals based on percent biomass. Sessile animals are prey items for many fish species and these sessile animals normally live in the crevices of shells and on the surfaces of reefs. In contrast to phytal animals, the percentage of arthropods (i.e., Malacostraca) among sessile was small on these artificial reef (Ito et al., 2008a).
5. **Biomass and composition of macrobenthos (annual average).** Figure 15.22 presents the composition of macrobenthos at distances of 0, 5, and 10 meters from the artificial reef. As shown (Figure 15.22), the most macrobenthos, by volume, lives 0 meters from the artificial reef. Presumably, this indicates that phytal animals and sessile animals have fallen from the vegetation or have moved next to the reef. In addition, the abundance of macrobenthos next to the reef also indicates that the

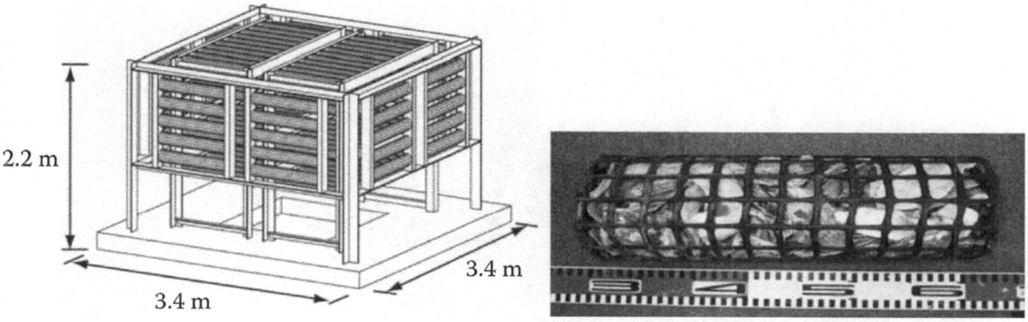

Figure 15.17 Illustration of the artificial reef (left) and shell substrate (right) for the marbled sole study.

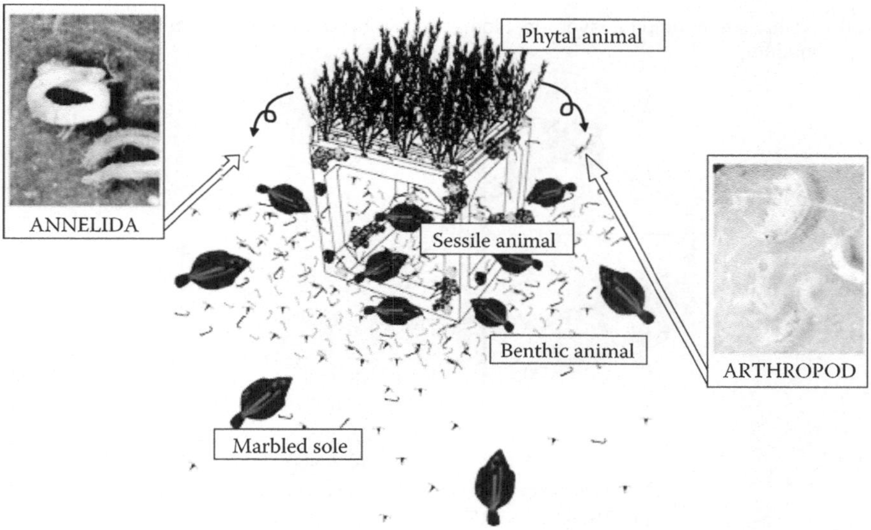

Figure 15.18 Distribution of prey animals for marbled sole around the artificial reef module.

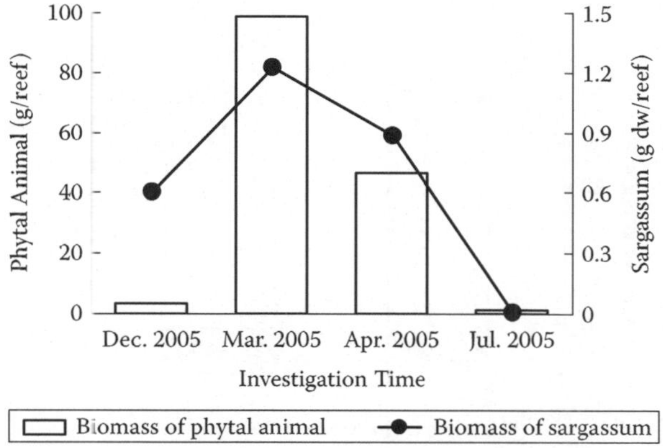

Figure 15.19 Graph indicating the relationship between seaweed biomass and phytal-associated animals relative to time.

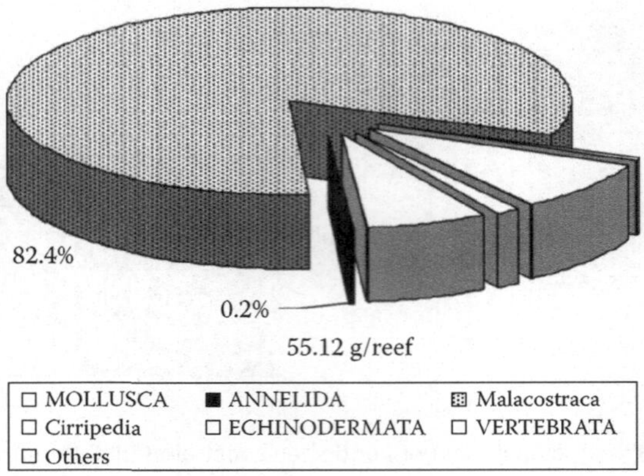

Figure 15.20 Composition (annual average biomass) of phytal-associated animals on the artificial reef modules.

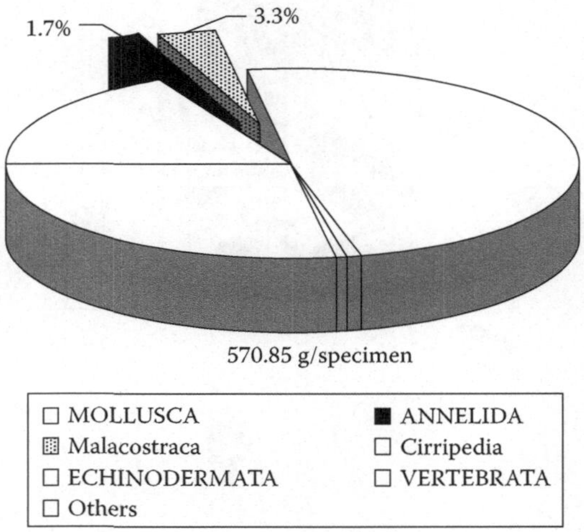

Figure 15.21 Composition (annual average biomass) of sessile animals associated with the artificial reef modules.

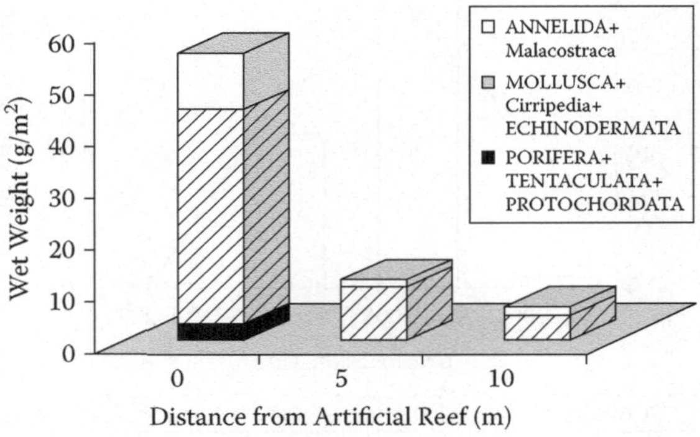

Figure 15.22 Annual average biomass and composition of macrobenthos associated with the artificial reef modules.

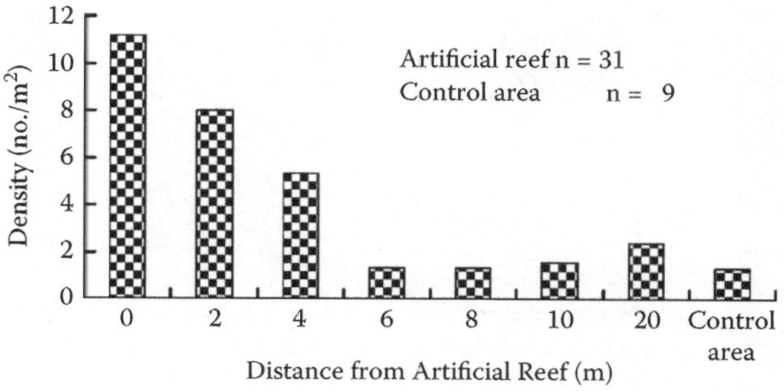

Figure 15.23 Horizontal distribution of juvenile marbled sole from the artificial reef modules.

habitat may have improved due to changes to the sediment caused by the flow of ocean water (Ito et al. 2008a).

6. **Horizontal distribution of juvenile marbled sole.** The density of juvenile marbled sole is highest at 0 m away from the artificial reef (Figure 15.23), and decreases in proportion to their distance from the artificial reef. At distances 0 to 5 m from the artificial reef, populations of juvenile marbled sole are significantly more dense that those more peripheral to the reef. At distances from 6 meters away from the artificial reef or more, the density of juvenile marbled sole is the same as densities of sole in the control area (Ito et al., 2008b).
7. **Food habits of fish associated with artificial reefs in 2005.** Table 15.1 presents data on the relative abundance of benthic animals in the digestive organ of marbled sole relative to their distance from the reef. Arthropods, which accounted for 45% of individuals in the digestive organ of marbled sole, were concentrated in an area 0 meter from the fishing reef. This indicates that marbled sole feed near artificial reefs and may make use of artificial reefs as feeding grounds (Ito et al., 2008b).
8. **Growth of Juvenile Marbled Sole.** Juvenile marbled sole were small in length and weight at the beginning of March 2006 (Figure 15.24). In April, they begin to grow rapidly and by mid-April they increased both in weight and length (Ito et al., 2008b). The average total length of juvenile marbled sole is compared from the artificial reef area and the control area in Figure 15.25. Size comparisons were made for 2 years (2008–2009), and it is clear that juvenile marbled sole from the artificial reef area grew faster. This indicates that deploying an artificial reef improves the growth among sole,

Table 15.1 Comparison between Composition in Digestive Organ Contents and the Distribution of Benthic Animal at the Artificial Reef Area (in 2005)

			Number of Macrobenthos (no./m²)			
		Composition of Digestive Organ Contents (%)	Artificial Reef Distance			Control Area
Taxa			0 m	10 m	20 m	Average
Mollusca	Gastropoda	2.3	60	20	40	90
	Pelecypoda	4.5	920	30	40	123
Annelida	Syllidae	6.8	0	0	0	0
	Spionidae	13.6	20	100	90	80
	Cirratulidae	2.3	360	10	40	17
Arthropoda	Mysinae	2.3	0	0	0	0
	Synchelidium sp.	2.3	0	10	10	3
	Pontogeneia sp.	4.5	10	0	0	0
	Corophium sp.	9.1	0	0	0	0
	Gammaridea	45.5	1,370	0	0	17
Others		6.8				
Total		100.0				

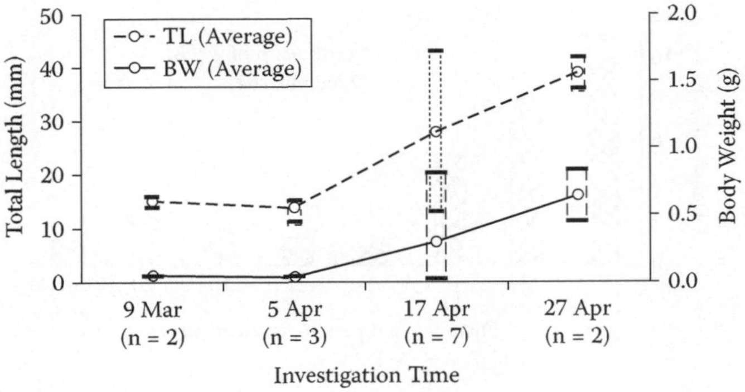

Figure 15.24 Growth (total length (TL) and biomass (BW)) of juvenile marbled sole on artificial reefs off Japan in 2006).

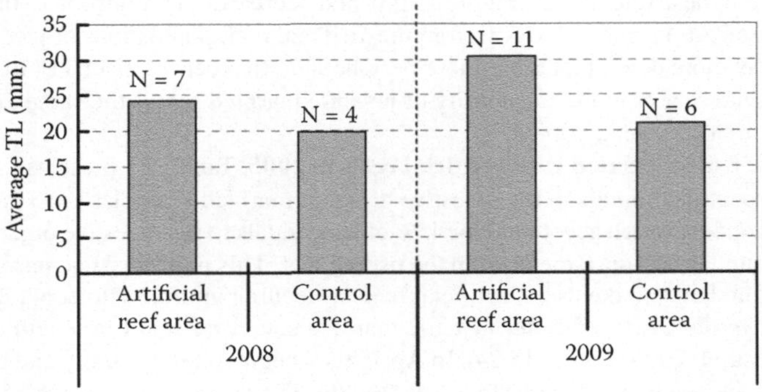

Figure 15.25 Growth of juvenile marbled sole compared between those associated with artificial reefs and the control area in 2008–2009.

most probably owing to an increase in the prey through an increase in a more hospitable environment for prey. Presumably, this leads to faster growth among juvenile marbled sole when they associate with artificial reefs.

Summary

In conclusion, (1) there was a higher density of benthic animals near the artificial reef in shallow waters; (2) there was a high abundance of marbled sole within 5 meters of the artificial reef; (3) it was confirmed that marbled sole feed on benthic organisms found near the artificial reef; and (4) it was confirmed that marbled sole grow faster when associated with artificial reefs.

Reef Affinity of Jack Mackerel

In this portion of the chapter the reef affinity of jack mackerel is examined relative to its life history features. A photograph (Figure 15.26) and a stylistic presentation (Figure 15.27) of jack mackerel in association with an artificial reef is offered to orient the reader to the mid-water association that jack mackerel have with artificial reefs.

Figure 15.26 School of about 50,000 jack mackerel associated with an artificial reef.

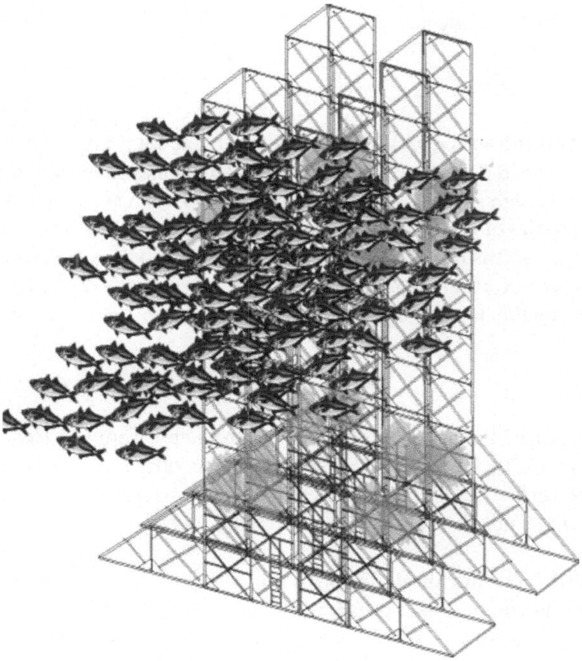

Figure 15.27 Schematic diagram of jack mackerel gathered around an artificial reef.

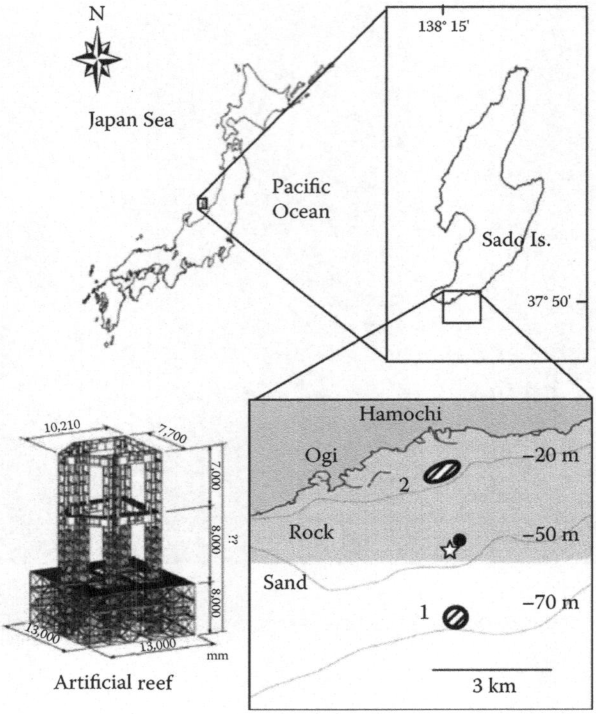

Figure 15.28 Map indicating the study site for jack mackerel off Japan and module (left).

Method

1. **Investigation point and observed artificial reef.** A map of the study areas is presented in Figure 15.28. This figure also indicates the type of artificial reef module deployed for this study. This single artificial reef was 21 meters high as was deployed on the ocean bottom at a depth of 45 meters.
2. **Behavior characteristics of jack mackerel (biotelemetry survey).** Behavior of jack mackerel in the proximity of artificial reefs is unknown. To determine their behavior in relation to artificial reefs, a biotelemetry device was attached to adult jack mackerel approximately 30-cm long. Fish movement was determined during the day and night by the ship using these telemetry data (Figure 15.29).

Results

1. **Horizontal migration of the jack mackerel.** Jack mackerel were released for four of the seven released fish (Figure 15.30). The battery life of the transponder is 4 days and this limited some of the observations. Nevertheless, the data indicated that jack mackerel move at least 3 km around the artificial reef. This indicates that mackerel tend to remain near (± 3 km) artificial and natural reefs (Ito et al., 2009).
2. **Swimming depth of jack mackerel.** The swimming depth of jack mackerel varies considerably between day and night (Figure 15.31). In Case 2 (Figure 15.31), the fish stayed near the artificial reef. Diurnally, the fish swam directly above the artificial reef and nocturnally, they migrated toward shallow water and swam at the surface. In Case 3 (Figure 15.31), the fish moved from the artificial reef to natural reefs where the water was deeper. Thus, jack mackerel remained around the artificial and natural reefs in the daytime and swam near the surface, unassociated with reefs at night.
3. **Frequency pattern of migration depth of jack mackerel.** Figure 15.32 shows the distribution of jack mackerel relative to their depth distribution diurnally and nocturnally. During daytime, jack

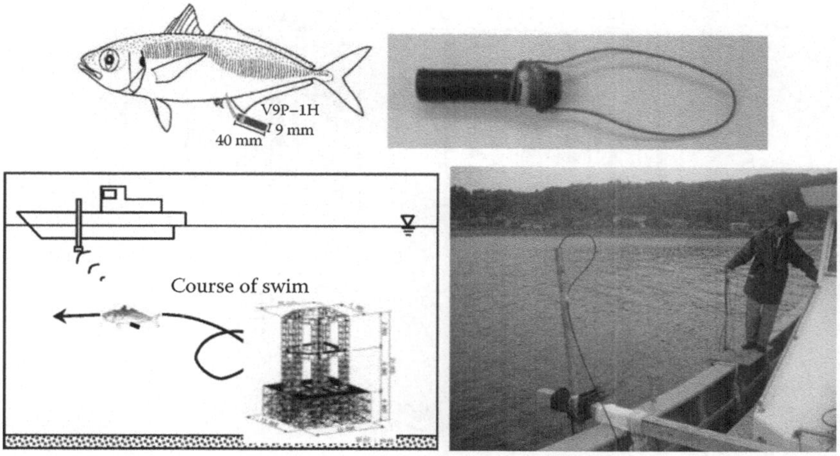

Figure 15.29 Diagram (above left) and photograph (above right) indicating the biotelemetry sender and receiver (ship diagram below left and photograph below right) for jack mackerel.

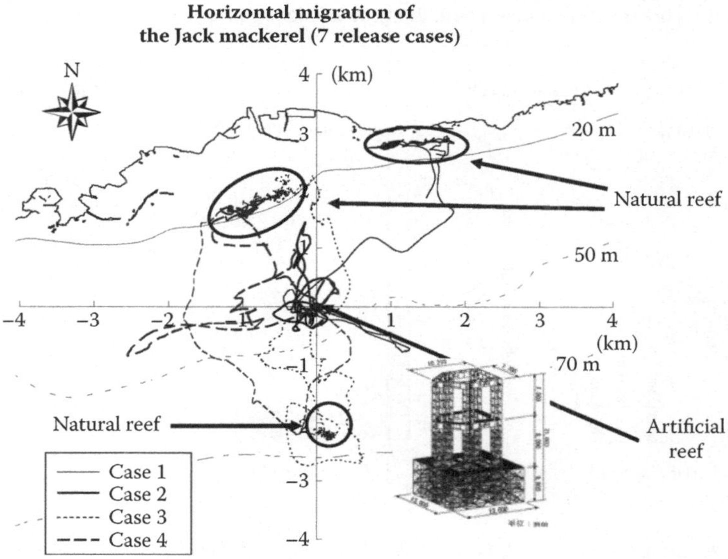

Figure 15.30 Map indicating horizontal migration paths of four jack mackerel near the study site.

mackerel remained around the artificial reef and natural reefs. Nocturnally, they swam near the water's surface (Ito et al., 2009), generally away from reefs.

4. **Migration frequency at artificial and natural reefs.** The frequency of migration to the artificial reef and natural reefs is presented in Figure 15.33. One mackerel (Case 2) remained near the same artificial reef during the daytime for 7 days. Another mackerel (Case 4) moved between the artificial and natural reefs. A characteristic of this fish was that, during the daytime, it remained around the artificial and natural reefs but during the nighttime, it was not associated with either type of reef (Ito et al., 2009).
5. **Distance of horizontal migration from the artificial reef.** The distance that the jack mackerel migrated during the nighttime away from the artificial reef varied (Figure 15.34). It is clear that

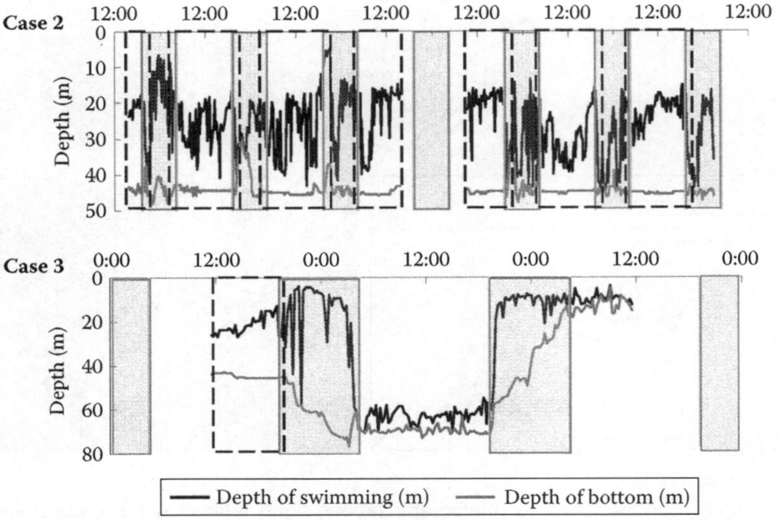

Figure 15.31 Swimming depth and time of two jack mackerel based on telemetry data. The black lines represent the swimming depth, and the gray lines represent the depth of the ocean. The white sections represent the daytime, and the gray sections represent the nighttime.

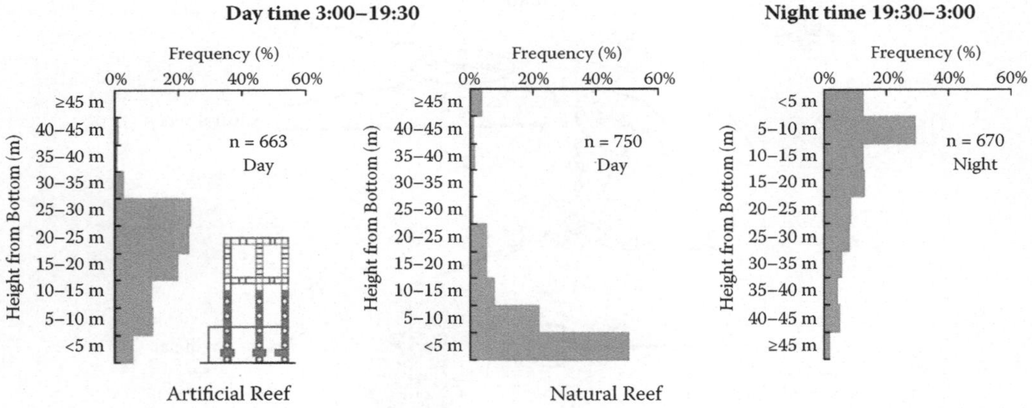

Figure 15.32 Frequency pattern of migration depth of jack mackerel released near the study sites (both artificial and natural reefs).

they migrated within a 3 km radius. In one instance (Case 2) a fish remained near an artificial reef for 7 days during the daytime. In the nighttime, however, it moved approximately 1 km away and returned. It is possible that jack mackerel are aware of an artificial reef located 1 km away even without a visual cue to its location (Ito et al., 2009).

Summary

To summarize: (1) jack mackerel remained around the artificial and natural reefs during the daytime; (2) during the nighttime, they withdrew from the artificial and natural reefs; (3) adult jack mackerel demonstrated daily movement; (4) fish remained near artificial reefs continuously for at

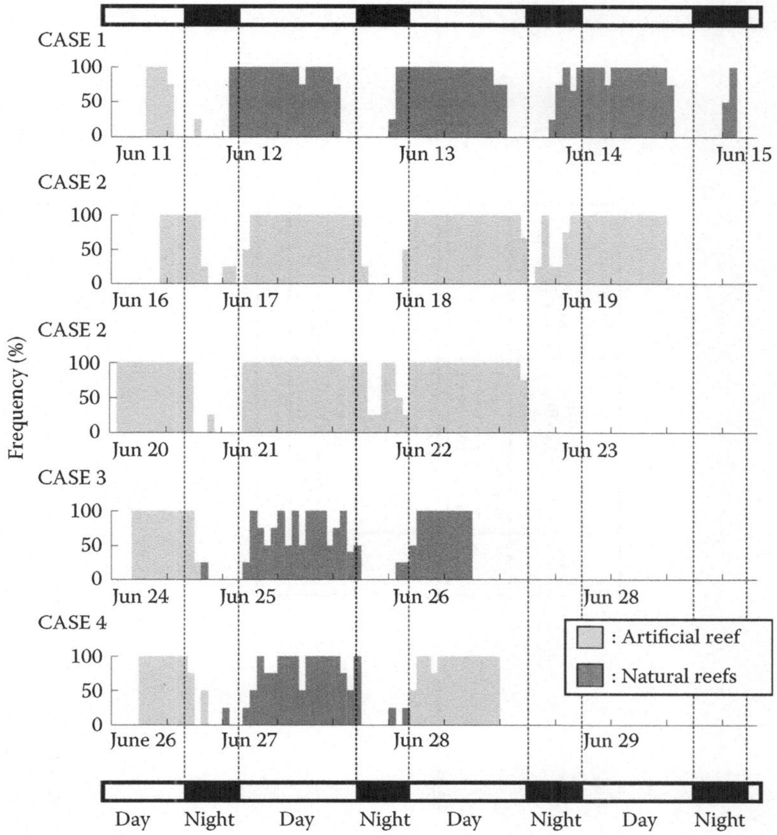

Figure 15.33 Migration frequency relative to time of day at artificial and natural reefs for jack mackerel. The white sections represent the daytime, and the black sections represent the nighttime. The light gray columns represent the artificial reef, and the dark gray columns represent the natural reefs.

least 7 days; and (5) jack mackerel were apparently aware of an artificial reef that was approximately 1 km away.

Function of the Protective Developmental Reef for Snow Crab

Snow crab live at a water depth of 200 to 500 meters (Figure 15.35). A survey was conducted at a depth of approximately 300 meters in the water (Figure 15.36) to examine the relationship that snow crab have with artificial reefs relative to the protective development feature of associating with artificial reefs.

Composition and Layout of the Protective Developmental Reef

The layout and structure of the protective developmental reefs studied here is presented in Figure 15.37. The artificial reef modules were deployed in an area 4 km × 4 km. Each reef module was 250 meters from other modules. Four different types of protective development reefs (modules) were deployed at the study site.

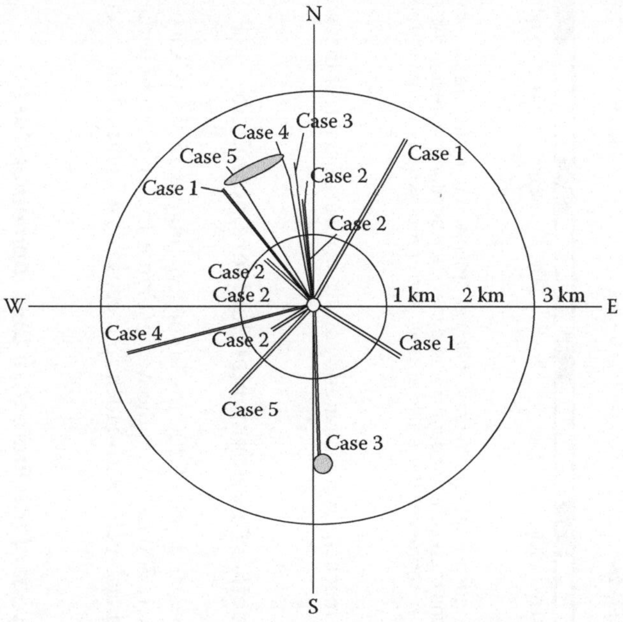

Case	Stay point in day time	Maximum distance from artificial reef at night (km)
2	Artificial reef	0.60
	Artificial reef	1.44
	Artificial reef	0.65
	Artificial reef	0.87
	Artificial reef	0.59
4	Natural reef offshore from Hamochi	2.68
Average		1.14

Figure 15.34 Distance of horizontal migration of jack mackerel from the artificial reef.

Figure 15.35 Snow crab associated with an artificial reef module.

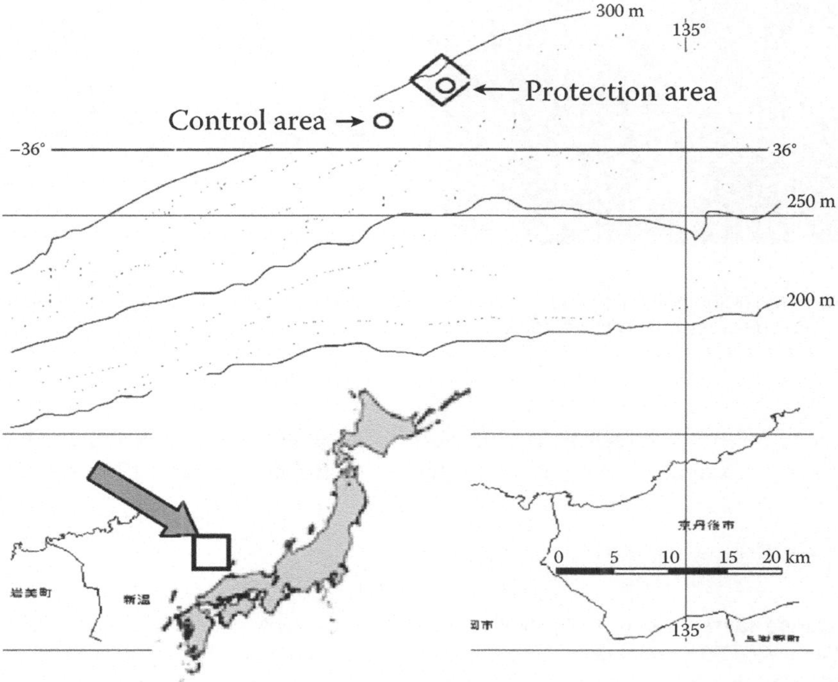

Figure 15.36 Study site for snow crabs off Japan.

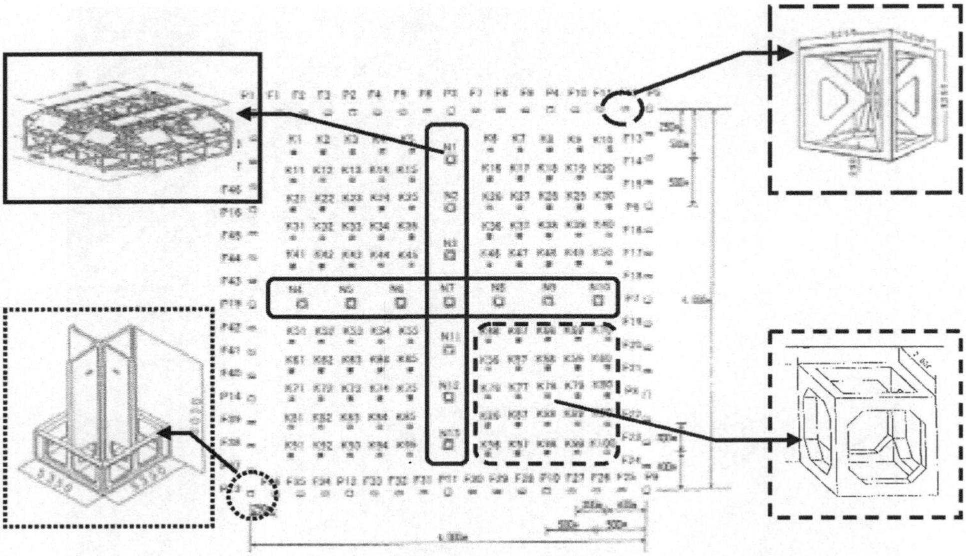

Figure 15.37 Composition and deployment profile of the protective developmental reef at the study site off Japan.

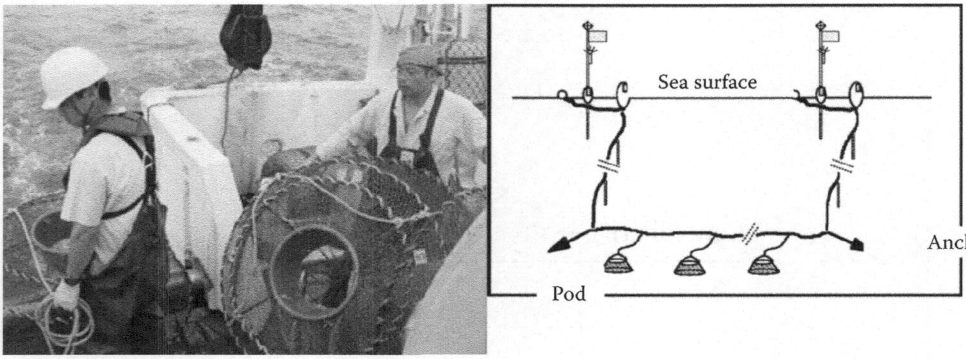

Figure 15.38 Trap (photograph left) and trap deployment schematic diagram (right) used to conduct the snow crab survey near artificial reefs.

Method

Comparisons of snow crab density were made between protection and control areas. Figures 15.38 and 15.39 shows the sampling gear (i.e., basket [pods or traps] nets and an ROV used to assess snow crab density).

Result

1. **Pod fishing survey.** Figure 15.40 shows the results of the snow crab survey. The results indicate that there were 212 snow crabs in the protection area compared to only 40 in the control area. With regard to sex, there were many females caught and most of these females were in the protection area. In contrast, there were few males caught, and there were approximately the same number of males in both the protection and control areas (Ito et al., 2008).

Figure 15.39 ROV (remotely operated vehicle; photograph right) and ROV tether arrangement used to survey for snow crabs.

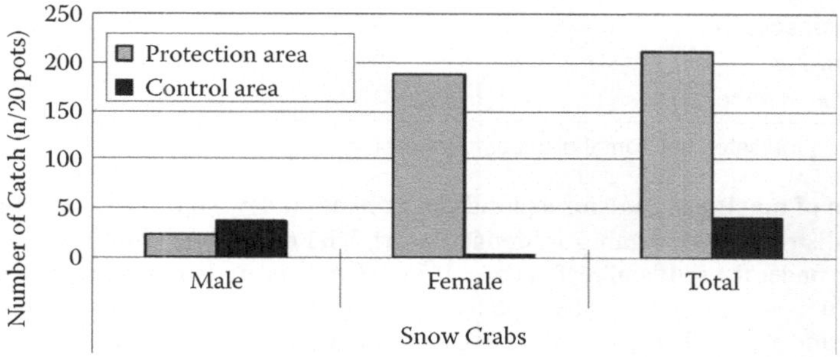

Figure 15.40 Snow crab catch (number of crabs per 20 pods or traps) from artificial reefs (protection area) and control sites.

Table 15.2 Result of ROV Observation

	Protective Area	Control Area	Total
Observed period (minute)	343	56	399
Depth (m)	283–289	289	—
Observed area (m^2)	9,376	1,484	10,859
Density (ind./1000 m^2)			
Species			
Chionoecetes Opilio (snow crab)	9.9	4.7	9.2
Pleuronectidae	2.6	4.7	2.9
Zoarcidae	51.6	41.1	50.2
Buccinidae	4.3	1.3	3.9
Macrura	7.7	0.0	6.6
Actiniaria	6.6	12.8	7.5

2. **ROV observation.** Table 15.2 shows the results of the ROV observations. Using the ROV, the amount of observed area was determined and fish abundance was converted to density. As a result, it was found that there were 9.9 snow crabs/1,000 m^2 in the protection area and 4.7 snow crabs/1,000 m^2 in the control area. These results indicate the benefits of setting up protective developmental reefs (Ito et al., 2008).

Table 15.3 Distribution of Marine Organisms Around the Protection Developmental Reef

	Protective Area (ind./1,000 m²)		
Taxa	**Inside the Reef (< 0.5 m)**	**Around the Reef (0.5 ~ 10 m)**	**Others (>10 m)**
Coelentelata			
Sea anemone	56.4	0	6.6
Mollusca			
Buccinidae	1848.3	80.5	4.3
Crustacea			
Macrura	846.6	5.4	7.7
Chionoecetes opilio (snow crab)	42.3	0	9.9
Echinodermata			
Ophiuroidea	3668.4	241.4	—
Crinoidea	3470.9	16.1	—
Pisces			
Pleuronectidae	0	0	2.6
Zoarcidae	0	16.1	51.6
Gadus macrocephalus	84.7	0	0

Note: ( ) Indicates horizontal distance from reef.

3. **Distribution of marine organisms around the protection developmental reef.** Table 15.3 shows the state of distribution of organisms around the artificial reef. Observations were divided into the area directly under the artificial reef, an area 0.5 to 10 meters around the artificial reef, and an area more than 10 meters away from the artificial reef. There were many organisms observed directly under the artificial reef. These organisms included snow crabs, prey, and fish.
4. **The "feeding ground" effect of the protective developmental reef.** The prime function of artificial is that they serve as a feeding ground (Figure 15.41). Swarms of plankton formed around these reefs. In addition, there was an extremely large number of starfish directly under the protective developmental reef. These observations indicate that protective developmental reefs serve as a feeding ground for fish.
5. **The "hideout" and "resting place" effect of the protective developmental reef.** It is thought that protective developmental reefs function as a "hideout" and "resting place." In Figure 15.42, a Pacific cod is located directly under the reef. It appears as though a snow crab is taking up residence with its back to the protective reef. Snow crabs are known for putting sea anemones and comatulids on their backs. To the snow crab, the protective reef may have resembled such organisms (Ito et al., 2008).
6. **The "spawning place" effect of the protective development reef.** As most of the female crabs caught around the protective developmental reef were incubating eggs, it could be assumed that such reefs can also serve as a spawning place or a place to lay eggs (Figure 15.43).

Figure 15.41 Photographs indicating the feeding-ground effect of the protective developmental reef.

Figure 15.42 The "hideout" (left) and "resting place" (right) for snow crabs at the protective developmental reef.

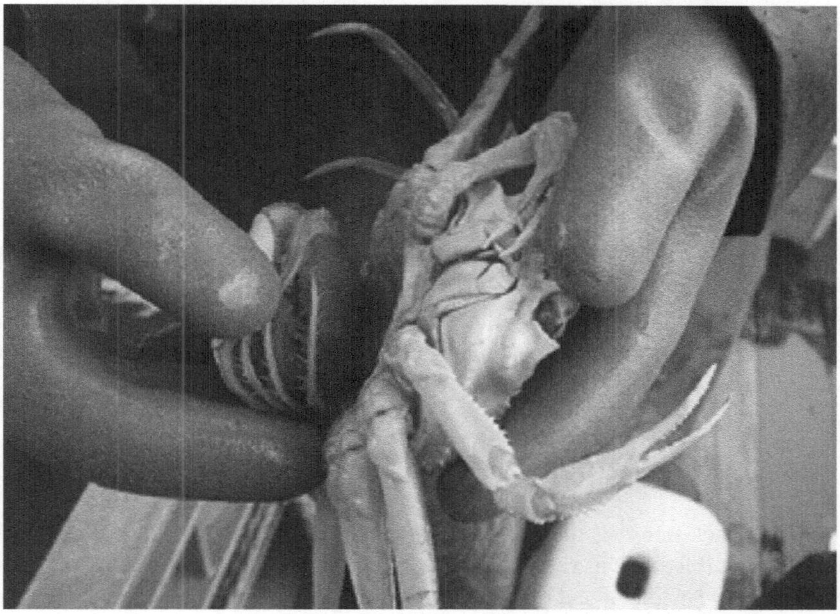

Figure 15.43 Evidence of spawning by snow crab at the protective developmental reef.

7. **Pattern diagram of marine organism in the area surrounding the reef.** The diagram in Figure 15.44 depicts the area surrounding a protective developmental reef according to type. Here, one can see that there is a high diversity of organisms living around the reef. The basic structure is similar to the ecosystem around an artificial fishing reef in shallow sea areas (Ito et al., 2008).

Summary

To conclude, the surveys (1) verified protective effects for snow crabs; (2) confirmed that snow crabs localize at artificial reefs and that the reefs have an effect as a place to live; (3) the area near the artificial reef is full of prey animals; and (4) a complex ecosystem similar to that of the coastline was formed near the artificial reef.

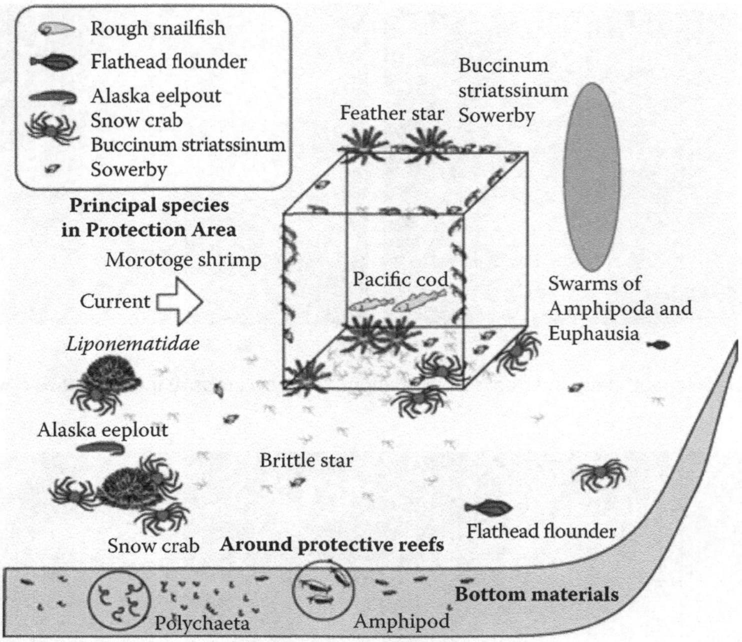

Figure 15.44 Schematic diagram of indicating marine organism associations and affinities around the artificial reef designed for snow crabs off Japan.

Artificial reefs, including those in shallow water areas to deep water areas, function as feeding grounds, hideouts (predator avoidance), and a place from which to orient. Moreover, artificial reefs create a suitable environment for fishes targeted by fisheries.

GENERAL CONSIDERATIONS (ARTIFICIAL REEF FUNCTIONS)

As for general conclusions, (1) there was an abundance of prey animals near the artificial reefs in shallow waters; (2) marbled sole use artificial reefs as feeding grounds; (3) anadromous jack mackerel repeatedly use the same artificial reef over long periods of time; (4) jack mackerel localize at artificial reefs during the daytime; (5) the protective effect of snow crab protection reefs was confirmed; and (6) as with shallow waters, prey animals are rich around artificial reefs in deep waters as well.

REFERENCES

Ito, Y., H. Miura, T. Nagahama, Y. Ohno, T. Yanase, and T. Nakamura. Evaluations of the functions of protective developmental reefs for snow crabs. *Chionoecetes opilio*. *Annual Journal of Civil Engineering in the Ocean, JSCE*, 24, 25–30, 2008.

Ito, Y., Y. Nakano, and S. Fujisawa. Distribution of small animal at and around artificial reefs: Study on propagating function of fisheries facilities-I. *Fisheries Engineering*, 45(2), 101–110, 2008a.

Ito, Y., Y. Nakano and S. Fujisawa. Distribution pattern and feeding habit of marbled sole *Pleuronectes yokohamae* at artificial reefs: Study on propagating function of fisheries facilities-II. *Fisheries Engineering*, 45(2), 129–138, 2008b.

Ito, Y., H. Miura, K. Nakamura, and T. Yoshida. Behavior properties of Jack mackerel *Trachurus japonicus* using ultrasonic biotelemetry in artificial reef off the coast of Hamochi, Sado Island, the Sea of Japan. *Nippon Suisan Gakkaishi*, 75(6), 1019–1026, 2009.

CHAPTER 16

Artificial Reefs to Induce Upwelling to Increase Fishery Resources

Takahiro Okano, Masanori Takeda, Yoshifumi Nakagawa, Kenji Hirata, Koji Mitsuhashi, Suetoshi Kawaguchi, and Junichi Ito

CONTENTS

ABSTRACT

Artificial upwelling has been used effectively in Japan through the deployment of artificial reefs or structures on the sea floor. These structures help move nutrient-rich bottom water toward the surface. The "induced-upwelling" brings new nutrients into the euphotic zone, increasing primary production and, hence, leading to increases in local fishery productivity. In 1991, the Japanese Government initiated a new subsidy program for local administrations interested in the construction of artificial underwater reefs to help create new fishing grounds. In this chapter we describe the steps or stages involved in this governmental program necessary for building artificial reefs to help induce upwelling. This includes (1) site selection, (2) design of structural shape, (3) construction and deployment of artificial upwelling systems, and (4) assessments to further assess fishery improvement and to evaluate the effectiveness of the financial investment.

FISHERIES BACKGROUND IN JAPAN

Japanese marine fisheries are generally organized into four categories: aquaculture, long distance, offshore, and coastal. The long-term variation in fishery production by each one of these

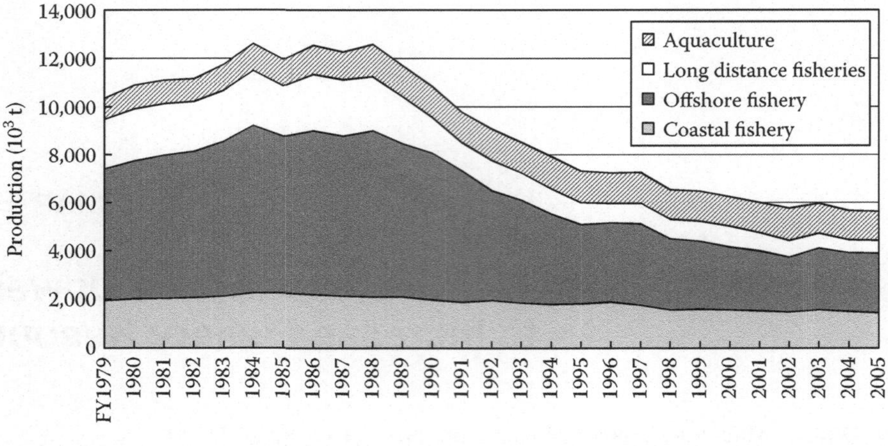

Figure 16.1 Fisheries production in Japan by fishery category.

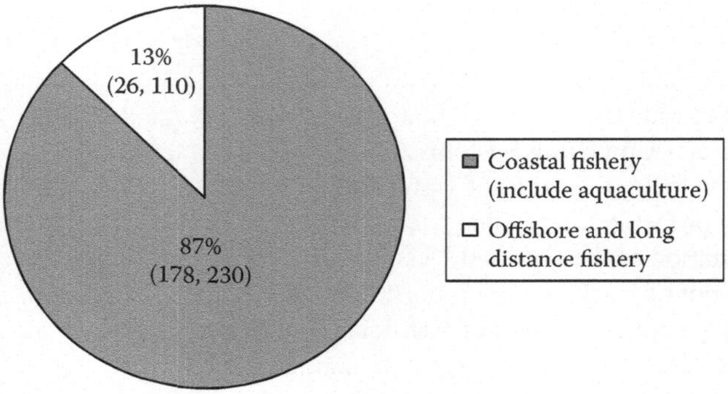

Figure 16.2 Number of fishery workers in Japan in 2007.

fishery categories is shown in Figure 16.1. Aquaculture production has been stable during the past few decades. The long-distant fishery operates in the open sea or in national exclusive economic zones under the agreements with the nations concerned. The offshore fishery operates outside the coastal fishery area to catch migratory fish schools. Its annual catch depends on the fluctuations in the recruitment of target species. The fishing yield of both these two categories has been decreasing steadily since late 1980. The coastal fishery takes place along the coast of Japan where many fishermen make their living on small-scale fisheries and aquaculture. The Japanese central and local governments have been working to enhance production by these two coastally based fishing categories because 87% of Japanese fishermen participate in these fisheries in coastal areas, and this corresponds to 64% of the total economic value of fish production in Japan (Figures 16.2 and 16.3). The artificial upwelling system represents one of the mechanisms adopted by the central government to enhance fishery resources around the coast of Japan.

THE PURPOSE OF ARTIFICIAL UPWELLING SYSTEMS

Upwelling from an artificial reef delivers nutrient-rich deep water to the upper euphonic zone. Natural upwelling zones around the world oceans are known as good fishing grounds, especially

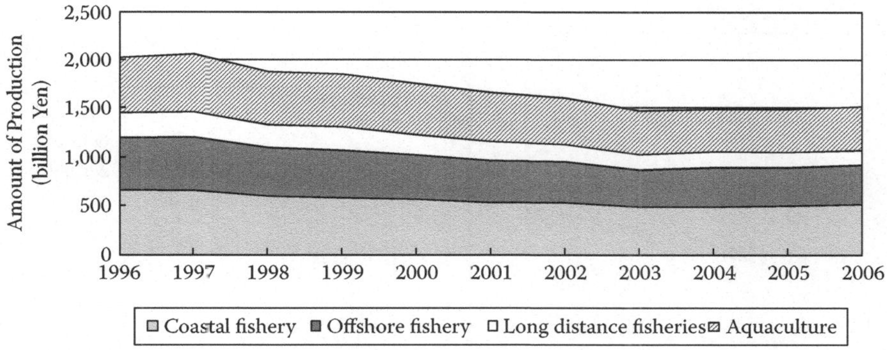

Figure 16.3 Economic value of fisheries production in Japan according to fisheries category.

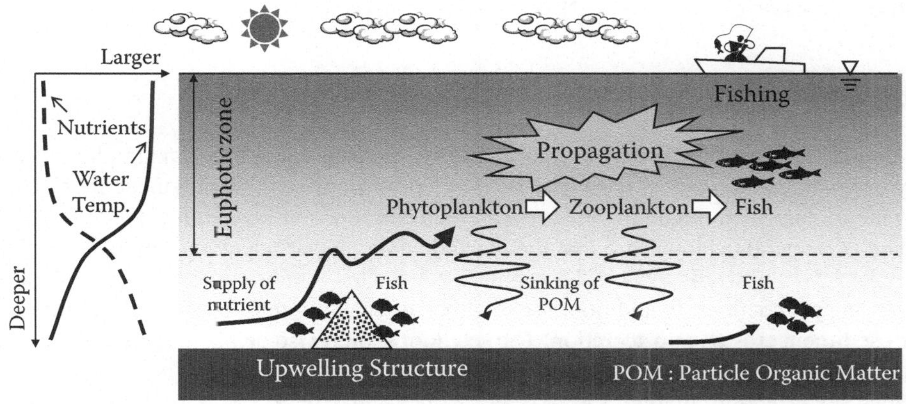

Figure 16.4 Expected local fishing enhancement based upon the construction and deployment of upwelling structures such as artificial reefs.

near islands, underwater banks, seamounts, and reefs. In addition, upwelling is expected to occur around rock mounds deployed on the sea floor; these will increase habitat complexity for fish assemblages and, subsequently, lead to an increase in local fish production. Figure 16.4 depicts how fishing production may be enhanced by artificial upwelling systems.

Figure 16.5 indicates the artificial upwelling systems in Nagasaki prefecture where fisheries are one of the major local industries. The Nagasaki prefecture administration has constructed artificial upwelling structures at five spots to create long-term fishing grounds based upon underwater manmade reefs.

THE GENERAL SEQUENCE FOR ARTIFICIAL REEF UPWELLING PROJECT

These artificial reef upwelling projects were implemented the following steps:

1. Selection of deployment sites
2. Determination of the optimal designs for reef structure
3. Deployment of rocks or concrete blocks supported by in situ monitoring of flow speed and direction
4. Postdeployment evaluation of artificial upwellings

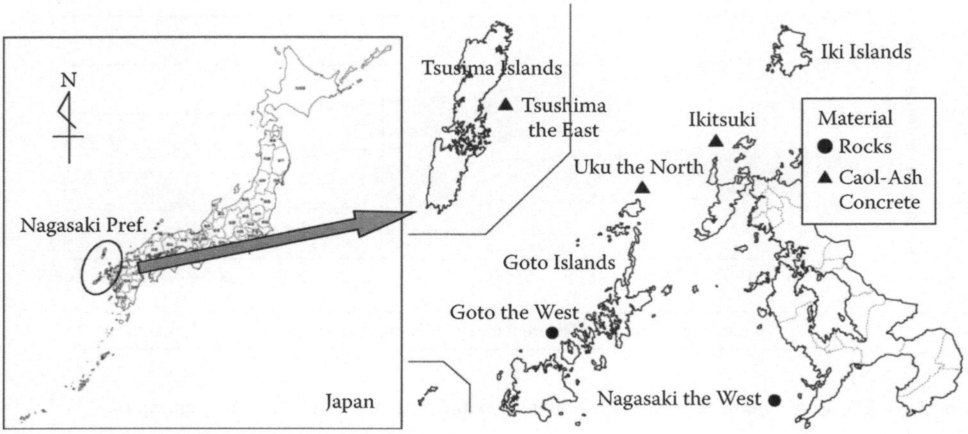

Figure 16.5 Upwelling reef structures in Nagasaki prefecture, Japan.

Step 1—Selection of Deployment Sites

Preferable sites are selected according to specific hydrographic and topographic criteria. The main hydrographic features that must be considered prior to material deployment are bottom circulation, water column stratification, and nutrient concentrations at bottom layers. Currents must be faster than adjacent areas to ensure that the inertial energy is sufficient to draw nutrient-rich water from deeper layers upwards toward the euphotic zone. This is more easily accomplished in areas without strong pycnoclines. In addition, the neighboring sea floor must be stable, to reduce the probability of sediment erosion and accretion. The sea floor should also be flat to facilitate building piles and assuring the future physical integrity of the artificial reef.

Besides site selection, there are important engineering aspects of the underwater mound construction that could restrict project development. The depth at the selected site must be in concurrence logistics and engineering realities. Japan has participated in the construction of upwelling structures on sea sites less than 90 m deep. Also the materials selected to build the mound should be readily available and transportable to the project site location to reduce project costs. Rocks and coal-ash concrete blocks have been used in Japan. Accessibility and the low cost of these materials make the reef construction economically feasible.

Step 2—Determination of the Optimal Designs for Reef Structure

Figure 16.6 diagrammatically depicts the systematic decision process used to determine the most effective design for future upwelling system construction. First, one has to determine the average bottom flow direction during high fish production seasons. And the artificial reef is placed at right angles against the bottom flow direction. The next step is modeling the upwelling that will be potentially induced by using some experimental simulations of reefs with various cross sections. The final reef design is determined after comparison with the model estimations of yearly upwelling volumes from various cross sections. The experimental modeling must also allow comparisons of calculated cost and benefit ratios of each experimental structure. Construction costs are mainly affected by the kind of materials used so that one has to consider the volume of material necessary for arriving at the final shape and size of the mound structure.

Tentative decisions regarding the most effective combination of cross sections are based on model tests made with historical data and in situ surveys. Cross sections are formed by two lateral peak

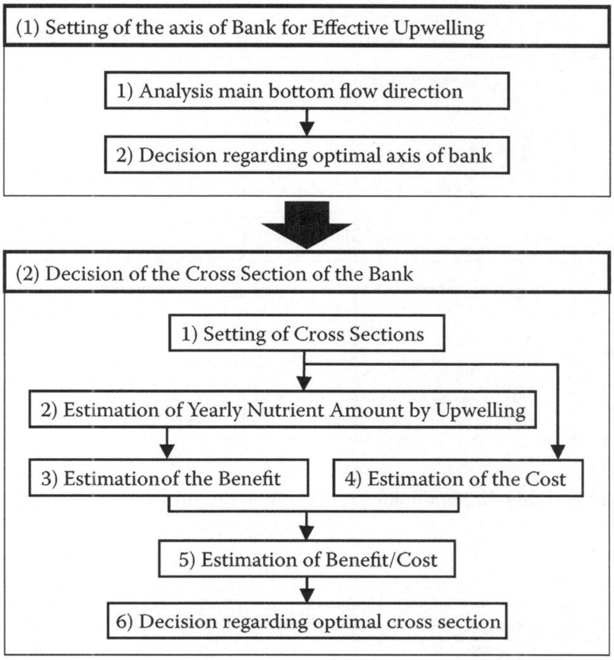

Figure 16.6 Flow diagram of the procedure necessary to design an effective upwelling structure.

Table 16.1 Ordinary Figures for Bank Planning

Material	Rocks	Concrete Blocks
Height (*H*)	*H* = 0.2 *D* (*D* = Depth)	
Distance between peaks (*L*)	*L* = 4*H* ~5*H*	
Height between peaks	*H*	0.875*H*
Slope	1:2.5	1:2

mounds connected by an embankment. Main variables to be considered are height of the peak mounds (*H*), distance between peak mounds (*L*), and the slope gradient of the structure; here defined as the ratio of vertical to horizontal dimensions of the mounds as shown in Table 16.1 and Figure 16.7.

For the standard design, *H* is 20% of local depth, *L* must be four to five times *H*, and the slope (inclination) is 1:2.5 for rocks or 1:2.0 for coal-ash concrete blocks. In the instance of using coal-ash concrete blocks, the reef should have higher peaks and a lower embankment connecting them. The height of the embankment is $0.875 \times H$.

A three dimension, nonhydrostatic pressure model (based on the steps outlined in Figure 16.8) has been used in Japan to calculate the annual load of nutrients expected from artificial upwelling systems. The model is based on eight environmental and design variables. Moreover, data on the general cross-section parameters identified in the previous section, the associated flow-field, water density, and nutrient distributions in each seasonal period (except winter) must be put into the model to calculate seasonal upward water flow rates. Results of this analysis are used to estimate the volume of water annually expected to be moved by upwelling.

An experimental artificial upwelling system was simulated by Kurihara et al. (2006). This model was based on variables in the cross-sectional design (Figure 16.9) and the resulting upwelling field

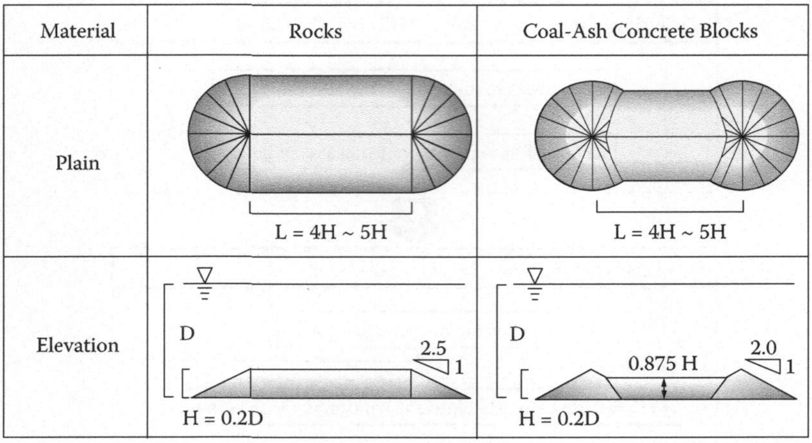

Figure 16.7 Typical plan and ratio of dimensions for planning during the development of an artificial reef used for upwelling.

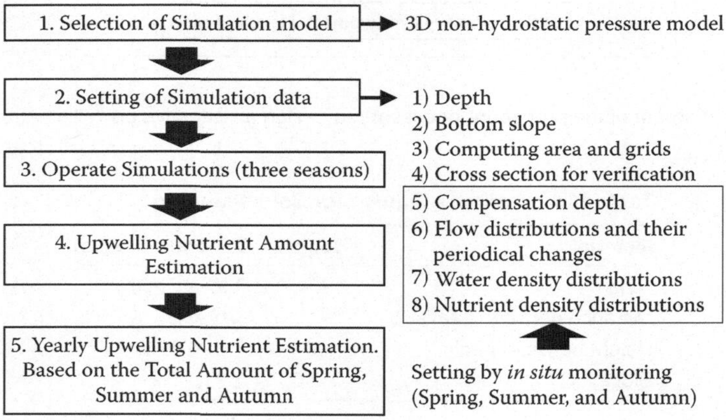

Figure 16.8 Estimation procedure to calculate annual amount of water moved via upwelling created from an artificial reef deployment.

(Figure 16.10) within 500 m from the artificial mounds. The model was run for three consecutive cycles of tidal flow to obtain stable results from simulated upwellings. The average volume of water moved during the last cycle was used to calculate the nutrient load moved to the euphotic zone via upwelling.

Step 3—Deployment of Rock or Concrete Blocks

After the best designs for artificial upwelling structures are determined, based on the results of computer model simulations, the units selected for building the underwater mounds (rocks or coal-ash concrete blocks) are loaded and transported to the selected deployment site on a hopper barge (Figure 16.11). The barge is precisely positioned at the deployment site using a geographic positioning system (GPS). Once in position, the bottom door of the hopper barge is opened, and the materials are released to bottom. The reef is formed by the release of materials, and no adjustment

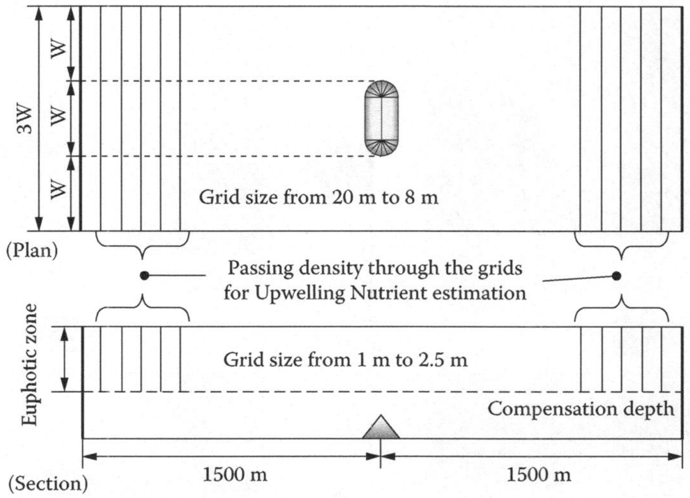

Figure 16.9 Diagrammatic representation used to determine potential upwelling effect of an artificial reef design and deployment configuration.

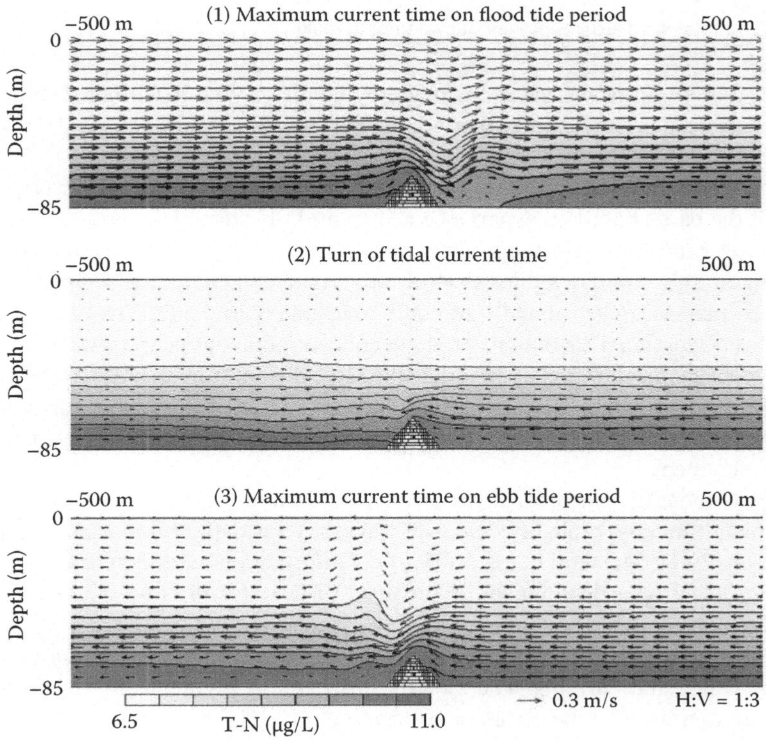

Figure 16.10 Numerical model simulation of the current field around artificial upwelling structure.

Figure 16.11 Example of a barge deployment cycle using rocks.

to the shape or dimension of the mound is made by divers or with machines. It often takes 2 to 3 years to fully construct an underwater mound, using up to 60 complete barge deployments for the complete project.

Three major steps must be carefully considered during block deployment: (1) Determine the best position of the barge based on experience using similarly shaped materials, local currents, and depth, (2) maneuver the barges during the material release to achieve desired mound configuration, and (3) confirm the shape of block piles to allow proper adjustment for subsequent barge releases.

The dumped mass of materials will gradually disperse with depth. The dispersal pattern may generally follow a Gaussian distribution (i.e., high concentration of materials in the idealized center, a progressively lower concentration of materials the more distant from the idealized center; Matsumi and Kishiguchi, 1990) in which deviation of the largest portion of the central mass depends on the density and number of blocks, depth, currents, etc. Figure 16.12 illustrates the final pile shape in case of rock deployment.

Prior to deployment, the currents at various depths between surface and bottom must be determined to calculate the most probably dispersal trajectory for deployed materials (Kumagai et al., 2004; Oono et al., 2004; Matsumi et al., 2005). This information helps determine the correct position of the barge for proper deployment. Figure 16.13 illustrates the final pile shape according to different current flow at different depths.

Deployment requires GPS instruments and tugboats for precise positioning of the barge due to the effect of winds and currents. The system illustrated in Figure 16.14 has been adopted one example in Japan to minimize the errors of positioning.

The last step in the deployment positioning process is the confirmation of the reef position and shape with GPS and multibeam side-scan sonar (Figure 16.15). Information from this procedure is used to determine the next deployment position.

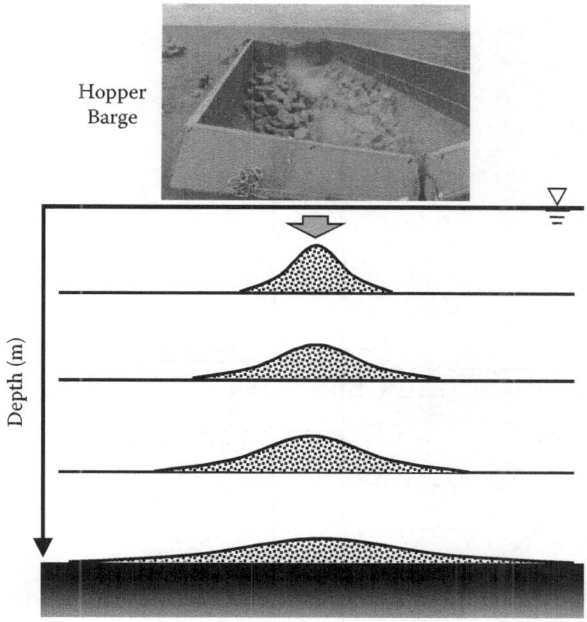

Figure 16.12 Possible patterns of rock dispersal upon deployment.

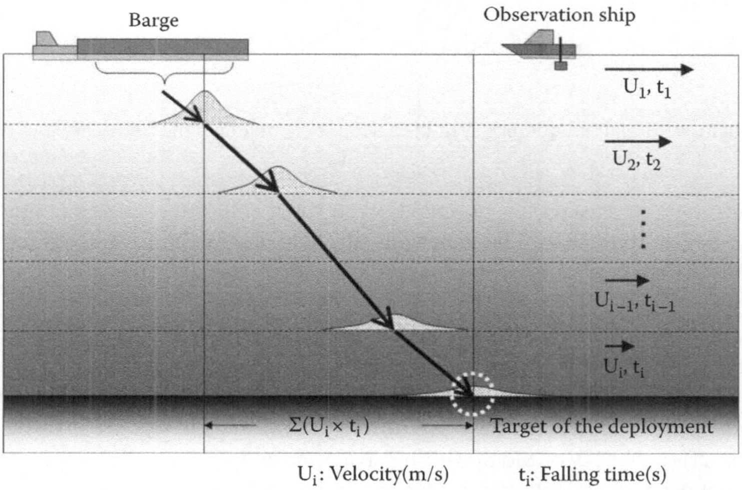

Figure 16.13 Trajectory of deployed mass of materials relative to currents and water depth.

Step 4—Postdeployment Evaluation of Artificial Upwellings

Monitoring environmental and biological variables is the last step necessary to confirm the effectiveness of an artificial reef deployed for upwelling. The most relevant variables to be measured are the variation in the water density (σ_t) field and the biological community (i.e., plankton, benthos, and nekton) around the artificial reef. Figure 16.16 indicates the water density of the water masses observed at the Uku artificial upwelling reef program along the north coast of Nagasaki. The water

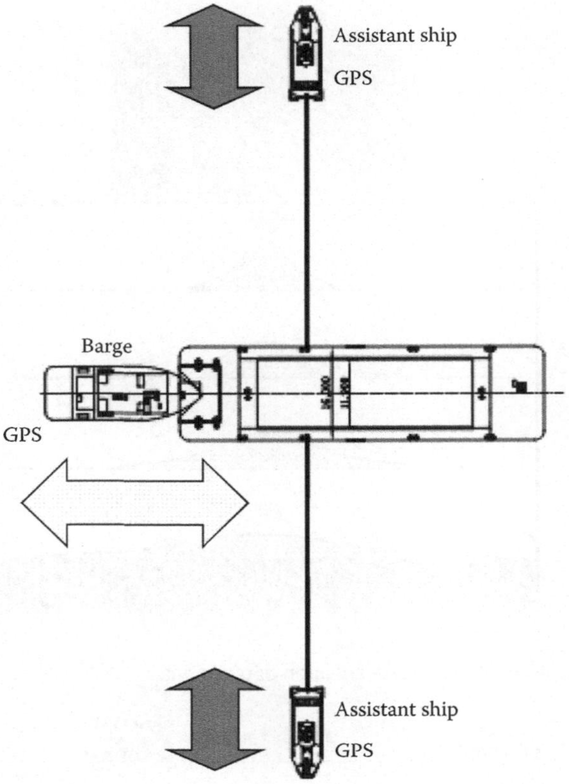

Figure 16.14 Guidance and management system to accurately maintain the barge deployment position.

density contour lines were based on a survey at seven water sampling stations (Figure 16.16) established along the line current flow. The water density field shows the 24.8 kg/m^3 isopycnal shifted upwards from 60 to 40 m depth; the water mass flowed above the structure at $x = 0$ m, thus confirming upwelling.

Plankton density before and after the construction of the upwelling structure is shown in Figure 16.17. Phytoplankton cell numbers were similar before and after the artificial upwelling project was undertaken. Interestingly, zooplankton abundance increased throughout the whole water after the reef deployment. However, these preliminary plankton surveys are insufficient in explaining the long-term effectiveness of the upwelling structure relative to enhancing local plankton production. Surveys must be conducted to continuously evaluate the effect of artificial structures on local ecosystem dynamics.

Benthic species composition, density, and biomass all increased (Figure 16.18) after the deployment of the upwelling structure. We suggest that changes in benthic community are due to the disturbance of the bottom flow that provides better conditions for the benthic assemblage the structures. Benthic assemblages may also be used as indices to estimate the effect of artificial upwelling structures as fish aggregate devices (FADs).

Figure 16.19 shows the aggregation of *Hyperoglyphe japonica*, *Trachurus japonicus*, *Epinephelus awoara*, and *Seriola lalandi* at the artificial upwelling systems of Tsushima on the east coast, Uku to the north, and Goto in the west. These fish are highly prized in Japanese fish markets. Total aggregated fish weight may be calculated from length/weight conversion formulas used in fishery research (Figure 16.20). Though results were different among seasons and sites, from 1 to 10 tons of

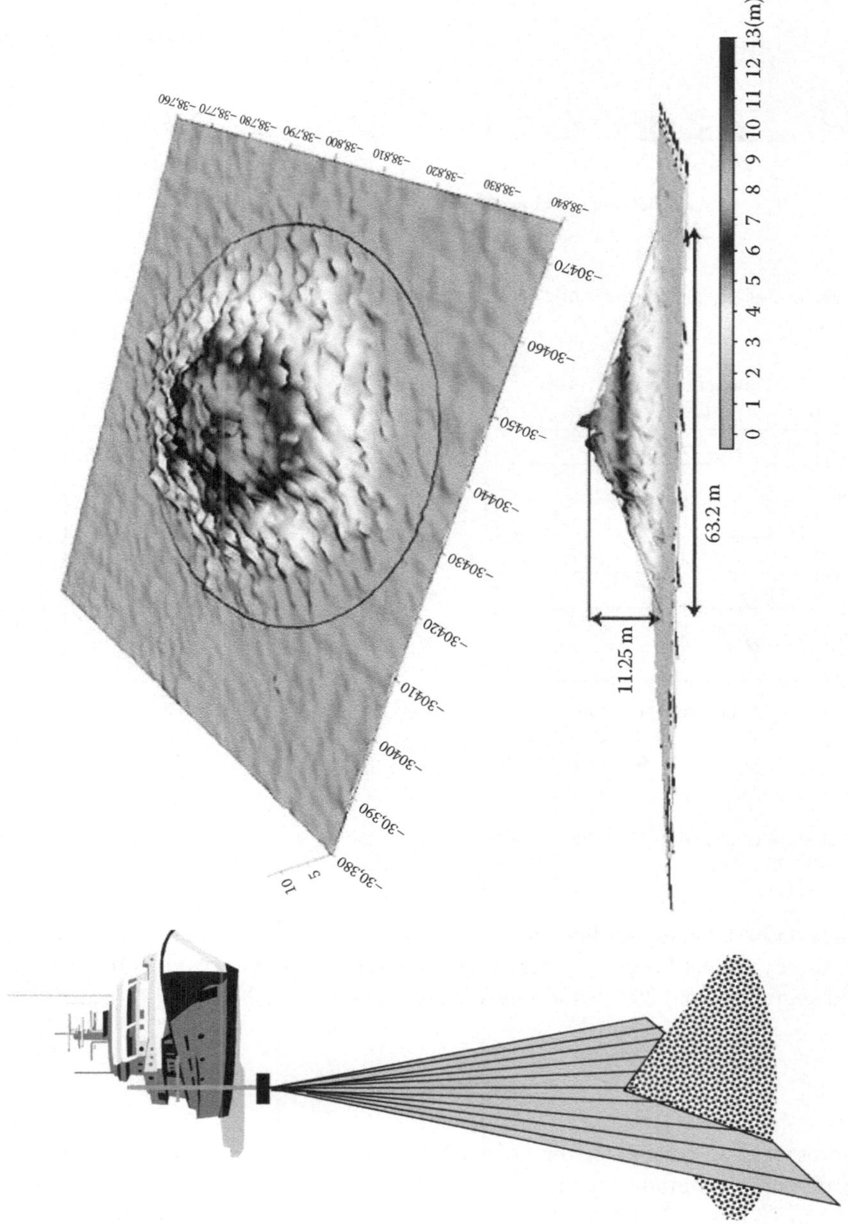

Figure 16.15 An example of a survey to determine the reef position, dimension, and configuration using multibeam side-scan SONAR.

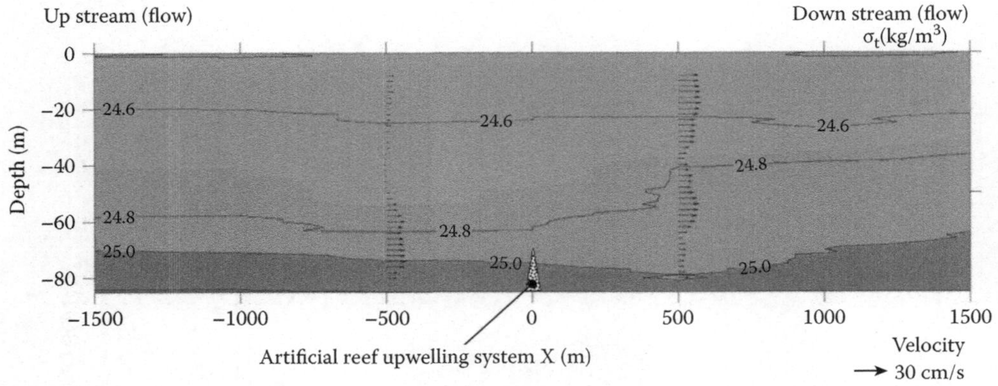

Figure 16.16 Water density distribution and flow vector (Uku, north coast, June 7, 2006).

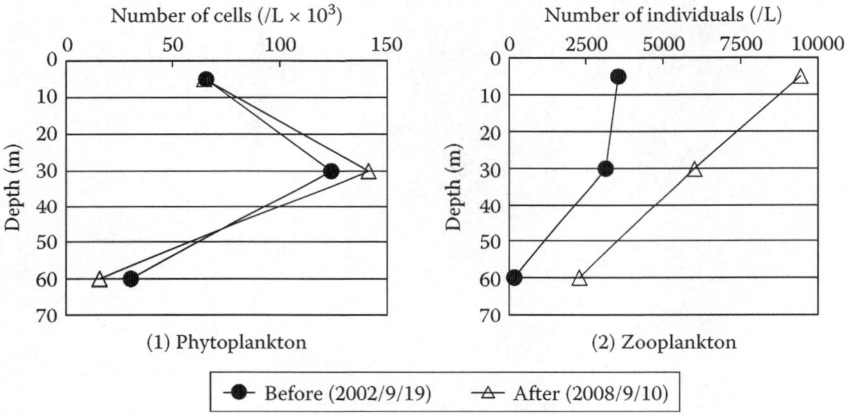

Figure 16.17 Change of plankton abundance before and after the construction of artificial upwelling systems (Goto the West).

fish biomass was observed at the artificial upwelling sites. Yearly fish catches using a purse seine on a 20 × 18 km^2 survey area at Ikituki artificial upwelling program confirmed that the annual fishing yield increased six times from 250 to 1500 tons (Suzuki and Izutsu, 2004).

CONCLUSIONS

Here, we presented an outline for the artificial upwelling program in Japan. Chiefly, this was based on the on-going programs in the continental shelf off Nagasaki prefecture, which were deployed less than 20 km from the coast in less than 100 m depth. With data obtained from these regional projects, we developed specific engineering systems based on in situ site decisions, planning and designs of rock piles, precise deployment procedures, and cost–benefit calculations. Improvement in the Japanese artificial upwelling program is occurring with the development of new simulation model (Nakayama et al., 2009) to more accurately estimate the enhancement of local

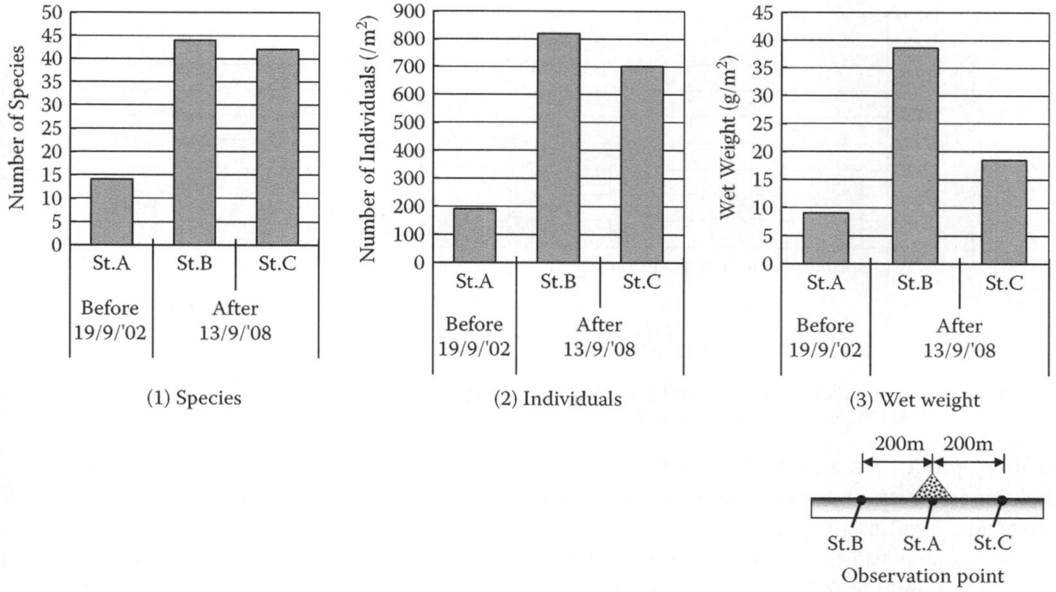

Figure 16.18 Changes of benthic community parameters before and after the construction of the artificial upwelling system (Goto the West).

Hyperoglyphe japonica (Tsushima the East)

Trachurus japonicus (Tsushima the East)

Epinephelus awoara (Uku the North)

Seriola lalandi (Goto the West)

Figure 16.19 Fish assemblages around artificial upwelling structures.

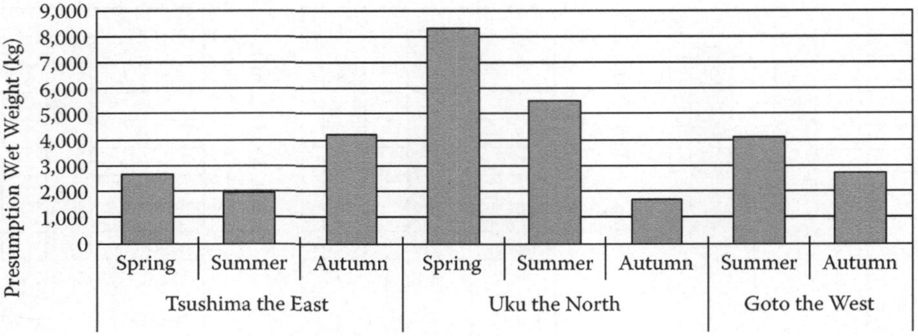

Figure 16.20 Estimation of aggregated fish biomass at different Japanese artificial upwelling programs as estimated from photographs.

primary production and, hence, the local fishery. This model requires validation with in situ data and this involves continuous surveys of environmental variables to fully assess the effectiveness of artificial upwelling programs.

In the near future, similar projects will be developed at deeper shelf locations for stable and long-term supply of fish resources in Japan.

REFERENCES

Kumagai, T., S.Yamamoto, T. Hirayama, T. Manabe, K. Kobayashi, H. Shimizu, T. Kawaguchi, K. Hirata, and K. Komiya. 2004. Movement cube blocks dropped from barge vessel to construct artificial marine ridge *Annual Journal of Coastal Engineering*. 51(2): 801–805 (in Japanese).

Kurihara, A., H. Sasaki, Y. Nakagawa, T. Aono, and A. Okayasu. 2006. Unsteady analysis of nutrients with upwelling flow due to artificial mound placed sea bottom. *Annual Journal of Coastal Engineering*. 53(2): 1306–1310 (in Japanese).

Matsumi, Y., and T. Kishiguchi. 1990. A study on barge position when constructed rubble mound. *Annual Journal of Coastal Engineering*. 37: 769–773 (in Japanese).

Matsumi, Y., K. Sumiya, and T. Asamuma. 2005. Development of management system on scheme of dumping block from barge in construting artificial fishing mound. *Annual Journal of Civil Engineering in the Ocean*. 21: 343–348 (in Japanese).

Nakayama, A., H. Yagi, Y. Fujii, K. Watanabe, T. Okano, M. Takeda, H. Watanabe, M. Takagi, and H. Ogawa. 2009. Field observation and development of numeric method for lower trophic production effect grasp by artificial upwelling producing structure. *Annual Journal of Coastal Engineering*. 56(2): 801–805 (in Japanese).

Oono, Y., M. Gomyo, T. Asamuma, T. Kawaguchi, and K. Hirata. 2004. Management on constructing artificial mound for upwelling generation with cubic blocks. *Annual Journal of Civil Engineering in the Ocean*. 20: 935–940 (in Japanese).

Suzuki, T. and T. Izutsu. 2004. The promising new market for coal ash use. *International Pittsburgh Coal Conference*.

CHAPTER 17

Estimating the Effective Wake Region (Current Shadow) of Artificial Reefs

Tae Gun Oh, Shinya Otake, and Moon Ock Lee

CONTENTS

ABSTRACT

The shape of artificial reefs can be characterized as having a very complicated structure that, in part, facilitates attracting fish by extending the reef's adhesion area or creating a large cavity. Drag coefficients are often used as one of the quantities to technically express this complexity of artificial reefs. For instance, it is well known that as a drag coefficient increases, larger eddies are generated downstream of artificial reefs so that plankton gathers more effectively around the reef. On the contrary, if the cavity of artificial reef is larger, the current flow easily passes through the reef and then plankton does not remain near the reef because the drag coefficient becomes small. This study contributes to our understanding of artificial reef deployment projects by clarifying the effect that a drag coefficient has on an artificial reef when the reef is subject currents of different direction. In addition, we elucidate the extent of vortex occurrence downstream of the artificial reef when this drag coefficient differs. The influence of the extent of this vortex was determined by comparing a drag coefficient to a swimming velocity of plankton. The maximum drag coefficient was similar to previous data obtained from a hydraulic experiment. The mean current velocity downstream of artificial reefs was faster than the swimming velocity of plankton, regardless of the drag coefficient. However, the extent of influence obtained from the distribution of vorticity magnitude was nearly similar to the vorticity obtained from potential theory. Results from this study permit more efficient design of artificial reef projects that are deployed to concentrate plankton and provide food resources for the associated fish assemblage.

Keywords: Artificial reefs, numerical experiments, wake region, drag coefficients, vorticity magnitude.

INTRODUCTION

Fish aggregate around artificial reefs for several reasons that include sound effects, light effects, flow effects, etc. Especially important in facilitating fish aggregation around artificial reefs is the presence of aggregated plankton that serves as food. Why does plankton aggregate around the artificial reef? The answer may be found by examining the flow or current characteristics around the artificial reef. The flow is generated by a weak stream called the "wake region," which is found down-current from the artificial reef. This wake region consists of counter-direction flow, weak flow, and many eddies that vary in scale. Plankton, which serve as an important food source for many fish species, aggregate in this wake region (Mann and Lazier, 1996). The wake region can attract fish to the reef by providing shelter and attracting food such as plankton. Moreover, turbulence at the edge of the wake region can attract certain pelagic species as well (Seaman et al., 2000). Japanese guidelines (Fisheries Agency, 2000) recognize this attractive effect on fish by the reef. However, more information is needed if the wake region effect is to be used in the designing of the artificial reefs. The effectiveness of an artificial reef can be increased by extending the wake region. Thus, it could be argued that clarifying of wake zone size is quite important in the future design of artificial reefs.

Generally, it is known that zooplankton, as food for fishes, are attracted to counter flow regions on the down current side of the artificial reef (Kakimoto et al., 1983). Moreover, the wake region size can be altered by flow direction over the reef. The irregular shape of a typical artificial reef means that current flowing from different directions leaves a different wake region shape and size relative to the reef size. Wake region size is determined by drag coefficient and projected surface area of the reef. Therefore, the objective of this study is to clarify counter and downstream flow distribution at artificial reef. Subsequently, the drag coefficient is calculated on artificial reef subjected to various flow directions. Also, vorticity magnitude is calculated to identify the effective wake region using a numerical solution.

It is well known that the region affected by an artificial reef is wide. For example, type B fishes, (as defined by Nakamura, 1985) group around the artificial reef in assemblages from 200 meters to 400 meters round the artificial reefs off Japan. This wide region cannot be explained by only using simple flow analysis. Often, artificial reefs include huge caves that permit fish to be able to swim through the reef. These caves (which include holes, crevices, and large pilings) often create complex flow patterns around the artificial reef. Under this condition, plankton often through the artificial reef and are not permitted to remain around the artificial reef because most plankton are weak swimmers. Generally, the flow and eddies generated around the artificial reef allow the plankton to stay in the current shadow and remain with the artificial reef. This relationship can explain the importance of understanding eddy diffusion coefficients and eddy dispersion theory. A quantitative examination of the effect of flow patterns around artificial reefs will help improve artificial reef designs in the future. Consequently, the objective of this study is to clearly examine the spatial scale of wakes on artificial reefs using drag coefficients and to understand the flow structure behind the artificial reef by examining eddy diffusion coefficients.

MATERIALS AND METHODS

Basic Equations

In this study the FLUENT modeling computer application package was used to simulate flow around the artificial reef. Also, a realizable k-ε model of the two-equation type was adapted to calculate turbulence. The transport equation of realizable k-ε model is defined by Equations 17.1 and 17.2.

$$\rho\frac{Dk}{Dt}=\frac{\partial}{\partial x_j}\left[\left(\mu+\frac{\mu_t}{\sigma_k}\right)\frac{\partial k}{\partial x_j}\right]+G_k+G_b-\rho\varepsilon+S_k \tag{17.1}$$

$$\rho\frac{D\varepsilon}{Dt}=\frac{\partial}{\partial x_j}\left[\left(\mu+\frac{\mu_t}{\sigma_\varepsilon}\right)\frac{\partial k}{\partial x_j}\right]+\rho C_1 S_\varepsilon-\rho C_2\frac{\varepsilon^2}{k+\sqrt{\nu\varepsilon}}+C_{1\varepsilon}\frac{\varepsilon}{k}C_3 G_b \tag{17.2}$$

where ρ is density of fluid, C_1 is max[0.43, η/(η + 5)], η is $S(k/\varepsilon)$, S is $\sqrt{2S_{ij}S_{ij}}$, G_k represents the generation of turbulence kinetic energy due to the mean velocity gradients, G_b is the generation of turbulence kinetic energy due to buoyancy, C_2 and $C_{1\varepsilon}$ are constants, σ_k and σ_ε are the turbulence Prandtl numbers for k and e, respectively, and S_k and S_ε are user-defined source term. After completing the flow analysis around an artificial reef, the drag coefficient can be calculated by considering the fluid total force that acts on an artificial reef body (Equation 17.3).

$$C_D=\frac{F}{0.5\ \rho u_i^2 A} \tag{17.3}$$

where F represents total force on the body, u_i^2 is the x-direction velocity of each cell, and A is the area projected against flow.

Also, vorticity is a measure of the rotation of a fluid element as it moves in the flow field, and is defined as the curl of the velocity vector (Equation 17.4).

$$\xi=\nabla\times\overline{V} \tag{17.4}$$

where $\overline{V}$ is velocity vector in each direction.

Eddy Energy Dispersion Theory Downstream of the Artificial Reef

A conservation equation of eddy energy downstream of the artificial reef can be shown as follows,

$$\frac{\partial W}{\partial t}=-\overline{u_a w}-\omega-\frac{\partial}{\partial k}\omega(k) \tag{17.5}$$

where W = eddy energy generated in the wake downstream of the artificial reef, t = time, u_a: horizontal velocity of x direction, w = vertical velocity of z direction, $u_a\ w$ = Reynolds stress, ω = eddy energy damping (=2ν k^2W), and k = wave number of eddies.

The first term on the right side of the equation is Reynolds stress and is related to eddy energy generated in the wake downstream of the artificial reef. The second term indicates eddy damping resulting from friction around the flow environment. The third term indicates apparent energy damping. It is known that the energy spectrum by wave number of eddy flow has great energy within a small wave number region which shows energy supply, and small energy within a large wave number region, which shows energy damping resulting from friction. Kolmogorov found the presence of an energy decrease without energy damping within this intermediate region of wave number of the eddy and named it the inertial sub-range. This third term shows this inertial sub-range.

Eddy scale becomes smaller in this sub-range without energy damping, and this change in eddy scale affects plankton movement. That is, the larger eddy scale with a small wave number can control the movement of larger plankton allowing them more swimming potential, and a smaller eddy scale with large wave number can only control the movement of smaller plankton with less swimming potential. According to Mann and Laizer (1996), plankton movement (swimming speed) is proportional to plankton size. This means the effect of eddy scale is related to plankton size. So an eddy generated downstream of the artificial reef can catch and release plankton inversely proportional to the eddy scale in this wake region. Therefore, this third term is related to plankton's movement.

We can show that the eddy energy represented by this third term changes using the eddy diffusion coefficient, eddy frequency, and distance from artificial reef as follows,

$$\frac{dW}{df} = -\frac{AK^{\alpha} f^{\beta}}{x/L} \tag{17.6}$$

where A is a proportional constant and is related to the Karman constant because the eddy strength is proportional to Prandle mixing length, f is eddy frequency, K is the eddy diffusion coefficient, x is the distance from the artificial reef, and L is the length of artificial reef.

That is,

$$A = f(\kappa)$$

So, we transform from Equation 17.6 to Equation 17.7 simply,

$$\frac{dW}{df} = -\kappa \frac{mKf^2}{x/L} \tag{17.7}$$

where m is a proportional constant.

By the dimension analysis, we get Equation 17.8

$$\frac{dW}{df} = -\kappa \frac{mK^{\alpha} f^{\beta}}{x/L} \tag{17.8}$$

From Equation 17.8 we can introduce the eddy diffusion coefficient K.

$$K = \frac{3(W_1 - W_2)}{m\kappa\left(f_{i+1}^3 - f_i^3\right)} \frac{L}{x} = \frac{3(W_1 - W_2)}{0.4m\left(f_{i+1}^3 - f_i^3\right)} \frac{L}{x} \tag{17.9}$$

This coefficient "K" can be obtained by calculating the energy spectrum using a hydraulic experiment, while measuring the concentration distribution using diffusion experiments. From these two experiments, we can verify this equation.

Table 17.1 Numerical Experiments Conditions for Flow Field Calculating

Method	Conditions
Solver	Pressure based, SIMPLEC
Formulation	Implicit
Time	Steady
Turbulence model	Realizable k-ε model
Near-wall treatment	Standard wall function
Model constant	C2-Epsilon : 1.8
	TKE Prandtl number 1
	TDR Prandtl number 1.2
Boundary conditions	Input: velocity inlet
	Output: pressure outlet (0 Pascal)
	Far boundary: symmetry

Experimental Conditions for Simulation Model

Numerical experiments were carried out using two methods. Firstly, numerical models on velocity distributions were verified by Schlichting's analytical solution with 2-D cylinder. Secondly, the change rates of a wake region by drag coefficients were calculated by 2-D square and inclined square.

Table 17.1 shows that the common experimental condition, which is applicable for all experiments.

Verification Experiments

To validate the realizable k-ε model, experiments were conducted using Schlichting's analytical solution with 2-D cylinder. Schlichting's (1979) analytical solution is the same as Equation 17.10. The width of the wake is represented by Equation 17.11.

$$\frac{u_D}{U} = \frac{\sqrt{10}}{18\beta}\left(\frac{x}{C_D D}\right)^{-1/2}\left\{1-\left(\frac{y}{B}\right)^{2/3}\right\}^2, \quad \frac{x}{C_D D} > 5 \tag{17.10}$$

$$B = \sqrt{10}\beta(xC_D D)^{1/2} \tag{17.11}$$

where u_D is velocity loss, U is the main flow velocity, β is constant (0.18), x is the length of the wake length direction, y is the length of wake width direction, C_D is drag the coefficient, and D is the diameter of the cylinder.

Figure 17.1 depicts the experiment for comparison velocity loss behind the cylinder in the modeling experiment and analytical solution. The Reynolds number was 300,000, and the velocity loss was compared at x/D = 15, 25, and 35, respectively.

The Change in the Effective Wake Region Using Drag Coefficient

To calculate changes of effective wake region by drag coefficient, flow fields were simulated on the 2-D squares and inclined square (Figure 17.2). Reynolds numbers were 300,000, 600,000, and

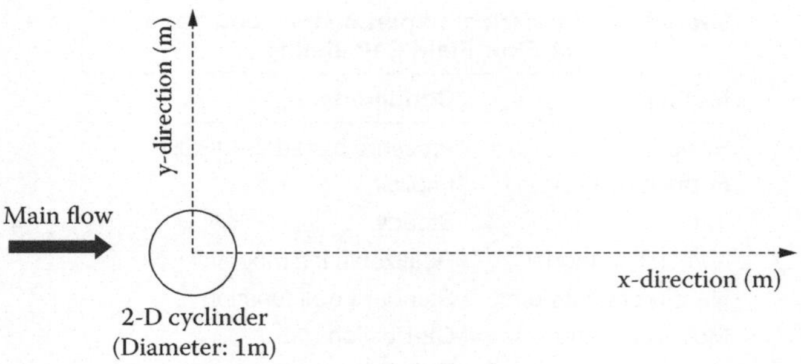

Figure 17.1 Definition of verification experiment.

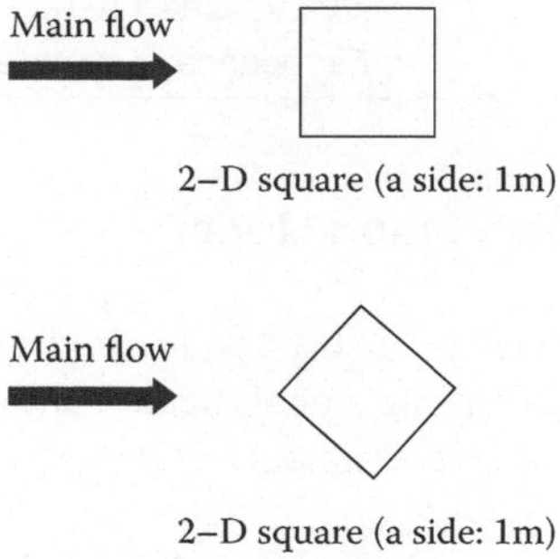

Figure 17.2 Numerical experimental conditions to simulate wake region on square and inclined square.

1,000,000, respectively. Also, vorticity magnitude was obtained using Equation 17.4, and the ecological effective wake region was calculated using vorticity magnitude using Equation 17.4.

Method of Hydraulic Diffusion Experiment

The hydraulic experiment was conducted using some flat plates placed instead of an artificial reef in the water to generate uniform eddies. Eddy scale was changed by changing the plate length. The flow velocity was constant at 2.4 cm/s. In the eddy diffusion experiments, scattered cylinders were made of nylon, 1 mm in length, 0.25 mm in diameter, and 1.14 kg/m^3 in density on the same scale as plankton. The cylinder concentrations were measured using a video camera. The diffusion coefficient was calculated from Equation 17.12 for comparison with Equation 17.9.

$$\frac{dC}{dt} = -u\frac{dC}{dx} + K\frac{\partial^2 C}{\partial x^2} \tag{17.12}$$

where C is concentration.

RESULTS AND DISCUSSIONS

The Result of Verification Experiments of Simulation Model

Figure 17.3 indicates the result of velocity distribution over the total flow domain of the verification experiment on the 2-D cylinder. Flow separation occurred in the back part of the cylinder due

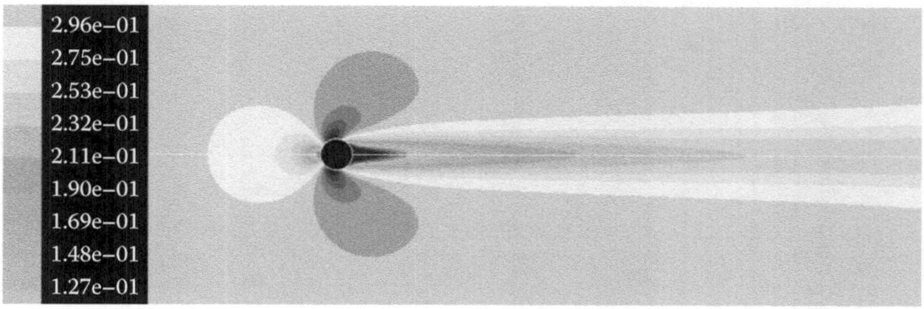

Figure 17.3 The result of verification experiment on total mean velocity distribution of wake region behind 2-D cylinder.

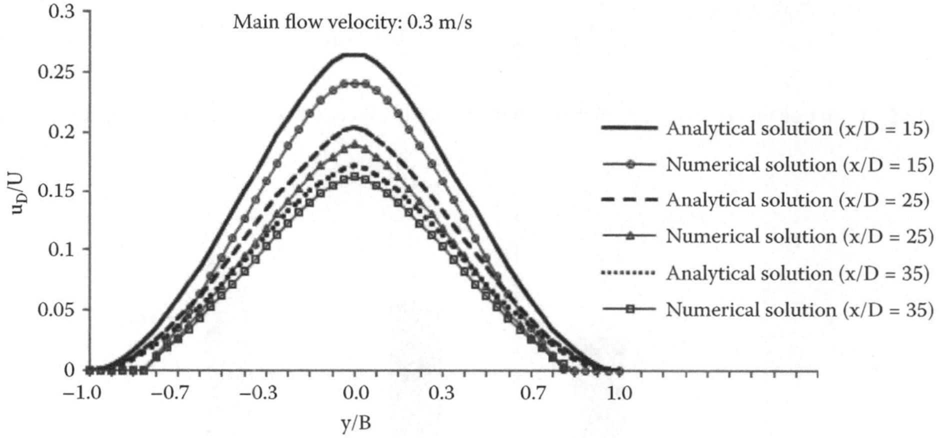

Figure 17.4 The result in comparison with numerical and analytical solutions by loss in the velocity rate behind a 2-D cylinder.

to turbulence flow. The length of the counter flow region was 1.5 times the cylinder diameter. By the law of mass conservation, velocity near both sides of the cylinder was being accelerated due to the slow velocity in the counter flow region. The mean velocity in the counter flow region and both sides of the cylinder (velocity accelerated region) were 0.05 m/s and 0.41 m/s, respectively.

Figure 17.4 compares the numerical and analytical solutions on the wake region velocity of the 2-D cylinder. In the case of $x/D = 15$, the loss velocity rate was 26%, and 17% at $x/D = 35$. The velocity loss rate in the numerical solution was low—5% less than that of the analytical solution—so that wake width in the numerical solution was narrower than that indicated by analytical solution. Values were determined using the analytical solution.

Changing the Effective Wake Region by Drag Coefficient

To understand the flow distribution down current of an artificial reef, the flow field was simulated with a realizable k-ε model. Square and inclined square artificial reefs were used to examine results from different projected length (or area), and also, by different Reynolds numbers (Figure 17.5). The drag coefficient of the square was 2.05, and inclined square was 1.81 at the Reynolds number of 300,000. Drag coefficients decreased at higher Reynolds numbers, and the drag coefficient of the square was 1.79, and the inclined square was 1.55 at the 1,000,000 Reynolds

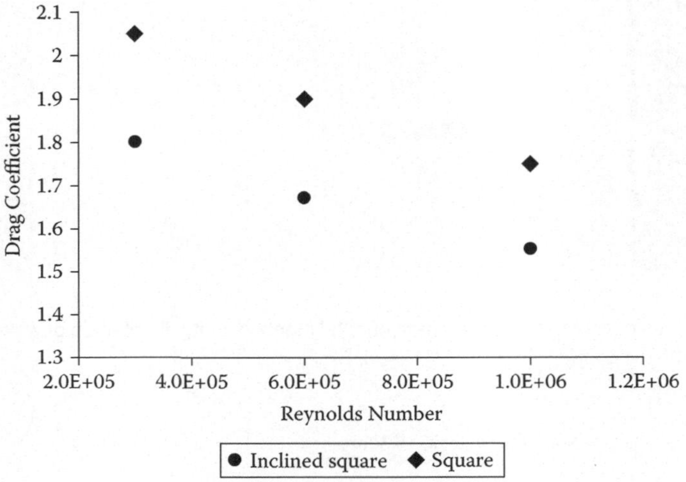

Figure 17.5 Changes in the drag coefficients of the square and inclined square body by Reynolds number.

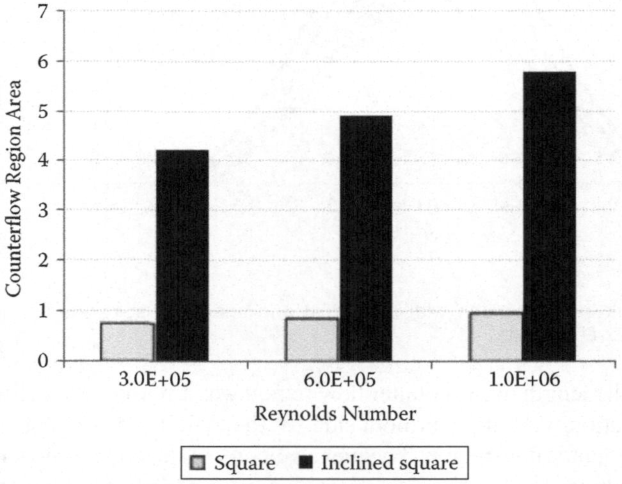

Figure 17.6 Comparison with counter flow region area of square and inclined square body by Reynolds number.

number. Figure 17.6 shows the counter flow area by various drag coefficients. The counter flow area is quite an important place on the reef to attract food organisms. Counter flow area of the square was 0.7 m^2, while that of the inclined square was 4.2 m^2, and it was 6 times larger than the counter flow area of the square.

Figure 17.7 indicates the area of ecological effectiveness according to the value calculated from the vorticity magnitude of Equation 17.4. Ecologically effective areas of the square and the inclined square were 8.03 m^2 and 15.37 m^2, respectively, with a range of Reynolds numbers from 300,000 to 600,000. However, at the 1,000,000 Reynolds number, the ecologically effective area of the square was 15% bigger than that of the inclined square. Figures 17.8 to 17.13 show streamlined distributions in the counter flow region of the square and the inclined square body. They show that the distribution of the streamlines differs by the Reynolds number. Figures 17.14 to 17.19 show

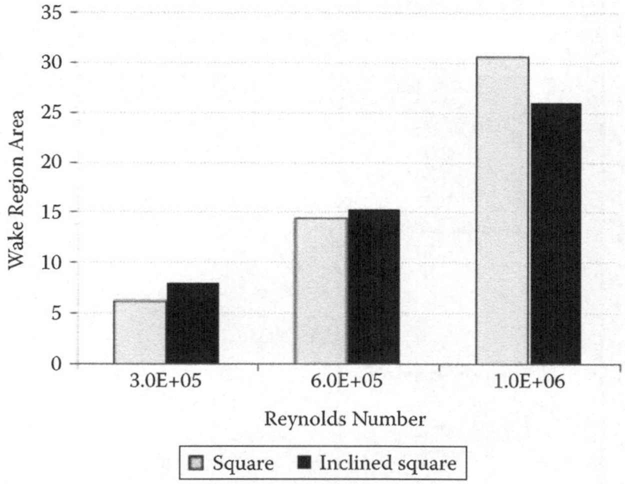

Figure 17.7 Comparison with wake region area of the square and the inclined square body according to Reynolds number.

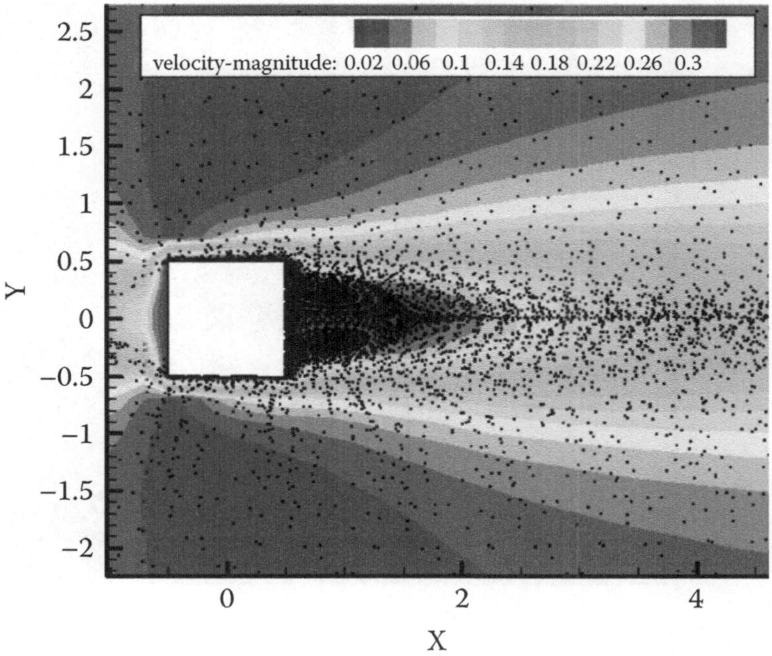

Figure 17.8 ***See color insert.*** Distribution of streamline by counter flow behind a square body (Re = 300,000).

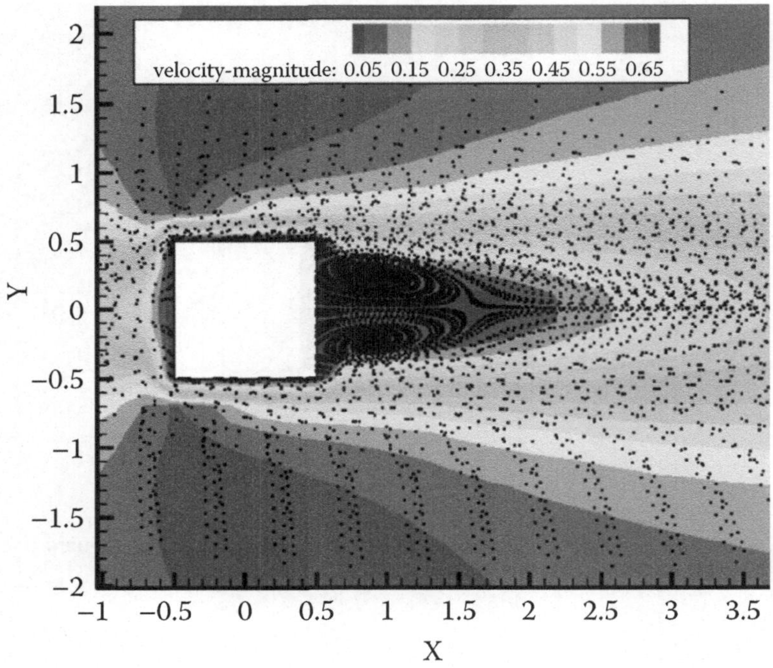

Figure 17.9 ***See color insert.*** Distribution of streamline by counter flow behind a square body (Re = 600,000).

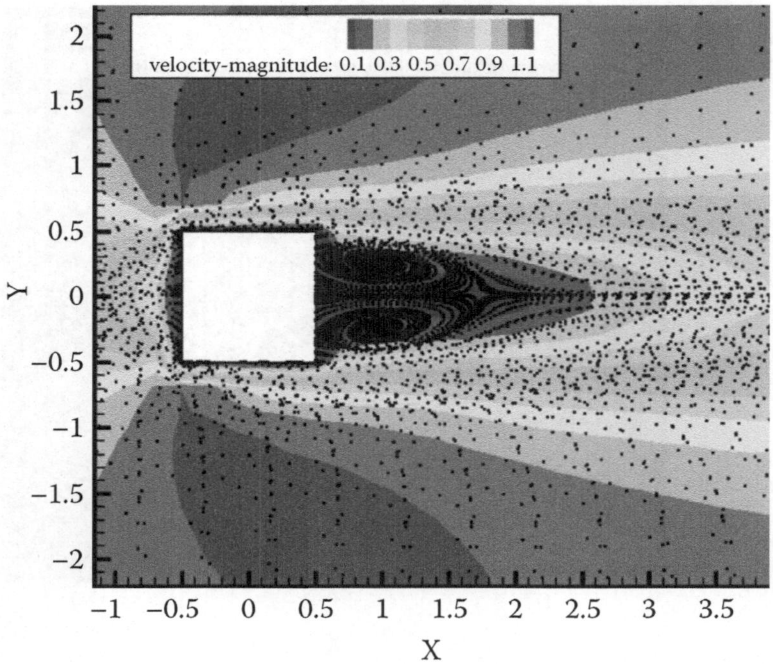

Figure 17.10 ***See color insert.*** Distribution of streamline by counter flow behind a square body (Re = 1,000,000).

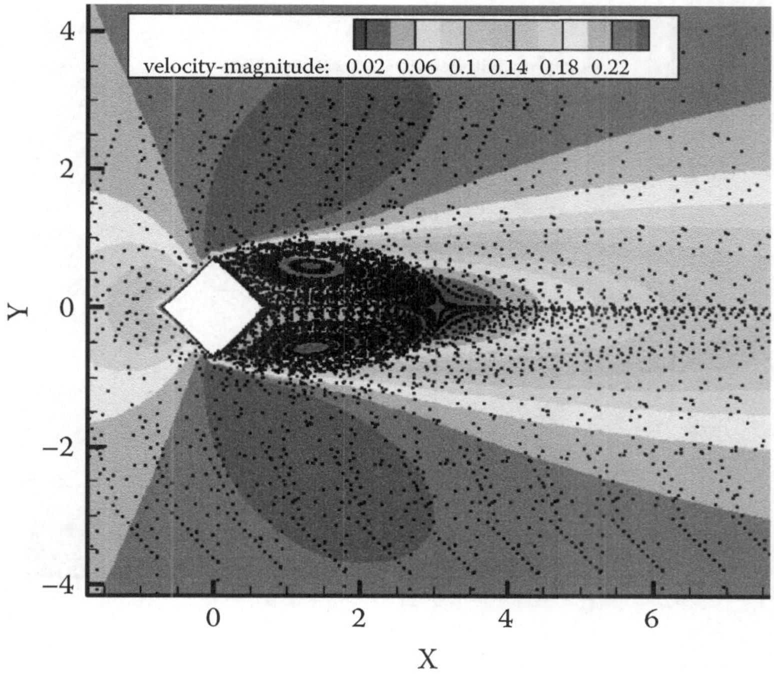

Figure 17.11 ***See color insert.*** Distribution of streamline by counter flow behind an inclined square body (Re = 300,000).

the distribution of the vorticity magnitude behind the square and the inclined square. These can be explained by the vorticity magnitude distribution according to the Reynolds number.

In this study, the ecologically effective areas were affected by the counter flow region and vorticity. The increase in drag coefficient did not lead to an increase in the ecologically effective area. Instead, however, we found out that an increase in the projected length (or area) led to an enhanced ecological wake region. Our study showed that differences of more than 6.0 times by projected length in the counter flow region attracted food organisms. This suggests that ecological effects can be increased or decreased by rearrangement of the same artificial reef.

Change in Eddy Scale and Eddy Diffusion Coefficient with Distance

Eddy strength decreased with distance from the plates. The eddy angular velocity and eddy radius decreased with the distance from plates (Figures 17.20 and 17.21). These results suggest that the eddy effect decreases and flow effect increases with distance from the artificial reef. Figure 17.22 shows the decrease in the eddy diffusion coefficient calculated in Equation 17.12. These results agreed with the suggestion that the eddy effect decreases with distance from the artificial reef. However, this effect does not disappear because this eddy scale is effectively large.

Relationships between Eddy Diffusion Coefficient and Eddy Energy Change

Figure 17.23 shows that the eddy diffusion coefficients calculated using the eddy energy dispersion theory presented in Equation 17.9 agree with a hydraulic diffusion experiment, if m = 0.5. This means that the eddy diffusion coefficient is proportional to the difference of eddy energy, and counter proportional to eddy frequency and distance from the artificial reef.

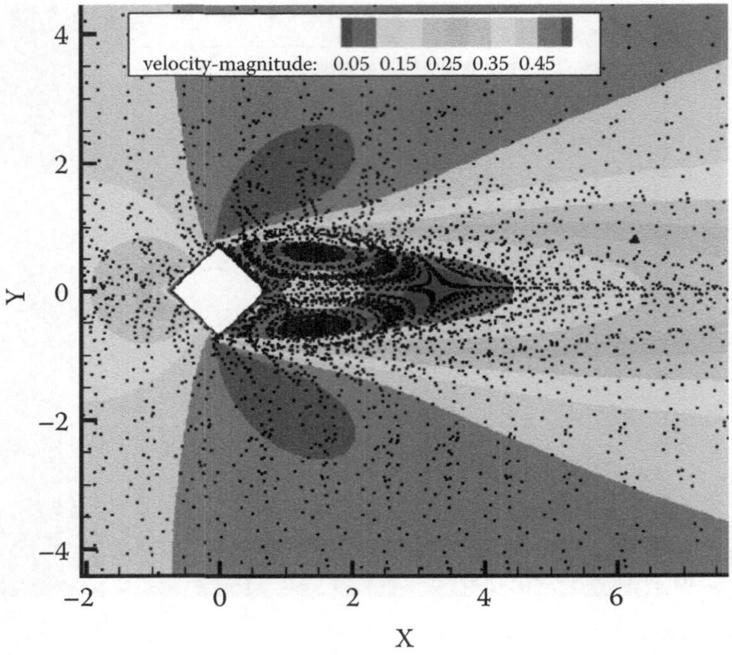

Figure 17.12 ***See color insert.*** Distribution of streamline by counter flow behind an inclined square body (Re = 600,000).

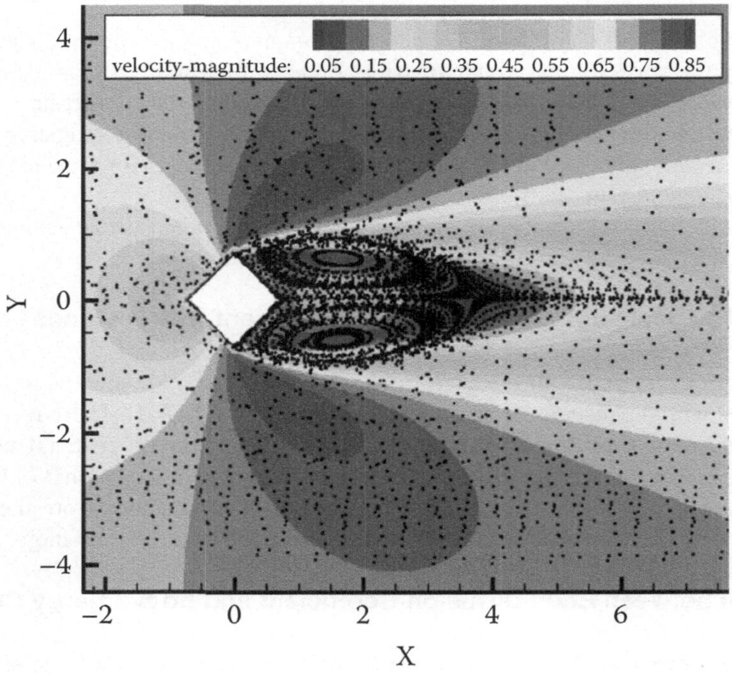

Figure 17.13 ***See color insert.*** Distribution of streamline by counter flow behind an inclined square body (Re = 1,000,000).

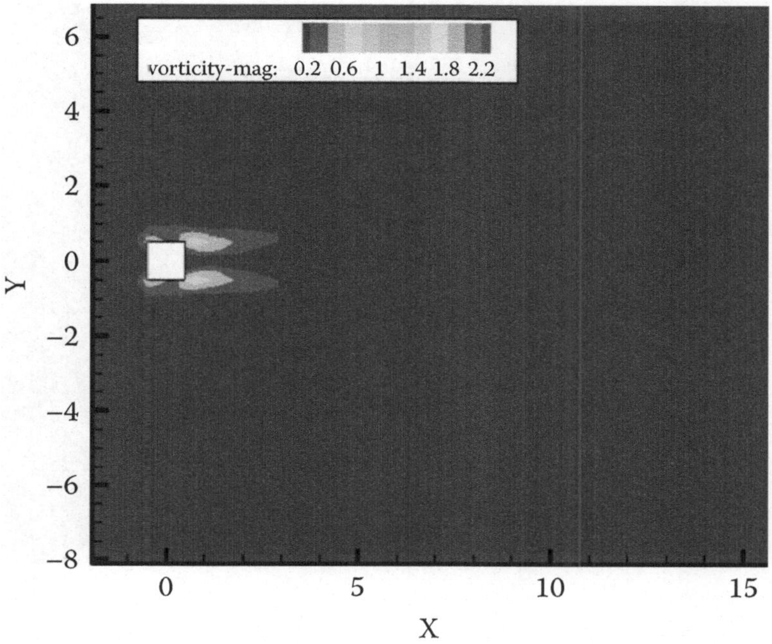

Figure 17.14 ***See color insert.*** Distribution of vorticity magnitude behind a square body (Re = 300,000).

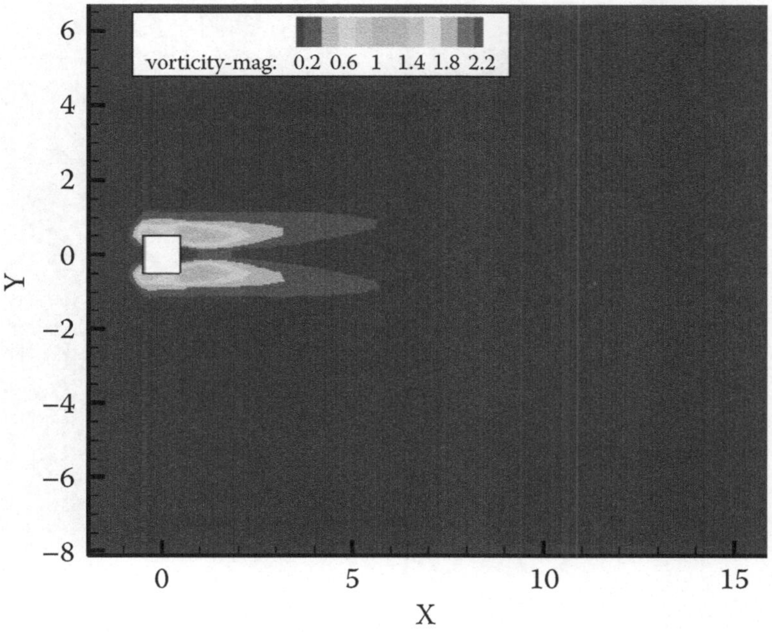

Figure 17.15 ***See color insert.*** Distribution of vorticity magnitude behind a square body (Re = 600,000).

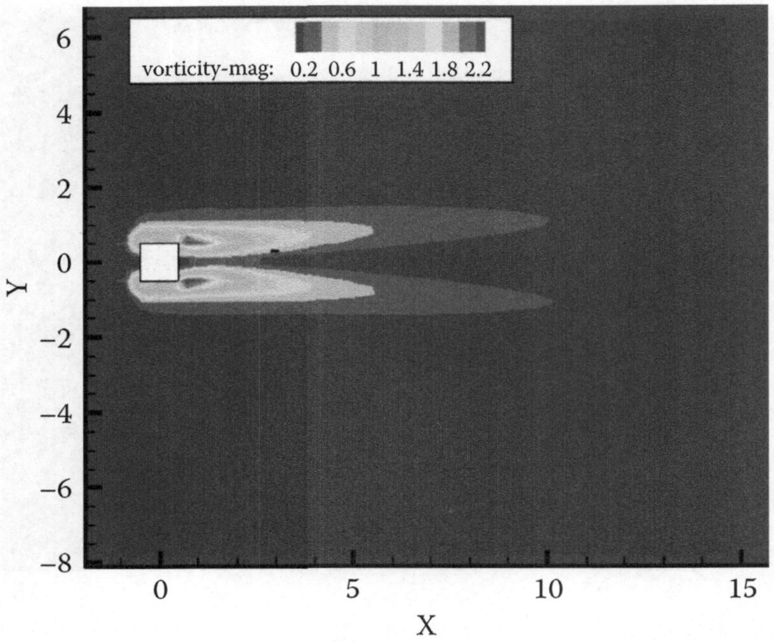

Figure 17.16 ***See color insert.*** Distribution of vorticity magnitude behind a square body (Re = 1,000,000).

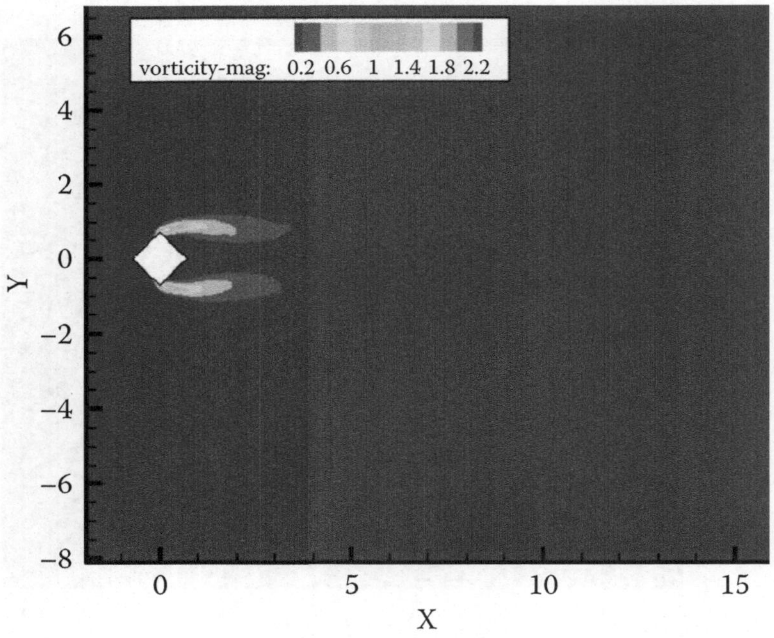

Figure 17.17 ***See color insert.*** Distribution of vorticity magnitude behind an inclined square body (Re = 300,000).

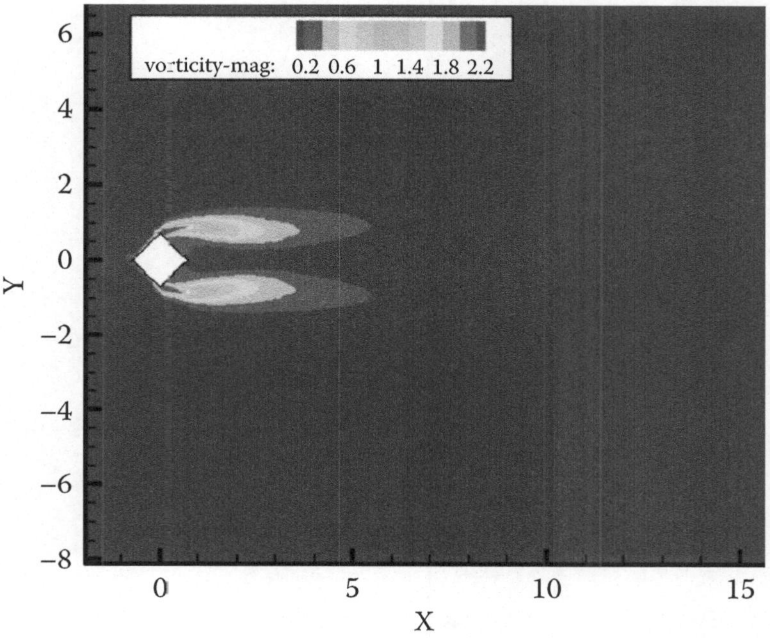

Figure 17.18 ***See color insert.*** Distribution of vorticity magnitude behind an inclined square body (Re = 600,000).

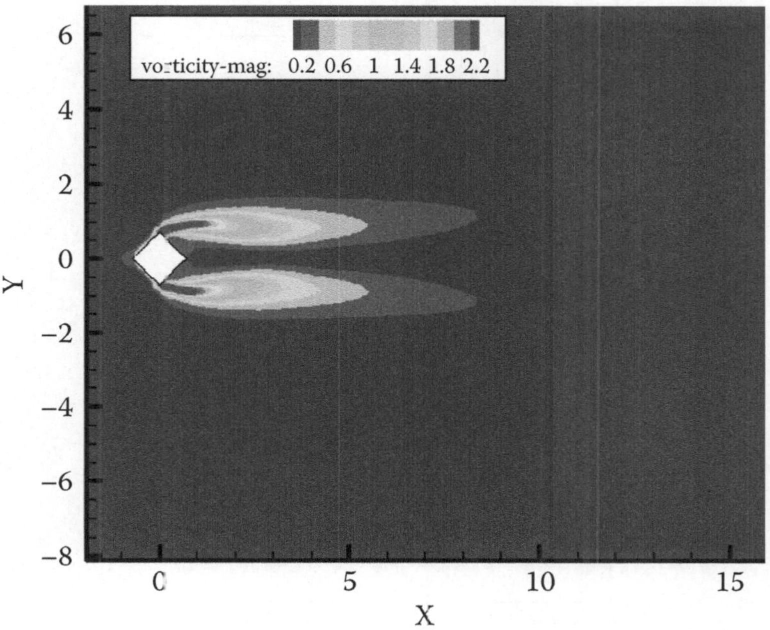

Figure 17.19 ***See color insert.*** Distribution of vorticity magnitude behind an inclined square body (Re = 1,000,000).

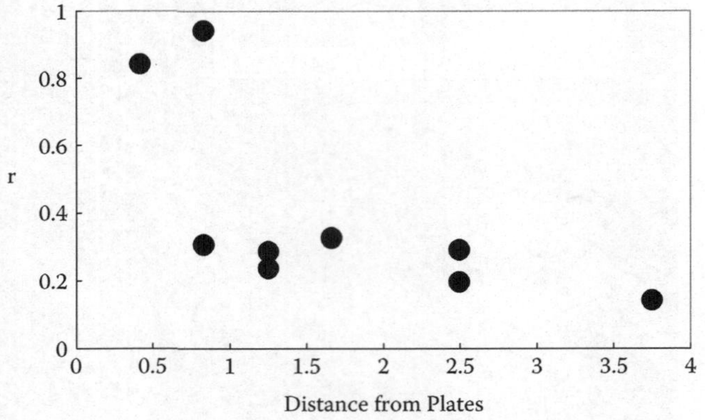

Figure 17.20 Decrease of the eddy radius with distance.

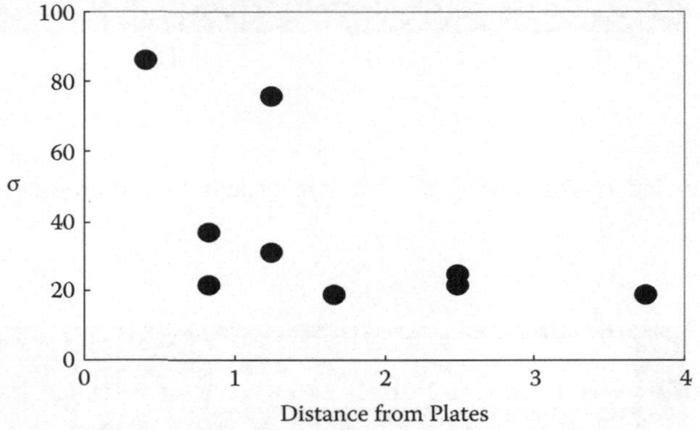

Figure 17.21 Decrease of the eddy angular velocity with distance.

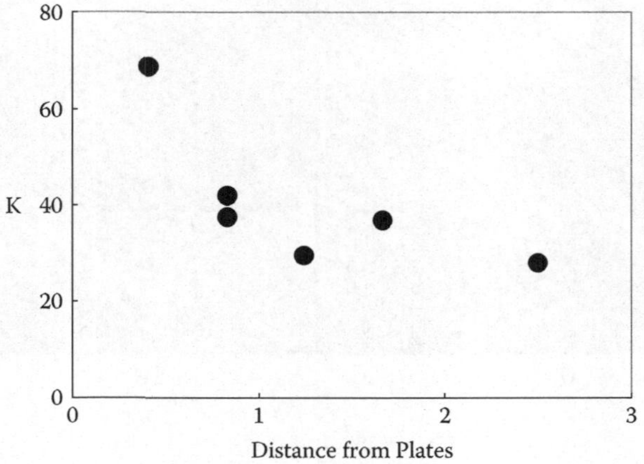

Figure 17.22 Decrease of the eddy diffusion coefficient with distance from plates.

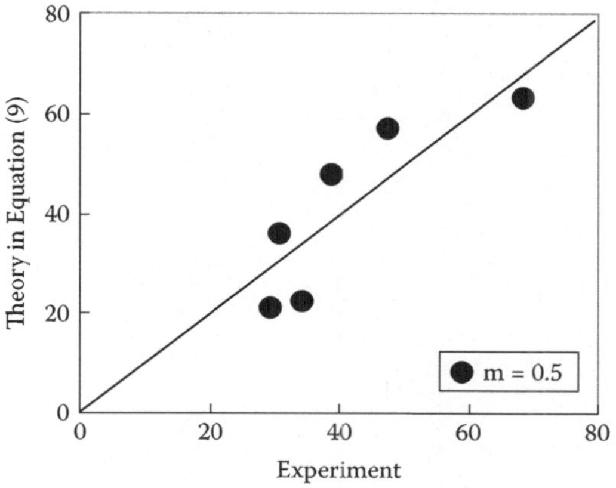

Figure 17.23 The verification of Equation 17.9 compared with the experiment.

CONCLUSIONS

In our examination of improving artificial reef designs, we looked at two physical features of flow around the reef. In describing the effect of the artificial reef, we considered two methods.

1. Complex flow in the wake can be expressed by the drag coefficient used in the hydro-dynamics. The drag coefficient indicates a magnitude of viscosity force generated by eddies in the boundary layers of the artificial reef and the pressure difference between upstream and behind the artificial reef. This feature can help determine the characteristics of the artificial reef shape
2. Change of eddies in the down current wake of the artificial reefs can be expressed by eddy diffusion coefficients used in hydro-dynamics. This is because the eddy diffusion coefficient includes the characteristics of the eddy strength and the eddy scale.

Our results indicate:

1. Drag coefficients are effective in indicating the wake scale generated by the artificial reef.
2. The larger the drag coefficient is, the larger the wake scale is.
3. The wake scale expressed by drag coefficient is small.
4. The longer the distance from the artificial reef, the smaller the eddy scale.
5. Eddy diffusion coefficients were theoretically plausible, and they were verified.
6. The eddy diffusion coefficient indicates the disappearance of eddy energy and is smaller with distance from the artificial reef.
7. The effective region of the artificial reef can be measured on site by using this method.

REFERENCES

Kakimoto, H. 1983. Study on the distribution of zoo plankton at artificial reefs (in Japanese). *Journal of Japanese Society of Fisheries Engineering* 19(2):21–28.

Mann, K.H. and J.R.N. Lazier, 1996. *Dynamics of Marine Ecosystems*, 2nd edition. Blackwell Science, Cambridge, MA: 394.

Nakamura, M., 1985. Evaluation of artificial reef concepts in Japan. *Bulletin of Marine Science* 37:271–278.

Schlighting, H., 1979. *Boundary Layer Theory*. New York: McGraw Hill, 24–95.

Seaman, W. Ed. 2000. *Artificial Reef Evaluation*. Boca Raton, Florida: CRC Press, 62.

CONCLUSIONS

REFERENCES

CHAPTER 18

Enhancing Food Production on the Continental Shelf by Artificial Seamounts

T. Suzuki and O. Hashimoto

CONTENTS

ABSTRACT

The Japanese government has been evaluating the possibility of enhancing fisheries production for food fish in the Japanese coastal region during the past quarter of a century. This has resulted in further promotion of artificial seamount (reef) development. There are several engineering issues that need to be resolved in order to continue to construct artificial seamounts on the continental shelf at deeper areas and at a larger scale. Based on our summary of engineering researches and developments, basic technological ideas are needed. This chapter summarizes the results of current engineering modeling studies and technical developments needed for future actions toward the improvement of the nearshore Japanese fisheries.

Keywords: Upwelling, artificial seamounts, fishing ground, continental shelf, density stratification

INTRODUCTION

To recover the current extreme low self-sufficiency of food resources in Japan, it is believed that it is important to provide good-quality protein by increasing fish and shellfish production off the coast of Japan. It is known that productivity in the ocean depends almost solely on phytoplankton productivity. Consequently, the fishery agency of Japan has been supporting research and development on artificial upwelling through the deployment of artificial reefs and structure since 1988 to increase primary productivity. Starting in 1995, a 6-year field experiment was conducted to evaluate the possibility of artificial enrichment of primary productivity by uplifting the lower water masses containing more nutrients and raise these water layers into the euphotic zone (Suzuki and Takahashi, 1997). The success of this project demonstrates the possibility for enhancing fish production in the ocean. Based on the results of this project, artificial seamounts were constructed at six different locations off the western Japanese coast (shallower than 100 m) as public enterprises Subsequently, in 2007, an agency, Ocean Fact of Japan, clearly demonstrated the effective enhancement of primary productivity in an offshore area. As a result, larger-scale artificial seamounts are being requested for construction on the continental shelf in the near future. Interestingly, there is no research supporting high upwelling efficiency using structure to facilitate upwelling in depths greater than 100 m. Since there is a significant density difference in seawater between surface and bottom water layers, resulting in a strong density stratification in the offshore region of the continental shelf, a larger artificial seamount is required.

The present chapter summarizes past studies conducted in Japan on the use of artificial structures to induce upwelling. Additionally, this chapter indicates possible engineering problems relevant to large-scale seamount deployments at depths over 100 m and the probable solutions for these problems.

REVIEW OF PAST STUDIES

Research and development on artificial reefs used for upwelling were conducted from 1988 to 2000 with financial support from the fishery agency of Japan. The research and development were initiated with a feasibility survey of the enterprise. As a result of these preliminary surveys, the suggested course of action was then undertaken in the form of public enterprises after thorough evaluation of their potential as business ventures. Below is a summary of these feasibility studies, the results of the preliminary test, an examination of the investment efficiency, and an examination of the performance of these enterprises.

Feasibility Studies on Artificial Upwelling

Artificially created upwelling currents were investigated from 1988 to 1994. Initially, vertical distributions of nutrient concentration from the surface to the bottom layer were determined in various seas in Japanese coastal waters. As a result, it was confirmed that nutrients are poor near the surface but become richer in deeper areas below the compensation depth throughout the Japanese coastal region during summer. This has also been observed in the other oceans around the world.

Subsequently, seawater samples taken from the surface and bottom layers were mixed after being filtered, incubated under light, and inoculated with phytoplankton seed cells. Figure 18.1 indicates the number of phytoplankton cells after incubation. Here we see that the nutrient concentration in seawater and the phytoplankton's cell yields are proportional.

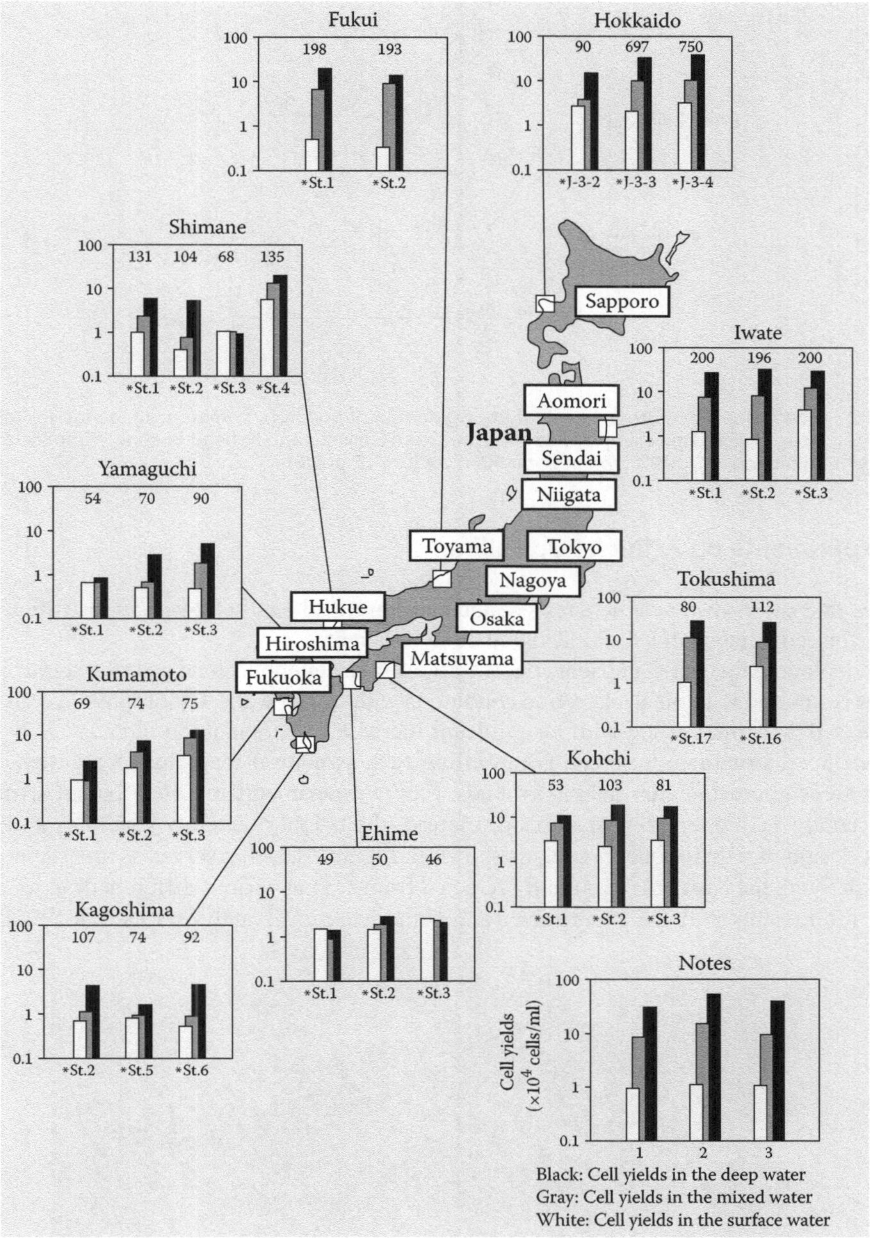

Figure 18.1 Final cell yields of *Chaetoceros gracillis* grown in filtered seawater of surface, bottom, and the 1:1 mixture of surface and bottom water collected from various coastal areas of Japan.

Based on this result, the possibility of enhancing phytoplankton growth was further evaluated by artificially mixing bottom water with upper compensation-depth water. Simultaneously, natural seamounts, well known as excellent fishing grounds, were evaluated for their capabilities of inducing upwelling. Moreover, flume experiments were also conducted to determine the most effective structure for generating upwelling vortex.

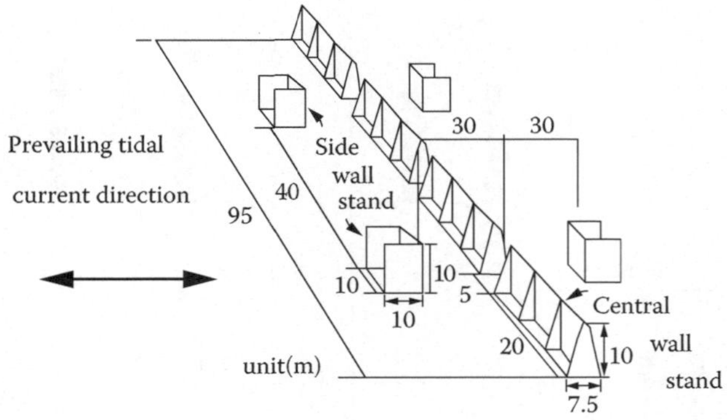

Figure 18.2 Size of the structure and actual arrangements of concrete cement wall stands for testing possible artificial upwelling in Uwa Strait between Shikoku and Kyushu Islands. (From Suzuki, T. and Takahashi, M. 1997. *Oceanology International* 97: 1–19.)

Field Experiments on Artificial Upwelling

Figure 18.2 shows results of field tests conducted from 1991 to 1994 for creating artificial upwelling using four concrete walls, four-sided and 10 m in height.

After deploying the walls, nutrient concentrations in the euphotic zone near the stands increased 2–4 times compared with the background conditions without the walls. Chlorophyll-a concentration also increased 2–3 times along with a significant increase of zooplankton biomass. Subsequently, suitable artificial structures for creating upwelling such as natural seamounts were then evaluated in deeper areas to create better fishing grounds. Flume experiments revealed that an artificial seamount structure with a length four times or more of the height efficiently generates an upwelling vortex. A design of a dual-cone mound, as shown in Figure 18.3, that serves as an efficient upwelling structure with the least cost was finally selected from among various different designs. Selection was done after testing in flume experiments and numerical model analyses (Suzuki, 1995). An arti-

Figure 18.3 ***See color insert.*** Depiction of artificial seamount constructed on the seabed at a depth of 82 m off Ikitsuki Island (height, 12 m; length, 120 m; conic bottom radius, 30 m; number of "Ashcrete" blocks, 4,860 (29,160 t)).

ficial seamount 12 m in height that extended 120 m in length was then constructed in 82 m of water off Ikitsuki Island in Nagasaki prefecture from 1997 to 2000.

Scientists cannot ignore potential environmental concerns for any new developmental activity of a large scale on fishing grounds. To reduce the potential damage to the natural environment attributable to the construction of the artificial seamount, a new engineering system was developed using a large number of huge sized concrete blocks made from coal ash, named Ashcrete® (a concrete of which all or part of the aggregate in plain concrete has been replaced with fly ash). Safety, durability, and the applicability of using Ashcrete in an artificial seamount were approved by the fishery agency of Japan. Accurate deployment of an artificial seamount made with Ashcrete blocks has been achieved using a differential global positioning system (DGPS) on an open-bottomed hopper barge.

Evaluation of Seamount Effects

Various investigations and evaluations concerning the upwelling effects of artificial seamounts were conducted from 1995 to 2001. Through oceanographic surveys and satellite image analyses, it was confirmed that phytoplankton in the surrounding area of artificial seamount increased.

Figure 18.4 shows the large fishing ground of 20 × 18 km that was created by constructing an artificial seamount. This fishing ground once supported a fishery that caught 250 t of fish annually. After enhancement through the addition of the artificial upwelling structure, the catch increased to 1500 t per year. Benefit by cost (B/C = (total convenience/rate of reduction during year)/total project

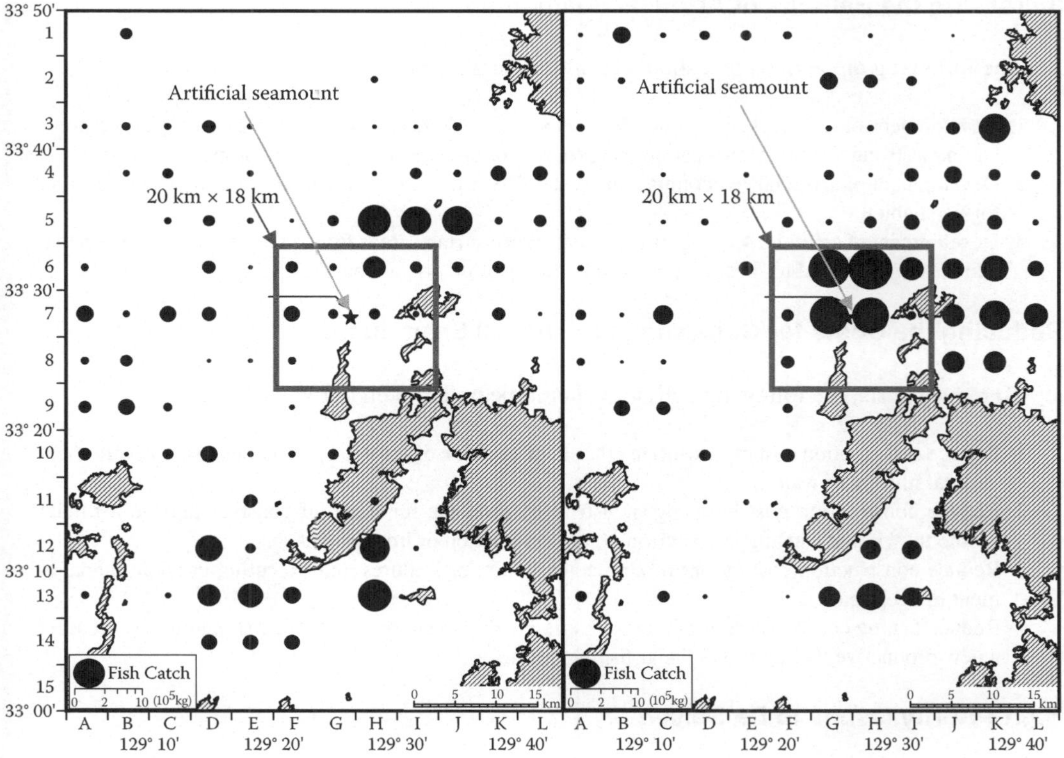

Figure 18.4 Fish catch increased six times in the area of 20 km × 18 km after the seamount was 70% complete (right) when compared to the catch before construction of the seamount (left).

cost) calculated from the results was 2.28. The Research Institute of Innovative Technology for the Earth in Japan estimated that there would be a potential increase in the fixation of CO_2 through the increase of primary production due to constructing the artificial seamount.

Expanding the Deployment of Artificial Seamounts

Based on the results of evaluating artificial seamount ventures, artificial seamounts, about 15 m in height and extending about 120 m long, were deployed at six different locations in 60–90 m depths in the coastal waters of Japan (Tsushima, Uku, Akune, west Goto, west Nagasaki, Maisaka and Iki) after 2005 as public works projects.

The Japanese government plans to construct a larger seamount in much deeper water on the continental shelf in the near future. This enterprise is expected to be on a gigantic scale; consequently additional consideration regarding the efficiency of the project needs to be included as part of the evaluation phase of the project.

ISSUES INVOLVING ARTIFICIAL SEAMOUNTS AT GREATER DEPTHS

It is necessary to improve benefit by cost ratio for businesses involved with large-scale artificial seamount deployments through a higher rate of upwelling with lesser costs for construction and management.

Increasing the Benefits of Artificial Seamounts

The following are actions that should result in increased benefits:

1. Development of artificial seamounts that support an effective vertical mixing in deeper sea areas having a strong vertical stratification and progress of effect evaluation technology.
2. Development of artificial seamounts that create diversified environments in water flow and sunlight for fish habitat.
3. Development of artificial seamounts that create more surface areas for the attachment of periphyton, which serves as food for benthic fishes and other benthic organisms.

Reducing the Costs for Constructing Artificial Seamounts

To reduce costs, the following activities should be undertaken to reduce costs:

1. Reduce construction cost by optimizing the shape and size of artificial seamounts to create sufficient vertical mixing of water.
2. Reduce construction cost by using industrial by-products for material constructing the artificial seamount without causing any environmental destruction or impairment.
3. Reduce construction cost by optimizing construction procedures and executing efficient management in deep sea.
4. Reduce fishing cost by shortening the navigation time and energy used by a fishing boat by creating a new productive fishing ground near fishing ports.

High-Priority Issues to Be Solved

The most essential issue, as determined in this study, is with regard to the upwelling of nutrients in water depths over 100 m into the euphotic zone. One plausible solution for this problem is to determine a seamount shape that efficiently creates upwelling of nutrients from deeper water.

Because it is difficult to accurately evaluate nutrient upwelling from deeper water into the euphotic zone under natural condition, a model using flow analysis will be the most practical method for determining the suitable shape and size of seamount. It is difficult to construct a mound having a long straight peak near the top in deep water. A cone-shaped mound can be more easily and exactly constructed. The cone-type mound was selected for deployment in deep water in the present study for comparison with the ordinary, levee-type mound.

Offshore seas are strongly vertically stratified due to temperature differences that are higher at the surface and lower in deeper water. Consequently, it is difficult to mix deeper water containing high concentrations of nutrients with the shallow waters in the euphotic zone. It is also difficult to determine vertical mixing at sea because of various microscale complex phenomena occurring in such a vast area. Consequently, an analysis that made use of an extended time series of vertical mixing and flow analyses was chosen in the present study. The following are the results of flow analysis relative to the shape and the vertical mixing efficiency of an artificial seamount.

UPWELLING FLUX OF NUTRIENTS AND SEA BOTTOM DEPTH ON AN ARTIFICIAL SEAMOUNT

In a strong flow field with a flat bottom, strong vertical stratification prohibits the vertical mixing of water. An artificial seamount constructed in such an area creates internal waves associated with the seamount. This has been confirmed by flow analysis and field observations. It seems that vertical flow generated by internal waves does not contribute directly to vertical mixing due to the counterbalancing of the positive and negative flows. However, vertical flow passing through each horizontal depth section, created by the internal waves, initiates upwelling of deeper water and downwelling of shallower water. Such vertical flows of water could transfer nutrients from deeper water to shallower water due to mixing. This is because nutrient concentrations are always higher in deeper water and lower in shallower water. For simulating this process, a three-dimensional LES (large eddy simulation) model was applied. In this model, water mixing does not occur when there is no artificial seamount. This is because there is no internal wave created when a seamount is absent. In the presence of a seamount, various types of internal waves could be created. This depends on the differences in water depth, flow speed, and the strength of vertical stratification. The internal waves create upwelling and gently move deeper water containing more nutrients to shallower water. Since internal waves are created only in the situation of constant flow with stratification, a vertical density gradient of 0.025 $kg/m^3/m$ and flow rate of 0.3 m/s were assumed in the present study. For investigating the vertical flux of nutrients at all depths, vertical nutrient distribution was assumed to be linear with 0 at surface and 1 at bottom. The upwelling flux was defined as the nutrient flux upwelled through a given horizontal cross section during a certain long period of time. Under the condition of uneven, vertical distribution of nutrients that was assumed in the present study, an upwelled flux of nutrients was expected to be created by return flow with the same flow rate. The total upwelled nutrients flux divided by the total volume of seamount in the target area was defined as the upwelling flux of nutrient per a unit volume. This represents the nutrient upwelling efficiency by the artificial seamount.

The condition of flow analysis was as follows: coverage horizontal area: 1074 × 1074 m (179 × 179 meshes); coverage depth: 80–200 m (20–50 meshes); flow velocity: 0.3 m/s; density gradient: 0.025 $kg/m^3/m$; and gradient of nutrient concentration: 0 relative unit at sea surface and 1 relative unit at sea bottom. Since wavelength and wave height of internal waves change due to the differences in sea bottom and sea surface, the vertical profile of upwelling flux change depends on the depth of the seabed, flow velocity, and vertical distributions of density.

Figure 18.5 shows the changes in the vertical profile of upwelling nutrient flux when the artificial seamount consists of eight cones with 20 m in height and 246,000 m^3 in volume deployed

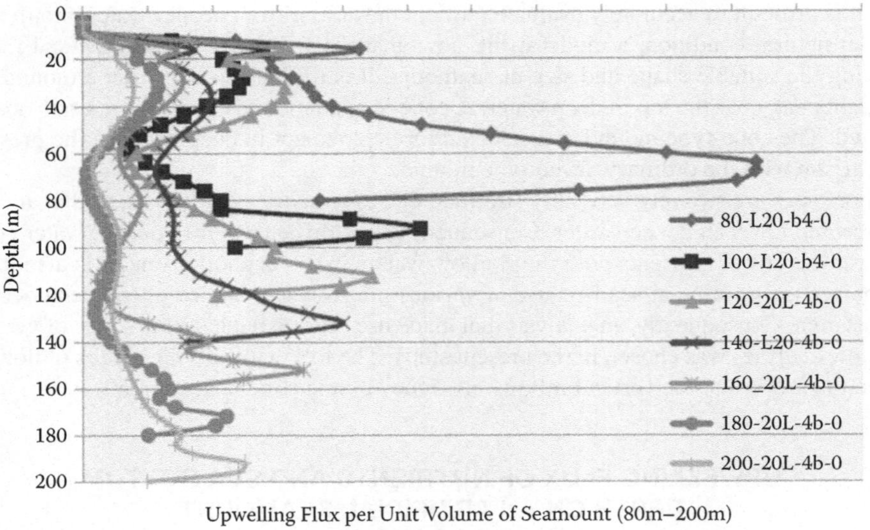

Figure 18.5 ***See color insert.*** Vertical profiles of upwelling flux of nutrients per volume versus construction depth of the artificial seamount. The lowest point of each line shows the depth where the seamount was constructed.

in water 80 m to 200 m deep. Vertical profiles of upwelling flux per unit volume changed greatly due to the differences in structure depth even with the same size and shape of artificial seamount. Upwelling flux was largest at the greatest depth of the artificial seamount, and the local maximum upwelling flux also appears in shallower depths. Since the shape of an artificial seamount relative to the effective upwelling flux under strong density stratification was in question, the value of upwelling flux per unit volume was not specified. The greater the depth of deployment, upwelling flux per unit volume near the seabed becomes smaller. Upwelling flux near the surface becomes smaller, but there also occurs a clear local maximum.

The unit upwelling flux around 30 m is maximal when an artificial seamount is installed at a depth of 80 m, and is minimal when it is deployed at 200 m. Under this analytical condition, with deployment at 120 m, the maximum value of the surface upwelling flux per unit volume was greater when compared to artificial seamounts deployed at 100 m. According to this flow analysis, the profile of an upwelling nutrient flux changes with the waveforms of the internal wave that is generated through the restrictions of the sea surface and seabed boundaries.

The Shape of Artificial Seamounts and Upwelling Nutrient Flux

To determine the shape of an artificial seamount that facilitates the most effective upwelling of nutrients, various shapes of seamounts were examined under the same conditions of bottom depth, flow speed of water, vertical stratification, and vertical distribution of nutrients. Upwelling flux per unit volume was evaluated by changing with two or more cones, each 20 m high and 40 m in bottom radius, along a line. If a vertex interval is expressed with the conic bottom radius of r, $2r$ represents two cones that are attached to each other, the volume of an artificial seamount changed with the intervals of the cone peaks L. The relationship between upwelling flux for each unit volume (unit upwelling flux) and the shape of artificial seamount was examined at a fixed depth of 160 m using the same conditions indicated earlier. Various parameters of artificial seamounts formed by cones are shown in Figure 18.6.

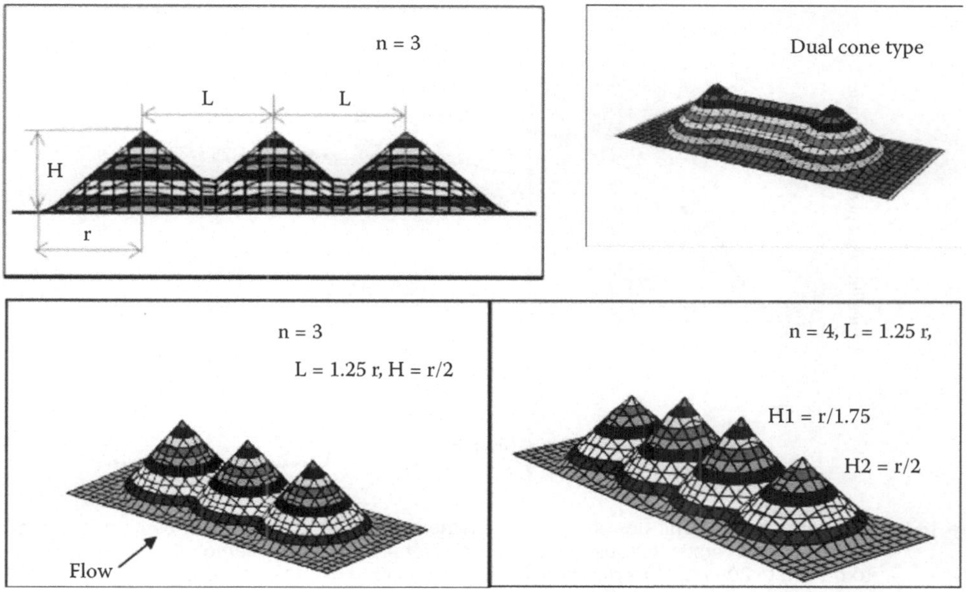

Figure 18.6 ***See color insert.*** Upper left are the parameters of an artificial seamount: *r*, conic bottom radius; *H* = *r*/2, conic height; *L*, conic top interval; and *n*, conic number. Lower left is a bird's-eye view of an artificial seamount consisting of three cones. Lower right is an artificial seamount consisting of four cones with deferent heights. Upper right is the previous dual-cone seamount.

The Number of Cones and Unit Upwelling Nutrient Flux

Upwelling nutrient flux per unit volume of artificial seamount changed with the number of cones, and the interval between cone tops arranged perpendicular to the flow direction. Figure 18.7 shows the relationship between upwelling nutrients flux per unit volume near the sea surface and cone numbers. If the number of cones constituting an artificial seamount increased from 1 to 5, there was a gradual increase in upwelled nutrient flux per unit volume, but nutrients attained saturation when the mound included more than four cones in all cases with cones having a radius of 3*r*, 2*r*, 1.5*r*, 1.25*r*, 1*r*, and 0.5*r*. The nutrient flux curve overlapped between cones having a radius of 1.5*r* and 1*r*. Upwelling flux of nutrients per unit volume changed with cone vertex interval *L*, as shown by six curves. Artificial seamounts composed of four or more cones, with a cone top interval *L* of 1 to 1.5 times compared to a cone bottom radius *r* can give a higher efficiency of upwelling than with previous dual-cone seamounts. Unit upwelling nutrients flux was largest with the cone vertex interval of 1.25 times the cone bottom radius. The unit upwelling flux (dots on the right side surrounded by the dashed line; Figure 18.7) of the conventional mound with a levee crown of 20 m in height changes with various distances between two peaks such as from 50 m (20*L*–50) to 200 m (20*L*–200). Although upwelling nutrients flux was large for 20*L*–150 and 20*L*–200, upwelling efficiency divided by volume of seamount was not very different due to the large volume. Artificial seamounts, having the vertex interval of 1.25*r* with four or more cones, showed a higher or similar upwelling efficiency to the conventional type. Even though structure of a straight peak contributed to upwelling flux under a continuous flow field with homogeneous density conditions, it has become obvious that a straight peak is not essential in a flow field with strong density stratification. Water flow with a strong density stratification tends to generate internal waves more effectively over the structure having three or more cones in a series than when compared to a structure with a straight long peak.

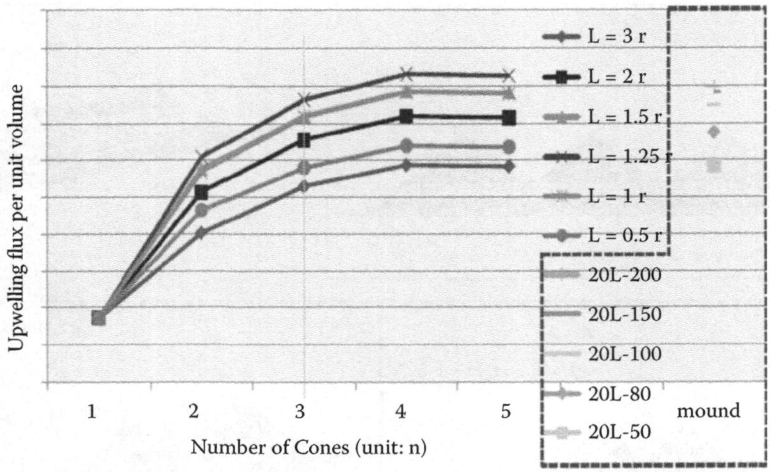

Figure 18.7 ***See color insert.*** Unit upwelling flux of nutrients versus number of cones composing artificial seamount and versus various types of conventional levee-type seamount varying peak distance from 50 m to 200 m. (Dots on the right side surrounded by the dashed line.)

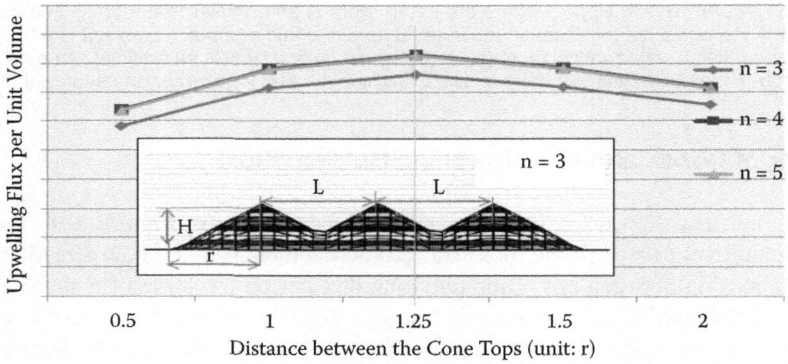

Figure 18.8 Unit upwelling flux near the sea surface versus distance between conic tops "*L*."

Distance between Cone Tops and Unit Upwelling Nutrient Flux

Figure 18.8 shows the upwelling nutrients flux per unit volume of seamount investigated by changing the vertex interval of *L* from 0.5*r* to 2*r* when the artificial seamount consisted of three to five cones. Upwelling nutrients flux per volume became large in the order of four, five, and three cones, and showed a maximum at 1.25*r* for the interval of cone top *L*.

Distance Ratio between Cone Tops (*La*/*Lb*) and Unit Upwelling Nutrient Flux

Change in upwelling nutrient flux was investigated when the cone vertex interval of the artificial seamount consisted of three cones, and the *La*/*Lb* as the distance of *La* and *Lb* was varied. Four curves from 1*r* to 2*r* cones with a vertex interval of *La* are shown in Figure 18.9. The vertex interval of three cones became equal, when the distance ratio between vertices *La*/*Lb* is 1. Upwelling nutrient flux for each unit volume of artificial seamount might become large, when top intervals

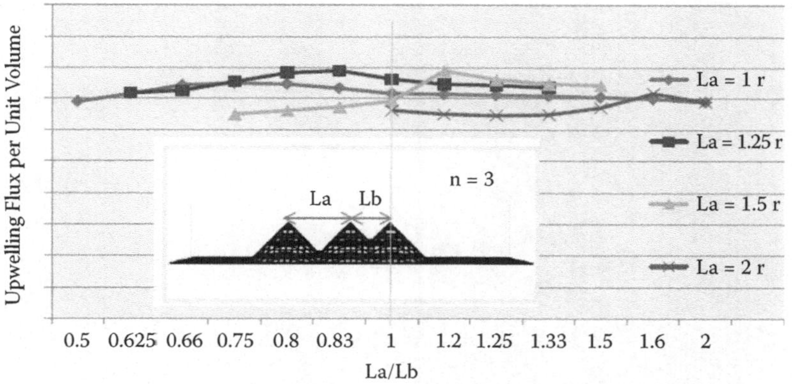

Figure 18.9 Unit upwelling flux of nutrients versus distance ratio between conic tops "*La*/*Lb*."

of *La* and *Lb* of three cones are slightly different. Consequently, there is an efficient arrangement even when the cone top intervals are unequal. There is a combination that became larger than the artificial seamount of a cone vertex interval $a = Lb = 1.25r$, and when it is $La = 1.5r$ and $Lb = 1.25r$, upwelling nutrient flux per unit volume served the maximum.

Distance between Seamounts and Unit Upwelling Nutrient Flux

To create a much larger upwelling flux of nutrients while keeping a large upwelling flux per unit volume, various combinations of two sets of artificial seamounts having a high efficiency were examined. Even if two or more artificial seamounts were deployed before and after a flow, upwelling efficiency of nutrients did not improve compared to a single artificial seamount. The upwelling nutrients flux per unit volume was investigated by changing the interval *Lm* of two sets of artificial seamounts consisting of four cones varied from $0.5r$ to $7r$.

Figure 18.10 indicates the upwelling nutrients flux response of two sets of artificial seamounts having four cones with vertex interval of $2r$, $1.25r$, and $1r$, respectively. When two sets of artificial seamounts with peak intervals *L* of $1.25r$ are arranged with $2r$ adjoining distance *Lm* between the peaks, the upwelling nutrients flux per unit volume was greatest.

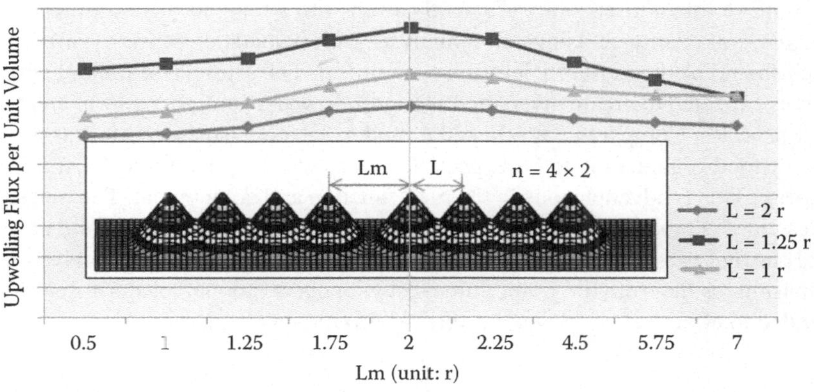

Figure 18.10 Unit upwelling flux of nutrients versus distance between two artificial seamounts "*Lm*."

Figure 18.11 ***See color insert.*** Depiction of cone-type artificial seamount made of Ashcrete blocks.

The upwelling nutrient flux per unit volume is larger when the seamount consists of four cones with the top interval $L = 1.25r$ by 10% or more than for a single artificial seamount. It is economically efficient to obtain the required upwelling nutrients flux by creating artificial seamounts with a large upwelling flux per unit volume. Figure 18.11 shows an image of cone-type artificial seamounts made of Ashcrete blocks with the top interval $L = 1.25r$ and the contiguity distance Lm of two sets of artificial seamounts to $2r$; height H: 20 m; total length: 460 m; conic bottom radius r: 40 m; conic number n: eight; conic interval L: 50 m; Lm: 80 m; seamount volume: 246,000 m^3; number of Ashcrete blocks: 37,300 (223,800 t).

CONCLUSIONS

Past efforts reveal that primary production in the sea can be enhanced by supplying nutrient- rich lower layer seawater to water above compensation depth using an artificial seamount. A structure that generates upwelling nutrients flux efficiently, even with strong density stratification on a continental shelf area, was examined using a flow analysis. This helped determine the artificial seamount design, which should be the most efficient and economical. The artificial seamount developed in the present study is expected to create diversified seawater flow and sunlight pattern, and these features could benefit fisheries. Compared with the older model, substrate surface area was greater, and the proliferation of food organism is to be expected. Blocks can be piled up to form a planned conic shape at its peak. This is advantageous in the both function and deployment. The most effective and certain forecasting method in the evaluation process is to determine the upwelling nutrients flux by the flow analysis. Future studies call for upgrading the method for numerical analysis on upwelling flux by comparing results from the water flume test with observational results at sea. This approach is recommended to effectively evaluate the vertical mixing mechanisms at sea.

REFERENCES

Suzuki, T. 1995. A study on the stimulation of coastal upwelling events created by artificial sea mounts. Doctor thesis. University of Tokyo, 127 pp.

Suzuki, T. and Takahashi, M. 1997. Enhancement of coastal upwelling by man-made sea-mount constructed by fly-ash concrete blocks for the increase of marine productivity. *Oceanology International* 97: 1–19.

CHAPTER 19

A Pathway to Resolving an Old Dilemma

Lack of Artificial Reefs in Fisheries Management

Stephen A. Bortone

CONTENTS

INTRODUCTION

In the first chapter, the argument was made that artificial reefs play almost no role in the management of any fishery. That is, they play almost no formal role, at least through an official management agency or group with management jurisdiction over fisheries. There are, of course exceptions, and the work presented by Kim et al. (Chapter 7 in this volume) is indicative of these exceptions. Nevertheless, the larger point is that artificial reefs have yet to achieve standing as a standard option in the management of fisheries. The plethora of research conducted over the past four decades indicates that there is interest among resource managers to include artificial reefs as a management alternative but there are many considerations yet to be resolved before true acceptance by managers is attained. The perception by the general public is that artificial reefs are already playing a role. Reality lies on the opposite side of this perception. Whether or not the premise is correct (i.e., artificial reefs play almost no role in the fisheries management), I offer a strategy for action leading to research results that are necessary to give fishery managers the information for proper evaluation of the use of artificial reef deployments. These will directly solve some fishery management issues that have been obstacles.

A SOLUTION TO THE DILEMMA

Simply stated, our current state of understanding of artificial reefs is insufficient in providing fishery managers with the information they need to make decisions to employ artificial reefs (or not). To play an effective and efficient role in management, our knowledge base needs to be at the level where there is a fair certainty of a positive effect on a fishery should an artificial reef be deployed. Steimle and Meier (1997) conducted a survey of resource managers and the respondents indicated that fisheries science needed to provide information on artificial reefs at a variety of levels to allow for the efficient development of effective reefs. It is not necessary to explicitly argue individual failures or how and why we got to this point after at least 35 years of deliberate research on artificial reefs. It is sufficient, nonetheless, to say we (i.e., the artificial reef researchers) have far to go in making a cogent case for fishery managers to be comfortable in recommending artificial reefs for inclusion as a fishery management alternative.

Following is an outline of the types of information required and a plan as to how researchers might design projects to better obtain the information necessary to justify the incorporation of artificial reefs into a fishery management plan. The outline has two basic parts: (1) directed decision making regarding research toward identifying the most appropriate species and their life history features that would benefit a fishery through using artificial reefs, and (2) a conceptual research model to directly apply this information in solving confronting fishery managers relative to artificial reefs.

Avenues of Artificial Reef Research

It is tempting to posit that we need to know all things about every facet of artificial reefs, the species, and features of the surrounding environment before fisheries science can proceed to fully appreciate the role that artificial reefs can play in management. This impractical and unrealistic perspective would only serve to further delay the implementation of artificial reefs as part of management plans. A more realistic and useful approach is to direct fishery research to specific areas of data gathering that would be directed toward resolving the dilemma. Following are offered two directions that should prove fruitful in achieving this larger goal.

Focusing on Appropriate Species

Jared Diamond in his book *Guns, Germs, and Steel* (Diamond, 1997), based on data from others, eloquently argued that of the potential number of species available for cultivation and efficient utilization by humans, only a small number of species are actually employed in agriculture. For a variety of reasons (i.e., physical attributes, tolerances, temperament, growth rate, ease of handling, social interactions, etc.) the vast number of species on our planet are essentially unmanageable—at least from a practical standpoint. Fishery managers might heed the examples from the history of human civilization and realize that not all fish species can be managed either. More to the point, perhaps not all species can be effectively managed by a particular management strategy. The extended argument here is that perhaps artificial reefs can only be implemented as a useful management option to a limited suite of species and not the entire community.

Too often, artificial reef research tries to target the fish community (more appropriately, fish assemblage; sensu Bohnsack et al., 1991) as a whole. Most research studies make assessments using species richness and species diversity parameters as standards for measuring success. While noble, and in some instances appropriate, this approach will most likely misdirect research efforts toward achieving species-specific management goals. These assemblage-based parameter objectives are useful if the goals for building the reef are to enhance recreational diving, esthetically improve an area, and/or increase the ecological services of a region, etc.

Our attention here focuses on an effort to develop artificial reef assessments or demonstration projects that help identify the species that are most appropriately managed, when they comprise a fishery, through the deployment of artificial reefs. The quest is to identify species that are more appropriate as targets for artificial reef deployments and/or those species that will have a higher probability of being manageable through manipulations involving artificial reefs.

One way to address this issue is to identify species based on their affinity to artificial reefs as structure. In simple terms, a species may be "preadapted" to artificial reefs, in that it may have an innate affinity, through life history attributes or characteristics, which is facilitated or enhanced by the individuals being present on, or associated with, an artificial reef. The demonstrative negative example is that if a species is repulsed by artificial reefs, a manager would be wasting considerable effort in trying to implement artificial reefs as a management option for this particular species.

Fish species have been classified according to their affinity to artificial reef structure (Nakamura, 1985; but see modifications to this scheme in Figure 19.1). In a practical sense, species with a closer spatial affinity to artificial reefs may give managers better reason to justify using artificial reefs to manage those particular species than managing other species with less affinity for artificial reef attributes. With this spatial relationship concept in mind, Bortone (2008) expanded on this concept of species-specific, spatial affinity and coupled it with the attraction/production concepts of artificial reef ecology (see Bohnsack et al., 1997; Lindberg, 1997 for discussions about attraction versus production). In Figure 19.2, according to their respective reef affinity, fish species are arranged according to axes indicating attraction and production. A species that has both low attraction and

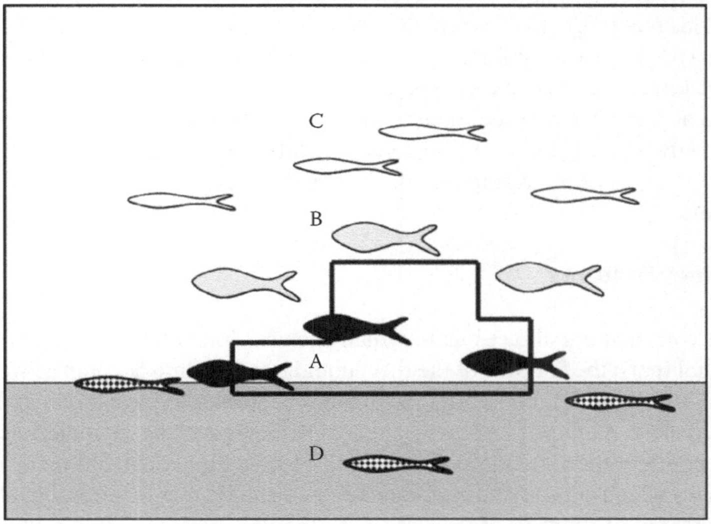

Figure 19.1 Classification of artificial reef fish based on their position relative to the reef: A-type species are directly on the reef or inside reef holes and crevices; B-type species are generally located directly around the reef but do not make contact with it; C-type species are found above the reef; D-type species are found on, in, or over the substrate next to the reef. (From Nakamura, M. 1985. *Bull. Mar. Sci.* 55:308–318 as modified by Bortone, S.A. 2008. Pages 917–924 in J. Nielsen, J.J. Dodson, K. Friedland, T.R. Hamon, J. Musick, and E. Verspoor (Eds.). *Reconciling Fisheries with Conservation: Proceedings of the Fourth World Fisheries Congress*. 1,946 pages/2 volumes, Symposium 49. American Fisheries Society. Bethesda, Maryland. With permission.)

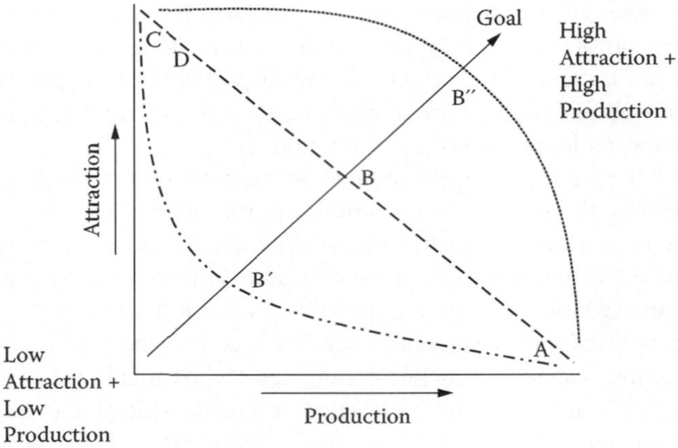

Figure 19.2 Model indicating the relationship of A, B, C, and D-type artificial reef species relative to attraction and production features of artificial reefs. The diagonal line from the origin indicates a position that species would occupy if they are balanced for attraction and production. B′ and B″ indicate the position of B-type species with different life history strategies. (From Bortone, S.A. 2008. Pages 917–924 in J. Nielsen, J.J. Dodson, K. Friedland, T.R. Hamon, J. Musick, and E. Verspoor (Eds.) *Reconciling Fisheries with Conservation: Proceedings of the Fourth World Fisheries Congress*. 1,946 pages/2 volumes, Symposium 49. American Fisheries Society. Bethesda, Maryland. With permission.)

production relationships with artificial reefs would clearly not be a candidate for management with an artificial reef in a fishery management plan. Fisheries that target species that are only attracted to artificial reefs and with little or no production benefit would get some benefit from artificial reef deployments but only through more efficient fishing pressure (i.e., resulting in an increase in fishing mortality). Fisheries that target species which only have a strong production relationship (e.g., spiny lobster or octopus) might gain a significant advantage through artificial reef deployments. Many reef fish fisheries that target reef-associated species (i.e., snapper, grouper, sparids, etc.) would likely benefit from both attraction and production relationships with artificial reef deployments.

While this is only one way to make decisions regarding which species to target with an artificial reef deployment, it serves as an example of the thought process necessary to give direction to reef planning activities.

Focusing on Reef Features

Concomitant with making choices as to which species should be targeted for management by employing artificial reefs; the features of the reef should be taken into account as well. Bortone et al. (2000: Figure 5.3 and Table 5.1) considered artificial reef studies to have biotic and abiotic components where the artificial reef itself had many abiotic attributes such as the materials of composition, dimensions, and configuration. In addition, it should be noted that artificial reefs have deployment controlled attributes which consist of the aforementioned abiotic reef features, but also include those environmental features selected in association with the deployment. These features are location-based and include depth, substrate type, general water conditions (i.e., those not weather related) and proximity to various in situ features such as coral reefs, harbor entrances, sea grass beds, etc.

The preamble above sets the stage for researchers to consider these abiotic features of reefs that can be controlled but also to which fish are potentially "preadapted." For example, if the target species prefers deeper areas, then the research design plan would include reef deployments in deeper areas. Similarly, placement-based decisions for other features that the target fish prefers would be included as well. This might include deployments near known nursery areas, in regions where fishing pressure is historically low, or in regions where upwelling may favor water column production for filter feeding species.

Life History Features

The importance of indicting the above aspects of deployment-controlled features of artificial reefs is that these attributes of artificial reefs can be controlled so that the reef positively affects the biotic response one is striving to achieve. Caddy (Chapter 4 in this volume) has indicated that many reef-associated species may have certain life history features that are limiting. An example would be where A-type species (Figure 19.1), which are totally dependent on the presence of a reef or reef-like substratum, are limited because of a lack of that substrate or substrate feature (e.g., hole size). Clearly, the species' abundance and even its presence are limited by habitat. Deploying more habitat would relieve this limiting feature (i.e., lack of structure) that presumably constrains the population size.

While this example is straightforward, an objective for fishery managers is to identify the specific life history features of particular species that a deployment may affect. This includes not only the mere deployment itself (i.e., presence/absence of reef-like substrate) but the specific life history features that can be enhanced by a deployment. For example, a species might be limited by a particular spawning substrate (e.g., flat rock surfaces for the attachment of eggs by the female pomacentrid garibaldi, *Hysypops rubicundus*) and increasing such habitat would increase its reproductive potential. Another example would be a situation where the high natural mortality in a species is relieved, and the artificial reef affords the life stage some respite from predation.

The objective for a fishery manager would be to identify the life history feature(s) that is being restricted through a "bottleneck" effect. A research objective would be to determine if an artificial reef deployment could help reduce the influence of a specific bottleneck on a particular life history feature that normally limits an important species' response characteristic. Identifying these relationships (i.e., life history bottlenecks and artificial reef attributes) should be the objective of much research prior to adopting artificial reefs as a management plan option. It is through this analytical process that managers eventually learn to deploy an artificial reef with specific design features (dimensions, materials, location, etc.) that can be used in a directed fashion to overcome the life history bottlenecks that limit a species' ability to meet its full potential relative to the fishery.

Employing a Logic Model to Resolve Fishery Management Problems

Resolutions to fisheries management problems can take many different paths—some random, some haphazard, and some purposeful—but eventually (and inevitably) most of them get resolved through the testing of hypotheses using field and laboratory experiments or conducting modeling exercises, The principle argument here is that most artificial reef research to date has not been purposefully driven, at least relative to fishery management, and that achievements and accomplishments toward resolving the issue concerning the use of artificial reefs in fishery management has lacked sufficient direction to allow resolution. Below is presented a conceptual scheme or model that should allow the eventual, but timely resolution of the dilemma. The pathway I have selected is borrowed from the social science literature and is termed a *logic model*. A generalized description of the logic model can be found in McCauley (2001), Hinchcliff (2004), and the Kellogg Foundation (2004). There are many other references that can be found on-line (e.g., http://www.uwex.edu/ces/lmcourse/) and in the published literature that describe the overall attributes and flow plan for

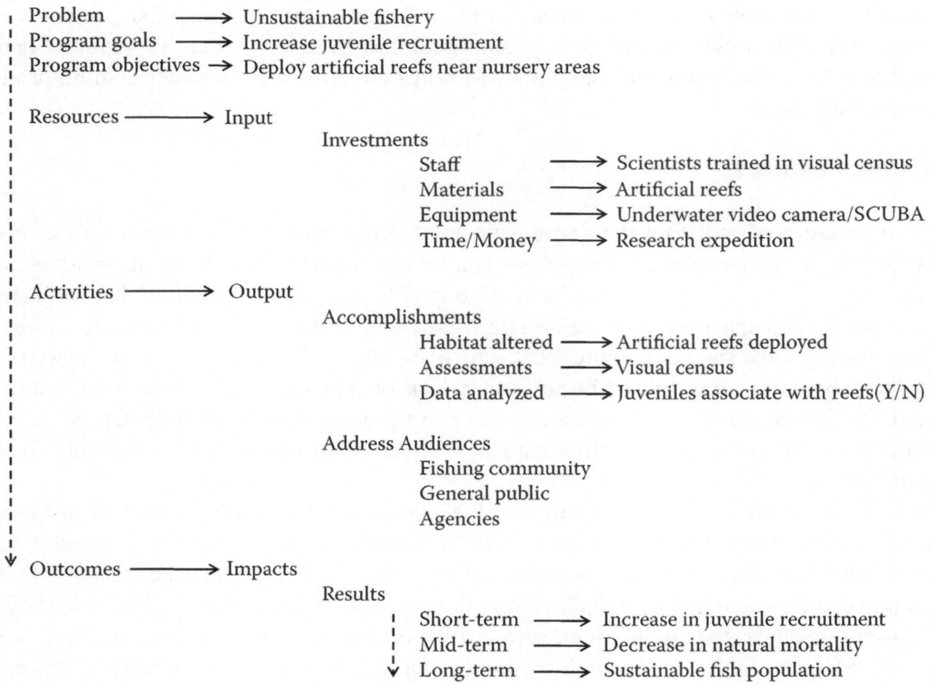

Figure 19.3 Flow diagram of resources, activities, and outcomes when using a Logic Model to design research that addresses the use of artificial reefs in management. Dashed arrows indicate sequence associations. Solid arrows indicate specific examples when the Logic Model is applied to artificial reef research.

the model. Each of these present slight variations in the logic model but the overall concept is the same. It should be noted that the logic model was designed to address social problems in complex situations. At first inspection, it would seem that the model is not particularly useful in resolving scientific problems as hypothesis testing is principally how scientific questions are resolved. The logic model helps pose the questions based on the problems and offers a framework to create study designs that answers the questions and, eventually, solves the problems identified. I believe it is the lack of appropriate questions and their resolution that has limited the directed application of artificial reefs in fishery management.

A Logic Model for addressing fisheries problems, as well as a Logic Model for the resolution for any problem begins succinctly with the identification of the problem (Figure 19.3). Once the problem has been identified, one or more program goals are listed. From the identified list of goals, one or more objectives are indicated for each goal. Once these foundations for the logic model are in place, the inputs by way of resources to be applied to the project are identified and, subsequently, the outputs or activities to be conducted are named. Lastly, the outcomes or impacts (i.e., results) are indicated and further subdivided into short-, mid-, and long-term results. In total, the logic model offers the research a direction in the form of a logical flow or pathway to problem resolution.

As indicated, logic models have been most often used to resolve complex social issues. One could easily come to the conclusion that answering the questions regarding the application of artificial reefs in fishery management is also a social issue. As an aside (and as every fishery management knows), managing fisheries is actually managing people! Below is a more specific description of the logic model as it can be applied to resolving issues in fishery science and, more to the point, the specific issue of artificial reef application in the management of fisheries.

Identifying the Problem in the Fishery

As with an investigation of any fishery, the first order of business is to determine the problem. Generally in fisheries, this comes down to identifying problems based on observations such as a declining fish stock, catch declining while effort is increasing, a reduction is occurring in the size of individuals, etc. Other problems in the fishery that deal with attributes of the fishery may be discovered as well. These problems are often concerned with fishing sector competition and exclusion issues, interference from competing jurisdictional regulations (i.e., local versus state or province versus federal or national), and regulations that supersede other regulations.

Selecting Appropriate Research Goals

Specific identification of the problems to be solved, as per the previous discussion, will give clarity of purpose to fishery management goals. These goals may or may not include the use of artificial reefs as part of the solution to the problem. If the problem is known, then achieving the program goal would solve the problem. For example, if the problem is identified as "landings are declining," then the goal becomes "increasing the landings" or maintain (sustain) the current landings.

Stating the Objectives Based on Goals

Objectives are identified to help achieve the goal or goals. Thus, with stated objectives, an artificial reef research program has the direction it needs to satisfy the major objections previously raised to using artificial reefs in fishery management. Generally, objectives are based on the rationale behind the suspected cause of the problem. Also, objectives should be offered that will help achieve the goal or goals. Moreover, objectives serve as a platform from which scientists offer hypotheses. In our argument to make use of artificial reefs a part of the solution to fishery management issues, hypotheses should be posed that are testable through the process of attempting to achieve stated objectives. These hypotheses could be one or a few among many but are generally directed (at least in the case of artificial reef deployment issues) toward relieving the life history bottlenecks that are suspected of limiting the species (or stock) in its normal ability to respond positively and recover.

The initial part of the logic model process involves determining the project goal, identifying the objectives and creating the hypotheses. It is during the creation of the hypotheses to be tested that it is important to recognize the assumptions that underlay the study design. Indicating or listing the assumptions is an important step in any scientific research project design, but this step is especially important in the scientific application of the logic model. The underlying assumptions need to be indicated and understood and even better if they are explicitly stated (Kellogg, 2004). Typical examples of assumptions are that the data gathered were done so without bias or that the populations of fish studied are representative of the entire population. Scientists often know that many assumptions are not testable, and may even be invalid, but they allow for the scientific process to proceed, nonetheless. Moreover, reexamination of the reality behind assumptions often becomes the basis for forming new hypotheses should we reject the initial first hypothesis.

Inputs or Resources

The next part of the logic model is a description of the inputs. Inputs are the resources available to the project. Another way of stating this is that inputs are what the researcher brings to bear on the problem to attain the goals by meeting the objectives. One could also think of these as the resources available or investments made in the project. Investments or resources include personnel such as scientists and technicians, materials such as the artificial reef materials, equipment such as

SCUBA gear to conduct visual surveys, and most importantly, the availability of funds as funds readily translate to personnel time, material, and equipment.

Outputs or Activities

Outputs in the logic model can be described as being the activities conducted to meet the objectives through use of the resources. These can be of two general types. The first type of outputs or activities relates to the accomplishments such as the habitats modified or altered. In the case of artificial reef studies, this normally would be the artificial reef or reefs being deployed. Another accomplishment would be data gathering or assessments. Accomplishments can also include the data analyses taking place on the data gathered, analyses, or synthesis, and comparisons to previous studies as well as model generation. The second type of activity is identification of the audience addressed (either directly or indirectly). When Logic Models are applied to social science situations, often the same community of individuals is being addressed as when the models are applied to natural science problems. In the situation with artificial reefs, the audiences can be specifically identified as the fishing communities (recreational and/or commercial), agencies, or the general public.

Outputs or Impacts

Once the activities in the project are conducted, there are some resultant outcomes or impacts. For fisheries research associated with artificial reefs, these are readily classified into short-, mid-, and long-term results. By way of example, short-term results would be some immediate change such as a change in the number of fish at a site. Mid-term results would be an achievement of some objective such as a change in an important population parameter (faster growth, higher catches, etc.). The last or long-term result would be the attainment (or lack thereof) of the project goal.

While Logic Models are generally not applied to resolving fisheries issues, the complete follow through of their application toward resolving the dilemma of showing how artificial reefs can, or cannot, be used in fisheries management is most appropriate as it gives direction to the research plan and provides a "blueprint" for actually conducting the project. In this sense, the Logic Model is bi-directional (Hinchcliff, 2004). For example, starting with the inputs and resources leads to the outputs or results, but going from the expected outcome to the resources available also allows one to formulate the research plan.

A CASE STUDY OF THE LOGIC MODEL IN THE APPLICATION OF ARTIFICIAL REEFS IN FISHERIES MANAGEMENT

A generalized description of the Logic Model might be sufficient to give fishery scientists an idea as to how to proceed with the model in assessing whether or not artificial reefs are appropriate for application to a particular fishery. Nevertheless, it is more intuitive to follow an example for a fishery. Below, is presented a scenario in which the Logic Model is used to formulate a research plan to resolve a real world fisheries issue. The flow diagram for this example can be seen in Figure 19.4.

In the example here, the problem is stated as "grouper declining." It could also be stated as "grouper unsustainable." This statement may be based on either fishery independent and/or fishery independent survey data or even an unsubstantiated impression. From that initial problem statement, the overall goal of the project is to attain and maintain sustainability of the grouper stock. Subsequent to the establishment of objectives for the project, a series of assumptions and hypotheses are generated. In this example using the grouper (in this case *Mycteroperca microlepis*, the gag serranid common to the shallow shelf areas of the western North Atlantic warm temperate waters) it is suspected that there may be a lack of recruitment to the adult populations that is leading to a

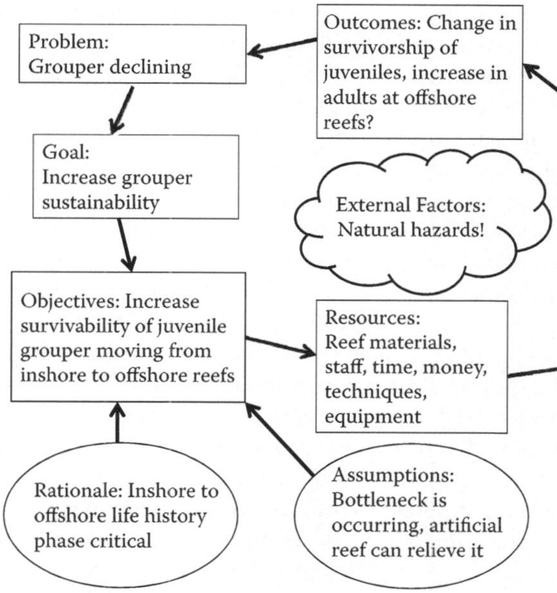

Figure 19.4 Flow diagram of a case study involving research plan development for a hypothetical scenario regarding a declining fish species. Rationale, assumptions, and external factors are added to the overall Logic Model process.

decline of the stock at current mortality levels. A reduction in recruitment of juveniles from inshore sea grass beds to offshore natural reefs has been offered as a possible explanation for the decline in adult stock. An assumption also could be offered that natural mortality among juveniles is high because juveniles at inshore grass beds must migrate dozens of kilometers offshore. Without protection during this migration, mortality is probably high. The hypothesis then becomes: the placement of artificial reefs at varying distances between the inshore sea grass beds to the offshore natural reefs would afford sufficient protection for juveniles to reduce natural mortality owing to predation. An assumption could also be that the trend being observed (i.e., the grouper population is declining) is correct and not an artifact of some poorly collected sets of observations.

With these assumptions and hypotheses in place, and concert with the overall goal for the fishery, the objectives of the project are to increase survival of juvenile grouper during their inshore to offshore movement. The rationale is that juvenile survivability during migration is a critical life history bottleneck and that artificial reefs can offer some degree of protection to juveniles and thus relieves the bottleneck.

The resources available to our hypothetical grouper case study are an abundance of reef construction materials in the form of relic bridge rubble. The nearby university has experienced staff and students trained in underwater visual assessments to document the affinity of juveniles recruitment to the "way station" or intermediate artificial reefs to be placed as "stepping stones" that lead out to deeper reefs. This study would involve several activities. First there are predeployment surveys of grouper for both juveniles and adults in grass beds, the open shelf, and at the offshore reefs. Activities also include the deployment of several sets of inshore to offshore artificial reefs at varying distances. The local fishing organizations would be informed of the project; project activities, goals, and objectives would be relayed to the fishing public. Surveys would be conducted to gather information of their opinions as to the best locations for the artificial reefs based on their experience. Output from these activities include the deployment of the artificial reefs, survey data assessing the population estimates of various life stages (juvenile and adult) of the target and other species associated with the artificial reefs. These data would be used to test the hypotheses related to the efficacy of artificial reefs in the reduction of juvenile gag mortality.

Outcomes from the study would be classified as short-term, mid-term, and long-term. A short-term outcome would be the presence of juvenile grouper on the various artificial reefs deployed. Mid-term outcomes would be an increase in the survival (i.e., decreased natural mortality) among juvenile grouper. Another mid-term outcome would be an increase in the number of small adult grouper on the natural offshore reefs. Long-term outcomes would be a higher number of adult grouper on offshore reefs. Moreover, these offshore populations remain higher after several years.

Once outcomes are observed, the overall project goal can be assessed as to being achieved or not. If the project goal is achieved, decisions can be made regarding a planned increase in the number of artificial reefs to achieve a predicted level of adult stock. Additionally, decisions could be made to modify the reef deployments slightly (i.e., alter the distance between reefs, etc.) to even further increase survival among juvenile grouper. If it is determined that juvenile survival (and hence number of offshore grouper) has not increased as a result of this project, then alternate hypotheses may be posed. For example, the type of reef may have to be changed under the assumption that the deployed reef type was not conducive to juvenile survival. Alternatively, the specific location of the deployments may need to be altered to facilitate transfer of juveniles from one reef to another.

Once the alternative hypotheses are offered (i.e., under the evidence that the goal was not achieved), the process in the Logic Model is repeated from the objective stage to the outcome stages of the model. This is because the logical flow of the project continues to proceed, but now is operating under a different set of assumptions.

The above outline of the implementation of the Logic Model to examine artificial reef application in the resolution of fishery management issues may seem inherently logical. This is because the model incorporates the role of hypotheses, testing (through data assessment), and rejection of the hypotheses and the offer of alternative hypotheses. On top of all this, however, researchers need to be mindful of the possible interference that external factors may have on the outcomes (and sometimes on the unencumbered execution of the activities). Density independent factors such as storms may interfere with reef construction to the point where the reef deployment is compromised (i.e., reef was moved because of strong currents, the reef was destroyed owing to wave action, etc.). Density independent factors may also lead to increased mortality of juveniles and adults despite the addition of artificial reefs that presumably helped reduce natural mortality. In the Gulf of Mexico, as well as in other parts of the world, harmful algal blooms may result in higher, shore-term mortality and thus interfere with the expected results.

NEW PARADIGM APPROACH TO FUTURE RESEARCH?

As offered as an argument at the outset of this chapter, there has been a plethora of research studies conducted on artificial reefs. While many of these purport to address the issue relative to the future application of artificial reefs in fishery management, the reality is that only a few notable situations exist where this has been the case. The future of artificial reefs as a fishery management option depends of several considerations that are currently unrealized, but the onus is chiefly on researchers involved with artificial reefs to resolve this current situation.

For the most part, artificial reef scientists have not conducted the type of research that provides results which are directly applicable to circumstances that allow fishery managers the luxury of selecting artificial reefs as one of the management options currently available. Artificial reefs remain an untested option with few data available that make their selection viable other than a hopeful suggestion of things to come. Fishery managers are not at fault. They must base decisions on a likelihood of success. To date, artificial reef scientists have not provided the background data to allow the directed implementation of artificial reefs. Given this scenario, it is not unreasonable that artificial reefs have remained an "art form" in the arsenal of fishery manager's choices.

For this circumstance to appreciably change, several things must occur and chief among these is purposefully directed research that gives fishery managers plausible options. Continuing down our current path will do little to relieve this situation. Offered above, is a guide to correcting this situation. Ironically, the outcome may be the same, that is, after all the research has been conducted, artificial reefs may still have virtually no role in the management of fisheries. That may be, but without an appreciable change in approach, the future of artificial reefs in fishery management may remain truly untested.

The plan of direction offered here is neither the only plan nor approach toward resolving the dilemma, but it at least offers a beginning at giving direction to the solution. State, provincial, and national funding agencies would do well to heed the call for directed research regarding artificial reefs and solicit proposals specific to its resolution. Hopefully, these solicitations would entice favorable responses to those proposals that follow a Logic Model or some similar approach that gives direction to problem resolution.

REFERENCES

Bohnsack, J.A., D.L. Johnson, and R.F. Ambrose. 1991. Ecology of artificial reef habitats and fishes. Pages 61–107 in W. Seaman, Jr. and L.M. Sprague (Eds.). *Artificial Habitats for Marine and Freshwater Fisheries*. Academic Press, San Diego.

Bohnsack, J.A., A.M. Eklund, and A.M. Szmant. 1997. Artificial reef research: Is there more than the attraction-production issue? *Fisheries* 22(4):14-16.

Bortone, S.A., M.A. Samoilys, and P. Francour. 2000. Fish and macroinvertebrate evaluation methods. Pages 127–164 in W. Seaman, Jr. (Ed.). *Artificial Reef Evaluation with Application to Natural Marine Habitats*. CRC Press. Boca Raton, Florida.

Bortone, S.A. 2008. Coupling fisheries with ecology through marine artificial reef deployments. Pages 917–924. In: J. Nielsen, J.J. Dodson, K. Friedland, T.R. Hamon, J. Musick, and E. Verspoor (Eds.). *Reconciling Fisheries with Conservation: Proceedings of the Fourth World Fisheries Congress*. 1,946 pages/2 volumes, Symposium 49. American Fisheries Society. Bethesda, Maryland.

Diamond, J. 1997. *Guns, Germs, and Steel: The Fates of Human Societies*. W.W. Norton and Company. Berkeley, California.

Hinchcliff, G. 2004. Logic Model Development. NOAA Coastal Learning Center, Silver Spring, Maryland.

Kellogg Foundation, W.K. 2004. Logic Model Development Guide. W.K. Kellogg Foundation, Battle Creek, Michigan (www.wkkg.org).

Lindberg, W.J. 1997. Can science resolve the attraction-production issue? *Fisheries* 22(4):10–13.

McCauley, P.F. 2001. The Logic Model for Program Planning and Evaluation. University of Idaho Extension, CIS 1097, Moscow, Idaho.

Nakamura, M. 1985. Evaluation of artificial reef concepts in Japan. *Bull. Mar. Sci.* 55:308–318.

Polovina, J.J. 1991. Fisheries applications and biological impacts of artificial habitats. Pages 154–176 in W. Seaman, Jr. and L.M. Sprague (Eds.). *Artificial Habitats for Marine and Freshwater Fisheries*. Academic Press, San Diego, California.

Steimle, F.W. and M.H. Meier. 1997. What information do artificial reef managers really want from Fishery Science? *Fisheries* 22(4):6–8.

Index

A

B

C

D

T